VOLUME SIX HUNDRED AND SIXTY SIX

METHODS IN ENZYMOLOGY

Advances in Biomolecular EPR

METHODS IN ENZYMOLOGY

Editors-in-Chief

ANNA MARIE PYLE
Departments of Molecular, Cellular and Developmental Biology and Department of Chemistry
Investigator, Howard Hughes Medical Institute
Yale University

DAVID W. CHRISTIANSON
Roy and Diana Vagelos Laboratories
Department of Chemistry
University of Pennsylvania
Philadelphia, PA

Founding Editors

SIDNEY P. COLOWICK and NATHAN O. KAPLAN

VOLUME SIX HUNDRED AND SIXTY SIX

METHODS IN ENZYMOLOGY

Advances in Biomolecular EPR

Edited by

R. DAVID BRITT
Department of Chemistry,
University of California, Davis,
CA, United States

ELSEVIER
ACADEMIC PRESS
An imprint of Elsevier

Academic Press is an imprint of Elsevier
50 Hampshire Street, 5th Floor, Cambridge, MA 02139, United States
525 B Street, Suite 1650, San Diego, CA 92101, United States
The Boulevard, Langford Lane, Kidlington, Oxford OX5 1GB, United Kingdom
125 London Wall, London, EC2Y 5AS, United Kingdom

First edition 2022

ISBN: 978-0-323-99974-8
ISSN: 0076-6879

For information on all Academic Press publications
visit our website at https://www.elsevier.com/books-and-journals

Publisher: Zoe Kruze
Developmental Editor: Federico Paulo S. Mendoza
Production Project Manager: James Selvam
Cover Designer: Christian Bilbow

Typeset by STRAIVE, India

Contents

Contributors

Aidin R. Balo
Department of Chemistry, University of California, Davis, CA, United States

Antonio Barbon
Department of Chemical Sciences, University of Padova, Padova, Italy

Arnau Bertran
Centre for Advanced Electron Spin Resonance and Inorganic Chemistry Laboratory, Department of Chemistry, University of Oxford, Oxford, United Kingdom

Squire J. Booker
Department of Chemistry; Department of Biochemistry and Molecular Biology; Howard Hughes Medical Institute, The Pennsylvania State University, University Park, PA, United States

Enrica Bordignon
Department of Physical Chemistry, University of Geneva, Genève, Switzerland

Alice M. Bowen
Centre for Advanced Electron Spin Resonance and Inorganic Chemistry Laboratory, Department of Chemistry, University of Oxford, Oxford; EPSRC National Research Facility for Electron Paramagnetic Resonance Spectroscopy, Department of Chemistry and Photon Science Institute, The University of Manchester, Manchester, United Kingdom

R. David Britt
Department of Chemistry, University of California, Davis, CA, United States

Gary K. Chow
Department of Chemistry, University of California, Davis, CA, United States

Marilena Di Valentin
Department of Chemical Sciences, University of Padova, Padova, Italy

Malte Drescher
Department of Chemistry and Konstanz Research School Chemical Biology (KoRS-CB), University of Konstanz, Konstanz, Germany

Gareth R. Eaton
Department of Chemistry and Biochemistry, University of Denver, Denver, CO, United States

Sandra S. Eaton
Department of Chemistry and Biochemistry, University of Denver, Denver, CO, United States

Laura Esteban-Hofer
ETH Zürich, Department of Chemistry and Applied Biosciences, Zürich, Switzerland

Laura Galazzo
Department of Physical Chemistry, University of Geneva, Genève, Switzerland

John H. Golbeck
Department of Biochemistry and Molecular Biology, and Department of Chemistry, The Pennsylvania State University, University Park, PA, United States

Russ Hille
Department of Biochemistry, University of California, Riverside, CA, United States

Gunnar Jeschke
ETH Zürich, Department of Chemistry and Applied Biosciences, Zürich, Switzerland

Wei Li
Department of Physics, Emory University, Atlanta, GA, United States

Andy LiWang
Department of Chemistry & Biochemistry; Health Sciences Research Institute, University of California, Merced, CA, United States

Glenn L. Millhauser
Department of Chemistry and Biochemistry, UC Santa Cruz, Santa Cruz, CA, United States

Benjamen Nforneh
Department of Physics, Emory University, Atlanta, GA, United States

Dimitri Niks
Department of Biochemistry, University of California, Riverside, CA, United States

Liliana Quintanar
Department of Chemistry, Center for Research and Advanced Studies (Cinvestav), Mexico City, Mexico

Katherine H. Richardson
Imperial College London, Molecular Sciences Research Hub, London, United Kingdom

Maxie M. Roessler
Imperial College London, Molecular Sciences Research Hub, London, United Kingdom

Maryam Seif-Eddine
Imperial College London, Molecular Sciences Research Hub, London, United Kingdom

Alexey Silakov
Department of Chemistry, The Pennsylvania State University, University Park, PA, United States

Adam Sills
Imperial College London, Molecular Sciences Research Hub, London, United Kingdom

Juliane Stehle
Department of Chemistry and Konstanz Research School Chemical Biology (KoRS-CB), University of Konstanz, Konstanz, Germany

Joshua Telser
Department of Biological, Physical and Health Sciences, College of Science, Health and Pharmacy, Roosevelt University, Chicago, IL, United States

Markus Teucher
MPI for Chemical Energy Conversion, Mülheiman der Ruhr, Germany

Art van der Est
Department of Chemistry, Brock University, St. Catharines, ON, Canada

Bo Wang
Department of Chemistry; Howard Hughes Medical Institute, The Pennsylvania State University, University Park, PA, United States

Kurt Warncke
Department of Physics, Emory University, Atlanta, GA, United States

Katie L. Whitcomb
Department of Physics, Emory University, Atlanta, GA, United States

CHAPTER ONE

Advances in rapid scan EPR spectroscopy

Gareth R. Eaton* and Sandra S. Eaton
Department of Chemistry and Biochemistry, University of Denver, Denver, CO, United States
*Corresponding author: e-mail address: gareth.eaton@du.edu

Contents

Abstract

Progress has been made in hardware for low frequencies, demonstrations of rapid frequency scans, hybrid instrumentation, and improved deconvolution software. The recent availability of the commercial Bruker BioSpin rapid scan accessory for their X-band EMX and Elexsys systems makes this technique available to a wide range of users without the need to construct their own system. Developments at lower frequencies are underway in several labs with the goal of facilitating in vivo and preclinical rapid scan imaging. Development of new deconvolution algorithms will make data processing more robust. Frequency scans have substantial promise at higher frequencies. New examples of applications show the wide applicability and advantages of rapid scan.

Methods in Enzymology, Volume 666
ISSN 0076-6879
https://doi.org/10.1016/bs.mie.2022.02.013

Comprehensive reviews of rapid scan electron paramagnetic resonance are given in (Eaton & Eaton, 2018; Eaton et al., 2014, 2017). A recent review provides a detailed discussion of the physical principles underlying rapid scan EPR (Tseytlin, 2021). In this chapter we provide an overview of the rapid scan experiment, discussion of recent hardware and software developments, and guidance on selection of experimental parameters to optimize information content for particular types of samples.

1. Over-view of the rapid-scan experiment

In a conventional continuous wave (CW) EPR experiment the microwave power is constant and the main magnetic field B_0 is scanned to achieve spin resonance. A second rapidly-oscillating field, designated as magnetic field modulation, is applied at a frequency up to 100 kHz. The EPR signal is recorded by phase-sensitive detection at the modulation frequency (Fig. 1). The output of the phase-sensitive detector is proportional to the difference in EPR signal amplitudes between the extremes of the modulation. If the modulation amplitude is sufficiently small relative to the spectral linewidth the signal from the phase-sensitive detector is proportional to the slope of the EPR lineshape. As the main magnetic field is stepped through resonance the phase-sensitive detected signal maps out the slope of the EPR spectrum, which gives the familiar first-derivative

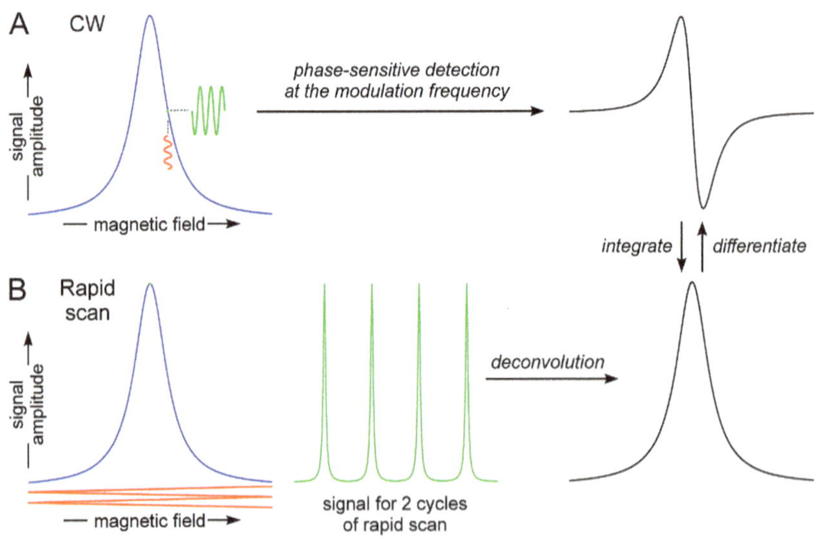

Fig. 1 Comparison of signal detection by CW and rapid-scan EPR.

display. If the amplitude of the modulation is too large relative to the linewidth the detected signal is broadened. If the amplitude of the modulation is small enough to accurately measure the slope of the EPR line, only a small fraction of the signal amplitude is detected at each point on the magnetic field axis. To obtain the most accurate signal intensity and lineshape the constant microwave power is selected to be in a region where the signal amplitude increases linearly with the square root of increasing power (Eaton, Eaton, Barr, & Weber, 2010). In these CW scans the spin magnetization vector is tipped only slightly by the microwave field and is only slightly perturbed from equilibrium.

In a field-swept rapid-scan experiment the main magnetic field B_0 is held constant near the center of the spectrum (Fig. 1). A rapidly-varying magnetic field is generated by an electrical current passing through coils. Provided that the rapidly-varying field is sufficiently large, the full EPR spectrum is recorded twice during each cycle of the modulation—once during the up-field portion of the scan waveform and again during the down-field portion (Fig. 1). The ability to detect the full signal amplitude in each pass through the spectrum substantially improves the signal-to-noise. The absorption and dispersion components of the EPR signal are recorded by direct detection with a quadrature mixer and both components are used to further improve the signal-to-noise. When the scan rate is faster than electron spin relaxation, the spin magnetic moment is perturbed substantially by the scan. After the scanning field has passed through resonance, the spins precess about B_0 at the instantaneous Larmor frequency that is determined by the sum of the B_0 field plus the scanned field. This frequency beats with the microwave or RF field, resulting in damped oscillations. The spacing between the oscillations decreases as the field scans away from resonance. This is different from the case for an FID, in which the oscillation frequencies are constant and define the static offset from resonance. The oscillations damp out with a time constant of T_2^*. Postprocessing software deconvolves the rapid scan oscillations and produces the EPR absorption spectrum. If the scan rate is not fast enough to generate oscillations the deconvolution software converts the experimental time axis to a magnetic field axis, but does not alter the spectrum. Thus the same signal processing algorithms can be used independent of scan rate.

Analogous to conventional CW experiments the microwave power is on continuously in a rapid scan experiment. Since the signal detection technique and signal processing is so different from the conventional experiment it is useful to use the designation "rapid scan" to distinguish from other experiments. Because the resonance condition is satisfied for such a short time in the rapid scan experiment, the spin system does not saturate as readily as in conventional CW. This means that higher microwave power can be used which increases signal intensity and improves signal-to-noise. The selection of microwave power for rapid scan experiments is discussed in Section 5.5.

As described previously (Eaton & Eaton, 2018) the primary advantages of rapid scan are improved signal-to-noise, ability to explicitly correct for passage effects that arise from scanning through the spectrum too rapidly, linewidths that are not broadened by overmodulation provided that bandwidth constraints are taken into consideration (see Section 5.4), and rapid data acquisition that can capture fast kinetics.

2. Recent hardware developments

2.1 Low frequency instrumentation

A rapid-scan EPR imager developed at West Virginia University is described in (Tseytlin et al., 2019). The system has several novel features. The EPR frequency is generated by mixing the output of a high-resolution arbitrary waveform generator (AWG) with a stable low-noise constant source frequency, which permits operation at a wide range of frequencies. Data are reported for operation at 800 MHz that was obtained by mixing the AWG output with a 700 MHz constant frequency source. The system has a digital automatic frequency control system that uses the AWG to sweep the RF frequency and lock to the frequency that provides the best resonator coupling which is defined as the minimum reflected power. The tuning sweeps are interleaved with data acquisition to adjust for animal motion. A digital automatic scan control monitors and corrects the amplitude and phase of the rapid sinusoidal magnetic field scans based on the voltage from a capacitor that is in-series with the scan coils. The digitized signal is compared with previous calibration to generate corrections, which produces highly accurate waveforms that are essential for rapid-scan deconvolution. A surface loop resonator with a 3D printed support structure was developed to reduce the amount of metal near the imaging region which is helpful in ensuring a uniform scanning magnetic field. The modular design from commercial components facilitates maintenance and upgrades of portions of the system. The performance was demonstrated with a phantom consisting of four tubes containing 1 mM solutions of trityl radicals with varying degrees of line broadening by a Gd^{3+} contrast agent (Tseytlin et al., 2019).

A tabletop spectrometer designed for rapid scan and pulsed EPR is based on an AWG and 25 mm cross-loop resonator (Buchanan, Rinard, Quine, Eaton, & Eaton, 2018). The first experiments were performed at 700 MHz, although only the resonator would need to be replaced to operate at frequencies up to 1 GHz. This frequency range was selected to permit studies in mice. The system is a first step toward design of compact systems for pre-clinical use. The compact 4-coil magnet has a uniformity of <25 ppm over the active volume of the resonator and weighs about 50 pounds (Fig. 2). The main magnet and the gradient coils are digitally controlled by compact CAEN power supplies. The AWG is used to create the source frequency and chirp pulses for resonator tuning. The system performance for 2-D spectral-spatial imaging and 4D-spectral-spatial imaging was demonstrated with phantoms consisting of trityl radicals and nitroxides.

A digital console for rapid-scan EPR that uses field programmable gate array (FPGA) technology has been demonstrated at 1.2 GHz (Sato-Akaba and Tseytlin 2019). The magnetic field scan waveform, spin excitation

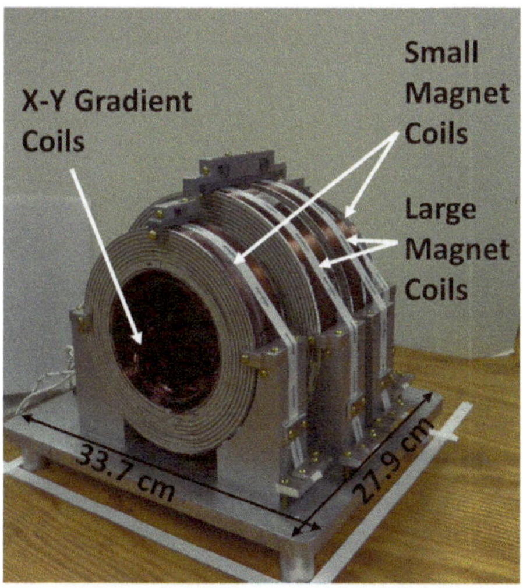

Fig. 2 Compact electromagnetic designed by rapid scan imaging of mice at frequencies between 700 MHz and 1 GHz. *Reproduced with permission from Buchanan, L. A., Rinard, G. A., Quine, R. W., Eaton, S. S. Eaton, G. R. (2018). Tabletop 700 MHz EPR imaging spectrometer.* Concepts in Magnetic Resonance Part B: Magnetic Resonance Engineering 48B. *e21384, doi:10.1002/cmr.b.21384.*

frequency, and digitizer sampling were run from a single internal clock. A digital feedback control system and voltage-tunable capacitors were used to perform automatic resonator tuning and matching adjustment with an electrically-tunable single loop-gap reflection resonator. A feedback system also was implemented to control the phase and amplitude of the rapid field scans which is very important for rapid-scan deconvolution. Data were accumulated continuously in the FPGA without time gaps, which eliminated the re-arm time of many digitizers. Band-pass sub-sampling was used to digitize the EPR signal at the carrier frequency. The narrow linewidths obtained for a sample of deoxygenated lithium octa-n-butoxyl-2, 3-naphthalocyanine (LiNc-BuO) tested the frequency stability, scan field, and reflection reduction controls. The system can also perform pulsed EPR measurements (Sato-Akaba & Tseytlin, 2019).

The rapidly-varying magnetic field of the rapid-scan experiments induces an oscillatory background signal that may be larger than the EPR signal from samples with low spin concentration. A background removal method was developed for use with bimodal cross-loop resonators (Buchanan, Woodcock, et al., 2018). For each spectrum two scans are acquired. In the second scan the main B_0 field is reversed and the acquisition trigger is offset by one half cycle of the scan field, relative to the first scan. Subtraction of the second scan from the first scan reinforces the EPR signal and cancels the background signal. This method has the advantage that no assumptions are made concerning the shape of the background. The method was demonstrated at 258 MHz for samples containing nitroxide and trityl radicals and for a nitroxide in the presence of a magnetic field gradient. The experiment depended on the ability of the CAEN power supplies to quickly reverse the polarity of the main magnetic field under computer control.

Hybrid systems that combine two imaging modalities provide the possibility of simultaneously monitoring two or more time-dependent physiological parameters. A prototype hybrid imager that combines PET and sinusoidal rapid-scan EPR has been reported (Tseytlin et al., 2018). Positron emission tomography (PET) uses radiotracers to monitor processes including glucose metabolism. EPR imaging experiments can be designed to measure local oxygen concentration and pH, for example. Operation at an EPR frequency of 800 MHz was selected to avoid interfering cell phone signals at ~750 MHz. The main magnetic field was generated with a permanent magnet. The rapid-scan coils were constructed from Litz wire on a polylactic acid cylinder and placed in the center of the PET ring.

A phantom consisted of four cylinders of 1 mM trityl and 25 μCi of ^{18}F-radiotracer with and without a Gd contrast agent that broadens the trityl signal. The phantom was placed inside an EPR surface loop resonator and that assembly was positioned inside the rapid-scan coils. EPR images co-registered with PET and MRI were in good agreement. There was no evidence of photon attenuation by the EPR resonator and scan coils.

2.2 Frequency scan

Most rapid-scan experiments have been performed by scanning the magnetic field at constant frequency with a resonant cavity that has a relatively narrow bandwidth. Rapid frequency sweep experiments have been reported in the frequency range of 170–250 GHz with scan rates up to 3×10^5 THz/s in a system without a microwave resonator (Laguta, Tucek, Van Slageren, & Neugebauer, 2018). These very fast sweep rates permit measurement of relaxation times as short as a few ns by analysis of the rapid-scan oscillations. A T_2 of 2.5–2.2 ns was measured for lithium phthalocyanine at fields of 7.225–8 T (Fig. 3). Sweeping the frequency has the advantage relative to field sweeps that eddy currents are not generated in conducting components of the resonator. There is, however, a strong background signal produced by standing waves. These background signals are independent of magnetic field and can be removed by subtraction of an off-resonance scan.

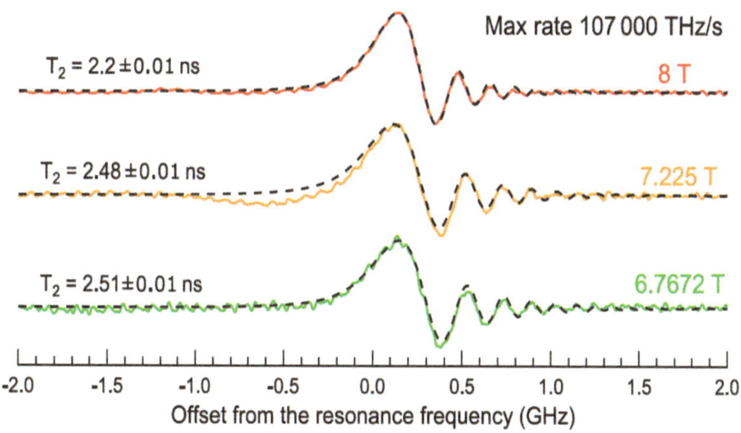

Fig. 3 Rapid scan spectra of LiPc recorded with 2 MHz modulation frequency (107,000 THZ/s at the center of the spectrum) at three magnetic fields. Dashed lines correspond to the spectra calculated using the modified Bloch equations. *Reproduced with permission from Laguta, O., Tucek, M., Van Slageren, J., Neugebauer, P. (2018). Multi-frequency rapid-scan HFEPR.* Journal of Magnetic Resonance, 296, *138–142.*

Although the sweeps were sinusoidal a small fraction of the sweep can be approximated as linear which permitted use of the simpler linear deconvolution routines to recover the slow scan lineshapes. The rapid scan experiments do not require the high microwave powers that are needed for pulse experiments and that are difficult to achieve at high frequency.

2.3 Rapid scan with electrically detected magnetic resonance

Characterization of point defects in solid-state electronic devices is key to optimizing performance and reliability (McCrory et al., 2019). Electrically-detected magnetic resonance (EDMR) is a highly-sensitive technique for studying those defects, but conventional EPR spectrometers are not well suited to performing those experiments on semiconductor wafers. A conventional semiconductor wafer probing station has been adapted to perform EDMR on a fully processed semiconductor wafer. A very small non-resonance near-field microwave probe is used in placed of a resonator. The use of a non-resonant detection system permits frequency-swept EDMR. Signals from a 4H-SiC MOSFET were acquired with frequencies in both the slow-scan and rapid-scan regime. The output from an AWG was used to control a voltage controlled oscillator and a voltage-controlled attenuator that was needed to generate a constant power as frequency was varied. Sweep rates up to ~100 GHz/s were produced. Signals were digitized with an oscilloscope with a sampling rate of 5 GS/s. The rapid frequency scan measurements have the potential to enhance the signal-to-noise for the EDMR measurements and reduce the overall measurement time (McCrory et al., 2019).

2.4 Bruker commercial system

Bruker BioSpin has developed an X-band rapid-scan accessory that is compatible with their EMX and Elexsys systems (Eichhoff & Hofer, 2020). For the dosimetric standard, irradiated alanine, the signal-to-nose was improved by a factor of about 17 for rapid scan relative to CW. The performance of this system for resolution of small hyperfine couplings (McPeak et al., 2020) and intracellular protein-lipid interactions (Braun et al., 2021) are discussed in Section 4.

3. Algorithms for deconvolution of rapid scans

In a sinusoidal scan the magnetic field passes through resonance twice during the full period of the sinusoidal scan, which makes deconvolution of the full cycle an ill-posed problem. In the original algorithms for sinusoidal deconvolution (Tseitlin, Mitchell, Eaton, & Eaton, 2012; Tseitlin, Rinard, Quine, Eaton, & Eaton, 2011) the deconvolution was performed on the up-field and down-field scans separately, which required that the EPR signal decay to zero by the end of each half cycle to avoid truncation errors. This limited the maximum scan rate. Faster scans are advantageous because higher microwave power can be used to improve signal-to-noise, as discussed in Section 5.5. A new deconvolution method was developed that is based on the additive properties of linear systems: the full cycle response is the sum of the responses in the two halves of the cycle. The constraint on scan frequency now is based on the need for the signal to decay during the full scan cycle. The method was demonstrated for the narrow lines of lithium octa-n-butoxy-2,3-naphthalocyanine (LiNC-BuO) (Tseytlin, 2017).

A general solution of the rapid-scan EPR deconvolution problem has been published (Tseytlin, 2020). The impact of the varying magnetic field on the spin magnetization, which is called the driving function, was shown to be related to the phase gained by the precessing spins, using the Bloch equations. The derivation provides a general framework for deconvolution algorithms, including a basis for the relationship between field-scanned and frequency-scanned rapid-scan signals. The one assumption is that the spin system can be approximated as linearly time-invariant. Previously developed algorithms are shown to be special cases of the general solution. In principle any arbitrary magnetic field waveform could be used for rapid scans. However, the mathematical realities of working with real-world noisy data make the numerical solutions ill-posed when working with full-scan data. Workarounds are required as have been discussed previously for sinusoidal scans (Tseitlin et al., 2012, 2011).

4. Examples of applications

The initial focus of rapid-scan EPR was on the improvement in signal-to-noise that could be achieved, especially for radicals with long electron spin relaxation times (Mitchell et al., 2013). Those experiments demonstrated lineshape fidelity for a range of species and accurate hyperfine splittings of nitroxide radicals. A related, and very important question, is how well closely spaced hyperfine lines that are key to defining electron spin density distribution can be resolved. The tradeoff between modulation amplitude and resolution of hyperfine splittings has been well documented for CW spectra. The analogous tradeoffs for rapid scan were examined in a study of 3-carbamoyl-2,2,5,5-tetramethylpyrrolidin-1-oxy (CTPO),

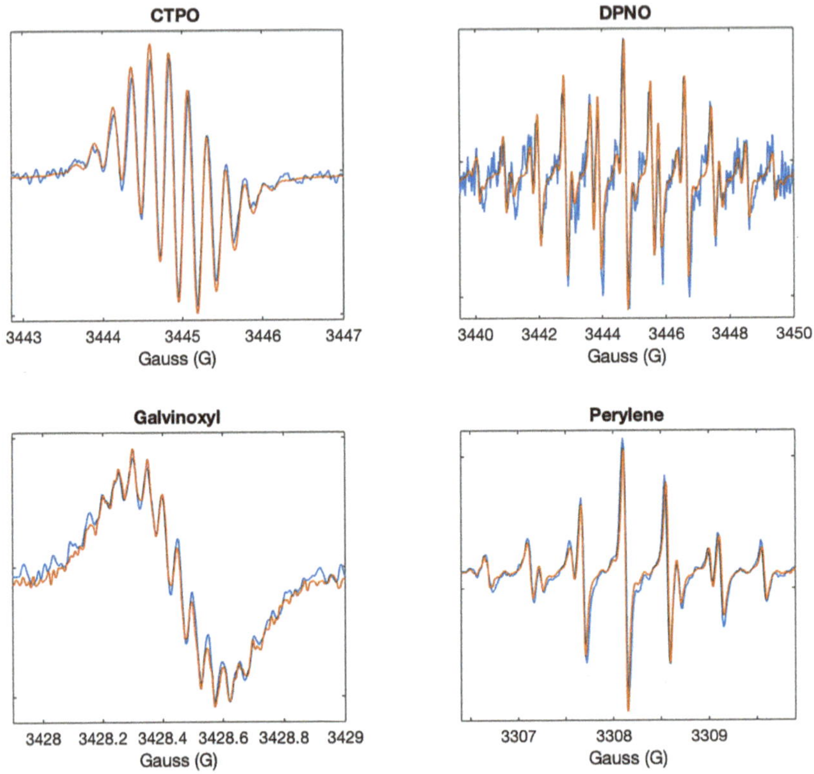

Fig. 4 Overlays of expanded segments of CW (blue) and rapid-scan spectra (orange) showing the agreement in lineshapes and improved signal-to-noise for the same data acquisition times. (A) CTPO, (B) DPNO, (C) galvinoxyl and (D) perylene. *Reproduced with permission from McPeak, J., Höfer, P., Kacprzak, S., Carl, P., Weber, R., Eaton, S. S. and Eaton, G. R. (2020). Rapid-scan electron paramagnetic resonance of highly resolved hyperfine in organic radicals.* ChemPhysChem, 21, *2564–2570.*

diphenyl nitroxide (DPNO), galvinoxyl, and perylene cation radical (McPeak et al., 2020). It was shown that hyperfine resolution in rapid scan can be as good as by CW while still achieving improved signal-to-noise (Fig. 4). In rapid scan there is a tradeoff between signal bandwidth and resonator bandwidth that can improve resolution at the expense of signal-to-noise. The detailed considerations of parameter selection for this tradeoff are discussed in Section 5.

Thiol redox status is a key factor in predicting the success or failure of cancer treatment (Epel et al., 2017). The rate of cleavage of the disulfide linkage in dinitroxides is proportional to the concentration of the major intracellular redox buffer, glutathione (Epel et al., 2017). Upon cleavage the spectrum of a dinitroxide with normal isotopic abundance changes from a 5-line spectrum to a 3-line spectrum, which can be monitored by EPR and EPR imaging. The higher signal-to-noise for rapid scan than for conventional CW is an important advantage for in vivo studies. The signal-to-noise can be further improved and the spectrum simplified by using isotopically substituted $^{2}H,^{15}N$ dinitroxide. Feasibility of the redox imaging experiments was demonstrated in a FSa fibrosarcoma tumor model in C3H mice (Fig. 5). Previous studies had shown that the intracellular glutathione concentration in mice is reduced by treatment with L-buthionine sulfoximine (BSO). The rate of cleavage of the dinitroxide in the tumor, k^{obs}, was decreased by BSO treatment, consistent with the hypothesis that the kinetics monitored by the imaging reflects changes in the glutathione concentration and local redox status. Integration of the spectra as a function of time also permitted measurement of the rate of clearance of the nitroxide, k^{clr} (Epel et al., 2017).

Alkaline phosphatase (ALP) catalyzes the dephosphorylation of substrates that contain a phosphate monoester and is reported to be a biomarker for some diseases. A spin-labeled substrate for ALP was synthesized, which contains a phosphate enol ether that is cleaved by ALP with high specificity

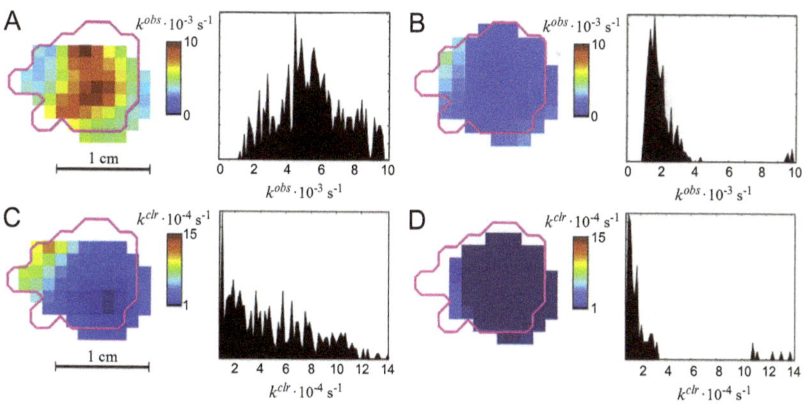

Fig. 5 Sagittal slices and histograms of the dinitroxide cleavage rate k^{obs} in FSa tumor (A) before and (B) 24h after application of BSO. Magenta outline shows the tumor border as obtained from a co-registered MRI image. The histograms shows k^{obs} in all voxels of the tumor. Sagittal slices and histograms of the clearance rate k^{clr} in FSa tumor (C) before and (D) after application of BSO. The histogram shows k^{clr} in all voxels of the tumor. *Reproduced with permission from Epel, B., Sundramoorthy, S. V., Krzykowska-Serda, M., Maggio, M. C., Tseytlin, M., Eaton, G. R., et al. (2017). Imaging thiol redox status in murine tumors in vivo with rapid-scan electron paramagnetic resonance.* Journal of Magnetic Resonance, 276, *31–36.*

(Sanzhaeva et al., 2018). The paramagnetic cleavage product is the well-known tempone radical, which has a nitrogen hyperfine coupling constant that is smaller than in the phosphate ester. As the cleavage reaction occurs the relative intensities of the well-resolved hyperfine lines from substrate and product change. Rapid scan at 800 MHz was used to image a phantom consisting of three tubes containing (i) substrate, (ii) substrate + ALP, and (iii) product. The image showed the time evolution of signal intensities in tube (ii) and the time invariance of the signals in tube (i) and (iii). This proof-of-principle experiment demonstrates the feasibility of monitoring activity of ALP *in vivo* using a specifically-designed substrate (Sanzhaeva et al., 2018).

Rapid-scan EPR of spin-labeled human α-synuclein (αS) has been used to study interaction with lipid membranes *in vitro* and *in vivo* in oocytes of *Xenopus laevis* (Braun et al., 2021). *In vitro* the spectra showed distinctive broadening in the presence of large unilamellar vesicles prepared from negatively charged 1-palmitoyl-2-oleoyl-*sn*-glycero-3-(phospho-rac-(1-glycerol)) (POPG) lipids, but not in the presence of vesicles prepared from uncharged 1-palmitoyl-2-oleoyl-*sn*-glycero-3-phosphocholine (POPC) lipids. The broadening is consistent with reduced rotational mobility that results from binding to the lipid. Prior attempts to study binding of αS to membranes *in vivo* had been limited by poor signal-to-noise. The improvement in signal-to-noise afforded by rapid scan permitted studies *in vivo*. The spin-labeled αS in the absence or presence of lipid vesicles was micro-injected into oocytes. Analogous to the observations *in vitro*, broadening of the EPR signal was observed only in the presence of the negatively charged lipids. The fast acquisition of rapid-scan EPR spectra permitted monitoring the extent of binding to lipid and endogenous membranes as a function of time, despite the relatively rapid reduction of nitroxide. Only a fraction of the αS is bound. The broadened signal from the bound αS is more readily distinguished from unbound αS in the absorption spectra obtained by rapid scan than in the first derivative spectra that are reported by CW spectroscopy (Braun et al., 2021).

Defects in thin-film silicon solar cell materials are present in low concentrations and have long electron spin relaxation times, which makes quantitation by CW EPR challenging. For a set of seven samples of hydrogenated amorphous silicon rapid-scan EPR was shown to improve the signal-to-noise by factors of between 10 and 50 relative to CW (Fig. 6) (Möser et al., 2017). Experiments were performed on a Bruker E580 spectrometer. The CW spectra were acquired with a Bruker high-quality factor SHQE

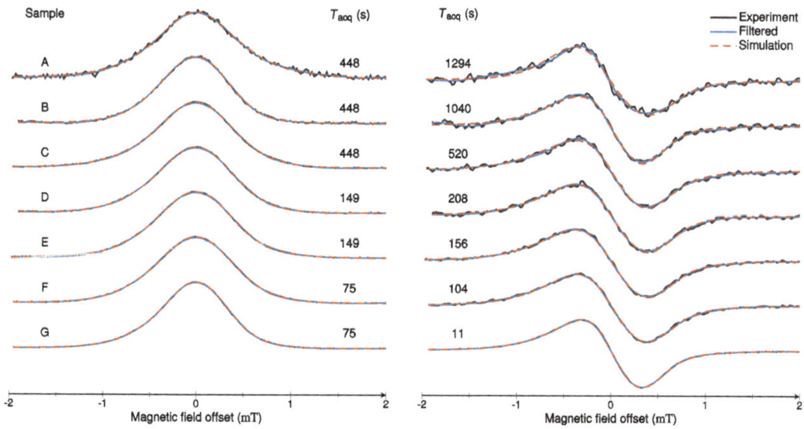

Fig. 6 Rapid scan (left) and CW (right) spectra of α-hydrogenated silicon samples, after baseline subtraction, obtained with the data acquisition time, T_{acq}. The plot includes the raw data (black), data after applying a low-pass filter (blue), and simulations obtained with EasySpin (red). *Reproduced with permission from Möser, J., Lips, K., Tseytlin, M., Eaton, G. R., Eaton, S. S., Schnegg, A. (2017). Using rapid-scan EPR to improve the detection limit of quantitative EPR by more than one order of magnitude.* Journal of Magnetic Resonance, 281, *17–25.*

resonator to obtain the best possible CW sensitivity. The rapid-scan spectra were acquired with a Bruker 4118X-MD5 dielectric resonator to minimize the effect of eddy current induced in metallic components of the resonator and take advantage of the resonator bandwidth. Sinusoidal scans at 37 kHz and 4 mT width were generated using the modulation coils built into the resonator. Under these conditions the signal bandwidth is less than the resonator and detection bandwidth, which is required to obtain undistorted lineshapes.

Many physiologic parameters including O_2 concentration, pH, and tissue redox status have been measured using nitroxide, trityl and lithium phthalocyanine (LiPc) radicals. The relaxation times and linewidths of these probes are quite different, so a fundamental question is how well these physiologic parameters can be measured simultaneously. A rapid-scan image at 250 MHz was obtained of a phantom consisting of solid LiPc and solutions of the trityl and nitroxide radicals in separate tubes (Shi, Eaton, Eaton, 2019). With a gradient of 10 G/cm the spatial resolution in 2D spectral-spatial images was about 0.1 mm for the narrow trityl and LiPc signals and about 1 mm for the broader nitroxide lines.

Oxygen is a polymerization inhibitor in many of the processes that are employed to convert liquid resins into solid structures for 3D stereolithography printing. EPR imaging of local oxygen concentration using LiNC-BuO as a probe has been used to monitor depletion of oxygen during photo-induced polymerization (Tseytlin, O'Connell, Tseytlin, Sivishankar, & Bobko, 2021). Three proof-of-principle experiments were demonstrated at 800 MHz: (1) spatial propagation of the polymerization in a bottle with an LED light source below the bottle, (2) oxygen depletion during postcuring of a series of layers, and (3) visualization of oxygen in a 3D printed spiral. It was proposed that high resolution oxygen imaging could be used to improve understanding and modeling of the polymerization process. The higher sensitivity of rapid-scan imaging, relative to CW imaging, permits lower probe concentration and/or faster image acquisition.

EPR imaging is usually performed in a uniform magnetic field plus intentionally added gradients. An inverse experiment can be performed in which the nonuniformity of the magnetic field is probed by monitoring shifts in the resonance position of a paramagnetic probe with a narrow EPR signal. Rapid-scan imaging at 800 MHz has been demonstrated using the narrow signal from deoxygenated LiNC-BuO microcrystals to probe the magnetic field in the vicinity of a magnetized assembly of iron oxide Fe_3O_4 silicon-coated nanoparticles (Tseytlin, Bobko, & Tseytlin, 2020). Rapid scan has the advantage that the images can be obtained more quickly than by CW, although there may be limitations on the magnitude of field offsets that can be measured. EPR imaging has the advantage over magnetic field cameras because the interior of an object can be scanned.

5. Guidance on parameter selection for field-swept rapid scan

As in the case of CW and pulsed EPR, one can observe a rapid-scan EPR signal with parameters far-removed from optimum values. The optimization depends on the information desired, and the amount of post-acquisition processing acceptable. In CW EPR if a line shape with less than 1% distortion is desired, one needs to be very conservative in selection of microwave power to avoid power saturation, in modulation amplitude to avoid broadening of the line, and in the choice of scan time and filter time constant. Factors impacting analogous parameter selection for rapid-scan experiments are discussed in this section. The scan rate should be chosen consistent with the bandwidth of the EPR signal relative to the bandwidths of the resonator and the detection system. Higher powers can be used for rapid scan than for CW without saturating the signal, but the power saturation depends on scan rate, so power should be selected after sweep width and sweep frequency have been selected.

5.1 Triangular vs sinusoidal scans

Methods are now available to deconvolve either linear or sinusoidal scans, so data processing does not favor one waveform over the other. From a hardware perspective it is easier to generate sinusoidal scans than triangular scans. For a sinusoidal scan the coils can be resonated at the scan frequency, which decreases the power that is needed in the circuit to create a scan. As a result, wider and faster scans can be performed with sinusoidal waveforms than triangular waveforms. For linear magnetic field scans, the system has to support at least the seventh harmonic of the nominal drive frequency. That is, a 10 kHz triangular scan requires a driver that can achieve at least 70 kHz bandwidth which requires a more sophisticated driver than for sinusoidal scans. Even with seven harmonics, only about 95% of the scan will be linear enough to use (Quine, Czechowski, & Eaton, 2009). Increasing B_1 can improve signal-to-noise, but only if the signal is not saturated. The maximum non-saturating B_1 increases with increasing scan rate (Section 5.5). In a sinusoidal scan the scan rate varies through the spectrum with the fastest rates in the center and slower rates toward the edges. This means that saturation is more severe toward the edges than in the center, which may require use of a smaller B_1 to avoid saturation of all regions of the spectrum. Combining multiple rapid-scan segments to form a wider scan currently is based on triangular scans (Yu et al., 2015).

5.2 Sample size relative to size of scan coils

To acquire spectra with accurate lineshapes and maximum spectral fidelity high uniformity of the scan field is required. The deconvolution software converts the original time axis to a magnetic field axis. If the magnetic field scan is not uniform over the volume of the sample, the relationship between time and magnetic field is not uniform, which causes spectral broadening in the deconvolved spectrum. The need for homogeneity of the scanned field means that the dimensions of the sample must be substantially smaller than the diameter of the scan coils. When the relatively small diameter modulation coils that are built into commercial CW resonators are used to perform a rapid-scan experiment the sample needs to be smaller than if larger purpose-built scan coils are used. The maximum length of the sample should be selected based on the dimensions of the scan coils for the system available.

5.3 Resonator selection

The rapidly-changing magnetic field induces eddy currents in conducting components of the resonator as well as other system components including the magnet and gradient coils. These eddy current may contribute to magnetic field inhomogeneity and to the background signal. To minimize these effects it is important to use a resonator with as little conductivity as possible. Of the standard Bruker X-band resonators, the dielectric resonator has been used for rapid scan experiments rather than the SHQE or split ring resonators that have much larger eddy currents.

The Q of a critically coupled X-band resonator often is too high for rapid-scan EPR. The Bruker X-band rapid-scan accessory includes a resonator with adjustable Q that is useful for matching resonator bandwidth to signal bandwidth as discussed in Section 5.4. The Q of the resonator is adjusted by introducing lossy material into the resonator. In the experiments in the Denver lab the resonator Q has been lowered by various means, usually by introducing water into the sample area of the resonator in one or more separate tubes or by using water or alcohols as the solvent for the sample. Since rapid scan is a continuously-driven experiment, lowering the Q by over-coupling, as is done in pulsed EPR, is not an option for a reflection resonator, because of the resulting increase in reflected power. However, overcoupling could be used to lower the Q of a cross-loop resonator, since high power is incident on one resonator and signal detection is from the second resonator.

At lower frequencies resonators have been custom built for rapid-scan experiments. These resonators have been designed to minimize eddy currents. Q values typically are lower than at X-band so lowering resonator Q to provide adequate resonator bandwidth typically is not needed. Cross-loop resonators have the advantage of decreasing the impact of source noise on the detected signal.

5.4 Selection of scan frequency

As in CW spectroscopy spectra usually are obtained first with a sweep width that is wide enough to encompass the full spectrum. For rapid scan the widest scan may be limited by the coil driver and the scan coils. In our laboratory the widest scans available are about 160 G at X-band (Yu et al., 2014) and about 200 G at L-band (Shi, Rinard, Quine, Eaton, & Eaton, 2016). Wider scans require recording spectra in segments and piecing segments together (Yu et al., 2015). For a given scan width there are tradeoffs involving scan rate and spectral resolution that are determined by the scan-rate dependent signal bandwidth and the Q-dependent resonator bandwidth (Eaton & Eaton, 2018; McPeak et al., 2020). These considerations are discussed in detail here.

5.4.1 Signal bandwidth

The scan rate in a sinusoidal scan varies continuously across the spectrum. The maximum rate in the center of the scan, a_s (Eq. 1) is given by

$$a_s = \pi f_s B_m \text{ G s}^{-1} \tag{1}$$

where a_s is the sinusoidal scan rate, f_s is the scan frequency, and B_m is the scan width in gauss. In triangular scans the scan rate a_t (Eq. 2) is independent of position in the scan.

$$a_t = 2 f_s B_m \text{ G s}^{-1} \tag{2}$$

The bandwidth of a rapid-scan signal is given by

$$BW_{sig} = \frac{N\gamma}{2\pi} a T_2^* \tag{3}$$

where the value of N depends on the acceptable extent of lineshape broadening and the scan rate a is either a_t (triangular) or a_s (sinusoidal). T_2* is the time constant for decay of the oscillations on the trailing edge of the signal, and is also the decay time constant that would be observed for a free induction decay (FID). Although the definition of the rapid-scan regime is that the magnetic field scans through the signal in a time that is less than T_2, the frequency bandwidth of the spectrum depends on T_2*. A value of N between 5 and 10 in Eq. (3) is a relatively conservative starting point for selecting scan parameters. The bandwidth required depends on the T_2* for the sample and the linewidth accuracy desired in an experiment. The larger the value of N, the less the signal is broadened.

For a Lorentzian lineshape

$$T_2^* = T_2 = \frac{2}{\sqrt{3}\gamma \Delta B_{pp}} \tag{4}$$

where ΔB_{pp} is the peak-to-peak first derivative linewidth. The bandwidth for the rapid-scan signal for a Lorentzian line is then

$$BW_{sig} = \frac{Na}{\sqrt{3}\pi\Delta B_{pp}} \tag{5}$$

The relationship between $T_2{}^*$ and ΔB_{pp} depends on the lineshape. Unresolved hyperfine structure results in EPR line broadening that is approximately Gaussian, and decreases the signal bandwidth which therefore changes the selection of parameters for rapid scan. For example, ^{15}N perdeutero tempone has $\Delta B_{pp} \sim 75$ mG with Gaussian broadening due to unresolved deuterium hyperfine coupling. Use of Eq. (4), which assumes a Lorentzian shape, gives $T_2{}^* = 380$ ns, but the experimental $T_2{}^*$ by pulse methods is about 430 ns. Eq. (5) is a useful starting point for estimating signal bandwidth and selecting scan rates consistent with a particular resonator Q or scan frequency.

5.4.2 Resonator bandwidth

A resonator acts as a filter for the rapid-scan EPR signal as was shown experimentally at X-band for scans in a high Q resonator (Joshi, Eaton, & Eaton, 2005). Resonator Q can be expressed in many ways, but one that is convenient for discussion of rapid-scan EPR is in terms of the frequency bandwidth of the resonator, $\Delta\nu$.

$$Q = \frac{\nu}{\Delta\nu} = \frac{\nu}{BW_{res}} \tag{6}$$

Since each half cycle of a rapid-scan experiment is recorded with either increasing or decreasing field/frequency, the relevant bandwidth that is available for a rapid-scan signal (BW_{RS}) is only half of the resonator bandwidth.

$$BW_{RS} = \frac{\nu}{2Q} \tag{7}$$

For an X-band resonator with $\nu = 9.3 \times 10^9$ Hz and $Q = 3000$, $\Delta\nu = 3.1$ MHz, and BW_{RS} is 1.55 MHz (Eq. 7). This bandwidth is too low for many rapid-scan experiments. If the Q is lowered to 300 then $\Delta\nu = 31$ MHz and BW_{RS} is 15.5 MHz. To take advantage of this bandwidth the spectrometer detection and amplification bandwidths should be at least 15.5 MHz, preferably greater. This detection bandwidth is available in pulse spectrometers but not in typical CW spectrometers.

5.4.3 Examples

Examples of the impact of scan rate on experimental spectra are shown in (McPeak et al., 2020) (Fig. 7) for data acquired in a resonator with Q = 1500. For the spectra shown in Fig. 7A,B the signal bandwidth is less than the resonator bandwidth and the lineshape is in good agreement with slow scan CW spectra. However for the faster scan rates used to acquire the data in Fig. 7C,D the signal bandwidth exceeds the resonator bandwidth and the signal is broadened. To avoid signal distortion a slower scan rate should be used. Alternatively, a smaller segment of the spectrum could be recorded with the higher scan rate, which would reduce the signal bandwidth and result in high resolution spectra.

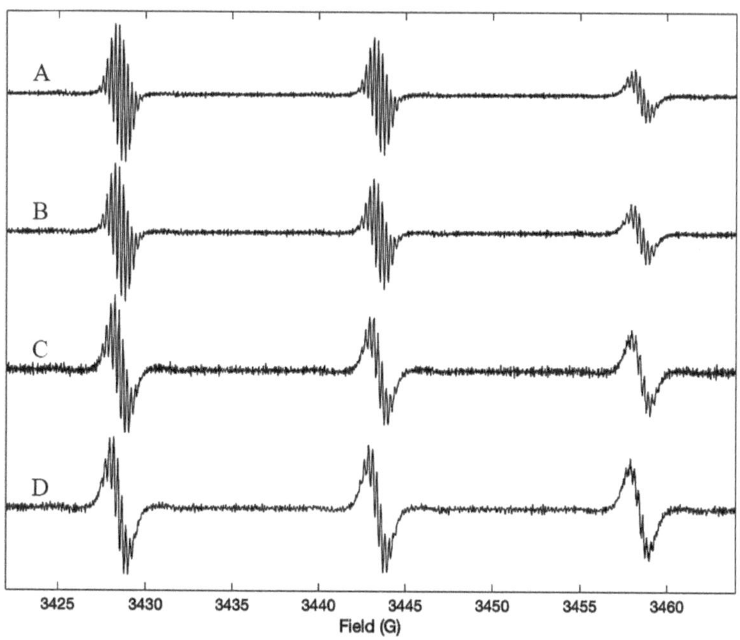

Fig. 7 First-derivatives of rapid-scan spectra of 0.2 mM CTPO in ethanol obtained with a sample height of 3 mm, scan width of 60 G and 20 mG B_1. (A) Triangular scan with frequency of 1.8 kHz and scan rate of 220 G/s. (B) Sinusoidal scan with frequency of 1.8 kHz and scan rate in the center of the spectrum of 350 kG/s. (C) Sinusoidal scan with frequency of 6.2 kHz and scan rate in the center of the spectrum of 1200 kG/s. (D) Sinusoidal scan with frequency of 10.5 kHz and scan rate in the center of the spectrum of 2000 kG/s. *Reproduced with permission from McPeak, J., Höfer, P., Kacprzak, S., Carl, P., Weber, R., Eaton, S. S. and Eaton, G. R. (2020). Rapid-scan electron paramagnetic resonance of highly resolved hyperfine in organic radicals.* ChemPhysChem, 21, *2564–2570.*

Consider calculations of the bandwidth requirements for two specific cases:

Case 1. For anoxic deuterated trityl radical $\Delta B_{pp} \sim 20$ mG. A 20 kHz sinusoidal scan with 5 G sweep width ($a_s = 3.4 \times 10^5$ G/s) results in a signal bandwidth of ~15 MHz, at $N = 5$. At 9.6 GHz, resonator Q (Eq. 7) should be less than about 320. If the Q cannot be lowered this far, then the scan rate would need to be reduced to avoid signal broadening.

Case 2. A nitroxide with $\Delta B_{pp} \sim 150$ mG, recorded with a 80 kHz sinusoidal scan and 40 G sweep width ($a_s = 1.0 \times 10^7$ G/s) has a signal bandwidth of ~62 MHz at $N = 5$. At 9.6 GHz, resonator Q would have to be less than about 78. Case (2) could be collected into a Bruker SpecJet which has a maximum bandwidth of 200 MHz, but the same signals would be distorted by the 20 MHz bandwidth of a Bruker SPU. In each case the video amplifier bandwidth in the bridge would need to be set to greater than the resonator and signal bandwidths. Most nitroxides have substantially broader lines so the constraint on Q would not be as severe. If the goal were to maximize signal-to-noise and some broadening was acceptable, a Q higher than 78 could be used. Alternatively, if lineshape fidelity was the highest priority, then a slower scan rate would be required if the resonator Q was greater than about 78 and the sweep width could be reduced to record only one of the nitrogen hyperfine lines.

For immobilized or slowly tumbling species anisotropies are not fully averaged and T_2^* may be much smaller than T_2, which makes the signal bandwidth requirement smaller than that determined by T_2. In practice, the scan rate is initially set to a conservatively slow value. The linewidth is monitored as the scan rate is increased, until the line broadening approaches the maximum acceptable value. The signal-to-noise can be improved, at the expense of line broadening, by acquiring data with smaller bandwidth, which permits use of higher resonator Q and increased post-processing filtering.

5.4.4 Constraints on scan rates imposed by rapid-scan oscillations.

If the relaxation times are long relative to the time to scan through the line, the transient effects on the trailing edge of the signal damp out with the time constant T_2^*. If the time between successive passes through resonance is less than about 5 T_2^*, oscillations have not fully damped out before the next spin excitation, which complicates interpretation of the data. This constraint requires that $1/f_s > \sim 20\ T_2^*$, and depends on the position of the signal in the scan. Since T_2^* usually is $< T_1$ this limitation on the scan frequency is less restrictive than the limitation on pulse repetition rates for pulsed EPR.

5.5 Microwave power

Higher powers can be used for rapid scan than for CW without saturating the signal. The scan rate should be selected based on the criteria discussed in the prior paragraphs prior to selecting the microwave power for a rapid scan experiment. A power saturation curve of signal amplitude as a function of microwave power should then be recorded at the selected scan rate. An example of the difference in power saturation for rapid scan and CW is shown in Fig. 8. In a sinusoidal scan the scan rate varies through the spectrum. Although signal bandwidth is calculated based on the maximum rate in the center of a sinusoidal scan, power saturation should be determined for peaks toward the edges of the sinusoidal scan because these signals will saturate at lower B_1 than peaks in the center of the scan.

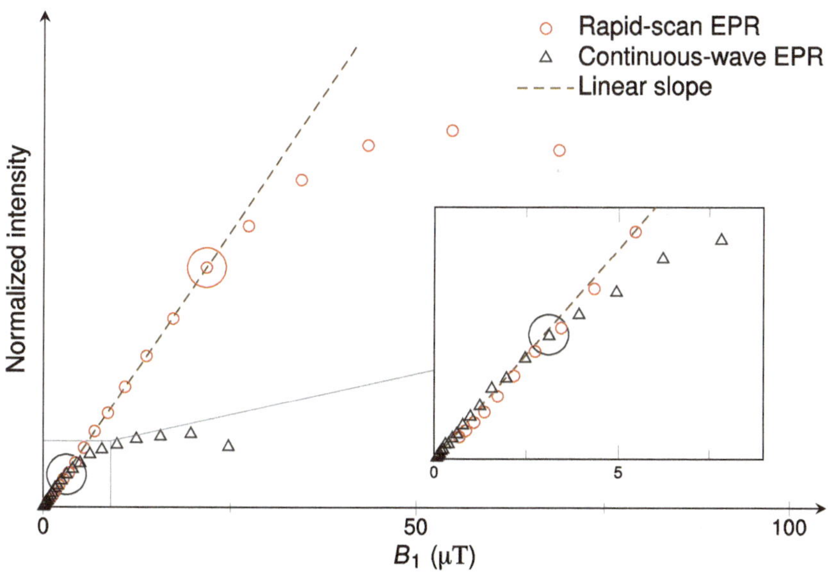

Fig. 8 Saturation curves for rapid scan (red circles) and CW (black triangles) EPR on a sample of α-hydrogenated silicon. Signal intensities are plotted as a function of the microwave magnetic field, B_1. Intensity values are normalized such that the slope in the linear regime (black dashed line) is equal for RS and CW. Circles mark the B_1 that was selected to acquiring data. The inset magnifies the low-power regime for the CW experiments. *Reproduced with permission from Möser, J., Lips, K., Tseytlin, M., Eaton, G. R., Eaton, S. S., Schnegg, A. (2017). Using Rapid-scan EPR to improve the detection limit of quantitative EPR by more than one order of magnitude.* Journal of Magnetic Resonance, 281, *17–25.*

5.6 Filtering in post-processing

Although the bandwidth required for acquisition of the rapid scan signal is greater than for CW, the bandwidths are the same after deconvolution of the rapid scan driving function. Thus the filtering that can be applied to a CW spectrum is the same as for rapid scan.

6. Summary

Progress has been made in multiple aspects of rapid scan EPR. The recent availability of the commercial Bruker BioSpin rapid scan accessory for their X-band EMX and Elexsys systems makes this technique available to a wide range of users without the need to construct their own system. Developments at lower frequencies are underway in several labs with the goal of facilitating in vivo and preclinical rapid scan imaging. Development of new deconvolution algorithms will make data processing more robust. Frequency scans have substantial promise at higher frequencies. New examples of applications show the wide applicability and advantages of rapid scan.

Acknowledgments

Support for work in our lab on rapid-scan EPR has been provided by NSF IDBR 0753018, NIH EB000557, NIH R01CA177744, and NIH R01CA262159, and by the University of Denver.

References

Braun, T. S., Stehle, J., Kacprzak, S., Carl, P., Hofer, P., Subramanian, V., et al. (2021). Intracellular protein-lipid interactions studied by rapid-scan electron paramagnetic resonance. *Journal of Physical Chemistry Letters*, *12*, 2471–2475.

Buchanan, L. A., Rinard, G. A., Quine, R. W., Eaton, S. S., & Eaton, G. R. (2018). Tabletop 700 MHz EPR imaging spectrometer. *Concepts in Magnetic Resonance Part B: Magnetic Resonance Engineering*, *48B*, e21384. https://doi.org/10.1002/cmr.b.21384.

Buchanan, L. A., Woodcock, L. B., Quine, R. W., Rinard, G. A., Eaton, S. S., & Eaton, G. R. (2018). Background correction in rapid scan EPR spectroscopy. *Journal of Magnetic Resonance*, *293*, 1–8.

Eaton, G. R., & Eaton, S. S. (2018). Rapid scan electron paramagnetic resonance. In D. Goldfarb, & S. Stoll (Eds.), *EPR spectroscopy: Fundamentals and methods* (pp. 503–520). N. Y., John: Wiley & Sons.

Eaton, G. R., Eaton, S. S., Barr, D. P., & Weber, R. T. (2010). *Quantitative EPR*. New York: Springer-Verlag/Wien.

Eaton, S. S., Quine, R. W., Tseitlin, M., Mitchell, D. G., Rinard, G. A., & Eaton, G. R. (2014). Rapid scan electron paramagnetic resonance. In S. K. Misra (Ed.), *Multifrequency Electron paramagnetic resonance: Data and techniques* (pp. 3–67). New York: Wiley.

Eaton, S. S., Shi, Y., Woodcock, L., Buchanan, L. A., McPeak, J., Quine, R. W., et al. (2017). Rapid scan EPR imaging. *Journal of Magnetic Resonance*, *280*, 140–148.

Eichhoff, U., & Hofer, P. (2020). 75 years of EPR. EPR milestones in 60 years of Bruker history. *Applied Magnetic Resonance*, *51*, 1723–1737.

Epel, B., Sundramoorthy, S. V., Krzykowska-Serda, M., Maggio, M. C., Tseytlin, M., Eaton, G. R., et al. (2017). Imaging thiol redox status in murine tumors in vivo with rapid-scan electron paramagnetic resonance. *Journal of Magnetic Resonance*, *276*, 31–36.

Joshi, J. P., Eaton, G. R., & Eaton, S. S. (2005). Impact of resonator on direct-detected rapid-scan EPR at 9.8 GHz. *Applied Magnetic Resonance*, *28*, 239–249.

Laguta, O., Tucek, M., Van Slageren, J., & Neugebauer, P. (2018). Multi-frequency rapid-scan HFEPR. *Journal of Magnetic Resonance*, *296*, 138–142.

McCrory, D. J., Anders, M. A., Ryan, J. T., Shrestha, P. R., Cheung, K. P., Lenahan, P. M., et al. (2019). Slow- and rapid-scan frequency-swept electrically detected magnetic resonance of MOSFETs with a non-resonance microwave probe within a semi-conductor wafer-probing stations. *Review of Scientific Instruments*, *90*, 014708.

McPeak, J., Höfer, P., Kacprzak, S., Carl, P., Weber, R., Eaton, S. S., et al. (2020). Rapid-scan electron paramagnetic resonance of highly resolved hyperfine in organic radicals. *ChemPhysChem*, *21*, 2564–2570.

Mitchell, D. G., Tseitlin, M., Quine, R. W., Meyer, V., Newton, M. E., Schnegg, A., et al. (2013). X-band rapid-scan EPR of samples with long electron relaxation times: A comparison of continuous wave, pulse, and rapid-scan EPR. *Molecular Physics*, *111*, 2664–2673.

Möser, J., Lips, K., Tseytlin, M., Eaton, G. R., Eaton, S. S., & Schnegg, A. (2017). Using Rapid-scan EPR to improve the detection limit of quantitative EPR by more than one order of magnitude. *Journal of Magnetic Resonance*, *281*, 17–25.

Quine, R. W., Czechowski, T., & Eaton, G. R. (2009). A linear magnetic field scan driver. *Concepts in Magnetic Resonance Part B: Magnetic Resonance Engineering*, *35B*, 44–58.

Sanzhaeva, U., Xu, X., Guggilapu, P., Tseytlin, M., Khramtsov, V. V., & Driesschaert, B. (2018). Imaging of enzyme activity by electron paramagnetic resonance: Concept and experiment using a paramagnetic substrate of alkaline phosphatase. *Angewandte Chemie, International Edition*, *57*, 11701–11705.

Sato-Akaba, H., & Tseytlin, M. (2019). Development of an L-band rapid scan EPR digital console. *Journal of Magnetic Resonance*, *304*, 42–52.

Shi, Y., Eaton, S. S., & Eaton, G. R. (2019). Rapid-scan EPR imaging of a phantom comprised of species with different linewidths and relaxation times. *Journal of Magnetic Resonance*, *308*, 106593.

Shi, Y., Rinard, G. A., Quine, R. W., Eaton, S. S., & Eaton, G. R. (2016). Rapid scan electron paramagnetic resonance at 1.0 GHz of defect centers in γ-irradiated organic solids. *Radiation Measurements*, *85*, 57–63.

Tseitlin, M., Mitchell, D. G., Eaton, S. S., & Eaton, G. R. (2012). Corrections for sinusoidal background and non-orthogonality of signal channels in sinusoidal rapid magnetic field scans. *Journal of Magnetic Resonance*, *223*, 80–84.

Tseitlin, M., Rinard, G. A., Quine, R. W., Eaton, S. S., & Eaton, G. R. (2011). Deconvolution of sinusoidal rapid EPR scans. *Journal of Magnetic Resonance*, *208*, 279–283.

Tseytlin, M. (2017). Full cycle rapid scan EPR deconvolution algorithm. *Journal of Magnetic Resonance*, *281*, 272–278.

Tseytlin, M. (2020). General solution for rapid scan EPR deconvolution problem. *Journal of Magnetic Resonance*, *318*, 106801.

Tseytlin, M. (2021). Continuous-wave rapid scan EPR. *Electron Paramagnetic Resonance*, *27*, 188–213.

Tseytlin, O., Bobko, A. A., & Tseytlin, M. (2020). Rapid scan EPR imaging as a tool for magnetic field mapping. *Applied Magnetic Resonance*, *51*, 1117–1124.

Tseytlin, O., Guggilapu, P., Bobko, A., Alahmad, H., Xu, X., Epel, B., et al. (2019). Modular imaging system: Rapid Scan EPR at 800 MHz. *Journal of Magnetic Resonance*, *305*, 94–103.

Tseytlin, O., O'Connell, R., Tseytlin, M., Sivishankar, V., & Bobko, A. A. (2021). Rapid scan oxygen imaging in photoactivated resin used for stereolithographic 3D printing. *3D Printing and Additive Manufacturing*, *8*(6). https://doi.org/10.1089/3dp.2020.0170.

Tseytlin, M., Stolin, A. V., Guggilapu, P., Bobko, A. A., Khramtsov, V. V., Tseytlin, O., et al. (2018). A combined positron emission tomography (PET)-electron paramagnetic resonance imaging (EPRI) system: Initial evaluation of a prototype scanner. *Physics in Medicine and Biology*, *63*, 105010.

Yu, Z., Liu, T., Elajaili, H., Rinard, G. A., Eaton, S. S., & Eaton, G. R. (2015). Field-stepped direct detection electron paramagnetic resonance. *Journal of Magnetic Resonance*, *258*, 58–64.

Yu, Z., Quine, R. W., Rinard, G. A., Tseitlin, M., Elajaili, H., Kathirvelu, V., et al. (2014). Rapid-scan EPR of immobilized nitroxides. *Journal of Magnetic Resonance*, *247*, 67–71.

CHAPTER TWO

Resolution and characterization of confinement- and temperature-dependent dynamics in solvent phases that surround proteins in frozen aqueous solution by using spin-probe EPR spectroscopy

Wei Li, Benjamen Nforneh, Katie L. Whitcomb, and Kurt Warncke*
Department of Physics, Emory University, Atlanta, GA, United States
*Corresponding author: e-mail address: kwarncke@physics.emory.edu

Contents

Abstract

Spin probe electron paramagnetic resonance spectroscopy is applied to characterize the dynamics of concentric hydration and mesophase solvent domains that surround proteins within the ice boundary in frozen aqueous solutions. The solvent dynamics are tuned by variation of temperature (190–265 K) and by the degree of ice boundary

Methods in Enzymology, Volume 666
ISSN 0076-6879
https://doi.org/10.1016/bs.mie.2022.02.009

confinement, which is modulated by the volume of added cryosolvent (0–$\sim$50 Å separation distance from protein surface). Goals are to: (1) characterize the protein-coupled solvent dynamics on correlation time scales of $\sim 10^{-10} < \tau < 10^{-7}$ s, and spatial scales from protein surface to periphery of the surrounding solution, from the perspective of a free, small-molecule ($\sim$7 Å diameter) probe, and (2) reveal properties of the solvent-protein coupling that can be correlated with protein functions, that are measureable under the same conditions. Rotational mobility of the nitroxide spin probe, TEMPOL, resolves and tracks two solvent components, the protein-associated domain (PAD; akin to hydration layer) and surrounding mesodomain, through their distinct temperature- and confinement-dependent values of τ and normalized weight. Detailed protocols are described for simulation of two-component nitroxide EPR spectra, which are categorized by line shape regime and guided by a library of template spectra and simulation parameters derived from two model soluble globular proteins. The order-disorder transition in the PAD, which is a universal feature of protein-coupled solvent dynamics, provides a well-defined, tunable property for elucidating mechanism in solvent-protein-function dynamical coupling. The low-temperature mesodomain system and EPR spin probe method are generally applicable to reveal solvent contributions to a broad range of macromolecule-mediated biological processes.

1. Introduction

Comprehensive, molecular-level description of the function of enzymes and other proteins requires characterization of the protein dynamics, or fluctuations and configurational (conformational) transitions (Hammes, Benkovic, & Hammes-Schiffer, 2011; Henzler-Wildman & Kern, 2007; Herschlag & Natarajan, 2013), that connect canonical structures portrayed by high-resolution structural techniques (Berman et al., 2000). Protein dynamics are intimately coupled to the dynamics of the surrounding solvent (Barnes et al., 2017; Lewandowski, Halse, Blackledge, & Emsley, 2015; Nucci, Pometun, & Wand, 2011; Orotny et al., 2017; Pal, Peon, Bagchi, & Zewail, 2002; Qin, Wang, & Zhong, 2016), consonant with the integral role of water in biological function (Ball, 2008). The solvent dynamics are influenced by the degree of confinement, and thus, external solvent confinement presents an incisive control property, that can be used to manipulate solvent-protein-function dynamical coupling, to discern mechanism. In general, confinement originates from a geometric restriction on the possible configurations of a system, that arises from volume excluded at boundaries, identified as static (Cerveny, Mallamace, Swenson, Vogel, & Xu, 2016; Rasaiah, Garde, & Hummer, 2008; Zhou, Rivas, & Minton, 2008). Physical confinement can be influenced by chemical interactions (Alcoutlabi & McKenna, 2005). A boundary that delimits the aqueous solvent around a protein creates confinement, that is incident on the protein. The surface of soluble proteins also exerts a confinement effect on the surrounding aqueous solvent, leading to structuring of water, that extends up to 30 Å from the protein surface (Ebbinghaus et al., 2007; Franck, Pavlova, Scott, & Han, 2013; Heyden & Tobias, 2013). This creates a distinct protein-associated solvent domain, or hydration layer. Measurements on protein powders show that significant internal protein motions are manifested when protein hydration achieves a mean thickness of approximately two water molecules, or 6 Å (Frauenfelder et al., 2009; Panagopoulou, Kyritsis, Shinyashiki, & Pissis, 2012). Confinement from boundaries external to the protein is conducted through the hydration layer to impact protein dynamics (Dantsker, Samuni, Friedman, & Agmon, 2005; Frauenfelder et al., 2009).

Here, we describe a system for control of solvent confinement, and an electron paramagnetic resonance (EPR) spectroscopy spin probe methodology, that resolves the solvent phases around proteins and characterizes their temperature- and mesophase volume-dependent dynamics (Fig. 1) (Ionescu, Li, Nforneh, & Warncke, 2021; Nforneh & Warncke, 2017, 2019). The system consists of protein in the polycrystalline ice-bounded, interstitial, aqueous-cryosolvent mesodomain in frozen samples, in broad analogy to

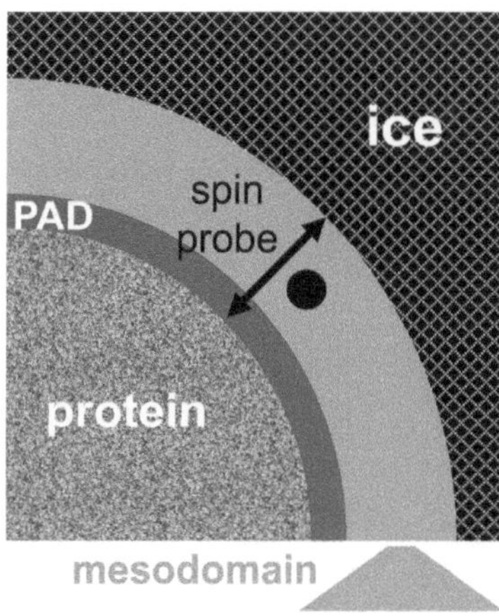

Fig. 1 Depiction of the fluid phases that surround a protein in frozen aqueous solution in the presence of a small volume (0.5–4% v/v) of cryosolvent, added prior to cryotrapping. The representative protein is ethanolamine ammonia-lyase and the cryosolvent is dimethylsulfoxide. The radial direction of the two-dimensional cross-section represents the relative mean dimensions for the condition of added 2% v/v dimethylsulfoxide. Protein (speckled) is surrounded by successive PAD (dark gray) and mesodomain (light gray) solvent layers, that are bounded by water ice (hatched). Range of mesodomain mean thickness variation is depicted by the trapezoid at bottom: minimum (top) to maximum (base). TEMPOL spin probe (black circle; mean diameter, 7 Å) ranges over PAD and mesodomain phases (arrow).

strategies of molecular confinement in amorphous or crystalline materials, under soft and hard conditions (Alcoutlabi & McKenna, 2005; Cerveny et al., 2016; Dunn & Zink, 1997; Wheeler, Nocek, & Hoffman, 2006) and in aerosols and reverse micelles (Arsane et al., 2021; Van Horn, Ogilvie, & Flynn, 2009). The formation of the system follows the phenomenon, that particles over an exceptionally broad size range, from several Ångstroms (molecules) to micrometers (polymer spheres) are excluded from the advancing ice fronts during the freezing of aqueous solutions, which leads to their collection in the freeze-concentrated liquid, or mesodomain, within the polycrystalline frozen aqueous solution (Körber, 1988). For the case of the model protein, lysozyme (1.4×10^4 g/mol), dispersal in the region of the maximum freeze concentration [MFC; the concentration at the eutectic point of the aqueous cryosolvent phase (Franks, 2003)], and depletion in the ice region, has been resolved by spatial imaging techniques (Curtis et al., 2012; Dong, Hubei, Bischof, & Aksan, 2009). This concurs with the venerable recognition that freezing of aqueous EPR samples in the presence of cryosolvents suppresses aggregation of the constituents (Leigh Jr. & Reed, 1971).

The low-temperature mesodomain system has the advantages for study of protein-solvent dynamical coupling of a fully-hydrated protein interior, a complete hydration layer, or protein-associated domain (PAD), and a surrounding solvent mesodomain (Ionescu et al., 2021; Nforneh & Warncke, 2017, 2019). Ice boundary confinement is tunable from strong to weak by addition of increasing amounts of cryosolvents. Cryosolvents are preferentially excluded from the PAD (Arakawa, Kita, & Timasheff, 2007; Timasheff, 2002), and thus contribute to a volume increase of only the mesodomain. The rotational mobility of the nitroxide EPR spin probe, TEMPOL, reports on the dynamical state of the PAD and mesodomain solvent phases over a detection window (Banerjee & Bhat, 2009) of correlation times of $\sim 10^{-10} \leq \tau_c \leq 10^{-7}$ s at the X-band microwave frequency of ~9.5 GHz (the range differs, for other probes and microwave frequencies) (Goldman, Bruno, Polnaszek, & Freed, 1972; Sato et al., 2008). TEMPOL is a particularly illuminating reporter of physiologically-relevant solvent dynamics, because its dimension (mean diameter, ~7 Å) is comparable to solvent molecules (water, ~3 Å; cosolvents, such as dimethylsulfoxide, DMSO, ~5 Å) and to protein surface relief features created by amino acid side chains. The line shape of the continuous-wave EPR spectrum provides a direct appraisal of spin probe mobility (Wertz & Bolton, 1986), that is quantified by using theory-based simulations (Stoll, 2014; Stoll & Schweiger, 2006) Fig. 2 shows characteristic profiles of the temperature-dependence

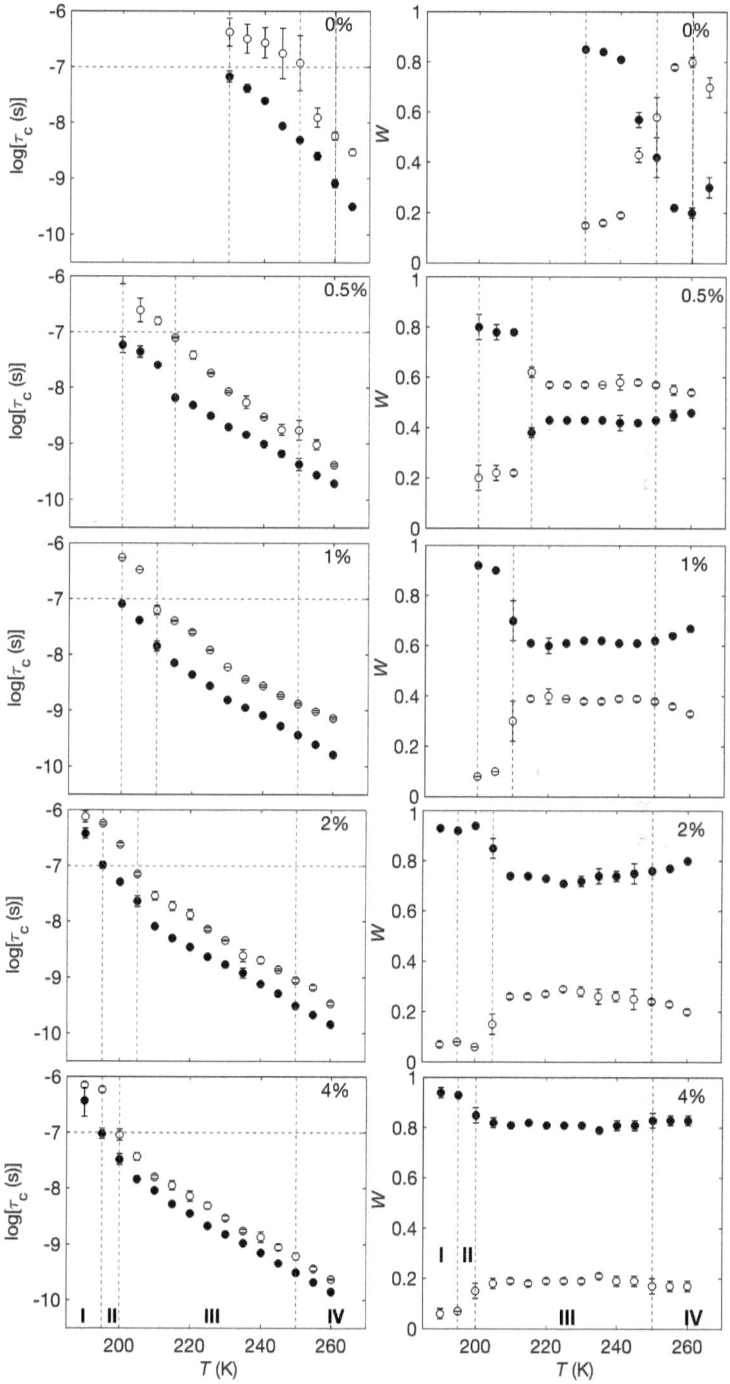

Fig. 2 See figure legend on next page.

of rotational reorientation of the EPR spin probe, TEMPOL, around a protein under conditions in which the mean mesodomain layer is increased from approximately 0 to 40 Å by using 0–4.0% v/v DMSO cryosolvent, added to the 0.3 mL EPR sample prior to freezing (Nforneh & Warncke, 2017, 2019). The "slow" mobility component, characterized by the rotational correlation time, $\tau_{c,s}$, and normalized weight, W_s, represents TEMPOL associated with the PAD, and the "fast" mobility component ($\tau_{c,f}$, W_f) corresponds to TEMPOL in the mesodomain, that envelops the PAD and is bordered externally by the polycrystalline ice (Fig. 1) (Nforneh & Warncke, 2017) Aqueous-cryosolvent systems generally preserve biological macromolecule structure at low temperature (Douzou, 1977).

The four regions (I–IV) (Nforneh & Warncke, 2019) distinguished by the temperature dependence of the spin probe mobilities indicate properties and events in the solvent dynamics that are poised for correlation with protein function. For example, a signature feature is the abrupt transition in W values and associated kink in $\log\tau_{c,f}$ (Fig. 2) (Nforneh & Warncke, 2019), and change in dielectric permittivity (Ionescu et al., 2021), that is caused by an order-disorder transition (ODT) in the PAD (transition from ordered to disordered state, with increasing temperature). The TEMPOL-detected ODT is associated with a fragile-strong crossover event involving a liquid-liquid phase transition in protein hydration water (Chen et al., 2006; Mallamace et al., 2006) and is a manifestation of transitions in the dynamics of protein and coupled solvent water observed in the general region of 220 ± 20 K by using different techniques (Doster, 2010; Lewandowski et al., 2015; Ngai, Capaccioli, & Paciaroni, 2013). Thus, the ODT represents a direct perturbation of a specific class (Ionescu et al., 2021) of PAD solvent-coupled protein dynamics, and therefore, a property that can be correlated with protein function, to reveal mechanism in solvent-protein-function coupling (Ionescu et al., 2021). Fig. 2, along with results from other

Fig. 2 Temperature dependence of the rotational correlation time of TEMPOL (column 1) and normalized mobility component weights (column 2) in the presence of protein, ethanolamine ammonia-lyase and dimethylsulfoxide cryosolvent, added at 0–4% v/v prior to cryotrapping (Nforneh & Warncke, 2017, 2019). In each panel, solid circles represent the fast component ($\log\tau_{c,f}$, W_f) and open circles represent the slow component ($\log\tau_{c,s}$, W_s) of TEMPOL spin probe rotational motion. Error bars represent standard deviations for three separate determinations. Regions I–III are defined by the detectability of spin probe motion and Region IV marks a trend of increasing W_f (indicated by dashed vertical lines). The upper limit of $\log\tau_c = -7.0$ for the detection of tumbling motion is indicated by horizontal dashed line.

conditions (Ionescu et al., 2021), shows that the ODT can be tuned from 245 to 200 K by increasing the amount of added cryosolvent.

Temperature-dependent transitions in solvent dynamics have been used to resolve contributions to ingress/egress and internal migration of diatomic gas molecules in heme proteins following iron-ligand photolysis in global-glass systems (Dantsker et al., 2005; Frauenfelder et al., 2009). Cryoreduction and low-temperature annealing has been used to characterize structures of metallo-center intermediates in heme enzymes (Blumenfeld, Davydov, Magonov, & Vilu, 1974; Davydov et al., 2002, 2005), and a multi-step annealing protocol used to measure reaction kinetics of intermediates in dihydrogen evolution in nitrogenase (Lukoyanov, Barney, Dean, Seefeldt, & Hoffman, 2007). The aqueous-cryosolvent mesodomain system allows cryotrapping of enzymes in the presence of substrates, and therefore, catalytic intermediate states. In the case of paramagnetic intermediates, such as organic radicals, and metal centers with relatively long spin-lattice relaxation times, full-spectrum, time-resolved EPR spectroscopy measures reaction kinetics continuously within the temperature range addressed by the TEMPOL spin probe (Kohne, Li, Zhu, & Warncke, 2019; Kohne, Zhu, & Warncke, 2017; Ucuncuoglu & Warncke, 2018; Zhu & Warncke, 2008). Thus, the reaction kinetics can be correlated with solvent dynamics.

The mesodomain system is well-suited for general application, because sample preparation and cryotrapping is straightforward, EPR spectroscopy can be conducted by using commercial EPR spectrometers with standard room to liquid nitrogen temperature control, and EPR simulations can be performed by using the open-source EasySpin (Stoll, 2014; Stoll & Schweiger, 2006) toolkit run on the MATLAB® (MathWorks, Natick, MA) software platform. The following detailed description of methodology provides a guide for application of the developed low-temperature mesodomain system (Ionescu et al., 2021; Nforneh & Warncke, 2017, 2019) to other protein and macromolecular systems.

2. Preparation of protein–mesodomain system in frozen solution

2.1 Sample preparation

Protein EPR samples are prepared in buffered aqueous solution. High buffer concentrations and added salts are avoided, to reduce the complexity of the frozen solution system. Other electrolytes or molecules with cryosolvent-like properties (Douzou, 1977) (for example, as a carry-over from stock solutions), unless added purposely for assessment, are avoided assiduously, because they alter the phase volumes and properties. Saccharides form heterogeneous crystalline domains, that create complexity in the mesodomain structure and properties (Chen, Sun, & Warncke, 2013; Malferrari, Savitsky, Lubitz, Möbius, & Venturoli, 2016). The conditions described here pertain to fluid cryosolvents, that create a detectably homogeneous two-component, PAD and mesodomain system. The standard protein concentration in the EPR sample for the weak confinement condition is determined by the amount of protein solvent accessible surface area. The goal is an aqueous-cryosolvent phase volume sufficient to surround the protein, in the absence of aggregation, while maintaining a protein concentration sufficient for detection of the PAD by the spin probe. The amounts of protein and cryosolvent added thus depend on the protein molecular mass, and on the mass and MFC of the cryosolvent.

Here, we describe the sample preparation parameters for the standard weak confinement condition in the mesodomain, achieved by adding 2% v/v DMSO to the sample prior to freezing (6.0 μL added to the sample, of total volume 0.3 mL). The DMSO-water phase diagram and properties have been reported (Rasmussen & MacKenzie, 1968; Shichman & Amey, 1972). For this condition, the target total protein solvent accessible surface area in the sample, $A_{S,prot,tot}$, is approximately 5×10^{20} Å^2. The solvent accessible surface area of an individual protein, $A_{S,prot}$, can be determined by analytical methods (Connolly, 1983; Wodak & Janin, 1980), and water with spherical radius, 1.4 Å. For a protein with available high-resolution structure, $A_{S,prot}$ can be determined by using VMD (Humphrey, Dahlke, & Schulten, 1996), PyMOL (DeLano, 2002; PyMOL, 2022) or other applications. For proteins of unknown structure, but known molecular mass, M_r (g/mol), the surface area of the equivalent sphere is calculated, by assuming a mean density for proteins, ρ_{prot}, of 1.35 g/mL (Fischer, Polikarpov, & Craievich, 2004; Quillin & Matthews, 2000). This method provides reasonable estimates for soluble proteins, over a wide molecular mass range. As an example, for the relatively large oligomeric protein, ethanolamine ammonia-lyase (EAL, *Salmonella enterica* Serovar typhimurium, M_r = 488,000 g/mol), the $A_{S,prot}$ determined from the protein structure is 1.3×10^5 Å^2 (Bovell & Warncke, 2013). The $A_{S,prot}$ from the equivalent sphere method is determined by using the following expression for the protein volume, V_{prot} (units, Å^3),

$$V_{prot} = \frac{M_r}{\rho_{prot} N_A} \times 10^{24} \quad (1)$$

where N_A is Avogadro's number (mol^{-1}). Thus, the radius of the equivalent sphere, r_{prot} (Å), is:

$$r_{prot} = \sqrt[3]{\frac{3\ V_{prot}}{4\pi}} \tag{2}$$

and

$$A_{S,prot} = 4\pi r_{prot}^2 \tag{3}$$

For EAL, the $A_{S,prot}$ estimated from M_r and d_{prot} is 3.4×10^4 Å^2, or approximately fourfold smaller than the structure-determined value, owing to the EAL oligomer surface relief (Bovell & Warncke, 2013). The concentration of EAL for the EPR sample of volume, V_{sample}, is calculated from

$$[EAL] = \frac{A_{S,prot,tot}}{A_{S,prot}\ N_A\ V_{sample}} \tag{4}$$

For the standard EPR sample volume of 0.3 mL and the structure-determined A_{prot}, the EAL concentration is 2×10^{-5} M or 20 μM (10 mg/mL).

TEMPOL is added to the sample volume, after addition of DMSO, and prior to addition of protein, from a stock TEMPOL solution in water. TEMPOL is soluble in water and buffered aqueous samples ($>10^{-1}$ M), and therefore does not require an organic solvent for dissolution. Preparation of TEMPOL in organic solvents, leads to significant carry-over of the organic solvent into the EPR sample, which adds to the mesodomain volume and changes its properties. A TEMPOL concentration of 2×10^{-4} M provides excellent signal-to-noise ratios over the full temperature range at relatively low microwave powers (0.2 mW). The low power averts power-broadening of the spectrum, and microwave heating. The protein samples described in this work included 10 mM potassium phosphate buffer (pH 7.5), protein at concentrations of 2.0×10^{-5} M (EAL) to 6.0×10^{-4} M (myoglobin, Mb; *Physeter macrocephalus*), and 0.2 mM TEMPOL (4-hydroxy-TEMPO, Sigma-Aldrich) spin probe in a final volume of 0.3 mL. Protein samples were prepared aerobically, at room temperature or on ice in small vials or Eppendorf tubes, mixed, and loaded into 4 mm outer diameter EPR tubes (707-SQ-250M, SP Wilmad-LabGlass).

2.2 Sample cryotrapping and storage

Following transfer to the EPR tube, the sample is frozen by plunging into $T=140$ K isopentane solution. The isopentane is chilled and maintained at 140 K in a dewar flask by adding liquid nitrogen to a thick glass cold finger, that is inserted into the flask. Temperature is determined by a thermocouple in the isopentane solution. This method has a characteristic cooling rate of 10 K/s for the contents of the 4 mm outer diameter quartz EPR tube (Wang, Zhu, Kohne, & Warncke, 2015). Methods for varying the cooling rate from 0.07 to 400 K/s have been described (Wang et al., 2015). A consistent freezing temperature and procedure are recommended, in order to achieve uniform properties over multiple samples. Following freezing, samples are transferred to a liquid nitrogen storage dewar.

3. Measurement of temperature-dependent EPR spectra

3.1 Continuous-wave EPR spectroscopy

Prior to temperature-dependence measurements, a spectrum of the sample is acquired at 120 K. This low-temperature spectrum serves as a rigid line shape standard, for later comparison, to assess possible changes in the spin probe EPR line shape or signal level incurred during the temperature-dependence measurements.

In the general measurement protocol, the sample is placed in the EPR cryostat, at an initial temperature, T, of 190 K, and held for at least 5 min. The initial EPR spectrum is obtained at 190 K. The controller is then set to the next higher T-value, and the system is allowed to equilibrate for 5 min at the new T-value. Temperature increments are typically 5 K. Toward the end of the 5 min period, the cavity is retuned, and at 5 min, a spectrum is acquired at the new T-value. This procedure is repeated for the entire T range, from 190 to 265 K (16 total spectra, for 5 K increments). In a separate series of measurements, a baseline sample (all components, but without TEMPOL) was subjected to the same measurement protocol. This completes one T-dependence series. Typically, a second experiment is then conducted, in the direction of decreasing T, starting from 265 K, by using the same procedure. This series provides a duplicate of the experiment, for the same sample. For soluble proteins examined so far, no hysteresis in correlation times or component amplitudes is observed. No hysteresis is observed for samples subjected to a second T-dependence experiment, after storage in liquid nitrogen. Therefore, equilibrium conditions at each

T prevail for soluble proteins. A second EPR sample, that is prepared independently, is carried through at least the T-increase protocol, to achieve at least three replications.

The X-band EPR spectrum acquisition parameters used for standard measurements in solutions that contain 0.2 mM TEMPOL spin probe are: microwave power, 2 mW; magnetic field modulation, 0.2 mT; modulation frequency, 100 kHz. For these parameter values, averaging of 4–8 spectra provides excellent signal-to-noise at each T value. The CW-EPR experiments described here were performed by using a Bruker E500 ElexSys EPR spectrometer and ER4123SHQE X-band cavity resonator. A Bruker ER4131VT temperature controller and cooling system, based on nitrogen gas flow through a coil immersed in liquid nitrogen, was used to establish T values of 190–265 K. For each temperature series (each day), the Bruker ER4131VT temperature readout was calibrated by using a 19180 4-wire RTD probe and Oxford Instruments ITC503 unit, as described (Wang et al., 2015). The uncertainty in T values is ± 0.5 K (Wang et al., 2015).

4. Simulation of temperature-dependent EPR spectra

4.1 EPR simulations: Goals and general background

The goal of the EPR simulations is to obtain motional parameters of the spin probe, that show temperature-dependent trends in the rotational correlation time (generally reported as the base-10 logarithm, $\log\tau_c$) and normalized component weights, W. A standard set of electron g-value, g, and ^{14}N hyperfine coupling, A, parameters was developed for use with spectra collected on our spectrometer, under all solution and temperature conditions, based on values that best-fit spectra of TEMPOL in aqueous-DMSO cryosolvent solution, in the rigid and fast motional limits. The g and A tensors are assumed coincident. In fact, the nitroxide g and A parameters display subtle dependences on solution environment (Clark, Sedhom, Elajaili, Eaton, & Eaton, 2018) and degree of motion (Hwang, Mason, Hwang, & Freed, 1975). Apparent g-values also depend on the accuracy of the spectrometer magnetic field calibration (Clark et al., 2018). These differences do not significantly influence the general trends in the motional simulation parameters, and the derived conclusions about the confinement-dependent mobility of the spin probe, that are obtained by using the methods reported, here. Absolute g and A parameters have been presented (Clark et al., 2018; Hwang et al., 1975), and parameters for different nitroxide (aminoxyl) radicals compiled (Berliner, 2014). Simulations of the CW-EPR spectra were performed by using the Chili algorithm in the program, EasySpin (Stoll & Schweiger, 2006) run in the MATLAB® (MathWorks, Inc.; www.mathworks.com) environment. EasySpin is available for download (www.easyspin.org).

The EPR spectrum can be calculated from theory, based on specification of the spin probe motion. In the opposite direction, certain specification of the spin probe motion, based on the EPR spectrum, is not possible. There are two levels of specification of free-solution spin probe motions: (1) resolution of different components, which correspond to different solvent phases, and (2) within a component, the relative freedom of the motion. Ambiguity in the spectrum–calculation correlation typically challenges the identification and resolution of different components. The temperature-dependence protocol alleviates this source of ambiguity for two-component systems, because temperature values are traversed that provide unequivocal two-component spectra. Thus, under the assumption that the system structure does not change dramatically in the interval 190–250 K, the two-component simulation can be confidently applied at all temperatures. Fig. 3 shows examples of one- and two-component simulations of the same two-component experimental spectrum, that illustrate the degrees of ambiguity and signature regions of mismatch.

The rotational reorientation is assumed to be isotropic in the simulation protocols presented here. This corresponds to equivalent freedom of motion about all three ^{14}N hyperfine principal axes, which are the dominant source of anisotropy at X-band microwave frequencies. This approximation is sound for $\log\tau_c < -8.0$, and tenable for $\log\tau_c \leq -7.0$ (Goldman et al., 1972; Sato et al., 2008). At $\log\tau_c$ values near and greater than -7.0, librational motions of the nitroxide moiety in spin labels contribute to the spectral broadening (Marsh, Bartucci, Guzzi, Sportelli, & Esmann, 2013). Tumbling motions outside of this range are accessible to EPR line shape analysis by using different microwave frequencies (Liang & Freed, 1999; Möbius et al., 2005) and detection schemes (Hyde & Thomas, 1980; Roozen & Hemminga, 1990). The upper limit on the detection bandwidth for the TEMPOL spin probe is therefore $\log\tau_c \approx -7.0$ (Banerjee, Bhat, Bhat, & Leporini, 2009). The lower limit on the detection bandwidth is dictated by the selected magnetic field modulation amplitude (here, 0.2 mT), and ultimately, the spin-spin relaxation time, T_2. Under the conditions used here, the lower limit of the detection bandwidth is $\log\tau_c \approx -10$. A single value of τ_c is attributed to each component, corresponding to spin probe occupancy of solvent phases that are homogeneous in their dynamics on the EPR time scale.

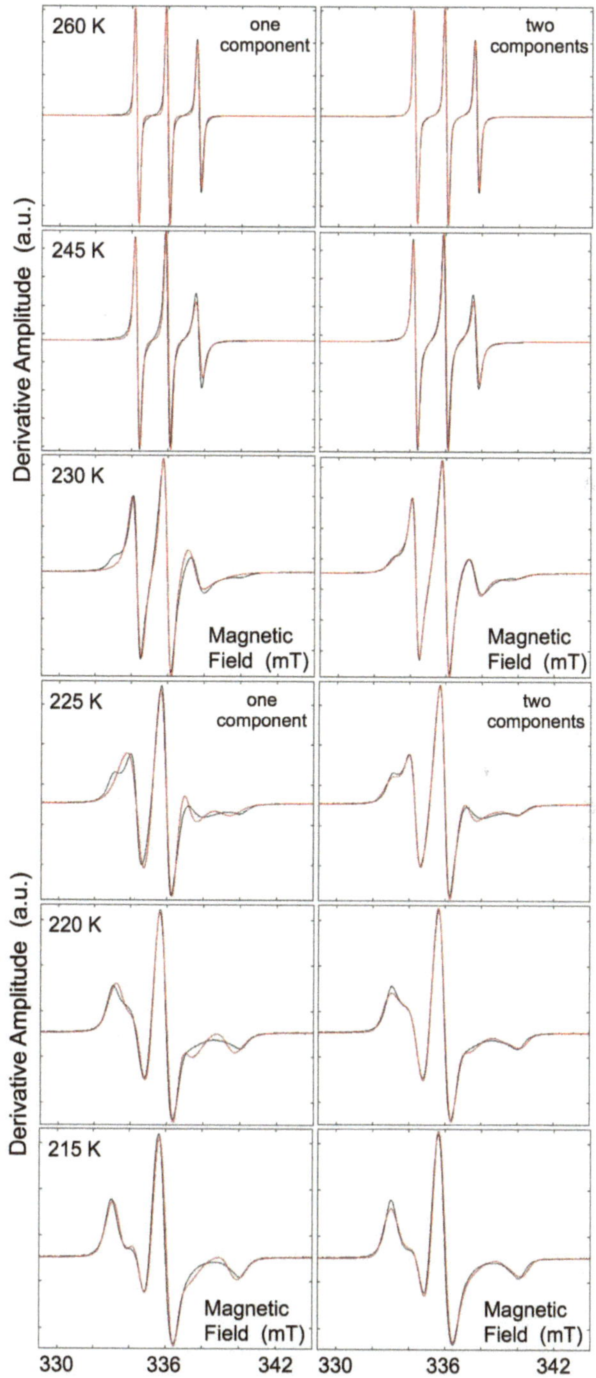

Fig. 3 See figure legend on next page.

4.2 EPR simulations: Systems

4.2.1 Mesodomain system in presence of protein and spin probe under weak confinement (2% v/v added DMSO)

Mesodomain systems with soluble, globular proteins ($M_r \geq 17{,}000\,g/mol$) are generally characterized by two-component spin probe EPR spectra (Li, Whitcomb, & Warncke, 2022). The approach to a simulation of a new experimental EPR spectrum is based on comparison with a template library of two-component spectra, as shown in Fig. 4. The library spectrum that matches most closely with the new spectrum is identified by inspection, and the simulation parameters from the library spectrum are used as starting parameters for the simulation. The library spectra at different temperatures are composed of the TEMPOL spectra for Mb in the presence of added 2% v/v DMSO (10 mg/mL protein in 10 mM KP_i buffer, pH 7.5). The line shape vs temperature dependence for the Mb spectra at added 2% v/v DMSO are essentially the same as for 2.0×10^{-5} M EAL protein (Nforneh & Warncke, 2019). For different added cryosolvent volumes or other solution conditions, one of the library line shapes will provide a best match, but not necessarily at the same temperature.

Specific simulation approaches are implemented according to the type of line shape, which in turn, depends on the temperature range. The simulation approaches are therefore sub-defined for different temperature intervals. The simulations corresponding to the library spectra are shown overlaid on the experimental spectra, and varied simulation parameters reported, in Fig. 4.

4.2.2 Spin probe in mesodomain system under other conditions, including no added cryosolvent

Starting parameters for simulations of EPR spectra obtained under other conditions are also determined by inspection of the line shape and the best match of the new spectrum with the library of TEMPOL spectra. For

Fig. 3 TEMPOL EPR spectra (black) and overlaid one-component (left column) and two-component (right column) EPR simulations (red) at selected temperature values (Nforneh & Warncke, 2019). Sample contains EAL protein and 0.5% v/v dimethylsulfoxide cryosolvent, added prior to cryotrapping. The selected spectra are representative of the different types of line shapes encountered for TEMPOL over the measured temperature range at different % v/v DMSO. Comparison of the one- and two-component simulations evinces the requirement for contributions of two components to the spectra, and illustrates regions of resolved contributions made by the "slow" motional component, relative to "fast" motional component, as described in Sections 4.5.2–4.5.5.

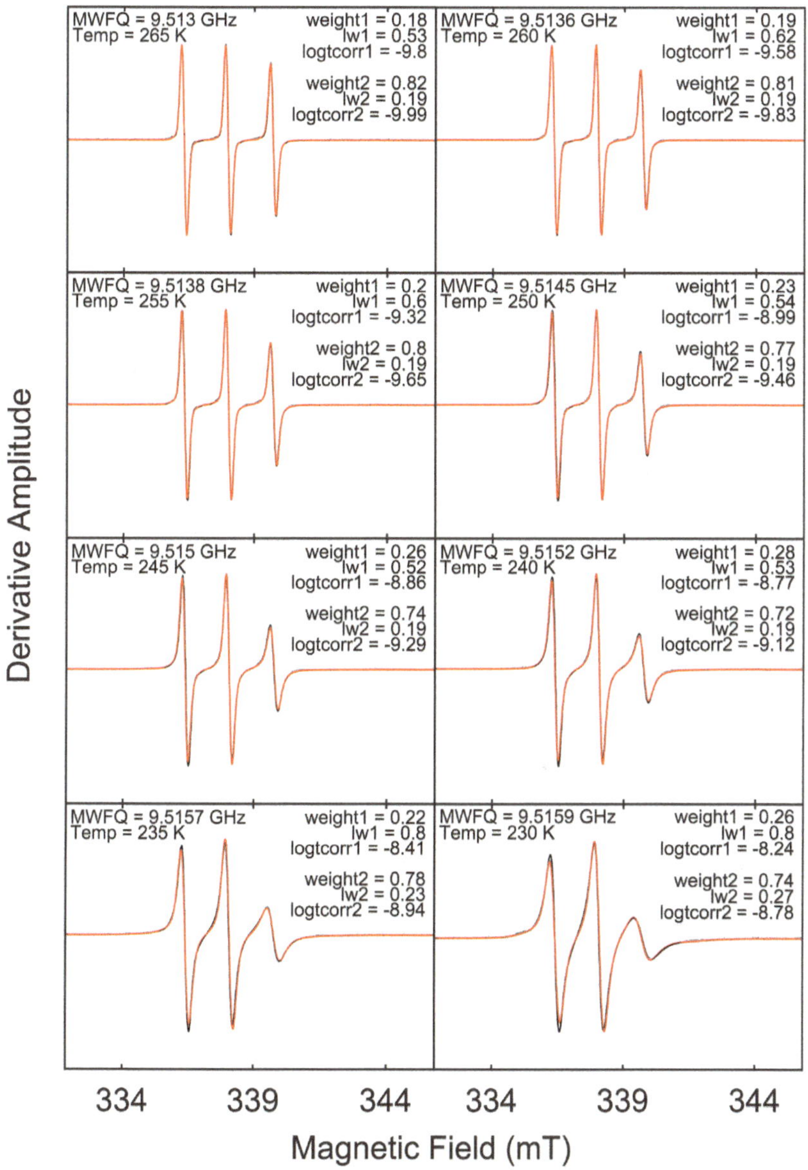

Fig. 4 Template library of TEMPOL EPR spectra (black) in the presence of the protein, myoglobin, in frozen aqueous solution at different temperatures. Samples contained 2% v/v dimethylsulfoxide cryosolvent added prior to cryotrapping. Two-component EPR simulations are overlaid (red), and the corresponding varied simulation parameters are listed in each panel. Spectra are normalized to the central peak-to-trough amplitude. Alignment along the magnetic field axis corresponds to the microwave frequency at 200 K.

(Continued)

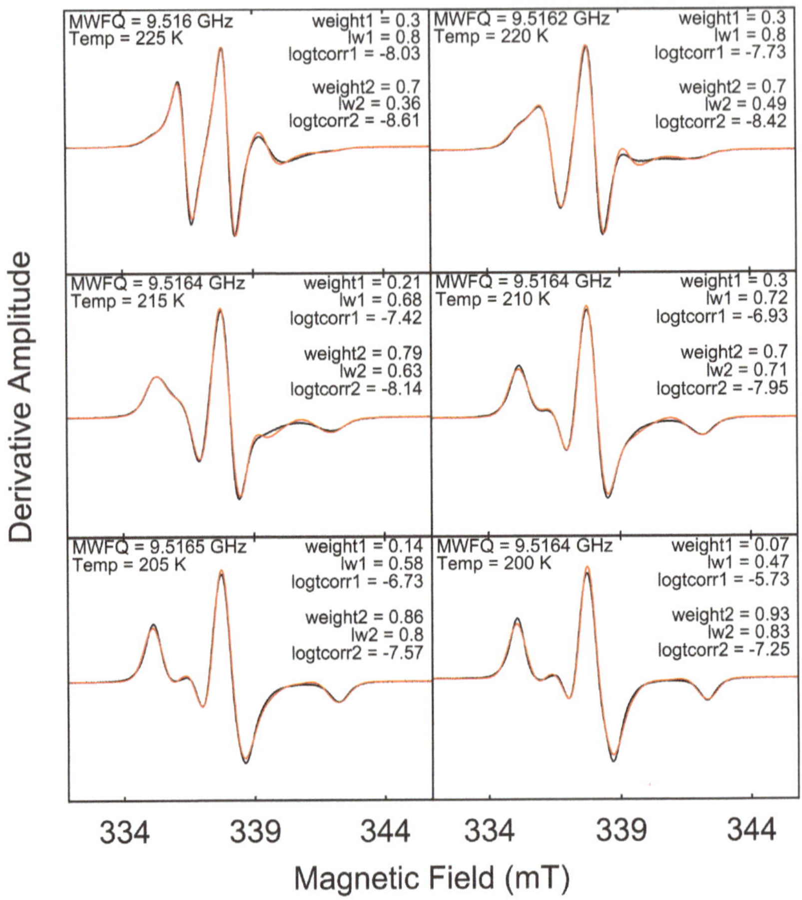

Fig. 4—Cont'd

example, for the strong confinement condition, attained in the absence of added cryosolvent, the EPR line shapes at the acquisition temperatures correspond to line shapes at lower temperature of the library in Fig. 4. Nitroxide spectra encountered for the aqueous-cryosolvent mesodomain systems in the absence of protein show a single or dominant component over the temperature range. In this case, the most general approach is to first achieve a one-component simulation of the single or dominant component, and then, if necessary, introduce the minor second component with 10-fold larger τ_c at a W_s value of 0.5.

4.3 Simulation parameters

The following set of parameters is held constant, through all simulations: principal values of the g tensor, $g_x=2.0130$, $g_y=2.0120$, $g_z=2.0073$ (historical; includes the magnetic field correction for our instrumentation; for absolute values, see description, Section 4.1) and principal values of the ^{14}N hyperfine tensor, $A_x=20.9$, $A_y=19.82$, $A_z=103.2$ MHz. The τ_c, W, and line width (lw) parameters were varied in the simulations. Random rotational motion of the spin probe was assumed. A fixed lower limit on the lw parameter of 0.2 mT was generally used, which corresponds to the experimental magnetic field modulation amplitude. Convergence of simulations was defined by the default, local least-squares fitting criteria.

4.4 Organization of simulations and documentation

It is assumed that data files have been converted to a format readable by MATLAB®, and that the data strings are present in a MATLAB® Workspace. EasySpin includes algorithms to accomplish this. An example MATLAB® "m-file" script for EPR simulations is shown in Fig. 5. Create a new m-file for each temperature. In this two-component m-file (Fig. 5), a one-component simulation is achieved by setting the weight and corresponding Vary weight parameters of one component equal to zero. Following convergence of the simulation, enter the final, optimized parameters, and save the m-file. This m-file can be used to efficiently replot the overlaid experiment and simulation, or to repeat the simulation with revised parameters or methods. Create a "gallery" file (by using Microsoft PowerPoint, for example), that holds the simulation-experiment spectrum overlay plots for each temperature series, copied directly from the simulation output plot. Enter the simulation parameters and conditions in a spreadsheet.

4.5 Detailed simulation protocols, by line shape type

Each experimental spectrum requires fine-tuning of the alignment with the output simulation spectrum, prior to applying the simulation protocol, which can be accomplished by adjusting the microwave frequency ("mw freq" parameter, in EasySpin). The reasons are manifold, including: (1) The microwave frequency recorded by the spectrometer software during acquisition does not represent the average frequency over multiple scans.

```
%   filename.m  Update: ##/##/##  ##:## pm
%
B1=B/10; % conversion of Gauss to mTesla
expSpc=expSpc_Mb_B_250;  % spectrum file
Params=Params_Mb_B_250; % parameters file
%
%   Enter static (g,Nucs,A) & dynamic(weight, lw, logtcorr) parameters
Sys1.g=[2.0130 2.0120 2.0073];
Sys1.Nucs='14N';
Sys1.A=[20.9 19.26 103.2]; % units, MHz
Sys1.weight=0;
Sys1.lw=0.5772906;  % units, mT
Sys1.logtcorr=-9.167811;  % units, tcorr, s
%
Sys2.g=[2.0130 2.0120 2.0073];
Sys2.Nucs='14N';
Sys2.A=[20.9 19.26 103.2];
Sys2.weight=3.148999;
Sys2.lw=0.1925866;
Sys2.logtcorr=-9.513441;
%
%   Enter experimental parameters
Exp.mwFreq =9.446110; % units, GHz
%   (Params freq mon: '9.450554'; correction, 9.446110)
Exp.Range = [330.0 344.0]; % apply field correction if desired
Exp.ModAmp = 0.2; % units, mT
Exp.Temperature = 250; % units, K
%
%   Enter simulation vary & fix info (# entry is Â ± range; fix=0)
Vary1.g=[0 0 0];
Vary1.A=[0 0 0];
Vary1.weight=0;
Vary1.lw=0.000000000001;
Vary1.logtcorr=0;
%
Vary2.g=[0 0 0];
Vary2.A=[0 0 0];
Vary2.weight=0.0;
Vary2.lw=0.0;
Vary2.logtcorr=0.0;
%
Options.LLKM=[28,14,12,4];
FitOpt.Method = 'fcn'; % select: 'fcn' 'int'
%   Run simulation
[BestSys,BestSpc]=esfit('chili',expSpc,...
    {Sys1,Sys2},{Vary1,Vary2},Exp,Options,FitOpt);
%   Plot scaled data and simulated spectrum
figure
plot(B1,expSpc,'k',B1,BestSpc,'r','LineWidth',1)
axis([330.0 344.0 ...
    min(expSpc)-0.1*(max(expSpc)-min(expSpc)) ...
    max(expSpc)+0.1*(max(expSpc)-min(expSpc))]);
xlabel('Magnetic Field  (mT)','FontSize',24)
ylabel('Derivative Amplitude (a.u.)','FontSize',24)
set(gca,'FontSize',20);
set(gca,'Box','on');
%
%   Title
title('date description','FontSize',14)
```

Fig. 5 MATLAB® script for simulation of experimental EPR spectra by using EasySpin. The Vary parameter values correspond to the initiation of the protocol for alignment of experimental and simulated EPR spectra (described in Section 4.5.1). This template file is cloned and repopulated for other temperature values and conditions.

(2) EPR spectrometers have different magnetic field offsets, which may not be known to the appropriate $\sim 2.0 \times 10^{-5}$ T accuracy. (3) Different quartz EPR tubes introduce shifts in the resonance frequencies. A tabular summary of the protocols corresponding to the different line shape types is presented in Table 1.

4.5.1 Optimize microwave frequency parameter to align spectra

The alignment of experimental and simulated EPR spectra is accomplished by a rigamarole involving optimization of the microwave frequency parameter, "mw freq." To optimize mw freq, use "fcn" (also known as "derivative," because it corresponds to the spectrum collected by using phase-sensitive detection with magnetic field modulation) as "FitOpt. Method" for the simulation. Choose the two-component parameters from the template library spectrum that most closely resembles the experimental spectrum. Enter the parameters in the simulation m-file. Set the "`Sys#.weight`" parameter for the lowest weight component equal to zero. Set the microwave frequency to the value in the Params extension file (Bruker spectrometer), or other format file that contains the data acquisition parameters. Set the Vary parameters for weight, lw, and logtcorr = 1.0 (search range of ±1.0, relative to entered value), for the largest weight component. "Run" the simulation. The simulation will end in several cycles, and the new best fit parameters will be displayed on the Command line. Enter the new, best fit parameters. To optimize the match of experiment and simulation microwave frequency, the one-component simulation is repeatedly executed with early truncation, while mw freq is systematically varied. To start, set all Vary parameters to "0," except for Vary of the lw parameter of the fast component, which is set as: Vary2.lw = 0.0000001. This provides a dummy Vary option, so that the simulation executes. Enter the microwave frequency from the acquisition file. For a Bruker file, use the "FrequencyMon" parameter (for example, "9.454222"). "Run" the simulation, note the value of the root-mean-square-deviation (rmsd) on the line in the Command Window. After several iterations, the simulation finishes and a figure is plotted, which shows a poor match. Change the mw freq by −0.001 GHz (−1.0 MHz). "Run" the simulation again. If the rmsd decreased, go to the next step. If the rmsd increased, repeat, but change mw freq by +0.001 GHz from original value. "Run" the simulation, and

Table 1 Simulation methods for different two-component nitroxide spin probe EPR spectra, according to line shape type, for the condition of protein[a] in frozen aqueous solution containing 2% v/v dimethylsulfoxide added prior to cryotrapping.

Observed line shape type[c]	T (K)	FitOptMethod	User input	Δ log [$\tau_{c,s}$ (s)]
Narrow, very rapid tumbling, A1 ≈ A2; A3 approaches A1, A2	260	fcn	Auto	–
Narrow, very rapid tumbling, A1 ≈ A2; A3 approaches A1, A2	255	fcn	Auto	–
Narrow, very rapid tumbling, A1 ≈ A2; A3 approaches A1, A2	250	fcn	Auto	–
Narrow, rapid tumbling; A1, A2 ≥ 2A3	245	fcn	Manual	0.3
Narrow, rapid tumbling; A1, A2 ≥ 2A3	240	fcn	Manual	0.3
Rapid tumbling with extension of A1 and A3	235	int	Manual	0.2
Rapid tumbling with extension of A1 and A3	230	int	Manual	0.2
Rapid tumbling with extension of A1 and A3	225	int	Manual	0.2
Slow component dominates rapid component	220	int	Manual	0.2
Slow tumbling; slow, fast components not well-resolved	215	int	Manual	0.1
Restricted motion; deep A2 trough	210	int	Manual	0.3
Restricted motion; shallow A trough	205	int	Manual	0.1–0.3
Rigid limit	200	fcn	Auto	–
Rigid limit	195	fcn	Auto	–
Rigid limit	190	fcn	Auto	–

[a]Myoglobin, 10 mg/mL; methods correspond to spectra and simulations in Fig. 4; T value–line shape correspondence and methods comparable for ethanolamine ammonia-lyase, 10 mg/mL.
[b]The T value and line shape correspondence will differ for other samples and conditions. For other samples and conditions, use the line shape, rather than T, to guide the initial simulation approach.
[c]The amplitudes of the $m_I = -1, 0, +1$, EPR line shape features are denoted A1, A2, A3, respectively. Temperatures correspond to the template library in Fig. 4.[b]

check that the rmsd decreased. Save time by performing a "quit" at the command line (Ctrl-c), to terminate the simulation after a few cycles. The rmsd does not change over multiple cycles, at this stage, so the first rmsd printed in the Command Window is the indicator.

Repeat the above procedure. "Run" the simulation, and compare the new rmsd to the previous rmsd value. If the rmsd is lower, continue to change mw freq in steps of 0.001 GHz (1.0 MHz). Continue iterating until the rmsd increases again, which signals that you have passed the optimal mw freq value. When the lowest rmsd is achieved, the corresponding "optimized" mw freq value for the simulation is identified. Enter this optimized value for mw freq.

4.5.2 Simulation protocol: T ≥ *250 K*

Defining features of this highest-temperature interval are a dominant "fast" component and minor (≤20%) "slow" component (Fig. 4). The disparity in apparent component amplitudes is accentuated by the extremely narrow lines, and hence large derivative amplitude, of the fast component. The minor slow component is visualized as a slight broadening of the three m_I features, around the base of the peak and trough contours. The fast feature has three clear, narrow low-, mid-, and high-field lines. The low-field and mid-field lines approach or achieve equal peak-to-trough amplitude. The "fcn" simulation is capable of reproducing both the "slow" and "fast" line shape contributions.

In the first stage, perform a one-component simulation. Modify the "title" in the m-file for the plot to include date, sample and temperature information. The template parameters are selected, and mw freq optimized, as described above. Set the "`Sys#.weight`" parameter and corresponding Vary parameter for the lowest weight component at 0. Set the Vary parameters for weight, lw, and logtcorr = 1.0, for the largest weight component. "Run" the simulation. The simulation will converge, and the new best fit parameters will be displayed on the Command line. Enter the best-fit parameters for weight, lw, and logtcorr, and save the m-file with file name indicating one-component simulation. Duplicate this m-file, and use it for the two-component simulation.

In the second stage, perform a two-component simulation. Modify the "title" for the plot: include date and sample information. Use the library (Vary B, Vary A) parameter set to perform a "Vary-all" simulation, that includes six adjustable parameters: The weight values for A and B, the lw values for A and B, and logtcorr values for A and B. The vary parameter range is ±0.5, for all. Following convergence of the simulation, check that the varied parameter values lie within the search range. If a parameter is at its upper or lower limit, replace with the limit value, and repeat the simulation. For example, if the initial logtcorr = −8.0 and the Vary for logtcorr is 0.5, then values at the end of the simulation of −7.5 or −8.5 indicate that the limit was reached. Thus, a repeat simulation with initial parameters −7.5 or −8.5, respectively, is performed.

4.5.3 Simulation protocol: 225 ≤ T ≤ 245 K

The component weights vary from comparable to dominant for the fast component, as the temperature increases from 225 to 245 K. The slow component is represented by a low-field "peak" or "shoulder" on the low-field feature of the fast component, and a shallow trough on the high-field side of the fast component feature. The narrower, fast component has three relatively resolved low-, mid- and high-field lines, and its amplitude grows, relative to the minor component features, with increase in temperature. The challenge for simulations in this *T*-interval is to characterize the slow component in the face of the simulation bias toward fitting the increasingly dominant fast component. The "fcn" simulation is generally not capable of reproducing the minor slow component features. The "int" simulation, which operates on the integral of the as-acquired derivative EPR spectrum, is better able to reproduce the slow component (in particular, the visualized low-field shoulder, and high-field trough features).

Follow the procedure to optimize mw freq and generate the single-component simulation, as described in the $T \geq 250$ K interval section.

The two-component simulation of line shapes in this *T*-interval may require two steps: "automatic" convergence followed by manual adjustment. Using the two-component parameters from the library spectrum, perform a "Vary-all" simulation, with parameter ranges set at 0.5. Alternatively, use the single-component parameters, obtained from the single-component simulation, in place of the library parameters for the "fast" component. "Run" the simulation. If visual inspection shows underrepresentation of the slow component in the simulation, then perform manual adjustment: In a separate sub-section of the m-file (sub-routine delineated by opening and closing "%%" on the line), reproduce the "automatic" simulation code lines, by using copy/paste, repeated 6 times, and separate each simulation line set by using "%" on the line. The m-file will now contain code for 6 simulations, and will run continuously to complete all 6 simulations. Starting with the first simulation, change the value oflogtcorr for the slow component in each successive simulation by $n \times (0.1)$ units, where $n = 1$–6. The final, 6th simulation will have a slow component logtcorr parameter value that is +0.6, relative to the first simulation. Set the "Vary" values in simulations 1–6 to zero, with the exception of the "Vary2.lw," which is set to 0.0000001, as a dummy vary parameter, to enable file execution. Set Fit.Method.Opt to "fcn," and "Run" this %%-delineated sub-section of the m-file. The individual simulations will execute, and the figures will be stacked-up on the desktop, at completion. Examine the simulations (zoom-in, for a better visual appraisal of the lowest-field and highest field regions). To be consistent in the assessment of the different conditions, use common x- and y-scale ranges for comparison of each logtcorr parameter value condition. The goal is to obtain the best match to the low-field shoulder or peak and high-field depression or trough, by visual inspection. Choose the slow logtcorr value that provides the best match to the low-field shoulder or peak and high-field depression or trough. The logging of the experiment and simulation overlay spectra used in the selection process, as well as recording all parameter values for the simulations, is performed in the Gallery file for each temperature value, at this stage. The results can be copied to a speadsheet file.

In a final simulation, the best manual slow logtcorr value, identified above, is fixed at the identified value, and all other parameters are set to Vary. This simulation gives the final parameter values for the particular temperature.

4.5.4 Simulation protocol: $205 \leq T \leq 220$ K

At the high-temperature end of this interval (220, 215 K), the slow and fast tumbling components are both visible in the spectrum. At the low-temperature end (210, 205 K), the spectrum approaches the rigid-limit line shape, and the distinction between the two separate components becomes subtle, or lost. Reproduction of the 220 K spectrum, which displays the lowest-field feature as a shoulder on the larger-amplitude, low-field derivative feature, absolutely requires two components, and is thus a hallmark for their presence. The 215 K spectrum, which shows the reverse in the low-field feature (larger-amplitude low-field feature relative to the derivative peak feature), can often be reproduced well by using a single component. This demonstrates the necessity of the temperature scan for experimental diagnosis of two-component behavior.

Use the two-component parameters from the library spectrum, that most closely resembles the new spectrum, and proceed with simulation as described for the previous entry. However, simulations in this T-interval generally do not require manual intervention.

4.5.5 Simulation protocol: 190 ≤ T ≤ 200 K

The overall line shape is very near, or in, the rigid limit. While subtle motional effects may be visible, the contributions of two species are generally not distinguishable by eye. Dominant contributions to line shape narrowing arise from librational motions (Marsh et al., 2013), which may differ about the Cartesian axes. The 200 K simulation includes one rigid component (logtcorr > −7.0) and one low-mobility component (logtcorr > −7.5). The 195 K simulation includes one rigid component (logtcorr > −7.0) and one low-mobility component very close to the rigid criterion. The 190 K simulation includes both components in the rigid limit (logtcorr > −7.0). Though the lowest temperature simulation will give logtcorr values that are both not defined as meaningful, in a simulation temperature series, it is necessary to identify this lowest temperature condition, which defines the end of the simulation series.

Use the two-component parameters from the library spectrum, that most closely resembles the target spectrum, and proceed with simulation as described for the previous *T*-interval.

5. Protein environment in the frozen aqueous solutions

The homogeneous spin probe mobilities in PAD and mesodomain, as indicated by the singular values of $\log\tau_{c,s}$ and $\log\tau_{c,f}$ at each temperature, and the facile equilibration of TEMPOL between PAD and mesodomain, with absence of temperature-directional hysteresis, led to the model that globular proteins are surrounded by successive PAD and mesodomain solvent layers (Fig. 1) (Nforneh & Warncke, 2019). This model is also supported by the uniform effects of confinement under each condition, from no added cryosolvent to added 4% v/v DMSO. Gathering of protein in large reservoirs of aqueous-cryosolvent would lead to a gradient of ice boundary confinement, from periphery to interior, which is expected to generate heterogeneity in $\log\tau_c$ values. We now describe how to estimate the phase volumes and dimensions in the frozen solution systems, under different confinement conditions, and use the EAL protein as an example. Under weak-to-moderate confinement, corresponding to added cryosolvent, and therefore, a defined mesodomain, the volumes of the mesodomain and PAD, V_{meso} and V_{PAD}, are related to the measured W values by (Nforneh & Warncke, 2019):

$$\frac{W_s}{W_f} = P\frac{V_{PAD}}{V_{meso}} \quad (5)$$

where P is the partition coefficient for TEMPOL (mesodomain→PAD). The V_{meso} is dependent on the volume of added DMSO, V_{DMSO}, as:

$$V_{meso} = \gamma_{meso} V_{DMSO} \tag{6}$$

where the volume enhancement factor, γ_{meso}, is the inverse of the fractional cryosolvent MFC. Substitution of Eq. (6) into Eq. 5 yields (Nforneh & Warncke, 2019):

$$\frac{W_s}{W_f} = P \frac{V_{PAD}}{\gamma_{meso} V_{DMSO}} \tag{7}$$

The measured linear dependence of W_s/W_f on V_{DMSO} has a slope of $PV_{PAD}/\gamma_{meso} = 1.9\,\mu L$ (Nforneh & Warncke, 2019). For DMSO, the MFC is 65% v/v (Rasmussen & MacKenzie, 1968), and $\gamma_{meso} = 1.54$. The value of P is unity to within ±5% (Nforneh & Warncke, 2019). Therefore, $V_{PAD} = 2.9\,\mu L$ (2.9×10^{21} Å^3). This value of V_{PAD}, distributed over the planar equivalent of the solvent accessible surface area of EAL ($A_{S,prot} = 1.3 \times 10^5$ Å^2; total EAL oligomers in sample, 4.0×10^{15}; see Section 1), creates a PAD layer of mean thickness, 6 Å, which is equivalent to two water molecules, and is consistent with dimensions determined for the protein hydration layer (Frauenfelder et al., 2009; Panagopoulou et al., 2012). A V_{meso} value of 7.9 μL for 2% added DMSO is determined from Eq. (5), by using the value of $V_{PAD} = 2.9\,\mu L$, $P = 1$, and the mean W_s/W_f ratio of 0.37 from Region III of the temperature dependence (Nforneh & Warncke, 2019). This value compares well with a V_{meso} of 9.2 μL, calculated directly from the 6.0 μL addition of DMSO to the sample, and γ_{meso}. Thus, approximately 90% of the added DMSO is accounted for in the mesodomain. The V_{meso} values for 2% added DMSO, when applied to the planar-equivalent, accessible surface area of EAL, indicate a mesodomain of mean thickness 15–18 Å, corresponding to 5–6 water molecules. This thickness is scaled, for other added % DMSO values (Nforneh & Warncke, 2019).

The strong confinement condition corresponds to the absence of added cryosolvent. Nevertheless, a fast motional component is observed, which emulates the temperature-dependent trends observed in the presence of cryosolvents. We refer to this fast component, in the absence of cryosolvent, as the "mesophase." Eq. 5, with V_{meso} assumed to represent the mesophase ($\gamma_{meso} = 1$), $V_{PAD} = 2.9\,\mu L$ as calculated above, $P = 1$, and the mean W_s/W_f value of 3.8 in Region III (Nforneh & Warncke, 2017), leads to an estimate for the PAD-external, TEMPOL-detected mesophase volume of 0.77 μL. This relatively small mesophase volume is commensurate with only partial coverage of the PAD area (for example, ~40% PAD surface coverage by water molecules of diameter, d_{H2O}=2.8 Å). Thus, the mesophase is proposed to arise from clusters or pools of water on the PAD surface, that permit a common, relatively high degree of spin probe mobility. The pools may be associated with structural disjunctions (defects) at the PAD-ice interface (Banerjee et al., 2009), and may also contain freeze-concentrated potassium phosphate buffer (Heller, Carpenter, & Randolph, 1997). The direct, vicinal ice-boundary confinement of the PAD under the no-cryosolvent condition leads to the highest *T* values observed for the ODT, 245–250 K. This range is approached by the ODT of 235–240 K observed for a system with minimal added cryosolvent (0.06% 2-aminopropanol) (Ionescu et al., 2021).

Overall, the variation of the mesodomain volume by using added cryosolvent from 0% to 4% v/v DMSO leads to scaled expansion of the mean mesodomain thickness from approximately 0 to 40 Å, and the consequent reduction in ice boundary confinement leads to a corresponding decrease in the ODT *T* value from 250 to 200 K (Ionescu et al., 2021; Nforneh & Warncke, 2017, 2019).

6. Conclusions and future directions

The EPR spin-probe perspective on solvent dynamics and structure in the low-temperature, frozen aqueous mesodomain systems concurs with the structures of polycrystalline ice and interstitial mesodomain in frozen aqueous solutions of protein, that are imaged at micrometer resolution by scattering techniques (Curtis et al., 2012; Dong et al., 2009). The EPR spin-probe methods extend this view to the nanoscale, by resolving the PAD or hydration water volume, that is directly associated with the proteins (Figs. 1 and 2). The temperature- and confinement-dependent spin probe dynamics provide the contrast, that enables high-fidelity reporting of both PAD and mesodomain properties.

Approaches used to address protein-hydration solvent coupling are distinguished by their temporal and spatial scales. On the shortest time scales of 1–100 ps, the dielectric relaxation in the region of a photoexcited, protein-associated fluorophore is detected by the Stokes shift, and reports on neighbor water within a few Ångstroms (Pal et al., 2002). A broader time scale of 0.1–300 ps was achieved by adding time-resolved fluorescence anisotropy (Qin et al., 2016). Overhauser dynamic nuclear polarization (ODNP) (Barnes et al., 2017; Franck et al., 2013; Orotny et al., 2017) uses a paramagnetic species (typically, a nitroxide) attached at a specific protein site to magnetically polarize water protons within approximately 10 Å, and the measured water translational diffusivity of freely exchangeable surface water reports on motional events on the 200–900 ps time scale. Both of these methods detect the dynamics of a network of water molecules around each label site, and report protein surface site-specific differences of up to 5-fold in motional time scale. NMR studies of ubiquitin also detect site heterogeneity in protein surface dynamics (Nucci et al., 2011), and motions on and near the protein surface from pico- to milliseconds (Lewandowski et al., 2015).

In comparison, the rotational reorientation of the spin probe measured by the standard X-band continuous-wave EPR detection scheme reports on motions over the 100 ps to 100 ns time range (Banerjee & Bhat, 2009; Goldman et al., 1972; Sato et al., 2008), and the free probe samples a spatial range around the protein that is limited only by the ice boundary-imposed confines of the system (up to approximately 50 Å from the protein surface in systems examined, thus far) (Nforneh & Warncke, 2019). Thus, relative to label techniques, the spin probe reports on, and distinguishes, the solvent beyond the hydration layer. The ODNP technique and terahertz spectroscopy (Ebbinghaus et al., 2007) directly address the solvent water molecules in the network, whereas probe molecules are proxies for solvent motions. Activation energies for rotational reorientation of TEMPOL (Nforneh & Warncke, 2019) are comparable to those of some solvent fluctuations, which suggests that the TEMPOL motion represents the energy surface for protein-coupled solvent motion (Fenimore, Frauenfelder, McMahon, & Young, 2004; Qin et al., 2016). The EPR spin probe method detects that PAD and mesodomain are dynamically homogeneous, in contrast to the heterogeneity observed in protein surface and coupled water hydration network dynamics (Barnes et al., 2017; Franck et al., 2013; Lewandowski et al., 2015; Nucci et al., 2011; Orotny et al., 2017; Pal et al., 2002; Qin et al., 2016). This is proposed to arise predominantly from both spatial and temporal averaging of the solvent dynamical inputs to the spin probe motion. For protein functional processes that occur on time scales of >1 ns, the EPR spin probe may represent well the influence of protein-coupled solvent fluctuations on function. The probe is similar in dimension to amino acid side chain and peptide backbone moieties, that respond to the solvent motions and conduct them to the protein interior.

The use of the mesodomain system to identify and quantify solvent-protein-function dynamical coupling is illustrated by the correlation of the ODT (Nforneh & Warncke, 2019) with the bifurcation (Kohne et al., 2019, 2017; Zhu & Warncke, 2008) in the kinetics of the substrate radical decay reaction in the EAL enzyme (Ionescu et al., 2021). In the direction of decreasing temperature, the ODT at 220 K causes the emergence of a barrier to interconversion of the substrate radical "capture" (stabilization following formation by the radical pair separation) and "enable" (for C—N bond cleavage and 1,2-amino group migration) configurational substrates (Ucuncuoglu & Warncke, 2018), that define the reaction progress coordinate within the substrate radical macrostate minimum. In parallel with dielectric permittivity measurements (Ionescu et al., 2021), the spin probe EPR results specify the origin of the ODT in the attenuation of the collective-cluster class of fluctuations (Capaccioli, Ngai, Ancherbak, & Paciaroni, 2012; Cerveny, Alegria, & Colmenero, 2008). Thus, coupling of protein functions to this class of fluctuations, which are also shown to be critical for small-molecule migration in heme proteins (Dantsker et al., 2005; Frauenfelder et al., 2009), can now be characterized for more complex processes, such as enzyme reactions.

The relatively hospitable nature of the mesodomain system portends application to a wide range of proteins. Study of soluble globular proteins, with M_r values of 2,000 to 500,000 g/mol, under weak and strong confinement, reveals common, protein mass and topology-independent trends in the temperature-dependent dynamics of the phases, with PAD ODT temperatures set by the degree of ice boundary confinement through the mesodomain (Li et al., 2022). A future direction is application of the mesodomain system to the study of the PAD and mesodomain dynamics in the classes of intrinsically disordered protein (IDP) and proteins with an intrinsically disordered region (IDR) (Gallat et al., 2012). Preliminary results for the IDR protein, α-synuclein (Meade, Fairlie, & Mason, 2019), reveal enhanced mobility in the solvent phases (shorter $\tau_{c,s}$ and $\tau_{c,f}$) and a lower-temperature ODT, relative to globular proteins. A fecund application area is the mechanistic characterization of protein-associated liquid-liquid phase separation and condensate formation (Banani, Lee, Hyman, & Rosen, 2017; Klosin et al., 2020; Mathieu, Pappu, & Taylor, 2020; Shin & Brangwynne, 2017). The variation of temperature and confinement present avenues for induction, manipulation, and time-resolution of phase formation.

Acknowledgments

Research reported in this publication was supported by the National Institutes of Diabetes and Digestive and Kidney Diseases (R01 DK054514) and National Institute of General Medical Sciences (R01 GM142113) of the National Institutes of Health. The purchase of the Bruker E500 EPR spectrometer was funded by the National Center for Research Resources of the National Institutes of Health (RR 17767), and by Emory University.

References

Alcoutlabi, M., & McKenna, G. B. (2005). Effects of confinement on material behaviour at the nanometer size scale. *Journal of Physics. Condensed Matter*, *17*, R461–R524.

Arakawa, T., Kita, Y., & Timasheff, S. N. (2007). Protein precipitation and denaturation by dimethyl sulfoxide. *Biophysical Chemistry*, *131*, 62–70.

Arsane, M.-L., Raut, I., Calin, M., Jecu, M.-L., Doni, M., & Gurban, A.-M. (2021). Versatility of reverse micelles: From biomimetic models to nano (bio)sensor design. *PRO*, *9*, 345.

Ball, P. (2008). Water as an active constituent in cell biology. *Chemical Reviews*, *108*(1), 74–108.

Banani, S. F., Lee, H. O., Hyman, A. A., & Rosen, M. K. (2017). Biomolecular condensates: Organizers of cellular biochemistry. *Nature Reviews. Molecular Cell Biology*, *18*, 285–298.

Banerjee, D., & Bhat, S. V. (2009). Vitrification, relaxation and free volume in glycerol-water binary liquid mixture: Spin probe ESR studies. *Journal of Non-Crystalline Solids*, *355*, 2433–2438.

Banerjee, D., Bhat, S. N., Bhat, S. V., & Leporini, D. (2009). ESR evidence for 2 coexisting liquid phases in deeply supercooled bulk water. *Proceedings of the National Academy of Sciences of the United States of America*, *106*, 11448–11453.

Barnes, R., Sun, S., Fichou, Y., Dahlquist, F. W., Heyden, M., & Han, S. (2017). Spatially heterogeneous surface water diffusivity around structured protein surfaces at equilibrium. *Journal of the American Chemical Society*, *139*, 17890–17901.

Berliner, L. J. (2014). Compilation of hyperfine splittings and g-factors for aminoxyl (nitroxide) radicals. In S. K. Misra (Ed.), *Multifrequency electron paramagnetic resonance: Data and techniques* (pp. 287–296). Singapore: Wiley-VCH.

Berman, H. M., Westbrook, J., Feng, Z., Gilliland, G., Bhat, T. N., Weissig, H., et al. (2000). The protein data bank. *Nucleic Acids Research*, *28*, 235–242.

Blumenfeld, L. A., Davydov, R. M., Magonov, S. N., & Vilu, R. O. (1974). Studies on the conformational changes of metalloproteins induced by electrons in water-ethylene glycol solutions at low temperatures. Cytochrome c. *FEBS Letters*, *45*, 256–258.

Bovell, A. M., & Warncke, K. (2013). Structure model of Salmonella typhimurium ethanolamine ammonia-lyase directs a rational approach to the assembly of the functional $[(EutB\text{-}EutC)_2]_3$ oligomer from isolated subunits. *Biochemistry*, *52*, 1419–1428.

Capaccioli, S., Ngai, K. L., Ancherbak, S., & Paciaroni, A. (2012). Evidence of coexistence of change of caged dynamics at T_g and the dynamic transition at T_d in solvated proteins. *The Journal of Physical Chemistry. B*, *116*, 1745–1757.

Cerveny, S., Alegria, A., & Colmenero, J. (2008). Universal features of water dynamics in solutions of hydrophilic polymers, biopolymers, and small glass-forming materials. *Physical Review E*, *77*, 031803.

Cerveny, S., Mallamace, F., Swenson, J., Vogel, M., & Xu, L. (2016). Confined water as a model of supercooled water. *Chemical Reviews*, *116*, 7608–7625.

Chen, S.-H., Liu, L., Fratini, E., Baglioni, P., Faraone, A., & Mamontov. (2006). Observation of fragile-to-strong dynamic crossover in protein hydration water. *Proceedings of the National Academy of Sciences of the United States of America*, *103*(24), 9012–9016.

Chen, H., Sun, L., & Warncke, K. (2013). Heterogeneous ordered-disordered structure of the mesodomain in frozen sucrose-water solutions revealed by multiple electron paramagnetic resonance spectroscopies. *Langmuir*, *29*, 4357–4365.

Clark, A., Sedhom, J., Elajaili, H., Eaton, G. R., & Eaton, S. S. (2018). Dependence of electron paramagnetic resonance spectral lineshapes on molecular tumbling: Nitroxide radical in water:glycerol mixtures. *Concepts in Magnetic Resonance Part A*, *45A*, e21423. 1–13.

Connolly, M. L. (1983). Solvent-accessible surfaces of proteins and nucleic acids. *Science*, *221*(4612), 709–713.

Curtis, J. E., Nanda, H., Khodadadi, S., Cicerone, M., Lee, H. J., McAuley, A., et al. (2012). Small-angle neutron scattering study of protein crowding in liquid and solid phases: Lysozyme in aqueous solution, frozen solution, and carbohydrate powders. *The Journal of Physical Chemistry. B*, *116*, 9653–9667.

Dantsker, D., Samuni, U., Friedman, J. M., & Agmon, N. (2005). A hierarchy of functionally important relaxations within myoglobin based on solvent effects, mutations and kinetic model. *Biochimica et Biophysica Acta*, *1749*, 234–251.

Davydov, R. M., Kofman, H., Fujii, T., Yoshida, M., Ikedo-Saito, M., & Hoffman, B. M. (2002). Catalytic mechanism of heme oxidase through EPR and ENDOR of cryoreduced oxy-heme oxygenase and its Asp 140 mutants. *Journal of the American Chemical Society*, *124*, 1798–1808.

Davydov, R. M., Perera, R., Jin, S., Yang, T.-C., Bryson, T. A., Sono, M., et al. (2005). Substrate modulation of the properties and reactivity of the oxy-ferrous and hydroperoxo-ferric intermediates of cytochrome P450cam as shown by cryoreduction-EPR/ENDOR spectroscopy. *Journal of the American Chemical Society*, *127*, 1403–1413.

DeLano, W. J. (2002). PyMOL: An open-source molecular graphics tool. *CCP4 Newletter, Protein Crystallography*, *40*, 82–92.

Dong, J., Hubei, A., Bischof, J. C., & Aksan, A. (2009). Freezing-induced phase separation and spatial microheterogeneity in protein solutions. *The Journal of Physical Chemistry. B*, *113*, 10081–10087.

Doster, W. (2010). The protein-solvent glass transition. *Biochimica et Biophysica Acta*, *1804*(1), 3–14.

Douzou, P. (1977). *Cryobiochemistry: An introduction*. New York: Academic Press.

Dunn, B., & Zink, J. I. (1997). Probes of pore environment and molecule-matrix interactions in sol-gel materials. *Chemistry of Materials*, *9*, 228–2291.

Ebbinghaus, S., Kim, S. J., Heyden, M., Yu, X., Heugen, U., Gruebele, M., et al. (2007). An extended dynamical hydration shell around proteins. *Proceedings of the National Academy of Sciences of the United States of America*, *104*, 20749–20752.

Fenimore, P. W., Frauenfelder, H., McMahon, B. H., & Young, R. D. (2004). Bulk-solvent and hydration-shell fluctuations, similar to alpha- and beta-fluctuations in glasses, control protein motions and functions. *Proceedings of the National Academy of Sciences of the United States of America*, *101*(40), 14408–14413.

Fischer, H., Polikarpov, I., & Craievich, A. F. (2004). Average protein density is a molecular-weight-dependent function. *Protein Science*, *13*(10), 2825–2828.

Franck, J. M., Pavlova, A., Scott, J. A., & Han, S. (2013). Quantitative cw Overhauser effect dynamic nuclear polarization for the analysis of local water dynamics. *Progress in Nuclear Magnetic Resonance Spectroscopy*, *74*, 33–56.

Franks, F. (2003). Scientific and technological aspects of aqueous glasses. *Biophysical Chemistry*, *106*, 251–261.

Frauenfelder, H., Chen, G., Berendzen, J., Fenimore, P. W., Jansson, H., McMahon, B. H., et al. (2009). A unified model of protein dynamics. *Proceedings of the National Academy of Sciences of the United States of America, 106*, 5129–5134.

Gallat, F.-X., Laganowsky, A., Wood, K., Gabel, F., van Eijck, L., Wuttke, J., et al. (2012). Dynamic coupling of intrinsically disordered proteins and their hydration water: Comparison with folded soluble and membrane proteins. *Biophysical Journal, 103*, 129–136.

Goldman, S. A., Bruno, G. V., Polnaszek, C. F., & Freed, J. H. (1972). Anisotropic rotational reorientation and slow tumbling in liquid and frozen media. *The Journal of Chemical Physics, 56*, 716–735.

Hammes, G. G., Benkovic, S. J., & Hammes-Schiffer, S. (2011). Flexibility, diversity, and cooperativity: Pillars of enzyme catalysis. *Biochemistry, 50*, 10422–10430.

Heller, M. C., Carpenter, J. F., & Randolph, T. W. (1997). Manipulation of lyophilization-induced phase separation: Implications for pharmaceutical proteins. *Biotechnology Progress, 13*, 590–596.

Henzler-Wildman, K. A., & Kern, D. (2007). Dynamic personalities of proteins. *Nature, 450*, 964–972.

Herschlag, D., & Natarajan, A. (2013). Fundamental challenges in mechanistic enzymology: Progress toward understanding the rate enhancements of enzymes. *Biochemistry, 52*, 2050–2067.

Heyden, M., & Tobias, D. J. (2013). Spatial dependence of protein-water collective hydrogen-bond dynamics. *Physical Review Letters, 111*, 2181011–2181015.

Humphrey, W., Dahlke, A., & Schulten, K. (1996). VMD—Visual molecular dynamics. *Journal of Molecular Graphics, 14*, 33–38.

Hwang, J. S., Mason, R. P., Hwang, L.-P., & Freed, J. H. (1975). Electron spin resonance studies of anisotropic rotational reorientation and slow tumbling in liquid and frozen media. III. Perdeuterated 2, 2, 6, 6-tetramethyl-4-piperidone N-oxide and an analysis of fluctuating torques. *The Journal of Physical Chemistry, 79*(5), 489–511.

Hyde, J. S., & Thomas, D. D. (1980). Saturation-transfer spectroscopy. *Annual Review of Physical Chemistry, 31*, 293–317.

Ionescu, A., Li, W., Nforneh, B., & Warncke, K. (2021). Coupling of ethanolamine ammonia-lyase protein and solvent dynamics characterized by temperature-dependence of the EPR spin probe mobility and dielectric permittivity. *The Journal of Chemical Physics, 154*, 175101.

Klosin, A., Oltsch, F., Harmon, T., Honigmann, A., Jülicher, F., Hyman, A. A., et al. (2020). Phase separation provides a mechanism to reduce noise in cells. *Science, 367*(6476), 464–468.

Kohne, M., Li, W., Zhu, C., & Warncke, K. (2019). Deuterium kinetic isotope effects resolve low-temperature substrate radical reaction pathways and steps in B_{12}-dependent ethanolamine ammonia-lyase. *Biochemistry, 58*, 3683–3690.

Kohne, M., Zhu, C., & Warncke, K. (2017). Two dynamical regimes of the substrate radical rearrangement reaction in B_{12}-dependent ethanolamine ammonia-lyase resolve contributions of native protein configurations and collective configurational fluctuations to catalysis. *Biochemistry, 56*, 3257–3264.

Körber, C. (1988). Phenomena at the advancing ice-liquid interface: Solutes, particles and biological cells. *Quarterly Reviews of Biophysics, 21*(2), 229–298.

Leigh, J. S., Jr., & Reed, G. H. (1971). Electron paramagnetic resonance studies in frozen aqueous solutions. Elimination of freezing artifacts. *The Journal of Physical Chemistry, 75*(9), 1202–1204.

Lewandowski, J. R., Halse, M. E., Blackledge, M., & Emsley, L. (2015). Direct observation of hierarchical protein dynamics. *Science, 348*, 578–581.

Li, W., Whitcomb, K. L., & Warncke, K. (2022). *Manuscript in preparation.*

Liang, Z., & Freed, J. H. (1999). An assessment of the applicability of multifrequency ESR to study the complex dynamics of biomolecules. *The Journal of Physical Chemistry. B*, *103*, 6384–6396.

Lukoyanov, D., Barney, B. M., Dean, D. R., Seefeldt, L. C., & Hoffman, B. M. (2007). Connecting nitrogenase intermediates with the kinetic scheme for N_2 reduction by a relaxation protocol and identification of the N_2 binding site. *Proceedings of the National Academy of Sciences of the United States of America*, *104*(5), 1451–1455.

Malferrari, M., Savitsky, A., Lubitz, W., Möbius, K., & Venturoli, G. (2016). Protein immobilization capabilities of sucrose and trehalose glasses: The effect of proteins/sugar concentration unraveled by high-field EPR. *Journal of Physical Chemistry Letters*, *7*, 4871–4877.

Mallamace, F., Broccio, M., Corsaro, C., Faraone, A., Wanderlingh, U., Liu, L., et al. (2006). The fragile-to-strong dynamic crossover transition in confined water: Nuclear magnetic resonance results. *The Journal of Chemical Physics*, *124*, 161102.

Marsh, D., Bartucci, R., Guzzi, R., Sportelli, L., & Esmann, M. (2013). Librational fluctuations in protein glasses. *Biochimica et Biophysica Acta*, *1834*, 1591–1595.

Mathieu, C., Pappu, R. V., & Taylor, J. P. (2020). Beyond aggregation: Pathological phase transitions in neurodegenerative disease. *Science*, *370*, 56–60.

Meade, R. M., Fairlie, D. P., & Mason, J. M. (2019). Alpha-synuclein structure and Parkinsons's disease - lessons and emerging principals. *Molecular Neurodegeneration*, *14*, 29.

Möbius, K., Savitsky, A., Wegener, C., Plato, M., Fuchs, M., Schnegg, A., et al. (2005). Combining high-field EPR with site-directed spin labeling reveals unique information on proteins in action. *Magnetic Resonance in Chemistry*, *43*, S4–S19.

Nforneh, B., & Warncke, K. (2017). Mesodomain and protein-associated solvent phases with temperature-tunable (200-265 K) dynamics surround ethanolamine ammonia-lyase in globally polycrystalline aqueous solution containing dimethylsulfoxide. *The Journal of Physical Chemistry. B*, *121*, 11109–11118.

Nforneh, B., & Warncke, K. (2019). Control of solvent dynamics around the B_{12}-dependent ethanolamine ammonia-lyase enzyme in frozen aqueous solution by using dimethylsulfoxide modulation of mesodomain volume. *The Journal of Physical Chemistry. B*, *123*, 5395–5404.

Ngai, K. L., Capaccioli, S., & Paciaroni, A. (2013). Nature of the specific relaxation in hydrated proteins and aqueous mixtures. *Chemical Physics*, *424*, 37–44.

Nucci, N. V., Pometun, M. S., & Wand, A. J. (2011). Mapping hydration dynamics in ubiquitin. *Journal of the American Chemical Society*, *133*, 12326–12329.

Orotny, J. H., Qiao, B., Newcomb, C. J., Keller, T. J., Palmer, L. C., Deiss-Yehiely, E., et al. (2017). Water dynamics from the surface to the interior of a supramolecular nanostructure. *Journal of the American Chemical Society*, *139*, 8915–8921.

Pal, S. K., Peon, J., Bagchi, B., & Zewail, A. H. (2002). Biological water: Femtosecond dynamics of macromolecular hydration. *The Journal of Physical Chemistry. B*, *106*, 12376–12395.

Panagopoulou, A., Kyritsis, A., Shinyashiki, N., & Pissis, P. (2012). Protein and water dynamics in bovine serum albumin-water mixtures over wide ranges of compositions. *The Journal of Physical Chemistry. B*, *116*, 4593–4602.

PyMOL. https://pymol.org/2/.

Qin, Y., Wang, L., & Zhong, D. (2016). Dynamics and mechanism of ultrafast water-protein interactions. *Proceedings of the National Academy of Sciences of the United States of America*, *113*(30), 8424–8429.

Quillin, M. L., & Matthews, B. W. (2000). Accurate calculation of the density of proteins. *Acta Crystallographica. Section D, Biological Crystallography*, *56*, 791–794.

Rasaiah, J. C., Garde, S., & Hummer, G. (2008). Water in nonpolar confinement: From nanotubes to proteins and beyond. *Annual Review of Physical Chemistry*, *59*, 713–740.

Rasmussen, D. H., & MacKenzie, A. P. (1968). Phase diagram for the system water-dimethylsulphoxide. *Nature*, *220*, 1315–1317.

Roozen, M. J. G. W., & Hemminga, M. A. (1990). Molecular motion in sucrose-water mixtures in the liquid and glassy state as studied by spin probe ESR. *The Journal of Physical Chemistry*, *94*, 7326–7329.

Sato, H., Bottle, S. E., Blinco, J. P., Micallef, A. S., Eaton, G. R., & Eaton, S. S. (2008). Electron spin-lattice relaxation of nitroxyl radicals in temperature ranges that span glassy solutions to low viscosity liquids. *Journal of Magnetic Resonance*, *191*, 66–77.

Shichman, S. A., & Amey, R. L. (1972). Viscosity and local liquid structure in dimethyl sulfoxide-water mixtures. *The Journal of Physical Chemistry*, *75*(1), 98–102.

Shin, Y., & Brangwynne, C. P. (2017). Liquid phase separation in cell physiology and disease. *Science*, *357*, eaaf4382.

Stoll, S. (2014). Computational modeling and least squares fitting of EPR spectra. In S. K. Misra (Ed.), *Multifrequency electron paramagnetic resonance: Data and techniques* (pp. 69–138). Singapore: Wiley-VCH.

Stoll, S., & Schweiger, A. (2006). EasySpin, a comprehensive software package for spectral simulation and analysis in EPR. *Journal of Magnetic Resonance*, *178*, 42–55.

Timasheff, S. N. (2002). Protein-solvent preferential interactions, protein hydration, and the modulation of biochemical reactions by solvent compounds. *Proceedings of the National Academy of Sciences of the United States of America*, *99*, 9721–9726.

Ucuncuoglu, N., & Warncke, K. (2018). Protein configurational states guide radical rearrangement catalysis in ethanolamine ammonia-lyase. *Biophysical Journal*, *114*, 2775–2786.

Van Horn, W. D., Ogilvie, M. E., & Flynn, P. F. (2009). Reverse micelle encapsulation as a model for intracellular crowding. *Journal of the American Chemical Society*, *131*, 8030–8039.

Wang, M., Zhu, C., Kohne, M., & Warncke, K. (2015). Resolution and characterization of chemical steps in enzyme catalytic sequences by using low-temperature and time-resolved, full-spectrum EPR spectroscopy in fluid cryosolvent and frozen solution systems. *Methods in Enzymology*, *563(Pt. A)*, 59–94.

Wertz, J. E., & Bolton, J. R. (1986). *Electron spin resonance*. New York: Chapman and Hall.

Wheeler, K. E., Nocek, J. M., & Hoffman, B. M. (2006). NMR spectroscopy can characterize proteins encapsulated in a sol-gel matrix. *Journal of the American Chemical Society*, *128*, 14782–14783.

Wodak, S. J., & Janin, J. (1980). Analytical approximation to the accessible surface area of proteins. *Proceedings of the National Academy of Sciences of the United States of America*, 77(4), 1736–1740.

Zhou, H.-X., Rivas, G., & Minton, A. P. (2008). Macromolecular crowding and confinement: Biochemical, biophysical, and potential physiological consequences. *Annual Review of Biophysics*, *37*, 375–397.

Zhu, C., & Warncke, K. (2008). Reaction of the Co^{II}-substrate radical pair catalytic intermediate in coenzyme B_{12}-dependent ethanolamine ammonia-lyase in frozen aqueous solution from 190 to 217 K. *Biophysical Journal*, *95*, 5890–5900.

CHAPTER THREE

Site directed spin labeling to elucidating the mechanism of the cyanobacterial circadian clock

Gary K. Chow[a], Andy LiWang[b,c], and R. David Britt[a,*]
[a]Department of Chemistry, University of California, Davis, CA, United States
[b]Department of Chemistry & Biochemistry, University of California, Merced, CA, United States
[c]Health Sciences Research Institute, University of California, Merced, CA, United States
[*]Corresponding author: e-mail address: rdbritt@ucdavis.edu

Contents

Abstract

Electron Paramagnetic Resonance (EPR) is a spectroscopic technique that provides structural and dynamic information on unpaired spins and their surrounding environments. Introduction of exogenous spin labels via site directed spin labeling (SDSL) enables characterization of systems of interests lacking intrinsic unpaired spins. This chapter describes the use of SDSL in quantifying KaiB-KaiC binding in the cyanobacterial circadian clock (Kai Clock), exploiting the changes in mobility of the local environment

Methods in Enzymology, Volume 666
ISSN 0076-6879
https://doi.org/10.1016/bs.mie.2022.02.011

around the spin label on KaiB-KaiC interactions. While the Kai system serves as our model system to demonstrate SDSL-EPR utility in quantifying protein-protein interactions, this technique is readily amenable to other systems of interest whenever specific protein-protein interactions need to be isolated. We first present a protocol for spin labeling KaiB. Then, we detail the sample preparation and acquisition processes to maximize signal-to-noise for downstream analysis. We close this chapter by highlighting recent advances in SDSL technology to incorporate spin labels into proteins of interest and in EPR technology to improve detection sensitivity that may allow greater flexibilities to the types of experiments possible.

1. Introduction

Electron Paramagnetic Resonance (EPR) spectroscopy is a powerful tool for studying the structure and dynamics of unpaired spins (Altenbach, López, Hideg, & Hubbell, 2015). The selectivity of EPR spectroscopy makes it ideal for studying the kinetics of radical-, transition metal- and triplet state-mediated reactions. Coupling EPR spectroscopy with site directed spin labeling (SDSL) enables interrogation of systems lacking intrinsic unpaired spins to provide structural information ranging from solvent accessibility (Altenbach, Flitsch, Khorana, & Hubbell, 1989) to conformational dynamics (Cafiso, 2014). It has long been recognized that the EPR spectrum of a spin label depends on its mobility (Freed, Bruno, & Polnaszek, 1971). Significant effort has been devoted to simulating EPR spectra in the nano- to microsecond regimes (Budil, Lee, Saxena, & Freed, 1996; Freed et al., 1971; Goldman, Bruno, & Freed, 1972; Meirovitch, Nayeem, & Freed, 1984; Polimeno & Freed, 1995) with the goal to extract dynamical information.

In theory, the changes in mobility in a spin label could be followed on the millisecond or slower timescales to track the reaction kinetics of a system. In practice, time-resolved EPR experiments are often limited to systems on millisecond (Bittl & Weber, 2005) to minute (Samouilov, Roubaud, Kuppusamy, & Zweier, 2004) timescales, whereas SDSL-based experiments on longer timescales have been hampered by the inherent instability of nitroxides (Bobko, Kirilyuk, Grigor'ev, Zweier, & Khramtsov, 2007) and/or the commonly used disulfide linker used for bioconjugation (Berliner, Grunwald, Hankovszky, & Hideg, 1982) toward reducing agents. On the other hand, the potential application of EPR spectroscopy in systems spanning hours or longer have not been adequately addressed from both theoretical and practical standpoints.

One class of systems that span timescales beyond typical EPR experiments is that of circadian oscillators. Circadian oscillators are intrinsic timekeeping systems in organisms that enable them to anticipate environmental changes originating from Earth's 24-h rotation (Vaze & Sharma, 2013; Woelfle, Ouyang, Phanvijhitsiri, & Johnson, 2004). As the only circadian oscillator that can be reconstituted *in vitro* by combining its constituent proteins, the cyanobacterial circadian oscillator (Kai oscillator) found in *Synechococcus elongatus* PCC 7942 serves as an important model for understanding circadian clocks in general (Chavan et al., 2021; Nakajima et al., 2005). The core Kai oscillator is composed of three proteins, KaiA, KaiB, and KaiC, in which when combined in vitro along with Mg^{2+} and ATP results in rhythmic oscillations in KaiC phosphorylation (Nishiwaki et al., 2007; Rust, Markson, Lane, Fisher, & O'Shea, 2007) and Kai protein complex (dis)association (Kageyama et al., 2006). The reader is referred to recent reviews on the inner workings of the Kai oscillator (Golden & LiWang, 2021; Swan, Golden, LiWang, & Partch, 2018). In this chapter, we begin by illustrating the utility of SDSL-EPR spectroscopy in giving biological insights complementary to orthogonal methods using the cyanobacterial circadian oscillator (Kai oscillator) as our model system while discussing the problem of reporter site design. Next, we describe the spin labeling protocol used for reproducible spin labeling and potential pitfalls. Then, we describe a strategy to quantifying protein-protein interactions and its associated data analysis strategies. Finally, we close the chapter by discussing recent advances in SDSL and EPR detection technologies that shall improve SDSL-EPR's utility in studying protein-protein interactions.

2. Experimental considerations

A key question to address prior to any SDSL-EPR experiment is to design the spin labeling/reporter site that when labeled will give spectral changes on protein complex formation that is relevant to the research question at hand. In this section, we give a brief overview of extractable information via SDSL-EPR using the Kai system as an example. Then, we discuss strategies for designing SDSL sites and other experimental considerations.

2.1 Information available from SDSL-EPR

In our experience with the Kai system, by labeling KaiB with 3-iodoacetamido-PROXYL (3IAP), we managed to resolve (i) free KaiB vs KaiC-bound KaiB by targeting the KaiB-KaiC binding interface (Chow et al., 2020) (Fig. 1) and (ii) "neighbored" versus "neighborless" KaiC-bound KaiB by targeting the KaiB-KaiB interface in the $KaiB_6C_6$ dodecamer (Chow et al., 2022) (Fig. 2). In (i), free KaiB-N19C-3IAP ($N19C_{3IAP}$ thereafter) exhibits a mobile spectrum due to solvent exposure of the spin labeled side chain whereas KaiC-bound $N19C_{3IAP}$ exhibits a spectrum with reduced mobility due to burial of the spin labeled side chain, irrespective of whether KaiA is bound or not. In (ii), free KaiB-K25C-3IAP ($K25C_{3IAP}$ thereafter) displays at least two motional components of intermediate mobility, with binding to KaiC resulting in an initial increase in mobility due to solvent exposure of the spin labeled side chain in "neighborless" KaiC-bound KaiB (B_-) followed by a decrease on further binding as the same side chain gets buried within the inter-KaiB interface (B_+). In both cases, the changes in mobility were confirmed by estimating the rotational correlation times of these spin-labeled species via spectral simulations implemented by EasySpin (Stoll & Schweiger, 2006).

The definitive spectral assignment between rotational dynamics states and their respective spectra gives credence to the relevance of the crystal structures in solution phase. More importantly, these assignments can be used for extracting quantitative real-time binding information. In terms of time resolution, we managed to obtain sufficient signal-to-noise by binning spectra as 7.5-min bins using 3.5 μM $N19C_{3IAP}$ while still being able to resolve free vs KaiC-bound KaiB (Chow et al., 2020). This resolution is comparable to fluorescence anisotropy (Chavan et al., 2021; Heisler, Chavan, Chang, & LiWang, 2019) and represents a significant improvement over native-PAGE (Kageyama et al., 2006) where samples have to be quenched at each time point. In terms of species resolvability, the magnetic field abscissa allows resolution of more than two components as illustrated by $K25C_{3IAP}$ (Chow et al., 2022). The fact that three spectral components, free KaiB, B_-, and B_+, were simultaneously quantified in a single experiment provided significant constraints to biochemical models that could explain the kinetics, a feature that was subsequently exploited to demonstrate the stepwise assembly of the KaiBC complex from $KaiB_0C_6$ to $KaiB_6C_6$. Thus, SDSL-cw-EPR offers improved time resolution and species resolvability and shall enable fitting to biochemical models to uncover previously unknown dynamics.

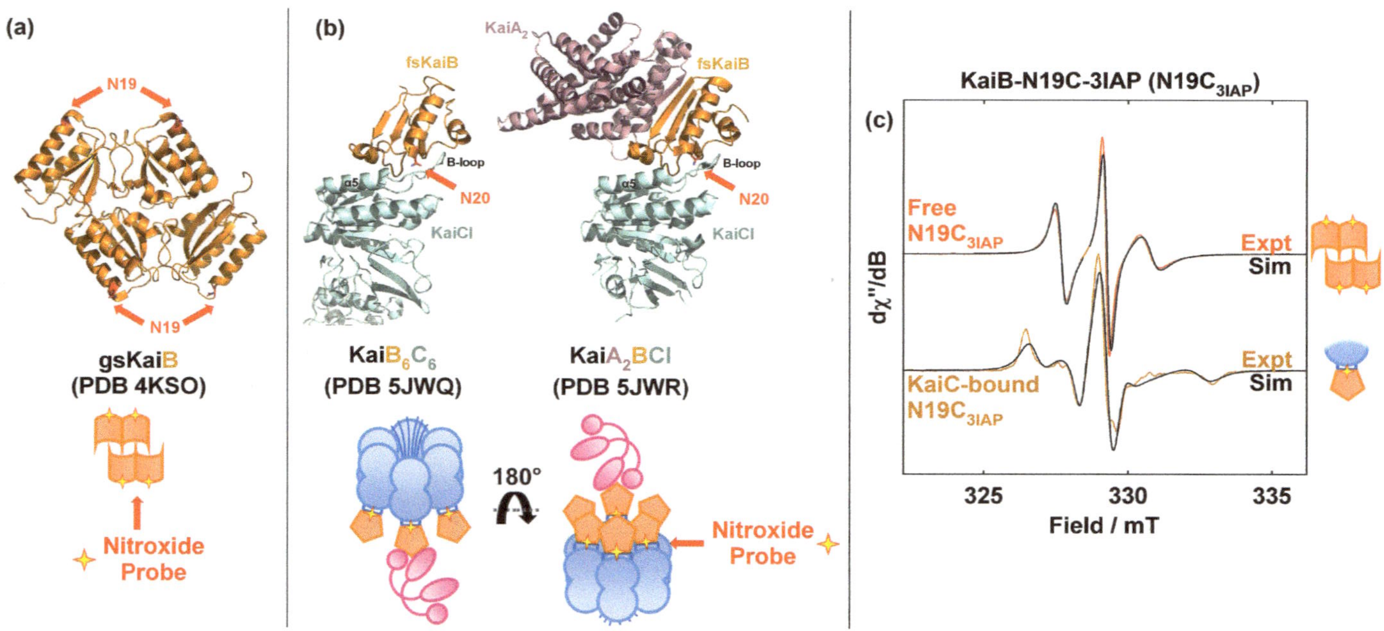

Fig. 1 Resolving free vs KaiC-bound KaiB using SDSL-EPR in X-band (9.5 GHz) cw-EPR by labeling at asparagine 19 (Chow et al., 2020). (A) and (B) Crystal structures (above) and cartoon representations (below) of (A) tetrameric ground-state (gs-) KaiB and (B) when bound to KaiC and KaiA. The coloring schemes are as follows: KaiA, magenta; KaiB, orange; and KaiC, cyan/blue. Spin labels are denoted as four-point stars. (C) Spectra of free (reddish orange) and KaiC-bound (light brown) KaiB-N19C-3IAP ($N19C_{3IAP}$) and their respective simulations (black). *Reprinted from Chow, G. K., Chavan, A. G., Heisler, J. C., Chang, Y.-G., LiWang, A., & Britt, R. D. (2020). Monitoring protein–protein interactions in the cyanobacterial circadian clock in real time via electron paramagnetic resonance spectroscopy.* Biochemistry, *59(26), 2387–2400. doi:10.1021/acs.biochem.0c00279 accessible at https://pubs.acs.org/doi/10.1021/acs.biochem.0c00279. Further permissions related to the material excerpted should be directed to the ACS.*

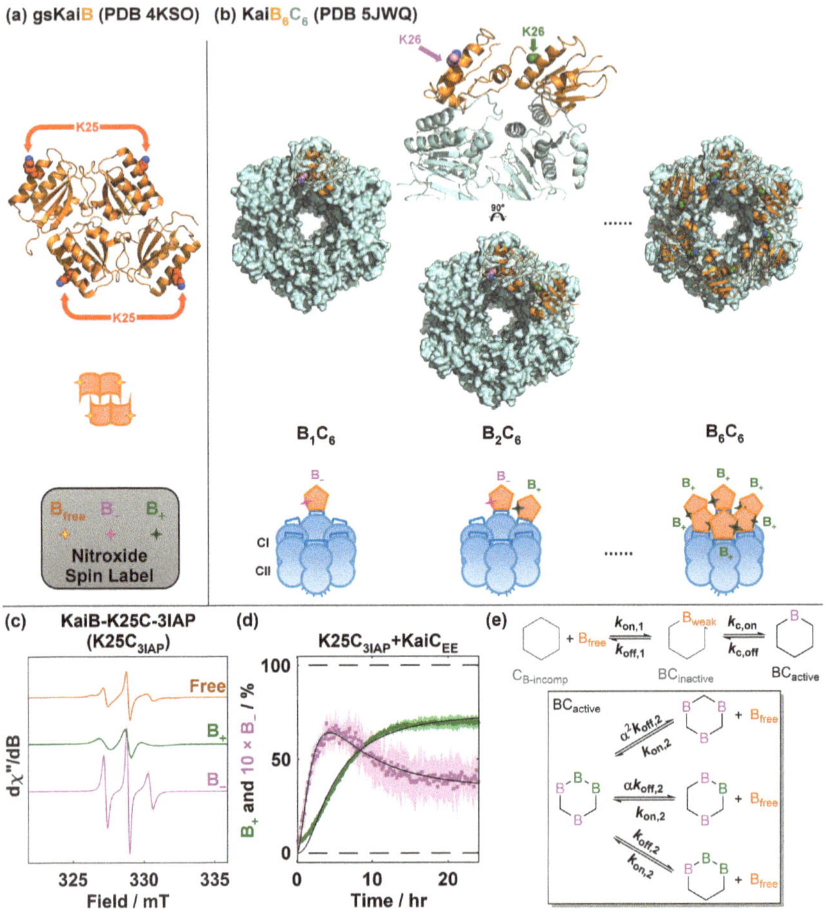

Fig. 2 Resolving "neighbored" versus "neighborless" KaiC-bound KaiB using SDSL-EPR in X-band (9.5 GHz) cw-EPR by labeling at lysine 25 (Chow et al., 2022). (A) and (B) Crystal structures (above) and cartoon representations (below) of (A) tetrameric ground-state (gs-) KaiB and (B) $KaiB_6C_6$ and presumed substoichiometric KaiBC intermediates. The coloring schemes are as follows: KaiB, orange; and KaiC, cyan/blue. The two types of KaiB-K25C-3IAP-resolvable KaiBC sites, "bulk" (B_+) and "edge" (B_-), are shown in green and light purple, respectively. Spin labels are denoted as four-point stars. The side chain atoms of K25 in (A) and K26 in (B) are shown as spheres. (C) Spectra of free KaiB (B_{free}, reddish orange) as well as estimated spectra of B_+ (green) and B_- (magenta). Shaded areas show 95% CI (visible at high magnification). (D) cw-EPR derived KaiB-$KaiC_{S431E, T432E}$ kinetics using $K25C_{3IAP}$ as a probe. B_+: Green circles. B_-: magenta squares. B_- is displayed at 10× concentration to facilitate visualization. Shaded areas show 95% CI ($n = 6$). Black solid lines show fits to model (E). *Adapted with permission from Chow, G. K., Chavan, A. G., Heisler, J., Chang, Y. G., Zhang, N., LiWang, A., & Britt, R. D. (2022). A night-time edge site intermediate in the cyanobacterial circadian clock identified by EPR spectroscopy.* Journal of the American Chemical Society 144 *(1), 184–194, doi:10.1021/jacs.1c08103. Copyright 2022 American Chemical Society.*

2.2 Toward a rational choice of labeling site

We note that $N19C_{3IAP}$ and $K25C_{3IAP}$ only represent two possible labeling sites out of the 100+ residues even when this discussion is confined to KaiB only. Other possible sites that on spin labeling can distinguish multiple states of KaiB without disrupting KaiB's function likely exist. For instance, it might be possible to target the KaiA-KaiB binding interface to quantify their interactions in real time and consequently contrast its kinetics against that of KaiB-KaiC interactions. Furthermore, spectral broadening originating from through-space spin exchange between two spins that are physically close to one another, a property that has previously been exploited in the annexin system (Langen, Isas, Luecke, Haigler, & Hubbell, 1998), might also be used to design spin probes.

The successful design of $N19C_{3IAP}$ and $K25C_{3IAP}$ in differentiating multiple modes of KaiC-bound KaiB hinged on the availability of published crystal structures (Iwase et al., 2005; Tseng et al., 2017) coupled with visual inspection of these structures. Unfortunately, these two aspects do not always apply to other systems of interest. On one hand, the lack of crystal structures might be partly resolved by recent advances in machine learning-based protein structure prediction (Jumper et al., 2021; Quadir, Roy, Soltanikazemi, & Cheng, 2021). On the other hand, the reliance on visual inspection and intuition to determine potential sites that undergo mobility changes begs the question as to whether a more objective means of site design exists. A more rigorous approach is to perform molecular dynamics (MD) simulations to simulate the spectra resulting from site directed spin labeling (SDSL). Multiple implementations exist to directly compute the cw-EPR spectra of spin labeled proteins from MD simulations (Budil, Sale, Khairy, & Fajer, 2006; DeSensi, Rangel, Beth, Lybrand, & Hustedt, 2008; Martin, Svensson, Thomas, & Stoll, 2019; Robinson, Slutsky, & Auteri, 1992). Alternatively, to design sites that rely on changes in interspin distances for spectral changes, one can rely on rotamer library predictions using the SDSL site scan module of Multiscale Modeling of Macromolecules (MMM) (Jeschke, 2018; Polyhach, Bordignon, & Jeschke, 2011). By performing an *in-silico* site scan, one can determine potential sites for labeling that will put spins near one another and hence experience spectral broadening on (hetero)dimerization.

We note in passing that, for conclusions drawn from spin labeling to stand, the functionality of the spin labeled protein must not be significantly different from the original system. For the Kai system, this meant performing parallel fluorescence anisotropy (Chavan et al., 2021; Heisler et al., 2019) and native-PAGE (Kageyama et al., 2006) experiments to determine respectively whether the spin labeled mutant sustains oscillations and have similar binding kinetics. In the case of $N19C_{3IAP}$, we observed an increase in steady-state $KaiB\text{-}KaiC_{S431E,T432E}$ binding as compared to wildtype KaiB even though $N19C_{3IAP}$ was capable of sustaining oscillations (Chow et al., 2020). Thus, despite the power of cw-EPR in discerning multiple states from a single protein of interest, we recommend researchers to test multiple spin labeling sites and validate their spin-labeled construct for functionality prior commencing cw-EPR experiments.

2.3 SDSL-EPR at other frequencies

In the previous sections, we only mentioned the use of X-band (9.5 GHz) cw-EPR for detection. X-band cw-EPR spectrometers are the most widely available EPR spectrometers and thus are the most accessible to researchers. However, the effect of motional broadening is dependent on the incident microwave frequency, and different frequencies are sensitive toward different motional regimes. The ability of cw-EPR in discerning two motional components with isotropic rotational correlation times (τ_c^{iso}) differing by a factor of two can be quantified by the normalized residual mean sum of squares deviation (NRMSD) of the difference spectra as shown in Fig. 3. When nitroxide motion is very fast (sub-nanosecond), Q-band cw-EPR gives the greatest difference in NRMSD between the two components. As the nitroxide motion slows down, X-band and eventually L-band cw-EPR become better at discerning the two motional components. Hence, it can be beneficial to perform quantitative cw-EPR spectroscopy at multiple microwave frequencies to distinguish more components than is possible at a single frequency. Furthermore, higher frequencies allow more accurate determinations of the g tensors as compared to X-band, changes of which have been associated with the microenvironment of the spin label (Gullá & Budil, 2001; Smirnov, Ruuge, Reznikov, Voinov, & Grigor'ev, 2004). Changes in the electron density parameters (g and A) due to local changes in pK_a have been documented at W-band (95 GHz) (Smirnov et al., 2004) and could potentially be adapted for kinetic assays. Thus, high-field EPR should not be disregarded despite the technical hurdles.

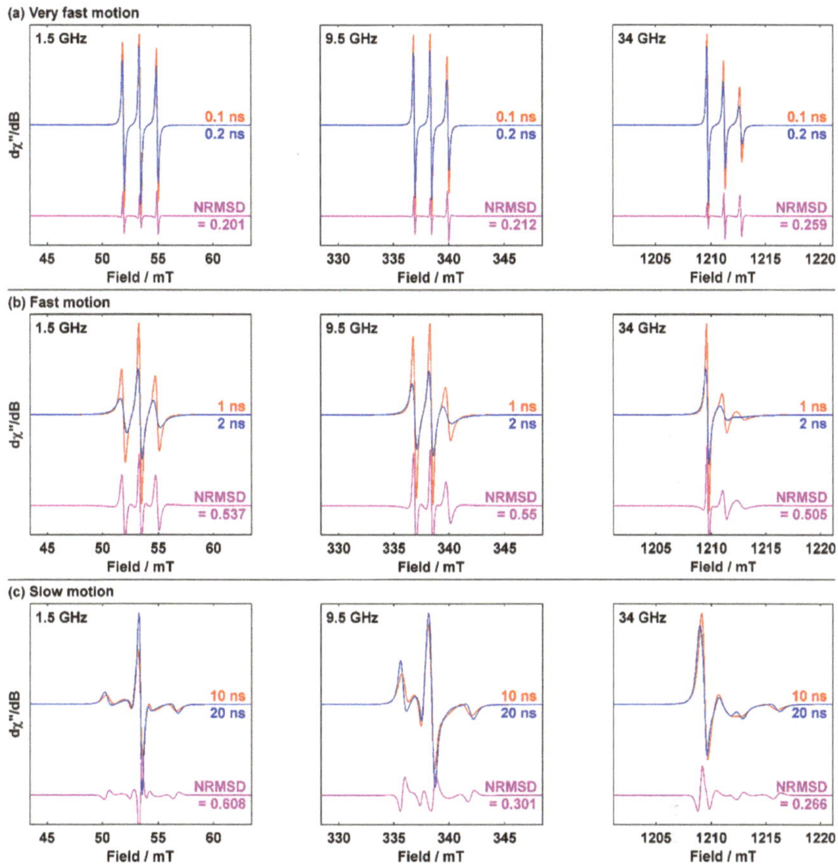

Fig. 3 Differential sensitivities of cw-EPR at various microwave frequencies toward multiple motional regimes. Simulation of nitroxide spectra at various motional regimes at three different microwave frequencies. (A) Very fast motion. (B) Fast motion. (C) Slow motion. From left to right: L-band (1.5 GHz), X-band (9.5 GHz) and Q-band (34 GHz). In each panel, the red trace possesses half the isotropic rotational correlation time (τ_c^{iso}) as the blue trace. The magenta trace at the bottom shows the difference spectrum and the normalized residual mean squared deviation. Simulations were performed using EasySpin (Stoll & Schweiger, 2006) with $g = [2.0086, 2.0060, 2.0023]$ and $A = [12, 16, 100]$.

3. Spin labeling of protein of interest

The spin labeling protocol described below assumes that the protein of interest has been purified by gel filtration. The buffer systems and reaction conditions listed below were optimized for labeling KaiB and/or carrying out the *in vitro* Kai oscillator reaction and should be adapted to one's system of interest as necessary. Details on mutagenesis and purification of Kai proteins have previously been described (Chang, Tseng, Kuo, & LiWang, 2012; Kim, Boyd, Espinosa, & Golden, 2015).

3.1 Equipment and materials

- Fast protein liquid chromatography (FPLC) system
- HiPrep 26/10 desalting column
- Spin desalting column
- UV-vis spectrophotomer
- Coomassie blue for Bradford assay
- Labeling buffer
  - Tris base, 20 mM, adjusted to pH 8.0 using hydrochloric acid[1]
  - Sodium chloride, 150 mM
  - Tris(2-carboxyethyl) phosphine (TCEP)[2]
  - 3-iodoactamido-PROXYL (3IAP)[2] dissolved in dimethyl sulfoxide (DMSO)
- TCEP stock solution, 500 mM in 20 mM Tris, 150 mM NaCl, pH 8.0
- 3IAP stock solution, 200 mM in DMSO
- Reaction buffer
  - Tris base, 20 mM, adjusted to pH 8.0 using hydrochloric acid
  - Sodium chloride, 150 mM
  - Magnesium chloride, 5 mM
  - Adenosine triphosphate (ATP) sodium salt, 1 mM
  - EDTA sodium salt, 0.5 mM

3.2 Spin labeling protocol

Duration: 2 days (more if re-labeling necessary)

1. Overnight labeling reaction (active: 1 h; inactive: overnight)

   Dilute protein to 100–200 μM in labeling buffer. Supplement with *fivefold excess*[2] TCEP from a 1 M stock solution as necessary. Incubate at room temperature for 15 min. Add *10-fold excess*[2] 3IAP. Incubate at 4 °C overnight. Next day, load sample onto desalting column pre-equilibrated in reaction buffer. Concentrate sample as appropriate. Perform Bradford assay to estimate the labeled protein's concentration.

2. Labeling efficiency determination (active: 1 h)

   Desalt sample into a mass spectrometry-compatible buffer using a spin desalting column. Determine the labeling efficiency by electrospray ionization high-resolution liquid chromatography mass spectrometry (ESI-HR-LCMS). Samples that were not at least 95% labeled should be re-labeled by repeating Step 1 and/or reoptimizing the labeling conditions.

3.3 Tips

1. It is the thiolate (R-S$^-$) that participates in the spin labeling reaction. As such, the rate of cysteine-iodoacetamide reaction depends on the pK_a of the cysteine thiol in question. Factors that influence the pK_a of the cysteine thiol (e.g., being at the N-terminus of an α-helix vs in a hydrophobic pocket) can impact the rate of reaction (Poole, 2015).
2. TCEP is used as a reducing agent to reduce disulfide bonds between cysteine residues introduced via mutagenesis and used for spin labeling. As TCEP reacts directly with iodoacetamide-based fluorescent dyes (Tyagarajan, Pretzer, & Wiktorowicz, 2003) and is inferred to do the same to 3IAP, excess 3IAP is necessary to ensure a stoichiometric excess of 3IAP to the protein of interest during spin labeling.

3.4 Stability of spin labeled adduct and choice of spin label

A prerequisite for using SDSL-EPR to study proteins is the stability of the spin labeled protein over the experimental timespan. Unlike typical SDSL-EPR experiments where samples are flash frozen for Double Electron-Electron Resonance (DEER) distance measurements, the goal of our experiments were to track the kinetics of protein-protein interactions at ambient temperatures. The Kai system oscillates with a period of approximately a day. Furthermore, five oscillatory cycles (120 h) have previously been recommended for reliable oscillation period determination (Zielinski, Moore, Troup, Halliday, & Millar, 2014). Thus, spin labeled adduct stability is of utmost importance for our experiments.

We chose 3IAP because we found that 3IAP-labeled KaiB was sufficiently stable for over a week at 30 °C (Chow et al., 2020). Prior our work, attempts have been made to label KaiB using MTSSL and 4-maleimido-TEMPO (4MT) by the Ishiura group (Mutoh, Mino, Murakami, Uzumaki, & Ishiura, 2011; Mutoh et al., 2010) but were met with multiple challenges. For the former, Mutoh et al. (2011) observed disulfide bond formation between KaiA and KaiB when KaiB-threonine 63 was mutated to cysteine and spin labeled using MTSSL. For the latter, they observed an increase in signal intensity when 4MT-labeled KaiB was

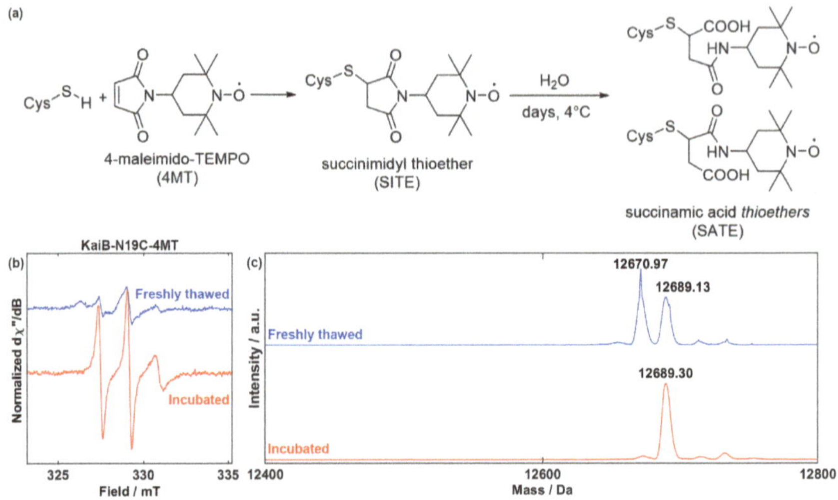

Fig. 4 4MT-spin labeled proteins may exhibit gradual hydrolysis as demonstrated by KaiB-N19C-4MT. (A) Chemical equation for hydrolysis. (B) Spectra of KaiB-N19C-4MT when freshly thawed (top, blue) vs after incubation at 4 °C for 2 weeks (bottom, red). Spectra were obtained from 3 × Kai reactions (3.6 μM KaiA, 10.5 μM KaiB-N19C-4MT, 10.5 μM KaiC, t = 0–7.5 min), background corrected, and normalized by double integrated intensity. (C) LC-HRMS of KaiB-N19C-4MT prior (top) and after (bottom) storage at 4 °C for 2 weeks. LCMS masses: KaiB-N19C-4MT (Calc: 12671.73; Found: 12670.97). KaiB-N19C-4SAT (Calc: 12689.73; Found: 12689.13 and 12689.30).

incubated at 40 °C for 24 h that could potentially be attributed to maleimide hydrolysis (Fontaine, Reid, Robinson, Ashley, & Santi, 2015) (Fig. 4A). In our experience, although KaiB-asparagine 19 when mutated to cysteine could be labeled successfully with 4MT to give KaiB-N19C-4MT, it was observed to slowly hydrolyze even when stored at 4 °C as determined by disappearance of a slow-motion feature in the EPR spectrum (Fig. 4B) and verified by mass spectrometry (Fig. 4C).

4. EPR sample preparation and data acquisition

The protocol below allows reproducible acquisition of SDSL-cw-EPR spectra for kinetics analysis. The reader is referred to the original publication (Chow et al., 2020) for details on data analysis.

4.1 Equipment and materials

- X-band cw EPR spectrometer (Bruker) equipped with an SHQE cavity
- EIP 578B frequency counter (National Instruments, Santa Clara, CA) or equivalent[1]

- Variable temperature cavity insert connected to an FTS XR401 Air-jet Crystal Cooler (SP Scientific, Stone Ridge, NY) or equivalent[2]
- Thermocouple and Universal Thermocouple Connector Direct USB to PC Connection (Omega Inc.)[2]
- Computer for logging microwave frequency[1] and temperature[2]
- Borosilicate rectangular miniature hollow tubing[3, 4] (VitroCom, Mountain Lakes, NJ, Part # 2540), cut into 12-cm long pieces and with one end flame-sealed
- Custom collet for holding sample cell[4]
- 1550 Rpm Centricone Centrifuge (Precision Scientific, optional)[3]
- Gel loading tip[3]
- Teflon tape
- Reaction buffer
  - Tris base, 20 mM, adjusted to pH 8.0 using hydrochloric acid
  - Sodium chloride, 150 mM
  - Magnesium chloride, 5 mM
  - Adenosine triphosphate (ATP) sodium salt, 1 mM
  - EDTA sodium salt, 0.5 mM
- (Spin labeled) Kai proteins in reaction buffer, listed below as final monomeric concentrations for the 1 × *in vitro* oscillator
  - KaiA, 1.2 μM
  - Spin labeled KaiB, 3.5 μM
  - KaiC, 3.5 μM
- 4-hydroxy-TEMPO (TEMPOL), 10 μM, in reaction buffer

4.2 Sample preparation and spectral acquisition protocol

Duration: 5 days (more if desired for period analysis)

1. Prepare spectrometer (active: 10 min; inactive: 30 min)

Turn on the spectrometer, including the magnet power supply and cooling module. Turn on the air/sample temperature controller. Allow the system to warm up and the cavity temperature to equilibrate for 30 min.[2] During equilibration, set up the following acquisition parameters:

Modulation frequency: 100 kHz
Modulation amplitude: 2 G
Conversion time: 40 ms
Time constant: 20 ms
Number of scans per spectrum: 10

Magnetic field sweep: 200 G

Set up an experiment for kinetics. Newer spectrometers often come with the Xepr software that allows one to select time as the second dimension. For older spectrometers (e.g., ECS 106), using the automation script editor, enter the following:

```
del
racq
write yourfilename_0.spc
loop 1 800
```

The script above will (i) delete the spectrum currently on display, (ii) acquire a new spectrum, (iii) write the file to yourfilename_0.spc and (iv) loop to line 1 for 800 times. The automation updates the file address automatically on execution and should result in 1600 files from yourfilename_0.spc and yourfilename_0.par to yourfilename_799.spc and yourfilename_799.par. Using the spectrometer settings listed above, the 800 spectra should correspond to approximately 4 days of acquisition.

2. Pre-tune the spectrometer (active: 15 min)

   Tune the spectrometer and determine the optimal magnetic field center position using the TEMPOL standard. The resonant frequency of an SHQE cavity with the variable temperature cavity insert is typically around 9.24 GHz when a 20–40 μL sample in reaction buffer is used, resulting in an optimal magnetic field center position of 3300 G.

3. Set up and load the oscillator reaction mixture. (active: 30 min)

   Calculate the volumes of proteins and buffer necessary to achieve the final desired volumes and concentrations. Prepare 1 μL excess of reaction to account for losses during pipetting. In a microfuge tube, combine the reaction mixture in the following order: buffer, KaiB, KaiA and KaiC. Then, using a gel loading tip, transfer the reaction mixture to the rectangular glass tube by pointing the tip's end at the edge of the opening of the cell. Allow the sample to settle at the bottom of the tube and remove any trapped bubbles using a low-speed centrifuge or by swinging the tube manually. Cover the opening of the tube with Teflon tape to prevent dust from entering the tube for the duration of the experiment.

4. Spectral acquisition (active: 10 min; inactive: 5 days)

   Load the sample cell into the spectrometer and secure it using the collet. Retune the spectrometer. Start spectral acquisition. Periodically inspect the spectrometer to ensure that the magnet power supply and cooling systems are in working condition throughout the experiment.

4.3 Tips

1. Older spectrometers (e.g., ECS 106) do not automatically record the microwave frequency at which the spectra were acquired. This can be omitted for newer models (e.g., ELEXSYS EMX).
2. The exact temperature settings for the air cooler depends on the length of the tubing between the outlet of the cooler and the cavity insert and the insulation between this tubing and ambient air. Use the thermocouple to monitor actual temperature.
3. Although most of the sample will settle into the bottom of the tube with little effort from the user, small amounts of sample may cling to the side that drip down slowly over the duration of the experiment. This drip can lead to volumetric changes in the sample and thus cavity resonance frequency drift over time. Occasionally, the sample may trap air pockets, resulting in the inability of the sample to settle at the bottom of the tube. Both problems are eliminated by centrifuging the sample prior loading onto the spectrometer. This step is more reproducible if a benchtop centrifuge is used.
4. Water absorbs microwaves strongly at X-band. A flat cell is used to minimize microwave absorption. In our experience, we can load up to 40 μL of sample into the cell without compromising the Q-factor and still be able to tune the spectrometer, with sample volumes larger than 40 μL being unable to produce better S/N. As the sample's temperature is controlled via an air jet, the sample needs to be centered and secured in the cavity via a custom collet that holds the sample in place for a stable tuning picture.

5. Concluding remarks

The dependence of cw-EPR spectra of spin labels on their mobility has long been recognized (Freed et al., 1971) but has rarely been put to bioanalytical use due to various technical challenges. In this chapter, we discussed experimental considerations, covering the information available from SDSL-cw-EPR and guidelines to site design. Then, we detailed the protocols to preparing spin labeled Kai proteins and setting up the acquisition parameters for studying the cyanobacterial circadian clock.

We have limited ourselves to performing site directed mutagenesis on KaiB owing to its lack of native cysteine residues. As native cysteine residues often possess varying functions from enzymatic active sites or recognition motifs for metal binding to structural stabilizers as disulfide bonds

(Marino & Gladyshev, 2012), the two main classes of strategies to bypass the limits posed by the use of native cysteine residues in SDSL are briefly discussed. First, alternative chemistries could be applied to label the protein of interest by making use of functional groups other than thiols. Ketone and alkyne functional groups have previously been engineered into recombinant proteins via amber codon suppression (Evans & Millhauser, 2015; Nguyen et al., 2009) whereas the aldehyde group could be introduced by the use of formylglycine generating enzyme (FGE) (Carrico, Carlson, & Bertozzi, 2007). Second, multidomain proteins that can be expressed separately could be labeled separately prior ligation to form the full-length protein. Common ligases used for segmental labeling followed by ligation include sortase or subtiligase (Antos, Truttmann, & Ploegh, 2016; Weeks & Wells, 2020).

We have also limited ourselves to using cw-EPR due to its relative accessibility. However, the current strategy has significant limitations as a quantitative bioanalytical assay due to its low throughput and relative sensitivity. Specifically, conventional cw-EPR spectrometers only allow a single sample to be studied at a time using samples in the micromolar range (3.5 μM $N19C_{3IAP}$ in our experience). This contrasts with fluorescence techniques where fluorescence platereaders can be used to collect data for hundreds of reactions simultaneously using submicromolar labeled protein (50 nM fluorescently labeled KaiB) (Chavan et al., 2021; Heisler et al., 2019). The difference in concentration is necessary to allow sufficient signal-to-noise in cw-EPR measurements and points toward the need for innovative EPR detection schemes for improved sensitivity. Significant advances in rapid-scan EPR by the Eaton group have been made in the last two decades and reviewed in a separate chapter (Eaton & Eaton, 2022). Rapid-scan EPR allows a much higher microwave power that can be used without saturating the sample, thereby improving the S/N of nitroxides by a factor of 40 (Mitchell et al., 2013). Rapid-scan technology for single samples has recently been commercialized by Bruker and will become more accessible to the research community in the near future. To improve throughput, rapid-scan can be coupled with imaging (Eaton et al., 2017) by the use of magnetic field gradients. Much of the rapid-scan imaging technology has been developed for *in vivo* applications and thus are performed at lower microwave frequencies. Nevertheless, spectral changes due to motion are still observed at lower microwave frequencies (Fig. 3). Thus, rapid-scan imaging can satisfy both requirements of increased S/N and throughput at physiological concentrations and will enable EPR spectroscopy to be a competitive bioanalytical tool for analyzing protein-protein interactions.

Acknowledgments

We thank Drs. William Jewell and Armann Andaya for his help in acquiring LCMS data. This work was supported by grants from the U.S. National Science Foundation (NSF-MCB-1615752 to R.D.B.; NSF-HRD-1547848 to CREST: Center for Cellular and Biomolecular Machines at UC Merced), National Institutes of Health (GM107521 to A.L.), and Army Research Office (W911NF-17-1-0434 to A.L.).

References

Altenbach, C., Flitsch, S. L., Khorana, H. G., & Hubbell, W. L. (1989). Structural studies on transmembrane proteins. 2. Spin labeling of bacteriorhodopsin mutants at unique cysteines. *Biochemistry*, *28*(19), 7806–7812. https://doi.org/10.1021/bi00445a042.

Altenbach, C., López, C. J., Hideg, K., & Hubbell, W. L. (2015). Exploring structure, dynamics, and topology of nitroxide spin-labeled proteins using continuous-wave electron paramagnetic resonance spectroscopy. *Methods in Enzymology*, *564*, 59–100. https://doi.org/10.1016/bs.mie.2015.08.006.

Antos, J. M., Truttmann, M. C., & Ploegh, H. L. (2016). Recent advances in sortase-catalyzed ligation methodology. *Current Opinion in Structural Biology*, *38*, 111–118. https://doi.org/10.1016/j.sbi.2016.05.021.

Berliner, L. J., Grunwald, J., Hankovszky, H. O., & Hideg, K. (1982). A novel reversible thiol-specific spin label: Papain active site labeling and inhibition. *Analytical Biochemistry*, *119*(2), 450–455. https://doi.org/10.1016/0003-2697(82)90612-1.

Bittl, R., & Weber, S. (2005). Transient radical pairs studied by time-resolved EPR. *Biochimica et Biophysica Acta*, *1707*(1), 117–126. https://doi.org/10.1016/j.bbabio.2004.03.012.

Bobko, A. A., Kirilyuk, I. A., Grigor'ev, I. A., Zweier, J. L., & Khramtsov, V. V. (2007). Reversible reduction of nitroxides to hydroxylamines: Roles for ascorbate and glutathione. *Free Radical Biology & Medicine*, *42*(3), 404–412. https://doi.org/10.1016/j.freeradbiomed.2006.11.007.

Budil, D. E., Lee, S., Saxena, S., & Freed, J. H. (1996). Nonlinear-least-squares analysis of slow-motion EPR spectra in one and two dimensions using a modified Levenberg–Marquardt algorithm. *Journal of Magnetic Resonance. Series A*, *120*(2), 155–189. https://doi.org/10.1006/jmra.1996.0113.

Budil, D. E., Sale, K. L., Khairy, K. A., & Fajer, P. G. (2006). Calculating slow-motional electron paramagnetic resonance spectra from molecular dynamics using a diffusion operator approach. *The Journal of Physical Chemistry. A*, *110*(10), 3703–3713. https://doi.org/10.1021/jp054738k.

Cafiso, D. S. (2014). Identifying and quantitating conformational exchange in membrane proteins using site-directed spin labeling. *Accounts of Chemical Research*, *47*(10), 3102–3109. https://doi.org/10.1021/ar500228s.

Carrico, I. S., Carlson, B. L., & Bertozzi, C. R. (2007). Introducing genetically encoded aldehydes into proteins. *Nature Chemical Biology*, *3*(6), 321–322. https://doi.org/10.1038/nchembio878.

Chang, Y. G., Tseng, R., Kuo, N. W., & LiWang, A. (2012). Rhythmic ring-ring stacking drives the circadian oscillator clockwise. *Proceedings of the National Academy of Sciences of the United States of America*, *109*(42), 16847–16851. https://doi.org/10.1073/pnas.1211508109.

Chavan, A. G., Swan, J. A., Heisler, J., Sancar, C., Ernst, D. C., Fang, M., et al. (2021). Reconstitution of an intact clock reveals mechanisms of circadian timekeeping. *Science*, *374*(6564), eabd4453. https://doi.org/10.1126/science.abd4453.

Chow, G. K., Chavan, A. G., Heisler, J. C., Chang, Y.-G., Zhang, N., LiWang, A., & Britt, R. D. (2022). A night-time edge site intermediate in the cyanobacterial circadian clock identified by EPR spectroscopy. *Journal of the American Chemical Society, 144*(1), 184–194. https://doi.org/10.1021/jacs.1c08103.

Chow, G. K., Chavan, A. G., Heisler, J. C., Chang, Y.-G., LiWang, A., & Britt, R. D. (2020). Monitoring protein–protein interactions in the cyanobacterial circadian clock in real time via electron paramagnetic resonance spectroscopy. *Biochemistry, 59*(26), 2387–2400. https://doi.org/10.1021/acs.biochem.0c00279.

DeSensi, S. C., Rangel, D. P., Beth, A. H., Lybrand, T. P., & Hustedt, E. J. (2008). Simulation of nitroxide electron paramagnetic resonance spectra from brownian trajectories and molecular dynamics simulations. *Biophysical Journal, 94*(10), 3798–3809. https://doi.org/10.1529/biophysj.107.125419.

Eaton, G. R., & Eaton, S. (2022). Advances in Rapid Scan EPR Spectroscopy. In R. D. Britt (Ed.), *666. Methods in Enzymology* Elsevier. In this issue.

Eaton, S. S., Shi, Y., Woodcock, L., Buchanan, L. A., McPeak, J., Quine, R. W., et al. (2017). Rapid-scan EPR imaging. *Journal of Magnetic Resonance, 280,* 140–148. https://doi.org/10.1016/j.jmr.2017.02.013.

Evans, E. G. B., & Millhauser, G. L. (2015). Genetic incorporation of the unnatural amino acid p-acetyl phenylalanine into proteins for site-directed spin labeling. *Methods in Enzymology, 563,* 503–527. https://doi.org/10.1016/bs.mie.2015.06.008.

Fontaine, S. D., Reid, R., Robinson, L., Ashley, G. W., & Santi, D. V. (2015). Long-term stabilization of maleimide-thiol conjugates. *Bioconjugate Chemistry, 26*(1), 145–152. https://doi.org/10.1021/bc5005262.

Freed, J. H., Bruno, G. V., & Polnaszek, C. F. (1971). Electron spin resonance line shapes and saturation in the slow motional region. *The Journal of Physical Chemistry, 75*(22), 3385–3399. https://doi.org/10.1021/j100691a001.

Golden, S. S., & LiWang, A. (2021). Mechanistic aspects of the cyanobacterial circadian clock. In C. H. Johnson, & M. J. Rust (Eds.), *Circadian rhythms in bacteria and microbiomes* (pp. 67–77). Cham: Springer International Publishing.

Goldman, S. A., Bruno, G. V., & Freed, J. H. (1972). Estimating slow-motional rotational correlation times for nitroxides by electron spin resonance. *The Journal of Physical Chemistry, 76*(13), 1858–1860. https://doi.org/10.1021/j100657a013.

Gullá, A. F., & Budil, D. E. (2001). Orientation dependence of electric field effects on the g factor of nitroxides measured by 220 GHz EPR. *The Journal of Physical Chemistry. B, 105*(33), 8056–8063. https://doi.org/10.1021/jp0109224.

Heisler, J., Chavan, A., Chang, Y. G., & LiWang, A. (2019). Real-time in vitro fluorescence anisotropy of the cyanobacterial circadian clock. *Methods Protocol, 2*(2), 42. https://doi.org/10.3390/mps2020042.

Iwase, R., Imada, K., Hayashi, F., Uzumaki, T., Morishita, M., Onai, K., et al. (2005). Functionally important substructures of circadian clock protein KaiB in a unique tetramer complex. *The Journal of Biological Chemistry, 280*(52), 43141–43149. https://doi.org/10.1074/jbc.M503360200.

Jeschke, G. (2018). MMM: A toolbox for integrative structure modeling. *Protein Science, 27*(1), 76–85. https://doi.org/10.1002/pro.3269.

Jumper, J., Evans, R., Pritzel, A., Green, T., Figurnov, M., Ronneberger, O., et al. (2021). Highly accurate protein structure prediction with AlphaFold. *Nature, 596*(7873), 583–589. https://doi.org/10.1038/s41586-021-03819-2.

Kageyama, H., Nishiwaki, T., Nakajima, M., Iwasaki, H., Oyama, T., & Kondo, T. (2006). Cyanobacterial circadian pacemaker: Kai protein complex dynamics in the KaiC phosphorylation cycle in vitro. *Molecular Cell, 23*(2), 161–171. https://doi.org/10.1016/j.molcel.2006.05.039.

Kim, Y. I., Boyd, J. S., Espinosa, J., & Golden, S. S. (2015). Detecting KaiC phosphorylation rhythms of the cyanobacterial circadian oscillator in vitro and in vivo. *Methods in Enzymology*, *551*, 153–173. https://doi.org/10.1016/bs.mie.2014.10.003.

Langen, R., Isas, J. M., Luecke, H., Haigler, H. T., & Hubbell, W. L. (1998). Membrane-mediated assembly of annexins studied by site-directed spin labeling. *The Journal of Biological Chemistry*, *273*(35), 22453–22457.

Marino, S. M., & Gladyshev, V. N. (2012). Analysis and functional prediction of reactive cysteine residues. *The Journal of Biological Chemistry*, *287*(7), 4419–4425. https://doi.org/10.1074/jbc.R111.275578.

Martin, P. D., Svensson, B., Thomas, D. D., & Stoll, S. (2019). Trajectory-based simulation of EPR spectra: Models of rotational motion for spin labels on proteins. *The Journal of Physical Chemistry. B*, *123*(48), 10131–10141. https://doi.org/10.1021/acs.jpcb.9b02693.

Meirovitch, E., Nayeem, A., & Freed, J. H. (1984). Analysis of protein-lipid interactions based on model simulations of electron spin resonance spectra. *The Journal of Physical Chemistry*, *88*(16), 3454–3465.

Mitchell, D. G., Rosen, G. M., Tseitlin, M., Symmes, B., Eaton, S. S., & Eaton, G. R. (2013). Use of rapid-scan EPR to improve detection sensitivity for spin-trapped radicals. *Biophysical Journal*, *105*(2), 338–342. https://doi.org/10.1016/j.bpj.2013.06.005.

Mutoh, R., Mino, H., Murakami, R., Uzumaki, T., & Ishiura, M. (2011). Thermodynamically induced conformational changes of the cyanobacterial circadian clock protein KaiB. *Applied Magnetic Resonance*, *40*(4), 525–534. https://doi.org/10.1007/s00723-011-0228-2.

Mutoh, R., Mino, H., Murakami, R., Uzumaki, T., Takabayashi, A., Ishii, K., et al. (2010). Direct interaction between KaiA and KaiB revealed by a site-directed spin labeling electron spin resonance analysis. *Genes to Cells*, *15*(3), 269–280. https://doi.org/10.1111/j.1365-2443.2009.01377.x.

Nakajima, M., Imai, K., Ito, H., Nishiwaki, T., Murayama, Y., Iwasaki, H., et al. (2005). Reconstitution of circadian oscillation of cyanobacterial KaiC phosphorylation in vitro. *Science*, *308*(5720), 414–415. https://doi.org/10.1126/science.1108451.

Nguyen, D. P., Lusic, H., Neumann, H., Kapadnis, P. B., Deiters, A., & Chin, J. W. (2009). Genetic encoding and labeling of aliphatic azides and alkynes in recombinant proteins via a pyrrolysyl-tRNA synthetase/tRNACUA pair and click chemistry. *Journal of the American Chemical Society*, *131*(25), 8720–8721. https://doi.org/10.1021/ja900553w.

Nishiwaki, T., Satomi, Y., Kitayama, Y., Terauchi, K., Kiyohara, R., Takao, T., et al. (2007). A sequential program of dual phosphorylation of KaiC as a basis for circadian rhythm in cyanobacteria. *The EMBO Journal*, *26*(17), 4029–4037. https://doi.org/10.1038/sj.emboj.7601832.

Polimeno, A., & Freed, J. H. (1995). Slow motional ESR in complex fluids: The slowly relaxing local structure model of solvent cage effects. *The Journal of Physical Chemistry*, *99*(27), 10995–11006. https://doi.org/10.1021/j100027a047.

Polyhach, Y., Bordignon, E., & Jeschke, G. (2011). Rotamer libraries of spin labelled cysteines for protein studies. *Physical Chemistry Chemical Physics*, *13*(6), 2356–2366. https://doi.org/10.1039/c0cp01865a.

Poole, L. B. (2015). The basics of thiols and cysteines in redox biology and chemistry. *Free Radical Biology & Medicine*, *80*, 148–157. https://doi.org/10.1016/j.freeradbiomed.2014.11.013.

Quadir, F., Roy, R. S., Soltanikazemi, E., & Cheng, J. (2021). DeepComplex: A web server of predicting protein complex structures by deep learning inter-chain contact prediction and distance-based modelling. *Frontiers in Molecular Biosciences*, *8*, 716973. https://doi.org/10.3389/fmolb.2021.716973.

Robinson, B. H., Slutsky, L. J., & Auteri, F. P. (1992). Direct simulation of continuous wave electron paramagnetic resonance spectra from Brownian dynamics trajectories. *The Journal of Chemical Physics*, *96*(4), 2609–2616. https://doi.org/10.1063/1.462869.

Rust, M. J., Markson, J. S., Lane, W. S., Fisher, D. S., & O'Shea, E. K. (2007). Ordered phosphorylation governs oscillation of a three-protein circadian clock. *Science*, *318*(5851), 809–812. https://doi.org/10.1126/science.1148596.

Samouilov, A., Roubaud, V., Kuppusamy, P., & Zweier, J. L. (2004). Kinetic analysis-based quantitation of free radical generation in EPR spin trapping. *Analytical Biochemistry*, *334*(1), 145–154. https://doi.org/10.1016/j.ab.2004.07.026.

Smirnov, A. I., Ruuge, A., Reznikov, V. A., Voinov, M. A., & Grigor'ev, I. A. (2004). Site-directed electrostatic measurements with a thiol-specific pH-sensitive nitroxide: Differentiating local pK and polarity effects by high-field EPR. *Journal of the American Chemical Society*, *126*(29), 8872–8873. https://doi.org/10.1021/ja048801f.

Stoll, S., & Schweiger, A. (2006). EasySpin, a comprehensive software package for spectral simulation and analysis in EPR. *Journal of Magnetic Resonance*, *178*(1), 42–55. https://doi.org/10.1016/j.jmr.2005.08.013.

Swan, J. A., Golden, S. S., LiWang, A., & Partch, C. L. (2018). Structure, function, and mechanism of the core circadian clock in cyanobacteria. *The Journal of Biological Chemistry*, *293*(14), 5026–5034. https://doi.org/10.1074/jbc.TM117.001433.

Tseng, R., Goularte, N. F., Chavan, A., Luu, J., Cohen, S. E., Chang, Y. G., et al. (2017). Structural basis of the day-night transition in a bacterial circadian clock. *Science*, *355*(6330), 1174–1180. https://doi.org/10.1126/science.aag2516.

Tyagarajan, K., Pretzer, E., & Wiktorowicz, J. E. (2003). Thiol-reactive dyes for fluorescence labeling of proteomic samples. *Electrophoresis*, *24*(14), 2348–2358. https://doi.org/10.1002/elps.200305478.

Vaze, K. M., & Sharma, V. K. (2013). On the adaptive significance of circadian clocks for their owners. *Chronobiology International*, *30*(4), 413–433. https://doi.org/10.3109/07420528.2012.754457.

Weeks, A. M., & Wells, J. A. (2020). Subtiligase-catalyzed peptide ligation. *Chemical Reviews*, *120*(6), 3127–3160. https://doi.org/10.1021/acs.chemrev.9b00372.

Woelfle, M. A., Ouyang, Y., Phanvijhitsiri, K., & Johnson, C. H. (2004). The adaptive value of circadian clocks: An experimental assessment in cyanobacteria. *Current Biology*, *14*(16), 1481–1486. https://doi.org/10.1016/j.cub.2004.08.023.

Zielinski, T., Moore, A. M., Troup, E., Halliday, K. J., & Millar, A. J. (2014). Strengths and limitations of period estimation methods for circadian data. *PLoS One*, *9*(5), e96462. https://doi.org/10.1371/journal.pone.0096462.

CHAPTER FOUR

Orthogonal spin labeling and pulsed dipolar spectroscopy for protein studies

Laura Galazzo[a], Markus Teucher[b], and Enrica Bordignon[a,*]
[a]Department of Physical Chemistry, University of Geneva, Genève, Switzerland
[b]MPI for Chemical Energy Conversion, Mülheiman der Ruhr, Germany
*Corresponding author: e-mail address: enrica.bordignon@unige.ch

Contents

Methods in Enzymology, Volume 666
ISSN 0076-6879
https://doi.org/10.1016/bs.mie.2022.02.004

Abstract

Different types of spin labels are currently available for structural studies of biomolecules both in vitro and in cells using Electron Paramagnetic Resonance (EPR) and pulse dipolar spectroscopy (PDS). Each type of label has its own advantages and disadvantages, that will be addressed in this chapter. The spectroscopically distinct properties of the labels have fostered new applications of PDS aimed to simultaneously extract multiple inter-label distances on the same sample. In fact, combining different labels and choosing the optimal strategy to address their inter-label distances can increase the information content per sample, and this is pivotal to better characterize complex multi-component biomolecular systems. In this review, we provide a brief background of the spectroscopic properties of the four most common orthogonal spin labels for PDS measurements and focus on the various methods at disposal to extract homo- and hetero-label distances in proteins. We also devote a section to possible artifacts arising from channel crosstalk and provide few examples of applications in structural biology.

Abbreviations

AWG	arbitrary waveform generator
DEER	double electron electron resonance
DQC	double quantum coherence
EDFS	echo-detected field sweep
EPR	electron paramagnetic resonance
FWHM	full width at half maximum
MTS	methanethiosulfonate
NO	nitroxide
NTA	nitrilotriacetic acid
PDS	pulse dipolar spectroscopy
PELDOR	pulsed electron-electron double resonance
RIDME	relaxation induced dipolar modulation enhancement
SDSL	site-directed spin labeling
SIFTER	single-frequency technique for refocusing
SNR	signal-to-noise ratio
Trityl	triarylmethyl
ZFS	zero-field splitting

1. Introduction

About a decade ago, Site-directed Spin Labeling (SDSL) EPR in combination with pulsed dipolar spectroscopy (PDS) initiated a transition from a nitroxide-only to a multi-label approach for protein studies. Three main driving forces guided this transformation: (i) method development (arbitrary waveform generators, increased microwave power at medium and high frequencies, development and optimization of pulsed dipolar techniques); (ii) synthesis of new labels with a higher degree of biocompatibility for the emerging field of in-cell EPR; (iii) the increased complexity of investigated protein systems (e.g., multi-protein-ligand complexes).

Pioneering work using Double Electron Electron Resonance (DEER aka PELDOR) on gadolinium chelates at Ka- and W-band frequencies (Raitsimring et al., 2007) paved the way to the development of the first Gd(III)-based spin labels targeting cysteine residues in proteins (Potapov et al., 2010). Shortly after, DEER experiments addressing the interactions between the spectroscopically "orthogonal" nitroxide and gadolinium spins appeared on model compounds at X and Q band (Lueders, Jeschke, & Yulikov, 2011) and on proteins at W and Q band (Garbuio et al., 2013; Kaminker et al., 2012). Notably, dual labeling with nitroxide and gadolinium labels was earlier identified as promising for distance measurements in proteins based on relaxation-enhancement experiments (Voss et al., 2001). Nowadays these two labels are routinely used to extract NO-Gd distances by DEER on protein complexes (Assafa et al., 2021; Galazzo et al., 2020; Gmeiner et al., 2017; Gmeiner, Dorn, Allain, Jeschke, & Yulikov, 2017; Kaminker et al., 2012, 2013; Lueders, Jäger, Hemminga, Jeschke, & Yulikov, 2013; Shah et al., 2019; Teucher et al., 2019).

Dual labeling can be used to increase the information content per sample, and it unravels its potential when multiple biomolecules interact and conformational changes occur, as in the case of oligomeric protein assemblies, protein-RNA (or DNA) complexes (Gmeiner, Dorn, et al., 2017), or protein interaction networks (Assafa et al., 2021). Dual labeling is more challenging when the two different labels target two cysteine residues in the same protein, however, unnatural amino acid strategies can be employed (see for example (Abdelkader et al., 2015; Garbuio et al., 2013)) or different target amino acids can be used (Gmeiner, Klose, et al., 2017).

Notably, orthogonal pairs can be chosen with optimal properties with respect to the investigated biomolecule and environment. For example, the properties of NO and Gd labels can be ad hoc tailored via chemical modifications. The steric hindrance and flexibility of the NO labels can be modified by adding substituents on the ring, or changing the linker, and their resistance to reducing agents can be increased when the common methyl flanking groups are exchanged with bulkier ones (Bleicken et al., 2019; Haugland, Lovett, & Anderson, 2018; Jagtap et al., 2015; Karthikeyan et al., 2018; Paletta, Pink, Foley, Rajca, & Rajca, 2012). The properties of Gd labels (zero field splitting, relaxation, rigidity) depend strongly on the structural properties of the chelator and on the linker used. Several gadolinium labels exist to date, as recently reviewed in (Giannoulis, Ben-Ishay, & Goldfarb, 2021).

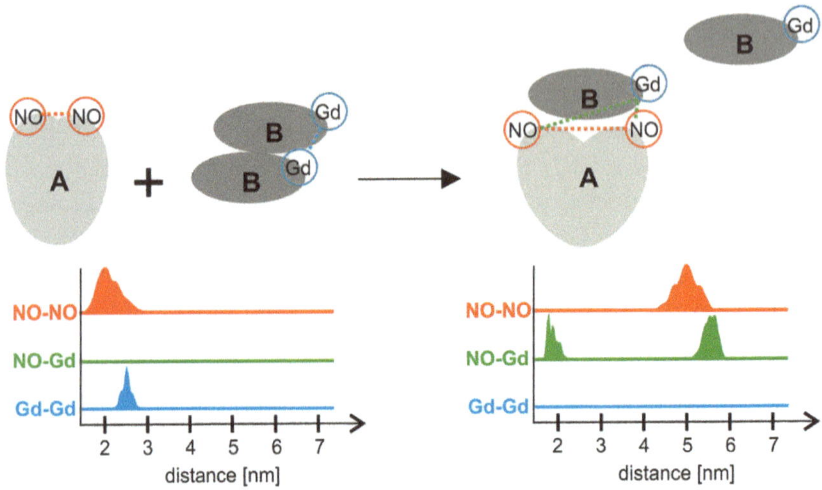

Fig. 1 Example of orthogonal spin labeling in structural biology. Sketch representation of the interaction between protein A, carrying two nitroxide labels (NO) and protein B, which is prone to dimerization, carrying one gadolinium label (Gd). In the depicted situation, A changes its conformation upon binding of B and the B–B dimer breaks. Combinations of distances that could be detected in the three channels before and after protein mixing are presented at the bottom.

The potential of a dual labeling strategy can be grasped immediately when considering an example that can be encountered quite often in structural biology (see sketch in Fig. 1): a protein A undergoes conformational changes when interacting with protein B. If protein A is labeled with two nitroxide labels (named NO for simplicity, as in Fig. 1) and protein B with one gadolinium label (named Gd in Fig. 1), it is possible to detect on the same sample the conformational change of A via NO-NO distances and to localize the binding site of B via NO–Gd distances (additionally, the oligomerization propensity of B can be monitored via the Gd–Gd distances). Therefore, with two orthogonal labels, three pairwise distances can be probed with three distinct optimized setups.

Ideally, increasing the number of independently addressable spin types from two to n, would quickly increase the number of measurable pairwise ($k=2$) distances to $\frac{(n+k-1)!}{k!(n-1)!}$ (multiset). Therefore, three labels would give access to 6 distinct pairwise interspin distances, and four labels already to 10, resulting in great enhancement of information content per sample. However, the higher the number of spins, the more challenging it becomes to selectively address all inter-spin distances avoiding channel crosstalk signals arising from the spectral overlap. A protein-DNA complex carrying three distinct labels (nitroxide, gadolinium and manganese) was already successfully investigated (Wu et al., 2017) and three independent distance constraints could be extracted. In this case, a nitroxide label was introduced in the DNA, one Gd^{3+} tag was attached to the protein with a genetically encoded *p*-azido-*L*-phenylalanine via Cu(I)-catalyzed click chemistry and the Mn^{2+} was chelated by and EDTA bound via a methanethiosulfonate (MTS) linker to a cysteine residue of the same protein.

Besides gadolinium spin labels, trityl- and copper-based labels have recently emerged as potential orthogonal probes for site-directed spin labeling EPR on proteins.

Additionally, biomolecules carrying natural metal cofactors or forming amino acid radicals under certain experimental conditions provide a complementary set of natural orthogonal labels. Some examples of PDS on pairs containing one intrinsic spin center can be found in (Abdullin et al., 2015; Kaminker et al., 2015; Motion et al., 2016; Guo et al., 2017; Russell et al., 2021; Yang et al., 2012). The chemical structures of the above-mentioned labels and their linkage to a protein (in orange) are depicted in Fig. 2.

Here we will address the optimal PDS setups and the properties of four orthogonal spin labels which have been widely used in structural studies of biomolecules, namely nitroxide (NO), gadolinium-based (Gd), trityl (Tr) and Cu(II)-NTA (Cu) spin labels. These orthogonal labels represent the state-of-the-art in structural biology studies, however, new combinations will surely emerge because the number of labels at disposal in the EPR community is growing, as demonstrated by the appearance of labels carrying triplet states (Di Valentin, Albertini, Zurlo, Gobbo, & Carbonera, 2014; Hintze, Bücker, Domingo Köhler, Jeschke, & Drescher, 2016; Williams, Tischlik, Scherer, Fischer, & Drescher, 2020) or different metals such as Mn^{2+} (Martorana et al., 2015).

2. Properties of four common orthogonal labels

In this section, we will describe the spectroscopic properties of the four most common labels (see Fig. 2), as a guide for the choice of the best combination for the desired experiment. The powder spectra simulated with EasySpin (Stoll & Schweiger, 2006) for these labels at X, Q and W band are compared in Fig. 3, highlighting their distinct spectral widths and frequency-dependent changes. An overview of the properties is presented in Table 1.

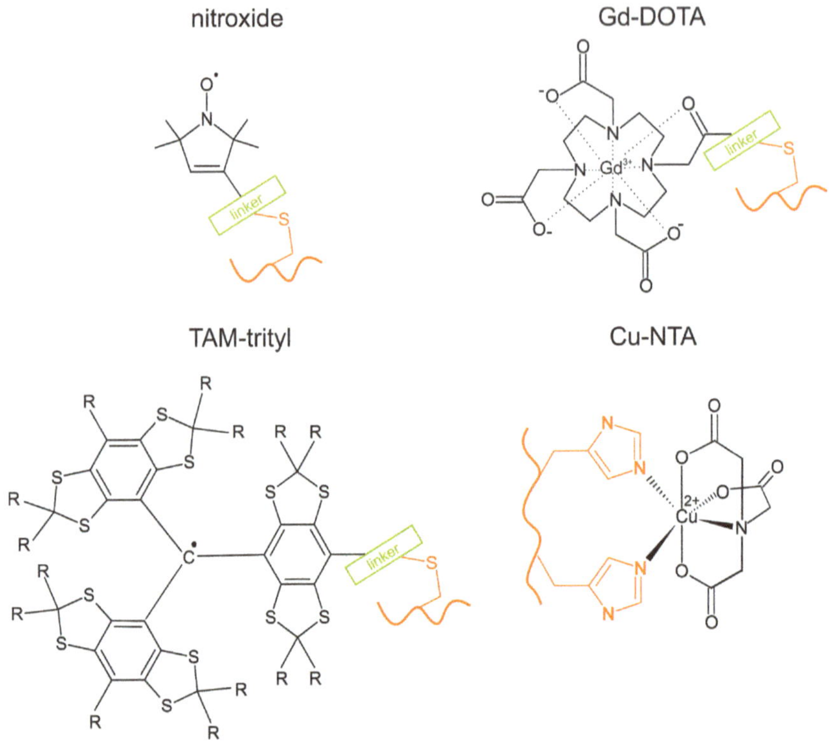

Fig. 2 Chemical structures of four common types of orthogonal labels bound to proteins. For the nitroxide, the five-membered pyrrolidine ring was chosen, as it is present in the common MTSL (methanethiosulfonate spin label). DOTA was chosen as representative chelator of the Gd(III) label. A TAM trityl is depicted, highlighting the variable R functional groups. Most labels are functionalized to target cysteines in proteins (depicted in orange in the structures), with exception of the copper label, which uses a double histidine motif (orange). Several functional groups targeting cysteines are available (methanethiosulfonate, maleimide, iodoacetamide, etc.) which create distinct linkers with tunable properties (green).

2.1 Nitroxide labels

Historically, nitroxide spin labels have been the first ones to be synthetized (Stone, Buckman, Nordio, & McConnell, 1965) and used for structural biology in combination with PDS. They are small molecules which are composed of a heterocycle (five- or six-member ring) carrying a N–O$^\bullet$ group ($S = 1/2$, $I = 1$) where the unpaired electron is delocalized. The paramagnetic moiety is shielded by alkyl groups, which confer stability and slightly prevent reduction of the N–O in mild reducing environments. Their small size and easy synthesis present a big advantage over other spin

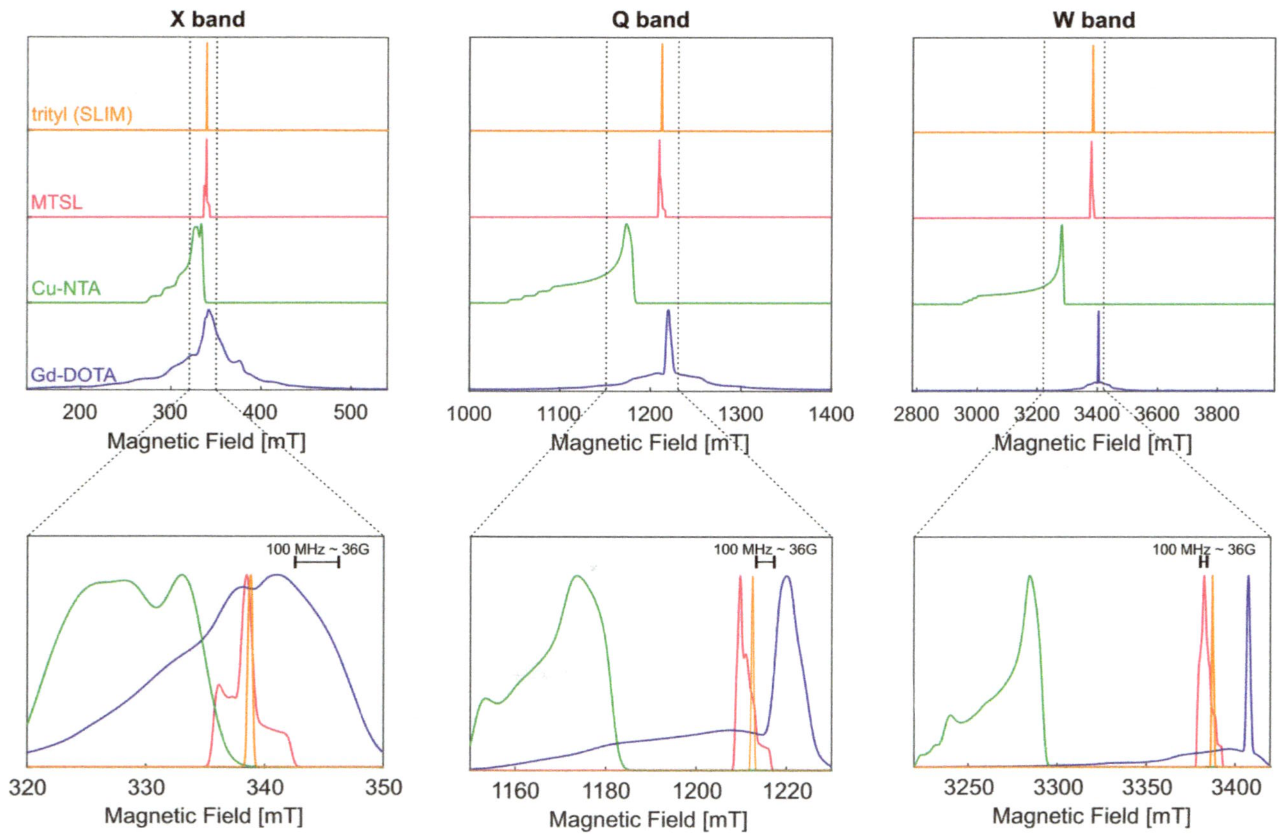

Fig. 3 See figure legend on next page.

labels and certainly contributed to their commercial availability and widespread use (a recent summary of their properties can be found in (Bordignon, 2017; Torricella, Pierro, Mileo, Belle, & Bonucci, 2021).

The EPR spectral shape of a nitroxide is mainly determined by the hyperfine coupling between the unpaired electron and the ^{14}N atom, while the interactions with nearby protons are usually unresolved within the spectrum. The lineshape of powder spectra is governed by the anisotropies of the magnetic tensors $\boldsymbol{g}$ and $\boldsymbol{A}$; in particular, at X-band frequencies, the hyperfine anisotropy (and especially A_{zz}) mostly determines the spectral breath, while the g-anisotropy is unresolved within the linewidth. As it can be seen for the MTSL spin label in Fig. 3 (magenta), W-band frequencies are sufficient to fully resolve the g-anisotropies.

Nitroxides are mainly covalently attached to cysteines via a disulfide, maleimide or iodoacetamide linker; due to the similarity in size to amino acids, these labels are unlikely to disturb the protein's structure and/or correct folding, and are suitable to probe their local microenvironment (Klare & Steinhoff, 2009). For all these reasons, nitroxides were used to label surface exposed and slightly buried positions, in all types of secondary structure elements. The most common linkers have several rotatable bonds, which can influence the broadening of the distance distributions in frozen samples. The maximal resolution of the interspin distances in capturing two distinct protein structures is related to the minimum width of the distance peaks. For nitroxides, the full width at half maximum (FWHM) varies from 0.5 nm to >5 nm depending on the intrinsic dynamics of the protein. However, even if two nitroxide labels are attached to rigid protein sites, the FWHM is rarely smaller than 0.5 nm (a comprehensive review on the distance distributions between nitroxide labels obtained by DEER can be found in (Schiemann et al., 2021)).

Fig. 3 Multifrequency comparison of four spin labels. Simulations were carried out with the EasySpin routine "pepper." The frequency was set to 9.5, 34 and 95 GHz for X, Q and W band, respectively. For the trityl label (orange), the simulation parameters have been chosen as reported for the SLIM label in (Fleck et al., 2020): g = [2.0028, 2.0036, 2.0040], A_C = [3.9, 6.5, 184.8] in MHz, A_{H1} = [3.6, 1.2, 2.4] in MHz, A_{H2} = [6.9, 5.1, 8.0] in MHz, lw = 0.2 mT. For the nitroxide label, MTSL was chosen (magenta) with the simulation parameters as reported in (Bordignon, 2017): g = [2.0085, 2.0061, 2.0022], A = [13, 13, 92] in MHz, lw = 0.5 mT. For Cu-NTA (green), the simulation parameters have been chosen as reported in (Jarvi, Ranguelova, Ghosh, Weber, & Saxena, 2018): g = [2.064, 2.064, 2.227], A = [70, 70, 490] in MHz, lw = 3 mT. The chelator chosen for the Gd label is DOTA (blue) with the simulation parameters as reported in (Clayton et al., 2018): g_{iso} = 1.992, D = [715, 150] in MHz, D_{strain} = 200 MHz, lw = 1 mT.

Table 1 Properties of spin labels for proteins.

	Trityl	NO	Cu(II)	Gd(III)
Target amino acid/s	Mainly Cys[a]	Mainly Cys[a]	d-His motif	Mainly Cys[a]
Linker	MTS[b] Maleimide	MTS Maleimide Iodoacetamide	None	MTS Maleimide
Labeling position	Surface	All, except buried in micelles or protein	α-Helix or β-sheet	Surface
Biocompatibility	High	Low (can be tuned with functional groups)	Potentially high but not yet tested	High
Concentration sensitivity	Submicromolar	Submicromolar	Submicromolar	Submicromolar
T_{opt} for detection	20–298 K	50–80 K	20–30 K	10 K

[a]Unnatural amino acids also can be introduced in proteins which requires different functional groups in the labels.
[b]The commonly used methanethiosulfonate linker is abbreviated as MTS.

The main disadvantage of nitroxides is their sensitivity to reducing environments, that ultimately leads to the reduction of the N–O$^{•}$ group to the corresponding diamagnetic hydroxylamine. This characteristic poses limitations to their use in cellular environments. In the recent years, efforts have been undertaken to synthetize more robust nitroxides, i.e., with bulkier groups that can better protect the N–O moiety (some examples can be found in (Haugland et al., 2018), but also increase the steric hindrance of the label.

PDS techniques are sensitive to nitroxide concentrations down to the submicromolar range (Ackermann, Wort, & Bode, 2021; Kucher, Elsner, Safonova, Maffini, & Bordignon, 2021), which is favorable but limited to in vitro studies. Rotamer libraries are available in MtsslWizard (Hagelueken, Ward, Naismith, & Schiemann, 2012) or MMM (Jeschke, 2018; Polyhach, Bordignon, & Jeschke, 2011) for different types of nitroxide labels, which facilitate comparison between distances and biomolecular structures.

2.2 Gadolinium labels

The quest for biological investigations in physiological environments has led to the development of more biocompatible labels that can survive longer in the presence of mild to strong reducing agents. Gd(III) complexes were shown to be ideal for such task as proven in (Martorana et al., 2014) since Gd(III) ions can accommodate up to nine ligands, which allows for the synthesis of very stable complexes, kinetically and thermodynamically. In such labels, a Gd(III) ion is complexed by a bulky chelator (an extensive review on the different types of Gd labels can be found in (Giannoulis et al., 2021)) and the complex is covalently attached to the protein via a linker, similar to the case of nitroxides. The length of the linker is generally bigger in gadolinium labels than in nitroxides, resulting in a higher number of rotational bonds, which ultimately leads to slightly broader interspin distance distributions. In contrast to nitroxide labels, only surface cysteines can be targeted in proteins.

Gd(III) is a high spin ion with seven unpaired electrons ($S=7/2$) and a half-filled *f*-shell, which gives rise to eight allowed transitions in the presence of a magnetic field, according to selection rules (Bleaney, 1970). The *g* tensor is isotropic and the main source of anisotropy in Gd-based labels comes from the ZFS (zero-field splitting) parameters D and E. The central transition (corresponding to the $-1/2 <-> +1/2$ transition) is dependent on the ZFS parameters only at second order, meaning that in a powder the width of this transition is proportional to D^2/ν (with ν the spectrometer frequency) (Meirovitch & Poupko, 1978). For this reason, this transition becomes narrower at higher frequencies, leading to a higher sensitivity at Q and W band (Clayton et al., 2018) (Fig. 3). The other transitions have widths that are proportional to *D* and are therefore very broad. Due to the fast phase memory time, samples containing Gd-labels need to be measured at very low temperatures (usually 10 K).

Gd-labels are nowadays the labels of choice to perform PDS experiments in cellular environments and it has been proven that PDS at submicromolar concentrations can be achieved at Q band with commercial setups (Kucher et al., 2021). Rotamer libraries are available for different types of Gd labels in MtsslWizard and MMM, which facilitate comparison between distances and biomolecular structures, as for the nitroxide case.

2.3 Copper labels

Cu(II) is a metal ion which has gained a lot of attention in PDS for structural biology in the recent years. It possesses an electron spin $S = 1/2$ and a nuclear spin $I = 3/2$. The EPR spectrum of Cu(II) is mainly governed by $\boldsymbol{g}$ and $\boldsymbol{A}$ anisotropy, and it encompasses 100 mT already at X band (Fig. 3). Due to the limited excitation bandwidth, the large spectral width has an impact on the sensitivity of the measurements. Copper can coordinate ligands similarly to Gd(III). The most used label is the Cu(II)-nitrilotriacetic acid (NTA) complex (Ghosh, Lawless, Rule, & Saxena, 2018; Ghosh, Saxena, & Jeschke, 2018) attached to the protein of interest via further chelation by an engineered double-histidine motif, which was previously introduced with the Cu(II)-IDA complex (Cunningham, Putterman, Desai, Horne, & Saxena, 2015). The two His residues must be either endogenously present in the protein or inserted site-specifically; in order to have a correct positioning of the complex, the two histidines must be in position i, $i+4$ in a α-helix and i, $i+2$ in a β-sheet. The resulting complex is very rigid, which translates in Cu(II)-Cu(II) distance distributions almost five times narrower than those obtained with MTSL (Cunningham et al., 2015; Ghosh, Lawless, et al., 2018; Ghosh, Saxena, & Jeschke, 2018). As a drawback, this labeling strategy requires the mutation of two residues per labeling position and cannot be used in loop regions. Notably, the non-covalent Cu(II) coordination might be vulnerable to low binding-affinity, changes in pH and external metal ions. However, at cryogenic temperatures low- to sub-μm dissociation constants at different pH values were demonstrated, and the presence of divalent metal ions in solution is tolerated (Ackermann et al., 2021; Wort, Arya, Ackermann, Stewart, & Bode, 2021).

Due to its fast relaxation times, the optimal temperatures to detect copper labels are 10–20 K. For the Cu(II)-Cu(II) distances, a sub-micromolar sensitivity has been reported (Ackermann et al., 2021). Rotamer libraries for the dHis-Cu-NTA labels are available (Ghosh, Lawless, et al., 2018; Ghosh, Saxena, & Jeschke, 2018;Hagelueken et al., 2012 ; Polyhach et al., 2011).

2.4 Trityl labels

Triarylmethyl (trityl, TAM) radicals, first reported by Nycomed Innovation in the late 90's have a carbon-centered $S = 1/2$ spin. The so-called Finland trityl radical (Andersson, Rydbeck, & Mahno, 1999) and in particular the tetrathioaryl-type of radicals (Reddy, Iwama, Halpern, & Rawal, 2002) fostered the creation of the spin labels for structural biology. Many types of trityl labels have been synthesized (see a few examples here (Jassoy et al., 2019; Kuzhelev et al., 2017; Shevelev et al., 2014; Yang et al., 2012)) and synthetic efforts have been made to facilitate their asymmetric functionalization (see for example (Hintz, Vanas, Klose, Jeschke, & Godt, 2019)).

A trityl label with a short linker and high reduction stability (called SLIM) (Fleck et al., 2020, 2021) recently enabled in-cell PDS at submicromolar concentrations. Therefore, SLIM is an optimal candidate to be employed in an orthogonal fashion with Gd labels for in cell measurements.

Trityls are characterized by a very small g anisotropy and hyperfine broadening. Therefore, they exhibit a narrow spectral line at both low and high frequency (Fig. 3). The phase memory time of these labels is in the microsecond range even at room temperature (Owenius, Eaton, & Eaton, 2005), which makes them suitable for PDS at experimental conditions that are usually not possible with other spin labels (Shevelev et al., 2014; Yang, Liu, et al., 2012).

If on the one side these labels bring the clear advantage of biocompatibility and long phase memory times, their bulkiness and hydrophobicity have been a limiting factor for their use in structural biology, together with the tendency of aggregation. Addition of hydrophilic substituents was shown to reduce the aggregation propensity in water and avoid non-specific binding to membranes (Tormyshev et al., 2020). Rotamer libraries are also available for a few trityl labels and will be further expanded in the future (Hagelueken et al., 2012; Polyhach et al., 2011).

3. Pulsed dipolar spectroscopy

The term pulsed dipolar spectroscopy (PDS) encompasses a variety of techniques to measure the dipolar coupling strength between electron spins to obtain distance information. In general, PDS techniques are distinguished according to whether they operate at a single or at two frequencies. Each technique is characterized by its specific strengths and weaknesses, which determine its applicability for a specific system under investigation. An elaborate comparison of the performance of different PDS methods on trityl model systems can be found in (Meyer, Jassoy, Spicher, Berndhäuser, & Schiemann, 2018). In this section we briefly discuss the characteristics of DEER (Double Electron Electron Resonance, aka PELDOR, Pulsed Electron Double Resonance), RIDME (Relaxation Induced Dipolar Modulation Enhancement) and DQC (Double Quantum Coherence).

3.1 DEER/PELDOR

DEER/PELDOR is the most used PDS technique. It was introduced in the form of a dead time limited 3-pulse sequence (Milov, Ponomarev, & Tsvetkov, 1984; Milov, Salikhov, & Shirov, 1981) which was later extended by a 4th pulse (Martin et al., 1998; Pannier, Veit, Godt, Jeschke, & Spiess, 2000) to become dead time free. This sequence, presented in Fig. 4, paved the way for the widespread adoption of DEER. It operates at two distinct frequencies, usually referred to as pump and observer. At the observer frequency, a refocused echo sequence is applied, while at the pump frequency, a single π-pulse, temporally positioned between the two observer π-pulses, is applied. If the pump pulse inverts spins that are dipolarly coupled to observer spins, the observer echo will experience a phase shift depending on the position of the pump pulse that manifests itself as a modulation of the echo intensity. By monitoring the observer echo intensity as a function of the pump pulse position T, a time trace can be recorded that is modulated by the sum of the dipolar coupling frequencies of all spin pairs excited by both the pump and the observer pulses. The amplitude of the modulation is referred to as modulation depth λ and depends on the inversion efficiency of the pump pulse. The length of t_2 can be adjusted to match the desired time trace length. The characteristic phase memory time of the observed spins defines the upper limit of the detectable distance range in a DEER experiment.

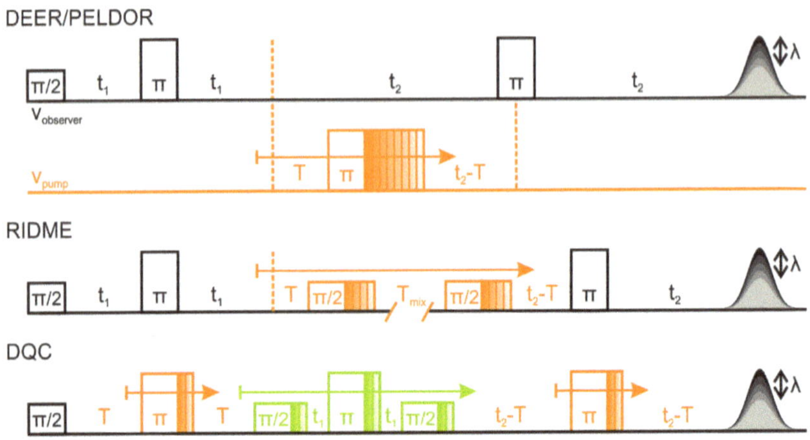

Fig. 4 PDS pulse sequences. 4-pulse DEER/PELDOR, the 5-pulse RIDME and the 6-pulse DQC sequences are discussed in the text. All sequences are constant time experiments where times t_1, t_2 and T_{mix} are constant, only time T is incremented during the experiment. λ is the modulation amplitude of the observer echo (gray).

Being a double frequency experiment, the bandwidth of the resonator plays an important role in defining which pairs of spins can be satisfactorily excited. Usually, frequency separations of 60–100 MHz can be easily achieved with commercial or home-made cavities at X and Q band (Polyhach et al., 2012), while dual mode resonators have proven to be superior for DEER at high field, especially for orthogonal spin pairs with a large frequency separation (Tkach, Sicoli, Höbartner, & Bennati, 2011).

The excitation profiles of pump and observer pulses should not overlap to avoid artifacts (Teucher & Bordignon, 2018). The shorter the pulse lengths and therefore the larger the excitation profile for both pulses is, the higher is the achievable signal-to-noise ratio (SNR) of the DEER time trace, which is usually defined as the modulation depth divided by the standard deviation of the noise. The SNR is dictated by the spin concentration, the excitation profiles the pulses and the phase memory time. A comprehensive discussion of all aspects related to the DEER/PELDOR method with nitroxide spin labels can be found here (Schiemann et al., 2021). The analysis of DEER time traces can be realized using different approaches, such as a traditional two-step analysis using Tikhonov regularization as in DeerAnalysis (Jeschke et al., 2006), a single step analysis as in Deerlab (Fábregas Ibáñez, Jeschke, & Stoll, 2020) or a neural network approach with DEERNet (Worswick, Spencer, Jeschke, & Kuprov, 2018).

DEER sequences with 5- or 7-pulses (Borbat, Georgieva, & Freed, 2013; Breitgoff, Polyhach, & Jeschke, 2017; Spindler et al., 2015) allow to prolong the coherence lifetime via dynamical decoupling (Soetbeer et al., 2021) and therefore provide an improved sensitivity. This gives access to longer distances but additional experiments (Breitgoff, Soetbeer, Doll, Jeschke, & Polyhach, 2017) or data processing methods (Fábregas Ibáñez et al., 2020) are required to reliably evaluate these data.

3.2 RIDME

The single frequency RIDME experiment was originally introduced as a 3-pulse experiment (Kulik, Dzuba, Grigoryev, & Tsvetkov, 2001), while the state-of-the-art sequence comprises 5 pulses (Milikisyants, Scarpelli, Finiguerra, Ubbink, & Huber, 2009).

The 5-pulse RIDME sequence, depicted in Fig. 4, is very similar to the 4-pulse DEER sequence. The difference is that the pump π-pulse operating at the second frequency in DEER is applied at the same frequency as the other pulse is split into two π/2-pulses with an inter-pulse delay called mixing time (T_{mix} highlighted in orange in Fig. 4). While the pump pulse in DEER directly inverts the interacting spins within the transverse plane, RIDME relies on the stochastic flips of the interacting spins during the mixing time. The first π/2-pulse of the moving mixing block moves the on-resonance observed spins out of the transverse plane into the longitudinal plane. Being subject to longitudinal relaxation with time constant T_1, some of the interacting spins will stochastically flip during the mixing time. The second π/2-pulse of the mixing block brings the observed spins back into the transverse plane and the refocused virtual echo signal is detected at the position $2t_1 + 2t_2 + T_{mix}$. As for the DEER case, the length of t_2 can be adjusted to match the required trace length and the phase memory time of the observed spins defines the upper limit of the detectable distance range. In contrast to DEER, the inversion efficiency of RIDME is thereby not limited by the pump pulse bandwidth but by the choice of the mixing time. The virtually infinite "pump bandwidth," which is not influenced by the resonance frequency offset between the two spins, and the consequent absence of orientation selection, make RIDME the method of choice for orthogonal labels with fast relaxing spins (which are "pumped") and slow relaxing spins (which are "observed"). However, RIDME experiments with just one spin type can also be successfully performed (see for example (Ackermann et al., 2021; Breitgoff et al., 2019; Keller et al., 2019)).

T_{mix} should be chosen in the order of the longitudinal relaxation time of the interacting "pumped" spins in the sample at the working temperature, however, the optimal value must be experimentally determined. The length of T_{mix} will determine the modulation depth and the background decay of the traces. Notably, the longitudinal relaxation of the observed spins must be sufficiently slow, so that the non-equilibrium magnetization survives during T_{mix}.

An important parameter to be considered to optimize the sensitivity of the method is the ratio between the phase memory time of the slowly relaxing observed spins ($T_{m,\ slow}$) and the longitudinal relaxation time of the fast relaxing "pumped" spins ($T_{1,\ fast}$). The first influences the intensity of the detected echo, while the second influences the sensitivity of the RIDME experiment. An optimal setup is achieved when $\frac{T_{m,slow}}{T_{1,fast}} \leq 0.2 - 0.3$ (Keller, Doll, et al., 2016; Keller, Zalibera, et al., 2016).

RIDME, compared to DEER, has a more complicated background decay function (Breitgoff et al., 2019) and overtones that are present in the time traces (Keller et al., 2017); however, methods to address these issues are available and software like DeerAnalysis can be used for evaluating RIDME time traces. Additionally, small artifacts present in the 5-pulse RIDME version (Ritsch, Hintz, Jeschke, Godt, & Yulikov, 2019) could be suppressed by the recently proposed 6-pulse RIDME sequence (Abdullin, Suchatzki, & Schiemann, 2021).

3.3 DQC

The DQC experiment operates at a single frequency and can be implemented using a variety of pulse sequences (Borbat & Freed, 1999; Saxena & Freed, 1996). The most used one is the 6-pulse version (Borbat & Freed, 1999) which exists in a symmetric (Borbat & Freed, 1999; Merz et al., 2014) and an asymmetric variant (Ruthstein, Ji, Shin, & Saxena, 2015; Yang, Ji, Cunningham, & Saxena, 2015), depending on how the pulses are stepped with respect to each other, which offers certain advantages and disadvantages. Fig. 4 shows the symmetric version of the 6-pulse DQC experiment, where time T is incremented so that pulses 2 and 6 (orange), as well as the block of pulses 3 to 5 (green), move in such a way that the echo remains at a constant time. This experiment works in such a way that the first three pulses are creating double quantum coherences which are refocused by the subsequent two pulses. The fifth pulse allows to evolve these into observable single quantum coherences which are then refocused by the last pulse into an echo.

The DQC traces can be analyzed with the same software used for DEER (Jeschke et al., 2006), however, with respect to DEER traces, the background decay is more difficult to fit.

3.4 Synopsis

In case of PDS experiments between orthogonal spin labels, DEER offers a high flexibility since it operates at two frequencies which allow to freely select the spectral region to pump and to observe. In orthogonal DEER experiments, the spin with the narrower spectrum is usually pumped, while the spin with the broader spectrum is observed to maximize the modulation depth of the time trace. However, DEER is technically demanding since it requires a large resonator bandwidth to host pulses operating at two different frequencies. The required frequency separation for specific types of spin pairs can easily become inaccessible at high frequencies. The maximum sensitivity is reached if pump and observer are positioned at the respective maxima of the spectra, but due to the resonator bandwidth limitations, a frequency separation smaller than optimal can still be used with satisfactory results. RIDME is favorable with respect to DEER for systems where the orthogonal spin species have very different relaxation times and/or spectra that are particularly broad and cannot be observed with high SNR or pumped efficiently in DEER. The spin with fast relaxing/broad spectrum is usually chosen to stochastically flip during the T_{mix} block, while the slow relaxing spin is detected. Thereby, the mixing block can be used as an efficient relaxation filter and the relaxation-driven inversion contributes to the signal based on its basically unlimited bandwidth. This is also what makes the RIDME experiment less prone to orientation selectivity in comparison to DEER and to DQC. DQC and, also to be mentioned, the single-frequency technique for refocusing (SIFTER) dipolar couplings technique (Jeschke, Pannier, Godt, & Spiess, 2000), perform better than DEER and RIDME on narrow spectra whose widths are on the order of the bandwidth of the used pulses.

4. PDS measurements with orthogonal labels

In this section, we will give an overview of the best practices to perform PDS experiments between pairs of orthogonal spin labels. The evolution of the experimental setups used for PDS goes hand in hand with the technological developments. Besides the never-ending quest for higher microwave powers at higher frequencies (Cruickshank et al., 2009;

Polyhach et al., 2012) and lower noise levels (Šimėnas et al., 2021), one major turning point in the past decade was the introduction of arbitrary waveform generators (AWGs) (Doll, Pribitzer, Tschaggelar, & Jeschke, 2013; Spindler et al., 2012) and the subsequent possibilities to freely shape pulses by varying their amplitude, frequency, and phase (Prisner, 2019; Spindler et al., 2015). These technical developments provide further degrees of freedom to tailor experiments to the specific properties of the utilized spin labels that can hardly be summarized. Here we focus on the capabilities of commercially available spectrometers. The following sections describe the optimal performance of specific PDS methods for each pair of labels chosen among the four selected ones (trityl, NO, Cu, and Gd). An overview of the optimal conditions, based on literature data, is given in Table 2.

Table 2 Possible combinations for homo- and hetero-DEER (in orange) and single frequency PDS (in green) with the description of optimal setup and experimental conditions (frequency and temperature of the experiment), pump (p) and observer (o) position for DEER. The gray boxes highlight the homo-experiments (same spins), the white ones the hetero-experiments (orthogonal spins).

	Trityl	NO	Cu	Gd
Gd	W band, 10 K p: Trityl, o: Gd ------------------------ Not yet explored	Q band, 10 K p: NO, o: Gd ------------------------ Not yet explored	Not yet explored	W band, 10 K ------------------------ W band, 10 K RIDME
Cu	Not yet explored	X band, 20K p: NO, o: Cu ------------------------ Q band, 50 K RIDME	X band, 20 K ------------------------ Q band, 20 K RIDME	
NO	Q band, 50-80 K p: Trityl, o: NO ------------------------ X band, 298 K RIDME	Q band, 50 K ------------------------ Q band, 50 K RIDME		
Trityl	W band, 50 K ------------------------ Q band, 50 K DQC			

4.1 NO–NO

The measurement of NO–NO distances using DEER dates to the early beginnings of the technique (Larsen & Singel, 1993; Milov et al., 1984; Milov, Maryasov, & Tsvetkov, 1998) and became particularly popular after the introduction of the dead time free 4-pulse sequence (Pannier et al., 2000). At present, the highest sensitivity can be reached with a high-power amplifier at Q band at 50 K (Ghimire, McCarrick, Budil, & Lorigan, 2009; Polyhach et al., 2012) using a separation of 100 MHz between pump and observer. W-band experiments suffer from power limitations (Bahrenberg et al., 2017) which can be partially overcome by using non-commercial equipment (Cruickshank et al., 2009; Motion et al., 2017). Notably, DEER can also provide orientation selection information, which allows quantitative extraction of the geometrical arrangement in rigid systems (as reviewed in (Endeward, Marko, Denysenkov, Sigurdsson, & Prisner, 2015)).

RIDME experiments have also been performed on nitroxide model systems (Kulik et al., 2001; Milikisyants et al., 2009). However, DEER clearly outperforms RIDME on this pair in terms of achievable modulation depth and SNR (Abdullin et al., 2021; Russell et al., 2021). Therefore DEER/PELDOR at Q band is the standard method used for NO-NO distances (Schiemann et al., 2021).

4.2 Gd–Gd

Due to the narrowing of the central transition of Gd towards higher frequencies, W band is the most suitable frequency to perform Gd–Gd DEER experiments (Goldfarb, 2014; Potapov et al., 2010; Raitsimring et al., 2007) and the use of the AWG capabilities can further increase sensitivity (Bahrenberg et al., 2017). High power Q band has established itself as a good alternative (see for example (Galazzo et al., 2020; Shah et al., 2019; Yulikov, Lueders, Warsi, Chechik, & Jeschke, 2012), especially when considering the purchasing and operating costs of a W-band spectrometer. Q band in combination with AWG can compensate for the lower sensitivity (Doll et al., 2015). Standard Q-/W-band Gd–Gd DEER experiments are performed at 10 K using a frequency separation of about 100 MHz.

Gd–Gd RIDME (Azarkh et al., 2019; Collauto et al., 2016; Keller et al., 2017; Razzaghi et al., 2014) can achieve much larger modulation depths compared to the analogous DEER experiment, highlighting the advantages of RIDME over DEER for spins with broad spectra. These experiments are best performed at W band and temperatures around 10 K. RIDME on Gd labels avoids broadening effects arising from a breakdown of the weak coupling approximation (Collauto et al., 2016) usually occurring in DEER experiments for short Gd–Gd distances (Cohen et al., 2016; El Mkami et al., 2020) without the need for any additional specialized equipment (Kaminker et al., 2013; Tkach et al., 2011).

4.3 Cu–Cu

Several Cu(II)–Cu(II) PDS experiments have been explored and optimized over the years on rulers and Cu-containing biomolecules. In fact, literature data can be found on Cu–Cu distances extracted via DEER (Bowen et al., 2016; Ghosh et al., 2020; Ghosh, Lawless, et al., 2018; Ghosh, Saxena, & Jeschke, 2018; Jarvi et al., 2018; van Amsterdam, Ubbink, Canters, & Huber, 2003; Yang, Becker, & Saxena, 2007), RIDME (Ackermann et al., 2021; Breitgoff et al., 2019; Engelhard, Meyer, Berndhäuser, Schiemann, & Clever, 2018; Stratmann, Kutin, Kasanmascheff, & Clever, 2021; Wort et al., 2019, 2021) and DQC (Becker & Saxena, 2005; Ruthstein, Ji, Mehta, Jen-Jacobson, & Saxena, 2013).

DEER measurements between copper labels suffer from the limited excitation bandwidth with respect to the large spectral width, resulting in a poor sensitivity already at X band. Moreover, orientation selection can complicate data analysis and it is likely to be present due to the limited spectral excitation at any frequency. Measurements are usually performed at X or Q band at 20 K with a separation between pump and observer of 60 or 100 MHz, respectively. At higher frequencies the spectrum becomes very broad and there is no real advantage in performing DEER to extract interspin distances. Due to the large spectral width, the use of shaped pulses has proven to help in gaining better sensitivity in DEER (Breitgoff et al., 2019).

RIDME has been successfully performed for Cu–Cu distance determination. It provides less orientation selectivity, given that the relaxation is uniform enough to guarantee similar relaxation of the spin packets during T_{mix}. Both DEER and RIDME methods have their advantages and disadvantages for the determination of Cu–Cu distances, which have to be taken into account to choose the optimal setup for the system under investigation (Breitgoff et al., 2019). RIDME experiments for Cu–Cu pairs introduced in proteins via the d-His motif have been performed down to 500 nM protein concentration (Ackermann et al., 2021), proving the high sensitivity of the RIDME technique.

4.4 Trityl–Trityl

DEER experiments between trityl spin labels at frequencies lower than W band suffer from low sensitivity due to the limited spectral width available (Kunjir, Reginsson, Schiemann, & Sigurdsson, 2013; Meyer et al., 2018; Reginsson, Kunjir, Sigurdsson, & Schiemann, 2012; Shevelev et al., 2014). In fact, up to Q band, single frequency techniques are clearly superior (Akhmetzyanov et al., 2015; Kunjir et al., 2013; Shevelev et al., 2014), while at higher frequencies such as W band (Fedin et al., 2016; Giannoulis et al., 2019; Yang et al., 2020) or G band (Akhmetzyanov et al., 2015) DEER experiments outperform single frequency techniques. W-band DEER experiments between trityls are best performed at 50 K with a frequency separation of about 20 MHz between pump and observer. Trityl labels have already been successfully used as biocompatible spin labels in cells, and DEER was used to extract interspin distances at W band (Yang et al., 2020).

RIDME experiments are rarely performed between trityl spin labels (Jassoy et al., 2017) because of their narrow spectral width and slow identical relaxation times making the detection/observation selectivity usually required for RIDME experiments quite difficult (Meyer et al., 2018). Single frequency experiments like DQC (Fleck et al., 2019, 2020, 2021; Kunjir et al., 2013; Reginsson et al., 2012; Shevelev et al., 2014; Yang, Liu, et al., 2012) or SIFTER (Akhmetzyanov et al., 2015; Jassoy et al., 2019) performed at Q-band frequencies and 50 K are the most sensitive PDS experiments for extracting distance information between trityl spin labels. The choice between both is a matter of taste; while SIFTER is usually more sensitive, DQC data are more reliable since their background correction is simpler (Akhmetzyanov et al., 2015; Meyer et al., 2018).

4.5 NO–Gd

DEER experiments between nitroxide and gadolinium spins have been performed at X-, Q- and W-band frequencies (Assafa et al., 2021; Galazzo et al., 2020; Garbuio et al., 2013; Gmeiner, Dorn, et al., 2017; Gmeiner, Klose, et al., 2017; Joseph, Korkhov, Yulikov, Jeschke, & Bordignon, 2014; Kaminker et al., 2012, 2013; Lueders et al., 2011, 2013; Shah et al., 2019; Teucher et al., 2019, 2020; Wu et al., 2017; Yulikov et al., 2012). While the separation between the spectral maxima of the two spins and the width of the NO spectrum increase proportionally to the microwave frequency, the width of the central transition of the Gd spectrum decreases at higher frequencies. These opposing effects make Q band the most suitable frequency for NO–Gd DEER experiments since it provides a good compromise between the spectral widths and the spectral separation (around 300 MHz) with respect to the available power and resonator bandwidth. Additionally, orientation selection is minimal at Q band for the nitroxide spin in most samples, but it can become problematic at W band. The optimal measuring temperature is 10 K. To maximize the modulation depth, the nitroxide label is pumped, and the observer is placed on the Gd. Additionally, the much shorter longitudinal relaxation time of the Gd spins with respect to the NO spins at 10 K (a few ms vs hundreds of ms) allows a fast shot repetition time while observing the Gd spins, which increases the SNR (Yulikov, 2015). W-band experiments are also possible (Kaminker et al., 2012; Wu et al., 2017) but only with specialized non-commercial equipment allowing 700 MHz frequency separation (Kaminker et al., 2013; Tkach et al., 2011).

DEER is the most established method for this pair because the achievable gain of a single frequency RIDME experiment is limited compared to its DEER counterpart. In fact, in DEER nearly the entire NO spectrum can be pumped providing a good modulation depth. However, a RIDME experiment is feasible because the Gd spins with the broader spectrum that should be "pumped" are characterized by much shorter longitudinal relaxation times than the NO at the same temperature.

4.6 NO–Cu

Together with the nitroxide-gadolinium pair, the nitroxide-copper orthogonal pair has been largely exploited in PDS. As clearly shown from the spectral simulations in Fig. 3, the g values of Cu(II) and nitroxides are very different, so that DEER experiments at frequencies higher than X band are not really permitted due to insufficient bandwidth of commercial resonators. In fact, the frequency separation between the maxima is already 150 MHz at X band.

Regarding DEER experiments, examples of applications at X band can be found in literature (Merz et al., 2014; Meyer, Abdullin, Schnakenburg, & Schiemann, 2016; Narr, Godt, & Jeschke, 2002; Russell et al., 2021; Sarver, Silva, & Saxena, 2013; Yang, Kise, & Saxena, 2010). Conventionally, for better sensitivity, the pump position is placed on the maximum of the nitroxide and the observer 200 to 500 MHz away from it. In DEER, by changing the frequency offset, orientation selection can be addressed

(Meyer et al., 2016). Optimal temperatures are around 15–20 K due to the fast relaxation of the observer copper spin. Concerning sensitivity, the large Cu(II) spectral width results in poor signal-to-noise ratios due to the limited excitation bandwidth.

The different longitudinal relaxation times of Cu(II) and NO spins favors the use of RIDME for this pair. The fast-relaxing Cu(II) spins are stochastically inverted during the mixing time, and the slowly relaxing NO spins are observed. Another advantage of RIDME with respect to DEER is the suppression of orientation selection, which facilitates the extraction of mean distances. Different studies have proven the superior concentration sensitivity of RIDME at Q band with respect to DEER for this pair (Giannoulis, Oranges, & Bode, 2017; Meyer et al., 2016; Ritsch et al., 2019; Wort et al., 2019).

4.7 Trityl–NO

Proof of principles studies on biradicals carrying a trityl and a nitroxide label (Kunjir et al., 2013; Reginsson et al., 2012) paved the way for DEER (Joseph et al., 2016; Ketter et al., 2021) and RIDME (Kuzhelev et al., 2018) on complex biomolecular systems.

DEER experiments are optimally performed at Q band due to the larger separation between the maxima of the respective spectra compared to X band, which provides higher sensitivity (see Fig. 3). Optimal sensitivity is achieved at 50–80 K with the pump placed on the narrow spectrum of the trityl (for optimal excitation and highest modulation depth) and with the observer placed on the nitroxide. The separation between the two maxima at Q band is around 100 MHz, a value that can be achieved with most commercial resonators. Modulation of the separation between pump and observer pulses allows orientation selective measurements to be performed, as shown in the studies on biradicals (Kunjir et al., 2013; Reginsson et al., 2012).

The long phase memory time of trityl spins allows detection with high sensitivity also at room temperature. At this temperature, DEER experiments are not feasible, due to the prohibitively fast relaxation of the nitroxides (observer spins). In contrast, X-band RIDME can be performed as shown in (Kuzhelev et al., 2018) after immobilization of the system in trehalose or water/nucleosil to avoid averaging of the dipolar coupling due to molecular motion.

4.8 Trityl–Gd

Only W-band DEER experiments involving these labels have been performed (Giannoulis et al., 2019) using a non-commercial resonator. The separation between the maxima of trityl and gadolinium is around 500 MHz at W band. The pump was placed on the trityl (due to its narrow spectrum) and the observer on the gadolinium. The temperature was chosen to be 10 K to optimize the detection sensitivity on the Gd spins.

With respect to spectral separation, the trityl–Gd pair is similar to the NO–Gd pair (Fig. 3), therefore DEER measurements using commercial Q-band spectrometers should be feasible with a frequency separation in the order of 200 MHz. Due to the narrower spectral width of the trityl (pumped spin), a larger modulation depth is expected for trityl–Gd than for NO–Gd.

Because of the difference in relaxation times, RIDME experiments are in principle also possible. The slowly relaxing trityl should be observed, while allowing stochastic inversion of the Gd spins. However, DEER should be in any case more favorable for this pair due to the large achievable modulation depth arising from the total excitation of the trityl spectrum.

4.9 Trityl–Cu

PDS experiments with this pair have not yet been shown. DEER experiments could be in principle performed at X band since the spectral separation is in the order of 100 MHz. At higher frequencies, the separation would be too large for DEER. For optimal sensitivity, the narrower trityl should be pumped and the detection would be placed on the copper, which poses the sensitivity problem highlighted before due to its large spectral width. Orientation selection effects should be tested by changing the frequency separation.

Both the large spectral separation and the presence of a slow (trityl) and a fast (Cu) relaxing spin, makes RIDME the optimal choice for this pair. The slower relaxing trityl spin label could be detected with high sensitivity and the stochastic inversion of the copper spins would prevent orientation selection.

4.10 Cu–Gd

To the best of our knowledge, no PDS experiments have been performed with this pair. The large spectral width of both spins and the increasing spectral separation towards higher frequencies (see Fig. 3) pose several challenges in performing PDS. These limitations make X band the most viable choice to perform DEER experiments. The choice of which spin to pump and which one to observe depends on their respective relaxation times. In case of DEER, the observer should have the longest phase memory time, and the shortest longitudinal relaxation time for optimal sensitivity. The differences in relaxation times should be addressed for each specific case. Due to the large *g* anisotropy of the Cu spins, orientation selection effects must also be considered.

RIDME experiments require the spin species with the longest longitudinal relaxation time to be observed. If copper is observed, low frequencies are more suitable due to the narrower spectral width of the Cu(II) spin, while if Gd is observed, high frequencies (W band) should be favorable for a high detection sensitivity.

5. Orthogonality and channel crosstalk

The strength of performing PDS experiments with orthogonal spin labels lies in the increased information content that can be obtained from a single sample, especially when assigning intra- and intermolecular distances. From a monomeric protein that is doubly spin-labeled with one label type, only a single distance can be obtained. The same protein doubly spin-labeled with an orthogonal label pair (spin A and spin B), will also provide only a single distance information but, in principle, the number of signal channels (aka setups to detect pairwise distances) that could be probed increases to three: two homo- (AA and BB) and one hetero-channel (AB). As previously addressed, for each additional orthogonal label added to the system, the number of signal channels is over-proportionately increasing, e.g., three orthogonal labels have six signal channels, while four orthogonal labels have already ten signal channels (unordered combination with repetition, or multiset). Already the simple example of one orthogonal label pair on a monomeric protein illustrates that not all these channels necessarily need to contain distance information. In fact, if the protein is monomeric, only the hetero-channel (AB) will contain a distance information, while the homo-channels (AA and BB) will show no dipolar oscillations.

Since in most cases it is a priori not known if the investigated protein could also form dimers or oligomers under certain conditions, it is advisable to check all available signal channels in real systems. However, as previously described, channel crosstalk signals may appear already when two orthogonal labels are present (Gmeiner, Dorn, et al., 2017; Shah et al., 2019;

Teucher et al., 2019, 2020; Wu et al., 2017) due to the non-perfect spectroscopic orthogonality of the spins under investigation. The higher the number of orthogonal spins in a sample, the more probable it is to have such unwanted effects in the different channels probed. For a reliable interpretation of the distances, crosstalk signals should be correctly identified and, if possible, suppressed.

The term "orthogonal" refers to pairs of spin labels that have at least a single property that makes them spectroscopically distinguishable from each other. These properties can be, e.g., the spin states with their connected transition moments, the relaxation times, and the positions of the central transitions. When performing PDS experiments in regions of spectral overlap between orthogonal spin labels, channel crosstalk signals will certainly appear since any experiment optimized for a specific signal channel (e.g., a hetero spin label experiment) will also act as a non-optimized experiment for another signal channel (e.g., the complementing homo spin label experiment). Therefore, one needs to be aware of this and consider the relative strength of the crosstalk signal in relation to the desired signal, which depends on the degree of orthogonality and the relative abundance of the respective labels in the sample. Crosstalk signals are most pronounced when they are the only signal which can be probed in the respective channel, but they can also appear alongside the desired signals.

The identification of crosstalk signals without performing additional experiments on samples in which one orthogonal spin has been removed can be challenging. Modulation depths can be used as an indicator since crosstalk signals tend to have small modulation depths in relation to actual signals since crosstalk signals are always recorded under sub-optimal conditions. Furthermore, a comparison of the suspicious signal with the signals of other channels can be very insightful. For example, if two channels recorded within the same region of overlap have the same signal, this should be further validated.

The direct identification and suppression of crosstalk signals can be accomplished by exploiting the orthogonal properties of the spin labels, as done for the Gd–NO DEER case (Teucher et al., 2020). Filtering for transverse relaxation can be realized to a certain degree by varying the sequence/time trace length but mainly by varying the experimental temperature which can be also used for longitudinal relaxation. Powerful for the identification of crosstalk signals can be to vary the pulse power to optimize the experiment for one or the other spin species if these have different spin states/transition moments. DEER offers in this scope the largest flexibility since it allows to pump and probe at different parts of the spectrum in contrast to the single-frequency techniques. This allows for example to exchange the positions of pump and observer which has proven to be a powerful tool (Shah et al., 2019; Teucher et al., 2020).

Crosstalk signals are also likely to be present in RIDME experiments. For example, if a slowly relaxing spin is coupled to two fast relaxing spins, the RIDME traces will show both contributions in the distance distributions. Identifying crosstalk signals could be possible by changing the mixing time and verifying changes in the distance distributions or modifying the temperature to increase the relaxation selectivity of the mixing time.

6. Applications

In this section, we will present four case studies where the dual orthogonal labeling approach has been exploited either to solve biological questions or as a proof of principle to pioneer its use for more complex structural biology problems. To prove the power and versatility of the technique, we show applications on different systems, using different types of PDS at various frequencies and under different experimental conditions.

6.1 NO–Gd on an ABC transporter

ABC transporters are transmembrane proteins that couple the binding and hydrolysis of ATP at dedicated nucleotide-binding domains (NBDs) with large conformational changes that switch the protein's conformation from an inward- to an outward-facing (IF to OF) state (or vice versa), resulting in substrate translocation. Our lab has been dedicated to the study of the heterodimeric exporter TM287/288, for which several key questions had to be addressed. One of those was the characterization of the OF state. The crystal structure of the OF state was solved thanks to a synthetic nanobody (i.e., the small fragment of an antibody) that binds specifically to the extracellular region of the protein, locking it in this conformational state (Hutter et al., 2019). To study the conformational dynamics of the nanobody-transporter complex, we spin labeled several pairs of amino acids across the transporter with MTSL (one pair shown as black dots in Fig. 5A) and the nanobody with Gd-maleimide-DOTA (green dot) and we followed the conformational changes during the ATP cycle. With this dual labeling strategy, we could investigate on the same sample the protein's conformation via NO–NO interspin distances and confirm the binding of the

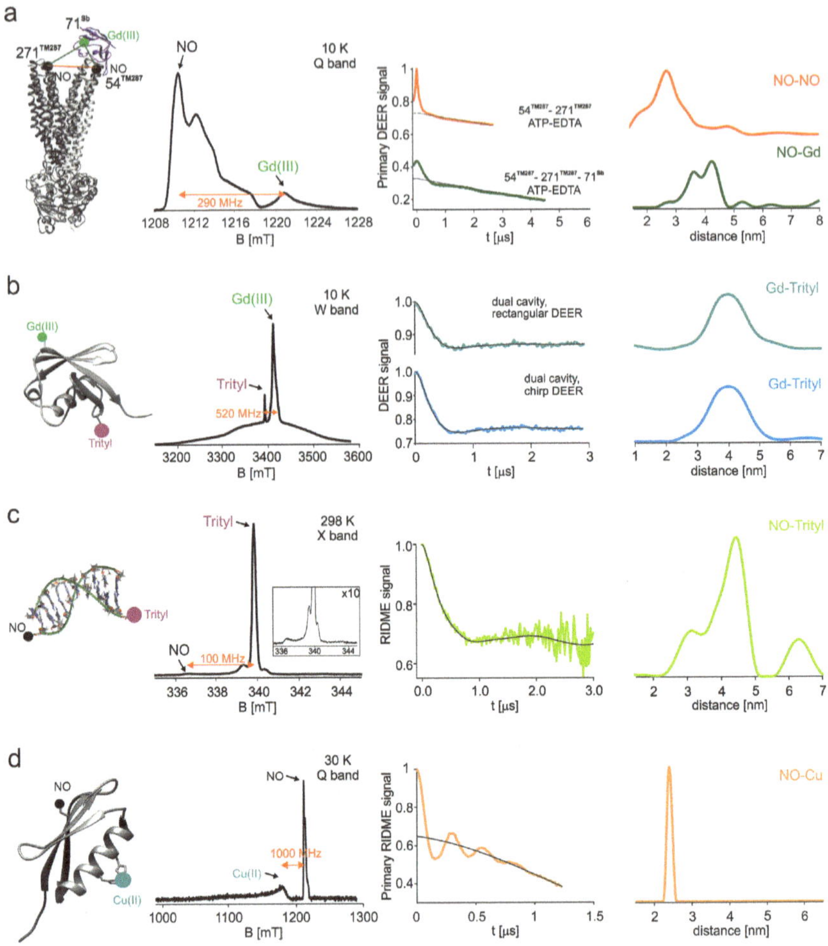

Fig. 5 Examples of orthogonal PDS in structural biology. For every panel (A–D), from left to right: representation of the system and the attached spin labels; echo-detected field sweep (EDFS) at the indicated temperature and frequency (the separation between the maxima of the two spectra is indicated); DEER or RIDME time traces (primary or background-corrected) and extracted distance distributions. *Panel A: Q-band DEER adapted from Galazzo, L., Meier, G., Timachi, M. H., Hutter, C. A. J., Seeger, M. A., & Bordignon, E. (2020). Spin-labeled nanobodies as protein conformational reporters for electron paramagnetic resonance in cellular membranes.* Proceedings of the National Academy of Sciences USA, 117 *(5), 2441–2448. doi:10.1073/pnas.1913737117; Hutter, C. A. J., Timachi, M. H., Hürlimann, L. M., Zimmermann, I., Egloff, P., Göddeke, H., . . . Seeger, M. A. (2019). The extra-cellular gate shapes the energy profile of an ABC exporter.* Nature Communications, 10 *(1), 2260. doi:10.1038/s41467-019-09892-6; Panel B: W-band DEER adapted from Giannoulis, A., Yang, Y., Gong, Y. J., Tan, X., Feintuch, A., Carmieli, R., . . . Goldfarb, D. (2019). DEER distance measurements on trityl/trityl and Gd(iii)/trityl labelled proteins.* Physical Chemistry Chemical Physics, 21 *(20), 10217–10227. doi:10.1039/c8cp07249c; Panel C: X-band RIDME adapted from Kuzhelev, A. A., Krumkacheva, O. A., Shevelev, G. Y., Yulikov, M., Fedin, M. V., & Bagryanskaya, E. G. (2018). Room-temperature distance measurements using RIDME and the orthogonal spin labels trityl/nitroxide.* Physical Chemistry Chemical Physics, 20 *(15), 10224–10230. doi:10.1039/c8cp01093e; Panel D: Q-band RIDME adapted from Wort, J. L., Arya, S., Ackermann, K., Stewart, A. J., & Bode, B. E. (2021). Pulse dipolar EPR reveals double-histidine motif CuII–NTA spin-labeling robustness against competitor ions.* The Journal of Physical Chemistry Letters, 12 *(11), 2815–2819. https://doi.org/10.1021/acs.jpclett.1c00211.*

nanobody via nitroxide-gadolinium distances. The third DEER channel (Gd–Gd distance, not shown in the figure) was also used to rule out dimerization of the nanobody in solution.

Fig. 5A shows the 4-pulse DEER experiments on the same sample conducted at 50K with the NO–NO channel pumping on the maximum of the nitroxide and detecting at 100MHz lower in frequency. At 10K the Gd–NO channel was probed by pumping on the maximum of the nitroxide and observing on the maximum of the gadolinium (around 280MHz separation). In the reported echo-detected field sweep (EDFS), the signal of the nitroxide is prevailing as the stoichiometry of NO:Gd was 2:1 (double mutant) and the setup (microwave power, shot repetition time) was chosen to optimize the nitroxide signal.

6.2 Trityl–Gd on ubiquitin

The first example present in literature for an orthogonal labeling of ubiquitin with a Gd-based label (BrPy-DO3MA-Gd(III)) and a trityl (CT02MA) (green and purple in Fig. 5B, respectively) is reported in (Giannoulis et al., 2019). In contrast to the previous case, where the different labels were attached to two different proteins and complex formation could be investigated, the hetero-labeling requires in the case of ubiquitin a particular labeling scheme. In principle, to achieve dual labeling on the same protein and assure specificity, orthogonal chemistry is required, but in this case the authors took advantages of the higher reactivity of one of the two engineered cysteines over the other to create a sample in which the trityl–Gd pair was present in a large fraction of the proteins (EDFS in Fig. 5B). In the presented W-band DEER experiment the pump was set to the maximum of the trityl (to obtain the highest modulation depth) and the observer on the Gd. For the setup with rectangular pulses a frequency separation of 520MHz was used, which could be achieved thanks to a dual mode resonator. Chirped pulses were used as pump pulses in the second setup tested.

The chirped pulses allowed a higher signal-to-noise ratio because of the higher excitation bandwidth and minimized possible orientation selection artifacts in the distance distribution.

The distance distributions obtained for this hetero-labeled protein are quite broad. One reason for this is the presence of proteins with swapped tags at the two labeling positions, which stems from the adopted statistical labeling scheme. Additionally, the MALDI-TOF analysis of the sample confirmed the presence of a fraction of doubly Gd-labeled proteins in the ensemble. This was corroborated by the detection of a distance in the Gd–Gd DEER channel (pump and observer pulses both placed on the Gd spectrum, not shown in Fig. 5B).

6.3 Trityl-NO on a DNA duplex

PDS is the technique of choice to interrogate the structure and function of biological systems, but most of the studies are carried out at cryogenic temperatures, as seen in the previous examples. In the recent years, thanks to the development of labels with longer relaxation times, room temperature PDS has become feasible. To perform PDS at room temperature, certain conditions must be fulfilled: (i) the phase memory time (T_m) of the spin labels has to be long enough to allow for dipolar time traces of appropriate length to extract distances with high fidelity; (ii) the rotational correlation time of the biomolecule and of the spin label needs to be slowed down to avoid averaging of the dipolar coupling. These conditions can be achieved with the use of trityl labels bound to biomolecules embedded in viscous media, as trehalose, sepharose or silica gels, as shown in this example.

The presented study (Fig. 5C) is a proof of principle that the combination of trityl and nitroxide labels with piperazine linkers (prepared according to (Shevelev et al., 2015)) attached to a DNA duplex can be used to perform PDS at room temperature. In this case, DEER is not the experiment of choice due to the short phase memory time of nitroxides at room temperature, but RIDME can serve the purpose. Trehalose was used to achieve optimal immobilization of the biomolecule without perturbing its structure. For trehalose, the phase memory time of the slow relaxing observer trityl is $T_{m,s} = 2.4$ μs, while the longitudinal relaxation time of the fast relaxing "pumped" nitroxide is $T_{1,f} = 15$ μs giving a ratio of $T_{m,s}/T_{1,f} = 0.16$ which is in a suitable range to perform RIDME experiments (Keller, Doll, et al., 2016; Keller, Zalibera, et al., 2016). The chosen mixing time ($T_{mix} = 20$ μs) gave the best compromise between time trace length, SNR and background function complexity. The X-band RIDME analysis is shown in Fig. 5C. The EDFS spectrum of the doubly-labeled DNA at 298 K presents the characteristic narrow peak of the TAM trityl with satellites due to the coupling with ^{13}C nuclei and a weak and broad signal originating from the nitroxide radical. The 5-pulse RIDME experiment led to a distance distribution which was in line with theoretical predictions, but probably broadened by some disorder arising from the immobilization in trehalose. The peak around 6–7 nm was found to be strongly dependent on the background correction, a delicate task in RIDME analysis, and therefore not interpreted.

6.4 NO–Cu to determine binding affinity

Since its early development, the use of PDS has been intertwined with protein studies to unveil structural details that are fundamental to gain a better understanding of the mechanisms regulating their function. However, distances are not the only observables achievable with dipolar techniques, as other information can be extracted from PDS. In the example reported in Fig. 5D, 5-pulse RIDME has been used as an assay for competitive binding, benefiting from the information derived from the modulation depth of the RIDME traces.

The protein used in this study, the *Streptococcus* sp. Group G. protein G, B1 domain (GB1), has been mutated to introduce a cysteine that can bind the nitroxide MTSL, and two histidines at positions i and $i+4$ in the α-helical portion of the protein, to bind Cu-NTA (Wort et al., 2021). The metal complex was added to the system with a titration, and RIDME traces were recorded at different copper concentrations. The obtained modulation depth of the RIDME traces is proportional to number of NO–Cu pairs formed and was used to determine the binding affinity of copper to the d-His motif ($K_D = 0.31$ μM). The accuracy of the RIDME-derived affinity constants was benchmarked with results obtained by isothermal titration calorimetry (ITC) experiments. The RIDME trace shown in Fig. 5D was recorded at Q band at 30K using the nitroxide as observer on a sample containing 5 μM protein incubated with 5 μM Cu(II). The very narrow distance distribution between the two labels is due to the rigidity of the Cu-NTA, an important property of the d-His labeling method, that offers great precision in the characterization of conformational changes of proteins. Notably, experiments carried out in the presence of large excess (around 1000-fold) of the diamagnetic competitor Zn–NTA also proved the robustness of the labeling against adventitious labeling, a characteristic which is very important for labels that are not covalently bound to the system of interest.

7. Outlook

In recent years, the number of systems that can be investigated via PDS has remarkably increased, thanks to significant developments in the different techniques and in the research on new spin labels with alternative properties. To this regard, the synthesis of model compounds carrying homo- and hetero- pairs of spin centers has been pivotal for the optimization of the experimental setups. The next challenge is to analyze in detail the spectroscopic complications arising from mixing several homo- and hetero-spin compounds and create model compounds with three and more spin centers to foster the development of optimized multi-frequency multi-spin detection protocols free of cross-talk signals. The study of large protein complexes and hetero-oligomeric systems is at reach, which will enhance the influential role of EPR in structural biology.

References

Abdelkader, E. H., Feintuch, A., Yao, X., Adams, L. A., Aurelio, L., Graham, B., et al. (2015). Protein conformation by EPR spectroscopy using gadolinium tags clicked to genetically encoded p-azido-l-phenylalanine. *Chemical Communications, 51*(88), 15898–15901. https://doi.org/10.1039/C5CC07121F.

Abdullin, D., Duthie, F., Meyer, A., Müller, E. S., Hagelueken, G., & Schiemann, O. (2015). Comparison of PELDOR and RIDME for distance measurements between Nitroxides and low-spin Fe(III) ions. *The Journal of Physical Chemistry B, 119*(43), 13534–13542. https://doi.org/10.1021/acs.jpcb.5b02118.

Abdullin, D., Suchatzki, M., & Schiemann, O. (2021). Six-pulse RIDME sequence to avoid background artifacts. *Applied Magnetic Resonance*. https://doi.org/10.1007/s00723-021-01326-1.

Ackermann, K., Wort, J. L., & Bode, B. E. (2021). Nanomolar pulse dipolar epr spectroscopy in proteins: CuII–CuII and nitroxide–nitroxide cases. *The Journal of Physical Chemistry B, 125*(20), 5358–5364. https://doi.org/10.1021/acs.jpcb.1c03666.

Akhmetzyanov, D., Schöps, P., Marko, A., Kunjir, N. C., Sigurdsson, S. T., & Prisner, T. F. (2015). Pulsed EPR dipolar spectroscopy at Q- and G-band on a trityl biradical. *Physical Chemistry Chemical Physics, 17*(37), 24446–24451. https://doi.org/10.1039/C5CP03671B.

Andersson, S., Rydbeck, A., & Mahno, R. S. (1999). *US Patent 5728370*.

Assafa, T. E., Nandi, S., Śmiłowicz, D., Galazzo, L., Teucher, M., Elsner, C., et al. (2021). Biophysical characterization of pro-apoptotic BimBH3 peptides reveals an unexpected capacity for self-association. *Structure, 29*(2), 114–124.e113. https://doi.org/10.1016/j.str.2020.09.002.

Azarkh, M., Bieber, A., Qi, M., Fischer, J. W. A., Yulikov, M., Godt, A., et al. (2019). Gd(III)-Gd(III) relaxation-induced dipolar modulation enhancement for in-cell Electron paramagnetic resonance distance determination. *The Journal of Physical Chemistry Letters, 10*(7), 1477–1481. https://doi.org/10.1021/acs.jpclett.9b00340.

Bahrenberg, T., Rosenski, Y., Carmieli, R., Zibzener, K., Qi, M., Frydman, V., et al. (2017). Improved sensitivity for W-band Gd(III)-Gd(III) and nitroxide-nitroxide DEER measurements with shaped pulses. *Journal of Magnetic Resonance, 283*, 1–13. https://doi.org/10.1016/j.jmr.2017.08.003.

Becker, J. S., & Saxena, S. (2005). Double quantum coherence electron spin resonance on coupled Cu(II)–Cu(II) electron spins. *Chemical Physics Letters, 414*(1), 248–252. https://doi.org/10.1016/j.cplett.2005.08.072.

Bleaney, B. (1970). *Electron paramagnetic resonance of transition ions*. Clarendon Press.

Bleicken, S., Assafa, T. E., Zhang, H., Elsner, C., Ritsch, I., Pink, M., et al. (2019). Gem-diethyl Pyrroline Nitroxide spin labels: Synthesis, EPR characterization, Rotamer Libraries and Biocompatibility. *ChemistryOpen, 8*(8), 1057–1065. https://doi.org/10.1002/open.201900119.

Borbat, P. P., & Freed, J. H. (1999). Multiple-quantum ESR and distance measurements. *Chemical Physics Letters, 313*, 145. https://doi.org/10.1016/s0009-2614(99)00972-0.

Borbat, P. P., Georgieva, E. R., & Freed, J. H. (2013). Improved sensitivity for long-distance measurements in biomolecules: Five-pulse double electron–electron resonance. *The Journal of Physical Chemistry Letters, 4*(1), 170–175. https://doi.org/10.1021/jz301788n.

Bordignon, E. (2017). EPR spectroscopy of Nitroxide spin probes. In *Vol. 6, issue 2. emigrees* Wiley.

Bowen, A. M., Jones, M. W., Lovett, J. E., Gaule, T. G., McPherson, M. J., Dilworth, J. R., et al. (2016). Exploiting orientation-selective DEER: Determining molecular structure in systems containing cu(ii) centres. *Physical Chemistry Chemical Physics, 18*(8), 5981–5994. https://doi.org/10.1039/C5CP06096F.

Breitgoff, F. D., Keller, K., Qi, M., Klose, D., Yulikov, M., Godt, A., et al. (2019). UWB DEER and RIDME distance measurements in Cu(II)–Cu(II) spin pairs. *Journal of Magnetic Resonance, 308*, 106560. https://doi.org/10.1016/j.jmr.2019.07.047.

Breitgoff, F. D., Polyhach, Y. O., & Jeschke, G. (2017). Reliable nanometre-range distance distributions from 5-pulse double electron electron resonance. *Physical Chemistry Chemical Physics, 19*(24), 15754–15765. https://doi.org/10.1039/C7CP01487B.

Breitgoff, F. D., Soetbeer, J., Doll, A., Jeschke, G., & Polyhach, Y. O. (2017). Artefact suppression in 5-pulse double electron electron resonance for distance distribution measurements. *Physical Chemistry Chemical Physics, 19*(24), 15766–15779. https://doi.org/10.1039/C7CP01488K.

Clayton, J. A., Keller, K., Qi, M., Wegner, J., Koch, V., Hintz, H., et al. (2018). Quantitative analysis of zero-field splitting parameter distributions in Gd(iii) complexes. *Physical Chemistry Chemical Physics, 20*(15), 10470–10492. https://doi.org/10.1039/c7cp08507a.

Cohen, M. R., Frydman, V., Milko, P., Iron, M. A., Abdelkader, E. H., Lee, M. D., et al. (2016). Overcoming artificial broadening in Gd3+–Gd3+ distance distributions arising from dipolar pseudo-secular terms in DEER experiments. *Physical Chemistry Chemical Physics, 18*(18), 12847–12859. https://doi.org/10.1039/C6CP00829A.

Collauto, A., Frydman, V., Lee, M. D., Abdelkader, E. H., Feintuch, A., Swarbrick, J. D., et al. (2016). RIDME distance measurements using Gd(iii) tags with a narrow central transition. *Physical Chemistry Chemical Physics, 18*(28), 19037–19049. https://doi.org/10.1039/C6CP03299K.

Cruickshank, P. A. S., Bolton, D. R., Robertson, D. A., Hunter, R. I., Wylde, R. J., & Smith, G. M. (2009). A kilowatt pulsed 94 GHz electron paramagnetic resonance spectrometer with high concentration sensitivity, high instantaneous bandwidth, and low dead time. *Review of Scientific Instruments, 80*(10), 103102. https://doi.org/10.1063/1.3239402.

Cunningham, T. F., Putterman, M. R., Desai, A., Horne, W. S., & Saxena, S. (2015). The double-histidine Cu^{2+}-binding motif: A highly rigid, site-specific spin probe for electron spin resonance distance measurements. *Angewandte Chemie International Edition English, 54*(21), 6330–6334. https://doi.org/10.1002/anie.201501968.

Di Valentin, M., Albertini, M., Zurlo, E., Gobbo, M., & Carbonera, D. (2014). Porphyrin triplet state as a potential spin label for nanometer distance measurements by PELDOR spectroscopy. *Journal of the American Chemical Society, 136*(18), 6582–6585. https://doi.org/10.1021/ja502615n.

Doll, A., Pribitzer, S., Tschaggelar, R., & Jeschke, G. (2013). Adiabatic and fast passage ultra-wideband inversion in pulsed EPR. *Journal of Magnetic Resonance, 230*, 27–39. https://doi.org/10.1016/j.jmr.2013.01.002.

Doll, A., Qi, M., Wili, N., Pribitzer, S., Godt, A., & Jeschke, G. (2015). Gd(III)-Gd(III) distance measurements with chirp pump pulses. *Journal of Magnetic Resonance, 259*, 153–162. https://doi.org/10.1016/j.jmr.2015.08.010.

El Mkami, H., Hunter, R. I., Cruickshank, P. A. S., Taylor, M. J., Lovett, J. E., Feintuch, A., et al. (2020). High-sensitivity Gd3+–Gd3+ EPR distance measurements that eliminate artefacts seen at short distances. *Magnetic Resonance, 1*(2), 301–313. https://doi.org/10.5194/mr-1-301-2020.

Endeward, B., Marko, A., Denysenkov, V. P., Sigurdsson, S. T., & Prisner, T. F. (2015). Advanced EPR methods for studying conformational dynamics of nucleic acids. *Methods in Enzymology, 564*, 403–425. https://doi.org/10.1016/bs.mie.2015.06.007.

Engelhard, D. M., Meyer, A., Berndhäuser, A., Schiemann, O., & Clever, G. H. (2018). Di-copper(ii) DNA G-quadruplexes as EPR distance rulers. *Chemical Communications*, *54*(54), 7455–7458. https://doi.org/10.1039/C8CC04053B.

Fábregas Ibáñez, L., Jeschke, G., & Stoll, S. (2020). DeerLab: A comprehensive software package for analyzing dipolar electron paramagnetic resonance spectroscopy data. *Magnetic Resonance*, *1*(2), 209–224. https://doi.org/10.5194/mr-1-209-2020.

Fedin, M. V., Shevelev, G. Y., Pyshnyi, D. V., Tormyshev, V. M., Jeschke, G., Yulikov, M., et al. (2016). Interaction of triarylmethyl radicals with DNA termini revealed by orientation-selective W-band double electron-electron resonance spectroscopy. *Physical Chemistry Chemical Physics*, *18*(42), 29549–29554. https://doi.org/10.1039/c6cp05904j.

Fleck, N., Hett, T., Brode, J., Meyer, A., Richert, S., & Schiemann, O. (2019). C–C cross-coupling reactions of trityl radicals: Spin density delocalization, exchange coupling, and a spin label. *The Journal of Organic Chemistry*, *84*(6), 3293–3303. https://doi.org/10.1021/acs.joc.8b03229.

Fleck, N., Heubach, C. A., Hett, T., Haege, F. R., Bawol, P. P., Baltruschat, H., et al. (2020). SLIM: A short-linked, highly redox-stable trityl label for high-sensitivity in-cell EPR distance measurements. *Angewandte Chemie International Edition English*, *59*(24), 9767–9772. https://doi.org/10.1002/anie.202004452.

Fleck, N., Heubach, C., Hett, T., Spicher, S., Grimme, S., & Schiemann, O. (2021). Ox-SLIM: Synthesis of and site-specific labelling with a highly hydrophilic Trityl spin label. *Chemistry - A European Journal*, *27*(16), 5292–5297. https://doi.org/10.1002/chem.202100013.

Galazzo, L., Meier, G., Timachi, M. H., Hutter, C. A. J., Seeger, M. A., & Bordignon, E. (2020). Spin-labeled nanobodies as protein conformational reporters for electron paramagnetic resonance in cellular membranes. *Proceedings of the National Academy of Sciences USA*, *117*(5), 2441–2448. https://doi.org/10.1073/pnas.1913737117.

Garbuio, L., Bordignon, E., Brooks, E. K., Hubbell, W. L., Jeschke, G., & Yulikov, M. (2013). Orthogonal spin labeling and Gd(III)-nitroxide distance measurements on bacteriophage T4-lysozyme. *The Journal of Physical Chemistry B*, *117*(11), 3145–3153. https://doi.org/10.1021/jp401806g.

Ghimire, H., McCarrick, R. M., Budil, D. E., & Lorigan, G. A. (2009). Significantly improved sensitivity of Q-band PELDOR/DEER experiments relative to X-band is observed in measuring the intercoil distance of a leucine zipper motif peptide (GCN4-LZ). *Biochemistry*, *48*(25), 5782–5784. https://doi.org/10.1021/bi900781u.

Ghosh, S., Lawless, M. J., Brubaker, H. J., Singewald, K., Kurpiewski, M. R., Jen-Jacobson, L., et al. (2020). Cu2+−based distance measurements by pulsed EPR provide distance constraints for DNA backbone conformations in solution. *Nucleic Acids Research*, *48*(9), e49. https://doi.org/10.1093/nar/gkaa133.

Ghosh, S., Lawless, M. J., Rule, G. S., & Saxena, S. (2018). The cu(2+)-nitrilotriacetic acid complex improves loading of α-helical double histidine site for precise distance measurements by pulsed ESR. *Journal of Magnetic Resonance*, *286*, 163–171. https://doi.org/10.1016/j.jmr.2017.12.005.

Ghosh, S., Saxena, S., & Jeschke, G. (2018). Rotamer modelling of Cu(II) spin labels based on the double-histidine motif. *Applied Magnetic Resonance*, *49*(11), 1281–1298. https://doi.org/10.1007/s00723-018-1052-8.

Giannoulis, A., Ben-Ishay, Y., & Goldfarb, D. (2021). Chapter eight—Characteristics of Gd(III) spin labels for the study of protein conformations. In J. A. Cotruvo (Ed.), *Vol. 651. Methods in enzymology* (pp. 235–290). Academic Press.

Giannoulis, A., Oranges, M., & Bode, B. E. (2017). Monitoring complex formation by relaxation-induced pulse Electron paramagnetic resonance distance measurements. *ChemPhysChem*, *18*(17), 2318–2321. https://doi.org/10.1002/cphc.201700666.

Giannoulis, A., Yang, Y., Gong, Y. J., Tan, X., Feintuch, A., Carmieli, R., et al. (2019). DEER distance measurements on trityl/trityl and Gd(iii)/trityl labelled proteins. *Physical Chemistry Chemical Physics*, *21*(20), 10217–10227. https://doi.org/10.1039/c8cp07249c.

Gmeiner, C., Dorn, G., Allain, F. H. T., Jeschke, G., & Yulikov, M. (2017). Spin labelling for integrative structure modelling: A case study of the polypyrimidine-tract binding protein 1 domains in complexes with short RNAs. *Physical Chemistry Chemical Physics*, *19*(41), 28360–28380. https://doi.org/10.1039/c7cp05822e.

Gmeiner, C., Klose, D., Mileo, E., Belle, V., Marque, S. R. A., Dorn, G., et al. (2017). Orthogonal tyrosine and cysteine site-directed spin labeling for dipolar pulse EPR spectroscopy on proteins. *The Journal of Physical Chemistry Letters*, *8*(19), 4852–4857. https://doi.org/10.1021/acs.jpclett.7b02220.

Goldfarb, D. (2014). Gd3+ spin labeling for distance measurements by pulse EPR spectroscopy. *Physical Chemistry Chemical Physics*, *16*(21), 9685–9699. https://doi.org/10.1039/C3CP53822B.

Guo, Y., Suess, D. L. M., Herzik, M. A., Iavarone, A. T., Britt, R. D., & Marletta, M. A. (2017). Regulation of nitric oxide signaling by formation of a distal receptor–ligand complex. *Nature Chemical Biology*, *13*(12), 1216–1221. https://doi.org/10.1038/nchembio.2488.

Hagelueken, G., Ward, R., Naismith, J. H., & Schiemann, O. (2012). MtsslWizard: In Silico Spin-Labeling and Generation of Distance Distributions in PyMOL. *Applied Magnetic Resonance*, *42*(3), 377–391. https://doi.org/10.1007/s00723-012-0314-0.

Haugland, M. M., Lovett, J. E., & Anderson, E. A. (2018). Advances in the synthesis of nitroxide radicals for use in biomolecule spin labelling. *Chemical Society Reviews*, *47*(3), 668–680. https://doi.org/10.1039/c6cs00550k.

Hintz, H., Vanas, A., Klose, D., Jeschke, G., & Godt, A. (2019). Trityl radicals with a combination of the orthogonal functional groups Ethyne and carboxyl: Synthesis without a statistical step and EPR characterization. *The Journal of Organic Chemistry*, *84*(6), 3304–3320. https://doi.org/10.1021/acs.joc.8b03234.

Hintze, C., Bücker, D., Domingo Köhler, S., Jeschke, G., & Drescher, M. (2016). Laser-induced magnetic dipole spectroscopy. *The Journal of Physical Chemistry Letters*, *7*(12), 2204–2209. https://doi.org/10.1021/acs.jpclett.6b00765.

Hutter, C. A. J., Timachi, M. H., Hürlimann, L. M., Zimmermann, I., Egloff, P., Göddeke, H., et al. (2019). The extracellular gate shapes the energy profile of an ABC exporter. *Nature Communications*, *10*(1), 2260. https://doi.org/10.1038/s41467-019-09892-6.

Jagtap, A. P., Krstic, I., Kunjir, N. C., Hänsel, R., Prisner, T. F., & Sigurdsson, S. T. (2015). Sterically shielded spin labels for in-cell EPR spectroscopy: Analysis of stability in reducing environment. *Free Radical Research*, *49*(1), 78–85. https://doi.org/10.3109/10715762.2014.979409.

Jarvi, A. G., Ranguelova, K., Ghosh, S., Weber, R. T., & Saxena, S. (2018). On the use of Q-band double Electron-Electron resonance to resolve the relative orientations of two double histidine-bound Cu2+ ions in a protein. *Journal of Physical Chemistry B*, *122*(47), 10669–10677. https://doi.org/10.1021/acs.jpcb.8b07727.

Jassoy, J. J., Berndhäuser, A., Duthie, F., Kühn, S. P., Hagelueken, G., & Schiemann, O. (2017). Versatile Trityl spin labels for nanometer distance measurements on biomolecules in vitro and within cells. *Angewandte Chemie International Edition*, *56*(1), 177–181. https://doi.org/10.1002/anie.201609085.

Jassoy, J. J., Heubach, C. A., Hett, T., Bernhard, F., Haege, F. R., Hagelueken, G., et al. (2019). Site selective and efficient spin labeling of proteins with a Maleimide-functionalized Trityl radical for pulsed dipolar EPR spectroscopy. *Molecules*, *24*(15), 2735. Retrieved from https://www.mdpi.com/1420-3049/24/15/2735.

Jeschke, G. (2018). MMM: A toolbox for integrative structure modeling. *Protein Science*, *27*(1), 76–85. https://doi.org/10.1002/pro.3269.

Jeschke, G., Chechik, V., Ionita, P., Godt, A., Zimmermann, H., Banham, J., et al. (2006). DeerAnalysis2006—A comprehensive software package for analyzing pulsed ELDOR data. *Applied Magnetic Resonance*, *30*(3), 473–498. https://doi.org/10.1007/BF03166213.

Jeschke, G., Pannier, M., Godt, A., & Spiess, H. W. (2000). Dipolar spectroscopy and spin alignment in electron paramagnetic resonance. *Chemical Physics Letters*, *331*, 243. https://doi.org/10.1016/s0009-2614(00)01171-4.

Joseph, B., Korkhov, V. M., Yulikov, M., Jeschke, G., & Bordignon, E. (2014). Conformational cycle of the vitamin B12 ABC importer in liposomes detected by double electron-electron resonance (DEER). *Journal of Biological Chemistry*, *289*(6), 3176–3185. https://doi.org/10.1074/jbc.M113.512178.

Joseph, B., Tormyshev, V. M., Rogozhnikova, O. Y., Akhmetzyanov, D., Bagryanskaya, E. G., & Prisner, T. F. (2016). Selective high-resolution detection of membrane protein-ligand interaction in native membranes using Trityl-Nitroxide PELDOR. *Angewandte Chemie International Edition English*, *55*(38), 11538–11542. https://doi.org/10.1002/anie.201606335.

Kaminker, I., Bye, M., Mendelman, N., Gislason, K., Sigurdsson, S. T., & Goldfarb, D. (2015). Distance measurements between manganese(ii) and nitroxide spin-labels by DEER determine a binding site of Mn2+ in the HP92 loop of ribosomal RNA. *Physical Chemistry Chemical Physics*, *17*(23), 15098–15102. https://doi.org/10.1039/C5CP01624J.

Kaminker, I., Tkach, I., Manukovsky, N., Huber, T., Yagi, H., Otting, G., et al. (2013). W-band orientation selective DEER measurements on a Gd3+/nitroxide mixed-labeled protein dimer with a dual mode cavity. *Journal of Magnetic Resonance*, *227*, 66–71. https://doi.org/10.1016/j.jmr.2012.11.028.

Kaminker, I., Yagi, H., Huber, T., Feintuch, A., Otting, G., & Goldfarb, D. (2012). Spectroscopic selection of distance measurements in a protein dimer with mixed nitroxide and Gd3+ spin labels. *Physical Chemistry Chemical Physics*, *14*(13), 4355–4358. https://doi.org/10.1039/c2cp40219j.

Karthikeyan, G., Bonucci, A., Casano, G., Gerbaud, G., Abel, S., Thomé, V., et al. (2018). A bioresistant nitroxide spin label for in-cell epr spectroscopy: In vitro and in oocytes protein structural dynamics studies. *Angewandte Chemie International Edition English*, *57*(5), 1366–1370. https://doi.org/10.1002/anie.201710184.

Keller, K., Doll, A., Qi, M., Godt, A., Jeschke, G., & Yulikov, M. (2016). Averaging of nuclear modulation artefacts in RIDME experiments. *Journal of Magnetic Resonance*, *272*, 108–113. https://doi.org/10.1016/j.jmr.2016.09.016.

Keller, K., Mertens, V., Qi, M., Nalepa, A. I., Godt, A., Savitsky, A., et al. (2017). Computing distance distributions from dipolar evolution data with overtones: RIDME spectroscopy with Gd(iii)-based spin labels. *Physical Chemistry Chemical Physics*, *19*(27), 17856–17876. https://doi.org/10.1039/C7CP01524K.

Keller, K., Qi, M., Gmeiner, C., Ritsch, I., Godt, A., Jeschke, G., et al. (2019). Intermolecular background decay in RIDME experiments. *Physical Chemistry Chemical Physics*, *21*(16), 8228–8245. https://doi.org/10.1039/C8CP07815G.

Keller, K., Zalibera, M., Qi, M., Koch, V., Wegner, J., Hintz, H., et al. (2016). EPR characterization of Mn(ii) complexes for distance determination with pulsed dipolar spectroscopy. *Physical Chemistry Chemical Physics*, *18*(36), 25120–25135. https://doi.org/10.1039/C6CP04884F.

Ketter, S., Gopinath, A., Rogozhnikova, O., Trukhin, D., Tormyshev, V. M., Bagryanskaya, E. G., et al. (2021). In situ labeling and distance measurements of membrane proteins in E. coli using Finland and OX063 Trityl labels. *Chemistry - A European Journal*, *27*(7), 2299–2304. https://doi.org/10.1002/chem.202004606.

Klare, J. P., & Steinhoff, H. J. (2009). Spin labeling EPR. *Photosynthetic Research*, *102*(2–3), 377–390. https://doi.org/10.1007/s11120-009-9490-7.

Kucher, S., Elsner, C., Safonova, M., Maffini, S., & Bordignon, E. (2021). In-cell double electron–electron resonance at nanomolar protein concentrations. *The Journal of Physical Chemistry Letters*, *12*(14), 3679–3684. https://doi.org/10.1021/acs.jpclett.1c00048.

Kulik, L., Dzuba, S., Grigoryev, I., & Tsvetkov, Y. (2001). Electron dipole–dipole interaction in ESEEM of nitroxide biradicals. *Chemical Physics Letters*, *343*, 315–324. https://doi.org/10.1016/S0009-2614(01)00721-7.

Kunjir, N. C., Reginsson, G. W., Schiemann, O., & Sigurdsson, S. T. (2013). Measurements of short distances between trityl spin labels with CW EPR, DQC and PELDOR. *Physical Chemistry Chemical Physics*, *15*(45), 19673–19685. https://doi.org/10.1039/C3CP52789A.

Kuzhelev, A. A., Krumkacheva, O. A., Shevelev, G. Y., Yulikov, M., Fedin, M. V., & Bagryanskaya, E. G. (2018). Room-temperature distance measurements using RIDME and the orthogonal spin labels trityl/nitroxide. *Physical Chemistry Chemical Physics*, *20*(15), 10224–10230. https://doi.org/10.1039/c8cp01093e.

Kuzhelev, A. A., Tormyshev, V. M., Rogozhnikova, O. Y., Trukhin, D. V., Troitskaya, T. I., Strizhakov, R. K., et al. (2017). Triarylmethyl radicals: An EPR study of 13C hyperfine coupling constants. *Zeitschrift für Physikalische Chemie*, *231*(4), 777–794. https://doi.org/10.1515/zpch-2016-0811.

Larsen, R. G., & Singel, D. J. (1993). Double electron–electron resonance spin–echo modulation: Spectroscopic measurement of electron spin pair separations in orientationally disordered solids. *The Journal of Chemical Physics*, *98*(7), 5134–5146. https://doi.org/10.1063/1.464916.

Lueders, P., Jäger, H., Hemminga, M. A., Jeschke, G., & Yulikov, M. (2013). Distance measurements on orthogonally spin-labeled membrane spanning WALP23 polypeptides. *The Journal of Physical Chemistry B*, *117*(7), 2061–2068. https://doi.org/10.1021/jp311287t.

Lueders, P., Jeschke, G., & Yulikov, M. (2011). Double electron – electron resonance measured between Gd3+ ions and nitroxide radicals. *The Journal of Physical Chemistry Letters*, *2*(6), 604–609. https://doi.org/10.1021/jz200073h.

Martin, R. E., Pannier, M., Diederich, F., Gramlich, V., Hubrich, M., & Spiess, H. W. (1998). Determination of end-to-end distances in a series of TEMPO Diradicals of up to 2.8 nm length with a new four-pulse double Electron Electron resonance experiment. *Angewandte Chemie International Edition English*, *37*(20), 2833–2837. https://doi.org/10.1002/(sici)1521-3773(19981102)37:20<2833::Aid-anie2833>3.0.Co;2-7.

Martorana, A., Bellapadrona, G., Feintuch, A., Di Gregorio, E., Aime, S., & Goldfarb, D. (2014). Probing protein conformation in cells by EPR distance measurements using Gd3+ spin labeling. *Journal of the American Chemical Society*, *136*(38), 13458–13465. https://doi.org/10.1021/ja5079392.

Martorana, A., Yang, Y., Zhao, Y., Li, Q.-F., Su, X.-C., & Goldfarb, D. (2015). Mn(ii) tags for DEER distance measurements in proteins via C–S attachment. *Dalton Transactions*, *44*(48), 20812–20816. https://doi.org/10.1039/C5DT04123F.

Meirovitch, E., & Poupko, R. (1978). Line shape studies of the electron spin resonance spectra of manganese protein complexes. *The Journal of Physical Chemistry*, *82*(17), 1920–1925. https://doi.org/10.1021/j100506a013.

Merz, G. E., Borbat, P. P., Pratt, A. J., Getzoff, E. D., Freed, J. H., & Crane, B. R. (2014). Copper-based pulsed dipolar ESR spectroscopy as a probe of protein conformation linked to disease states. *Biophysical Journal*, *107*(7), 1669–1674. https://doi.org/10.1016/j.bpj.2014.07.068.

Meyer, A., Abdullin, D., Schnakenburg, G., & Schiemann, O. (2016). Single and double nitroxide labeled bis(terpyridine)-copper(ii): Influence of orientation selectivity and multispin effects on PELDOR and RIDME. *Physical Chemistry Chemical Physics*, *18*(13), 9262–9271. https://doi.org/10.1039/C5CP07621H.

Meyer, A., Jassoy, J. J., Spicher, S., Berndhäuser, A., & Schiemann, O. (2018). Performance of PELDOR, RIDME, SIFTER, and DQC in measuring distances in trityl based bi- and triradicals: Exchange coupling, pseudosecular coupling and multi-spin effects. *Physical Chemistry Chemical Physics*, *20*(20), 13858–13869. https://doi.org/10.1039/C8CP01276H.

Milikisyants, S., Scarpelli, F., Finiguerra, M. G., Ubbink, M., & Huber, M. (2009). A pulsed EPR method to determine distances between paramagnetic centers with strong spectral anisotropy and radicals: The dead-time free RIDME sequence. *Journal of Magnetic Resonance*, *201*(1), 48–56. https://doi.org/10.1016/j.jmr.2009.08.008.

Milov, A. D., Maryasov, A. G., & Tsvetkov, Y. D. (1998). Pulsed electron double resonance (PELDOR) and its applications in free-radicals research. *Applied Magnetic Resonance*, *15*(1), 107–143. https://doi.org/10.1007/BF03161886.

Milov, A. D., Ponomarev, A. B., & Tsvetkov, Y. D. (1984). Electron-electron double resonance in electron spin echo: Model biradical systems and the sensitized photolysis of decalin. *Chemical Physics Letters*, *110*, 67. https://doi.org/10.1016/0009-2614(84)80148-7.

Milov, A., Salikhov, K., & Shirov, M. (1981). Application of ELDOR in electron-spin echo for paramagnetic center space distribution in solids. *Fizika Tverdogo Tela*, *23*(4), 975–982.

Motion, C. L., Cassidy, S. L., Cruickshank, P. A. S., Hunter, R. I., Bolton, D. R., El Mkami, H., et al. (2017). The use of composite pulses for improving DEER signal at 94GHz. *Journal of Magnetic Resonance*, *278*, 122–133. https://doi.org/10.1016/j.jmr.2017.03.018.

Motion, C. L., Lovett, J. E., Bell, S., Cassidy, S. L., Cruickshank, P. A. S., Bolton, D. R., et al. (2016). DEER sensitivity between Iron centers and nitroxides in heme-containing proteins improves dramatically using broadband, high-field EPR. *The Journal of Physical Chemistry Letters*, 7(8), 1411–1415. https://doi.org/10.1021/acs.jpclett.6b00456.

Narr, E., Godt, A., & Jeschke, G. (2002). Selective measurements of a nitroxide-nitroxide separation of 5 nm and a nitroxide-copper separation of 2.5 nm in a terpyridine-based copper(II) complex by pulse EPR spectroscopy. *Angewandte Chemie International Edition English*, *41*(20), 3907–3910. https://doi.org/10.1002/1521-3773(20021018)41:20<3907::Aid-anie3907>3.0.Co;2-t.

Owenius, R., Eaton, G. R., & Eaton, S. S. (2005). Frequency (250 MHz to 9.2 GHz) and viscosity dependence of electron spin relaxation of triarylmethyl radicals at room temperature. *Journal of Magnetic Resonance*, *172*(1), 168–175. https://doi.org/10.1016/j.jmr.2004.10.007.

Paletta, J. T., Pink, M., Foley, B., Rajca, S., & Rajca, A. (2012). Synthesis and reduction kinetics of sterically shielded Pyrrolidine Nitroxides. *Organic Letters*, *14*(20), 5322–5325. https://doi.org/10.1021/ol302506f.

Pannier, M., Veit, S., Godt, A., Jeschke, G., & Spiess, H. W. (2000). Dead-time free measurement of dipole-dipole interactions between electron spins. *Journal of Magnetic Resonance*, *142*(2), 331–340. https://doi.org/10.1006/jmre.1999.1944.

Polyhach, Y., Bordignon, E., & Jeschke, G. (2011). Rotamer libraries of spin labelled cysteines for protein studies. *Physical Chemistry Chemical Physics*, *13*(6), 2356–2366. https://doi.org/10.1039/C0CP01865A.

Polyhach, Y., Bordignon, E., Tschaggelar, R., Gandra, S., Godt, A., & Jeschke, G. (2012). High sensitivity and versatility of the DEER experiment on nitroxide radical pairs at Q-band frequencies. *Physical Chemistry Chemical Physics*, *14*(30), 10762–10773. https://doi.org/10.1039/C2CP41520H.

Potapov, A., Yagi, H., Huber, T., Jergic, S., Dixon, N. E., Otting, G., et al. (2010). Nanometer-scale distance measurements in proteins using Gd3+ spin labeling. *Journal of the American Chemical Society*, *132*(26), 9040–9048. https://doi.org/10.1021/ja1015662.

Prisner, T. F. (2019). Shaping EPR: Phase and amplitude modulated microwave pulses. *Journal of Magnetic Resonance*, *306*, 98–101. https://doi.org/10.1016/j.jmr.2019.07.023.

Raitsimring, A. M., Gunanathan, C., Potapov, A., Efremenko, I., Martin, J. M., Milstein, D., et al. (2007). Gd3+ complexes as potential spin labels for high field pulsed EPR distance measurements. *Journal of the American Chemical Society*, *129*(46), 14138–14139. https://doi.org/10.1021/ja075544g.

Razzaghi, S., Qi, M., Nalepa, A. I., Godt, A., Jeschke, G., Savitsky, A., et al. (2014). RIDME spectroscopy with Gd(III) centers. *The Journal of Physical Chemistry Letters*, *5*(22), 3970–3975. https://doi.org/10.1021/jz502129t.

Reddy, T. J., Iwama, T., Halpern, H. J., & Rawal, V. H. (2002). General synthesis of persistent trityl radicals for EPR imaging of biological systems. *The Journal of Organic Chemistry*, *67*(14), 4635–4639. https://doi.org/10.1021/jo011068f.

Reginsson, G. W., Kunjir, N. C., Sigurdsson, S. T., & Schiemann, O. (2012). Trityl radicals: Spin labels for nanometer-distance measurements. *Chemistry - A European Journal*, *18*(43), 13580–13584. https://doi.org/10.1002/chem.201203014.

Ritsch, I., Hintz, H., Jeschke, G., Godt, A., & Yulikov, M. (2019). Improving the accuracy of Cu(ii)–nitroxide RIDME in the presence of orientation correlation in water-soluble Cu(ii)–nitroxide rulers. *Physical Chemistry Chemical Physics*, *21*(19), 9810–9830. https://doi.org/10.1039/C8CP06573J.

Russell, H., Stewart, R., Prior, C., Oganesyan, V. S., Gaule, T. G., & Lovett, J. E. (2021). DEER and RIDME measurements of the Nitroxide-spin labelled copper-bound amine oxidase homodimer from Arthrobacter Globiformis. *Applied Magnetic Resonance*, *52*(8), 995–1015. https://doi.org/10.1007/s00723-021-01321-6.

Ruthstein, S., Ji, M., Mehta, P., Jen-Jacobson, L., & Saxena, S. (2013). Sensitive Cu2+–Cu2+ distance measurements in a protein–DNA complex by double-quantum coherence ESR. *The Journal of Physical Chemistry B*, *117*(20), 6227–6230. https://doi.org/10.1021/jp4037149.

Ruthstein, S., Ji, M., Shin, B. K., & Saxena, S. (2015). A simple double quantum coherence ESR sequence that minimizes nuclear modulations in Cu(2+)-ion based distance measurements. *Journal of Magnetic Resonance*, *257*, 45–50. https://doi.org/10.1016/j.jmr.2015.05.005.

Sarver, J., Silva, K. I., & Saxena, S. (2013). Measuring Cu2+-Nitroxide Distances Using Double Electron–Electron Resonance and Saturation Recovery. *Applied Magnetic Resonance*, *44*(5), 583–594. https://doi.org/10.1007/s00723-012-0422-x.

Saxena, S., & Freed, J. H. (1996). Double quantum two dimensional Fourier transform Electron spin resonance: Distance measurements. *Chemical Physics Letters*, *251*, 102–110.

Schiemann, O., Heubach, C. A., Abdullin, D., Ackermann, K., Azarkh, M., Bagryanskaya, E. G., et al. (2021). Benchmark test and guidelines for DEER/PELDOR experiments on Nitroxide-labeled biomolecules. *Journal of the American Chemical Society*, *143*(43), 17875–17890. https://doi.org/10.1021/jacs.1c07371.

Shah, A., Roux, A., Starck, M., Mosely, J. A., Stevens, M., Norman, D. G., et al. (2019). A gadolinium spin label with both a narrow central transition and short tether for use in double electron electron resonance distance measurements. *Inorganic Chemistry*, *58*(5), 3015–3025. https://doi.org/10.1021/acs.inorgchem.8b02892.

Shevelev, G. Y., Krumkacheva, O. A., Lomzov, A. A., Kuzhelev, A. A., Rogozhnikova, O. Y., Trukhin, D. V., et al. (2014). Physiological-temperature distance measurement in nucleic acid using triarylmethyl-based spin labels and pulsed dipolar EPR spectroscopy. *Journal of the American Chemical Society*, *136*(28), 9874–9877. https://doi.org/10.1021/ja505122n.

Shevelev, G. Y., Krumkacheva, O. A., Lomzov, A. A., Kuzhelev, A. A., Trukhin, D. V., Rogozhnikova, O. Y., et al. (2015). Triarylmethyl labels: Toward improving the accuracy of EPR nanoscale distance measurements in DNAs. *The Journal of Physical Chemistry B*, *119*(43), 13641–13648. https://doi.org/10.1021/acs.jpcb.5b03026.

Šimėnas, M., O'Sullivan, J., Zollitsch, C. W., Kennedy, O., Seif-Eddine, M., Ritsch, I., et al. (2021). A sensitivity leap for X-band EPR using a probehead with a cryogenic preamplifier. *Journal of Magnetic Resonance*, *322*, 106876. https://doi.org/10.1016/j.jmr.2020.106876.

Soetbeer, J., Millen, M., Zouboulis, K., Hülsmann, M., Godt, A., Polyhach, Y., et al. (2021). Dynamical decoupling in water–glycerol glasses: A comparison of nitroxides, trityl radicals and gadolinium complexes. *Physical Chemistry Chemical Physics, 23*(9), 5352–5369. https://doi.org/10.1039/D1CP00055A.

Spindler, P. E., Waclawska, I., Endeward, B., Plackmeyer, J., Ziegler, C., & Prisner, T. F. (2015). Carr–purcell pulsed electron double resonance with shaped inversion pulses. *The Journal of Physical Chemistry Letters, 6*(21), 4331–4335. https://doi.org/10.1021/acs.jpclett.5b01933.

Spindler, P. E., Zhang, Y., Endeward, B., Gershernzon, N., Skinner, T. E., Glaser, S. J., et al. (2012). Shaped optimal control pulses for increased excitation bandwidth in EPR. *Journal of Magnetic Resonance, 218*, 49–58. https://doi.org/10.1016/j.jmr.2012.02.013.

Stoll, S., & Schweiger, A. (2006). EasySpin, a comprehensive software package for spectral simulation and analysis in EPR. *Journal of Magnetic Resonance, 178*(1), 42–55. https://doi.org/10.1016/j.jmr.2005.08.013.

Stone, T. J., Buckman, T., Nordio, P. L., & McConnell, H. M. (1965). Spin-labeled biomolecules. *Proceedings of the National Academy of Sciences USA, 54*(4), 1010–1017. https://doi.org/10.1073/pnas.54.4.1010.

Stratmann, L. M., Kutin, Y., Kasanmascheff, M., & Clever, G. H. (2021). Precise distance measurements in DNA G-Quadruplex dimers and sandwich complexes by pulsed dipolar EPR spectroscopy. *Angewandte Chemie International Edition, 60*(9), 4939–4947. https://doi.org/10.1002/anie.202008618.

Teucher, M., & Bordignon, E. (2018). Improved signal fidelity in 4-pulse DEER with Gaussian pulses. *Journal of Magnetic Resonance, 296*, 103–111. https://doi.org/10.1016/j.jmr.2018.09.003.

Teucher, M., Qi, M., Cati, N., Hintz, H., Godt, A., & Bordignon, E. (2020). Strategies to identify and suppress crosstalk signals in double electron–electron resonance (DEER) experiments with gadoliniumIII and nitroxide spin-labeled compounds. *Magnetic Resonance, 1*(2), 285–299. https://doi.org/10.5194/mr-1-285-2020.

Teucher, M., Zhang, H., Bader, V., Winklhofer, K. F., García-Sáez, A. J., Rajca, A., et al. (2019). A new perspective on membrane-embedded Bax oligomers using DEER and bioresistant orthogonal spin labels. *Scientific Reports, 9*(1), 13013. https://doi.org/10.1038/s41598-019-49370-z.

Tkach, I., Sicoli, G., Höbartner, C., & Bennati, M. (2011). A dual-mode microwave resonator for double electron-electron spin resonance spectroscopy at W-band microwave frequencies. *Journal of Magnetic Resonance, 209*(2), 341–346. https://doi.org/10.1016/j.jmr.2011.01.012.

Tormyshev, V. M., Chubarov, A. S., Krumkacheva, O. A., Trukhin, D. V., Rogozhnikova, O. Y., Spitsyna, A. S., et al. (2020). Methanethiosulfonate derivative of OX063 Trityl: A promising and efficient reagent for side-directed spin labeling of proteins. *Chemistry - A European Journal, 26*(12), 2705–2712. https://doi.org/10.1002/chem.201904587.

Torricella, F., Pierro, A., Mileo, E., Belle, V., & Bonucci, A. (2021). Nitroxide spin labels and EPR spectroscopy: A powerful association for protein dynamics studies. *Biochimica and Biophysica Acta Proteins and Proteomics, 1869*(7). https://doi.org/10.1016/j.bbapap.2021.140653, 140653.

van Amsterdam, I. M. C., Ubbink, M., Canters, G. W., & Huber, M. (2003). Measurement of a Cu–Cu distance of 26 Å by a pulsed EPR method. *Angewandte Chemie International Edition, 42*(1), 62–64. https://doi.org/10.1002/anie.200390053.

Voss, J., Wu, J., Hubbell, W. L., Jacques, V., Meares, C. F., & Kaback, H. R. (2001). Helix packing in the lactose permease of Escherichia coli: Distances between site-directed nitroxides and a lanthanide. *Biochemistry, 40*(10), 3184–3188. https://doi.org/10.1021/bi002333e.

Williams, L., Tischlik, S., Scherer, A., Fischer, J. W. A., & Drescher, M. (2020). Site-directed attachment of photoexcitable spin labels for light-induced pulsed dipolar spectroscopy. *Chemical Communications, 56*(93), 14669–14672. https://doi.org/10.1039/D0CC03101A.

Worswick, S. G., Spencer, J. A., Jeschke, G., & Kuprov, I. (2018). Deep neural network processing of DEER data. *Science Advances*, *4*(8), eaat5218. https://doi.org/10.1126/sciadv.aat5218.

Wort, J. L., Ackermann, K., Giannoulis, A., Stewart, A. J., Norman, D. G., & Bode, B. E. (2019). Sub-micromolar pulse dipolar EPR spectroscopy reveals increasing Cu(II) -labelling of double-histidine motifs with lower temperature. *Angewandte Chemie International Edition English*, *58*(34), 11681–11685. https://doi.org/10.1002/anie.201904848.

Wort, J. L., Arya, S., Ackermann, K., Stewart, A. J., & Bode, B. E. (2021). Pulse dipolar EPR reveals double-histidine motif CuII–NTA spin-labeling robustness against competitor ions. *The Journal of Physical Chemistry Letters*, *12*(11), 2815–2819. https://doi.org/10.1021/acs.jpclett.1c00211.

Wu, Z., Feintuch, A., Collauto, A., Adams, L. A., Aurelio, L., Graham, B., et al. (2017). Selective distance measurements using triple spin labeling with Gd3+, Mn2+, and a nitroxide. *The Journal of Physical Chemistry Letters*, *8*(21), 5277–5282. https://doi.org/10.1021/acs.jpclett.7b01739.

Yang, Z., Becker, J., & Saxena, S. (2007). On Cu(II)–Cu(II) distance measurements using pulsed electron electron double resonance. *Journal of Magnetic Resonance*, *188*(2), 337–343. https://doi.org/10.1016/j.jmr.2007.08.006.

Yang, Z., Ji, M., Cunningham, T. F., & Saxena, S. (2015). Cu^{2+} as an ESR probe of protein structure and function. *Methods in Enzymology*, *563*, 459–481. https://doi.org/10.1016/bs.mie.2015.05.026.

Yang, Z., Kise, D., & Saxena, S. (2010). An approach towards the measurement of nanometer range distances based on Cu2+ ions and ESR. *The Journal of Physical Chemistry B*, *114*(18), 6165–6174. https://doi.org/10.1021/jp911637s.

Yang, Z., Kurpiewski, M. R., Ji, M., Townsend, J. E., Mehta, P., Jen-Jacobson, L., et al. (2012). ESR spectroscopy identifies inhibitory Cu2+ sites in a DNA-modifying enzyme to reveal determinants of catalytic specificity. *Proceedings of the National Academy of Sciences USA*, *109*(17), E993–1000. https://doi.org/10.1073/pnas.1200733109.

Yang, Z., Liu, Y., Borbat, P., Zweier, J. L., Freed, J. H., & Hubbell, W. L. (2012). Pulsed ESR dipolar spectroscopy for distance measurements in immobilized spin labeled proteins in liquid solution. *Journal of the American Chemical Society*, *134*(24), 9950–9952. https://doi.org/10.1021/ja303791p.

Yang, Y., Pan, B.-B., Tan, X., Yang, F., Liu, Y., Su, X.-C., et al. (2020). In-cell trityl–trityl distance measurements on proteins. *The Journal of Physical Chemistry Letters*, *11*(3), 1141–1147. https://doi.org/10.1021/acs.jpclett.9b03208.

Yulikov, M. (2015). Chapter 1 Spectroscopically orthogonal spin labels and distance measurements in biomolecules. In *Vol. 24. Electron paramagnetic resonance: Volume 24* (pp. 1–31). The Royal Society of Chemistry.

Yulikov, M., Lueders, P., Warsi, M. F., Chechik, V., & Jeschke, G. (2012). Distance measurements in Au nanoparticles functionalized with nitroxide radicals and Gd(3+)-DTPA chelate complexes. *Physical Chemistry Chemical Physics*, *14*(30), 10732–10746. https://doi.org/10.1039/c2cp40282c.

CHAPTER FIVE

Tracking protein domain movements by EPR distance determination and multilateration

Juliane Stehle and Malte Drescher*
Department of Chemistry and Konstanz Research School Chemical Biology (KoRS-CB), University of Konstanz, Konstanz, Germany
*Corresponding author: e-mail address: malte.drescher@uni-konstanz.de

Contents

Abstract

Electron paramagnetic resonance (EPR) spectroscopy in combination with site-directed spin labeling (SDSL) is a powerful approach to supplement the large toolbox of methods for protein structure determination. The strength of EPR spectroscopy is the ability to map flexibility of protein domains and conformational ensembles. EPR distance determination in the nanometer range and subsequent multilateration enables the three-dimensional visualization of the localization of a spin label in a protein domain. Therefore, multilateration can not only be used to represent the degree of flexibility of protein structural elements, but also track movements of whole domains. We report a detailed protocol for all needed steps, beginning with the choice of the labeling sites, the spin labeling reaction, the EPR distance measurement by double electron-electron resonance (DEER) and finally the multilateration exploiting the EPR distance restraints.

1. Introduction

To understand the complex mechanism of life, a profound knowledge of the intracellular communication and regulation between the thousands of proteins and other macromolecules, is required (Kim et al., 2014). A crucial

Methods in Enzymology, Volume 666
ISSN 0076-6879
https://doi.org/10.1016/bs.mie.2022.02.016

step towards that, is the determination of their three-dimensional structures. Subsequently those structures might allow to conclude about the protein function and role in the whole network of the cell. Several methods have been established over the years to resolve the structure of proteins, X-ray crystallography, cryo-electron microscopy (cyro-EM) and nuclear magnetic resonance (NMR) spectroscopy, being the most significant (Benjin & Ling, 2020). However, some proteins have no defined three-dimensional fold, being either completely disordered, containing unstructured domains or show a broad ensemble of different conformations in an equilibrium. This makes the investigation of those proteins by X-ray crystallography or cryo-EM challenging (Bonomi, Heller, Camilloni, & Vendruscolo, 2017). In recent years, electron paramagnetic resonance (EPR) spectroscopy in combination with site-directed spin labeling (SDSL) has proven to be an ideal complementary method to those, making it possible to overcome some of these limitations (Hett et al., 2021; McHaourab, Steed, & Kazmier, 2011; Weickert, Cattani, & Drescher, 2019). Most biomolecules are natively EPR silent. Upon attaching a spin label at specific sites in the protein structure, only this label is detected in the EPR experiment, without any background signals (Roser, Schmidt, Drescher, & Summerer, 2016). EPR spectroscopy thus is a very powerful tool to investigate proteins independent of their size and even in their natural environment, inside cells (Braun et al., 2021; Dalaloyan et al., 2019; Kucher, Elsner, Safonova, Maffini, & Bordignon, 2021). Among many other applications, EPR distance determination, complemented with computer based multilateration of those distance information, proved to be extremely useful to unravel the conformation of flexible or disordered domains within proteins (Assafa, Anders, Linne, Essen, & Bordignon, 2018; Steed, Stein, Mishra, Goodman, & McHaourab, 2013; Wingler et al., 2019).

Double electron-electron resonance (DEER) spectroscopy, also called PELDOR, is a pulsed EPR method, that relies on detection of the dipolar interaction between two single electron spins (Pannier, Veit, Godt, Jeschke, & Spiess, 2000). It allows distance determination between the two coupled electron spins in the nanometer range, typically between 1.5 and 8 nm. Under ideal circumstances even a distance range up to 17 nm can be covered (Schmidt, Wälti, Baber, Hustedt, & Clore, 2016). Moreover, it does not only provide a single averaged distance, but distance restraints, i.e. distance distributions, that allow the effective study of conformational distributions of biomolecules in a quantitative manner (Jeschke, 2002, 2012).

An established method to introduce the EPR sensitive spin labels into the protein is site-directed spin labeling (Cornish et al., 1994; Hubbell, Cafiso, & Altenbach, 2000). Usually a site-directed mutagenesis, substituting the site under study by a cysteine residue, is performed, followed by the reaction with a sulfhydryl-specific reagent. If there are native cysteines in the sequence they are often replaced by non-reactive residues, like serine or alanine. Commonly used are nitroxide labels with a thiol-reactive linker such as methanthiosulfonate, like MTSL ((1-oxyl-2,2,5,5- tetramethylpyrroline-3-methyl)methanethiosulfonate) (Altenbach, Marti, Khorana, & Hubbell, 1990). An often used alternative is 3-maleimido-PROXYL, that carries a maleimide linker, which has a higher reduction stability compared to MTSL (Griffith & McConnell, 1966). Both linkage types present a high reactivity and specificity towards the cysteine residues and therefore yield quantitative labeling under very mild conditions. Other established labeling approaches are bioorthogonal labeling after introduction of an unnatural amino acid by genetic code extension (Fleissner et al., 2010; Widder, Berner, Summerer, & Drescher, 2019) or the direct incorporation of a spin-labeled amino acid into the protein (Schmidt, Borbas, Drescher, & Summerer, 2014). Beside nitroxides as EPR sensitive species, also complexed metal ions, like Gd^{3+} and Cu^{2+}, or trityl radicals have been applied as spin labels for proteins (Cunningham, Putterman, Desai, Horne, & Saxena, 2015; Potapov et al., 2010; Yang et al., 2012).

A series of distance distributions each between two spin labels located in a protein can be used to localize structural elements of that protein, applying multilateration. One spin label has to be attached to the structural element to be localized, i.e. the site of interest. The other spin label has to be located in a rigid and structurally well-defined domain of the protein, so that it can be used as reference site. Using several distance distributions between the unknown site of interest and different reference sites, the location of the site of interest can be triangulated in a GPS-like approach (see Fig. 1). Three distance restraints still give an ambiguous result with two possible solutions, that are symmetrical to each other. At least four distance restraints are required for an explicit result, more distance restraints improve the accuracy of the determined location. In literature this approach is also referred to as triangulation or trilateration (Abdullin & Schiemann, 2021; Reichenwallner et al., 2019). As in most cases more than three distances are used, hereafter it is termed multilateration. In the easiest approximation the mean distance of the distribution is taken and interpreted as the radius of a sphere around the respective reference site. The intersection point of all spheres around

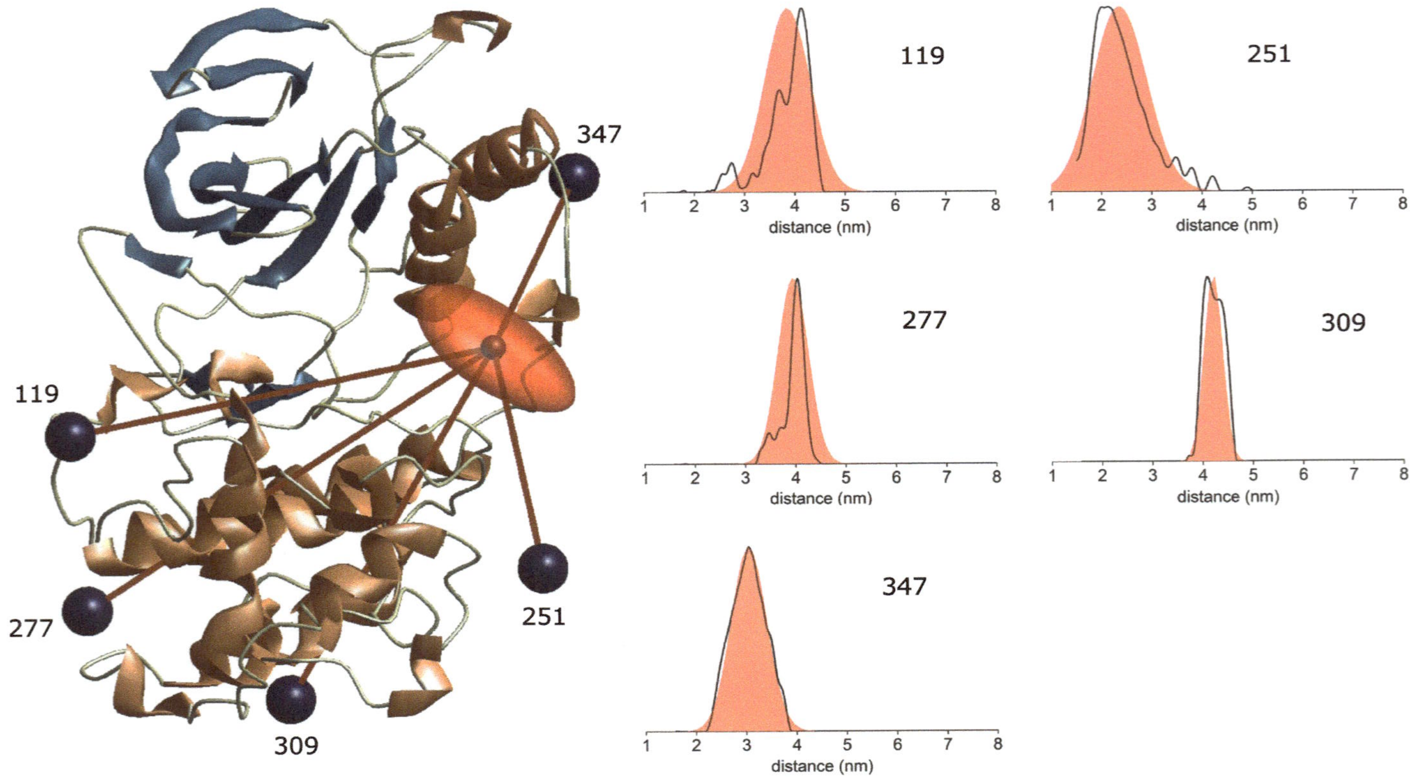

Fig. 1 See figure legend on opposite page.

the reference point, is the location of the site of interest. However, this approach neglects one of the main strength of distances obtained by DEER, the possibility to obtain distributions of distances. Therefore, a more elegant approach takes also the distribution into account, resulting in localization volumes, rather than only one intersection point (Fig. 1). The localization volume represents the probability density distribution of the spin label localization. The point of the maximum probability density is represented by the small black sphere within the red isosurface. The probability decreases towards the edges of the isosurface, while the displayed isosurface usually includes 50 % of the total probability. This direct translation of the full distance distribution to a three-dimensional probability density distribution makes multilateration an ideal tool for investigation of flexible or unstructured protein domains. However, some width of the distribution usually originates from the flexibility of the spin label linker, which also needs to be considered. The spin label at the reference sites as well as the spin label at the site of interest can contribute to that width. By performing a rotamer analysis, the possible spin label conformations can be predicted, taking into account the internal energetic states of the different conformations, as well as the interaction with the protein structure (Polyhach, Bordignon, & Jeschke, 2011). The resulting spin label rotamer distribution is then considered in the multilateration calculations. This complex appearing analysis is implemented in several software, for example MMM (Multiscale Modeling of Macromolecules, Jeschke, 2018), which runs in a Matlab environment or MtsslTrilaterate within MtsslSuite, a plugin for Pymol (Hagelueken, Abdullin, & Schiemann, 2015; Hagelueken, Abdullin, Ward, & Schiemann, 2013), making the analysis user-friendly.

Fig. 1 Multilateration of a spin label in the activation loop of the kinase p38α. Five distance distributions between the reference sites, labeled with MTSL, were used to localize the spin label at residue 172 in the activation loop. Left: The shown reference structure was obtained by X-ray diffraction (PDB: 3QUE). The dark purple spheres visualize the mean position of all possible spin label conformation of the reference sites. The red semi-transparent isosurface represents the localization volume of the spin label at position 172, which includes 50 % of the total probability. The numbers indicate the labeling positions of the reference labels. Right: The experimental distance distributions are shown in black together with a fitted Gaussian in red that was used for the multilateration. *Prepared with data provided from Roser, P., Weisner, J., Stehle, J., Rauh, D., & Drescher, M. (2020). Conformational selection: Vs. induced fit: Insights into the binding mechanisms of p38α MAP Kinase inhibitors.* Chemical Communications, *56(62), 8818–8821. 10.1039/d0cc02539a with permission from the Royal Society of Chemistry.*

One often used application of multilateration in literature is the localization of the copper ion in metal binding proteins. Cu^{2+} is a paramagnetic species, and in combination with site-specific spin labels introduced at different positions in the protein, the distance distribution can be determined via pulsed EPR measurements. In one of the first examples a triangulation between two Cu^{2+}-ions and one MTSL spin label was performed to identify the Cu^{2+}-binding residue in the EcoRi homodimer, a DNA-modifying enzyme (Yang et al., 2012). A first real multilateration, using MTSL spin labels at six reference sites to determine the Cu^{2+} localization in the blue copper protein azurin was performed by Abdullin et al. (Abdullin, Florin, Hagelueken, & Schiemann, 2015). The found position of copper was determined with only a shift of 1.9 Å compared to the crystal structure. In another study, labeling of an unnatural amino acid with nitroxide spin labels at four different positions was applied to determine the copper binding site in the prion protein (Evans, Pushie, Markham, Lee, & Millhauser, 2016), revealing an insight into its regulation mechanism. A significant improvement of the multilateration approach has been shown recently, by using a two armed linker. The spin label is based on a double histidine (dHis) motif, that complexes a Cu^{2+} ion. The paramagnetic center is conformationally unambiguous and therefore results in much narrower distance distributions than conventional spin labels (Gamble Jarvi, Bogetti, Singewald, Ghosh, & Saxena, 2021). With only four distance restraints, the location of the copper ion was determined with a 2 Å resolution (Gamble Jarvi, Cunningham, & Saxena, 2019). No metal ion, but the location of a lipid in the lipoxygenase enzyme SBL1 (soybean seed lipoxygenase-1) was investigated (Gaffney et al., 2012). The distance restraints between five different MTSL label positions and a spin-labeled lipid were determined with DEER. After multilateration of those, the polar-end of the lipid was localized at the entrance to the active site of the lipoxygenase SBL1.

Multilateration was not only used to localize a small ligand or metal ion in the protein, but also to find the relative orientation of whole protein domains towards each other. One example was presented for the membrane-binding ATPase EHD2 (Shah et al., 2014). One residue in the N-terminus of the protein was pinpointed by 4 distance restraints, showing that the N-terminus is folded into the hydrophobic pocket of the GTPase domain of EHD2. In a second example the conformation of the membrane-bound Bax dimer was investigated (Bleicken, Jeschke, Stegmueller, & Salvador-Gallego, 2014). Four different spin label positions in the C-terminal helices of the dimerization domain were localized, using between four and six distance restraints for each site of interest. The resulting spatial extension of the localization volumes indicates a significant flexibility of the C-terminal part, which might be important for Bax to flexibly bind to differently composed membranes. Multilateration was also applied to describe the flexibility of a protein loop by Peter et al. (2019). One spin label was attached in the activation loop of the kinase YopO and was multilaterated with respect to several fixed anchor points in the rigid and structurally resolved parts of the protein. The activation loop showed a very high flexibility in its location, but still resides mainly on top of the active site of the kinase domain, involved in the regulation of kinase activity.

Furthermore, the multilateration method is applicable to track conformational changes within proteins, which was shown successfully for the conformational change of the phytochrome photosensory module from the photoreceptor Cph2 (Assafa et al., 2018). The localization of two residues in the photosensory module was multilaterated based on four and five distance restraints. The conformations of two conformational states, Pr and Pfr, whose transition is triggered by red and far-red light, have been compared. Two distinguishable localization volumes were shown for the two states, indicating a conformational change, which might regulate the downstream signaling of the phytochrome receptor. Another study investigated the conformational changes in the Angiotensin II type I receptor, induced by different types of ligands (Wingler et al., 2019). For the multilateration only the main distance was considered, not the whole distribution. Nevertheless, they found a conformational ensemble of the intracellular region of the receptor, whose equilibrium is balanced by the different ligands.

One recently published study used the multilateration approach to investigate the activation loop in the p38α kinase. Not only the degree of flexibility of a structural element of the protein was shown, but the multilateration approach enabled additionally to resolve conformational changes (Roser et al., 2020). The activity of the kinase p38α, but also other kinases like the Aurora-A kinase and YopO, is regulated by conformational rearrangements of the activation loop. The activation loop contains a conserved motif at the N-terminus, the Asp-Phe-Gly (DFG) motif (Treiber & Shah, 2013). Two conformations of the activation loop were distinguished by X-ray crystallography, named after the orientation of this DFG motif. The catalytically active DFG-in state, where the aspartate residue points into the ATP-binding pocket to coordinate Mg-ATP and the inactive DFG-out state, where a phenylalanine residue is flipped and blocks the active site

(Roskoski, 2016). Crystal structures show specific conformational states with a very high resolution, but are not ideal to display flexibility in structures or conformational equilibria. Therefore, these conformational rearrangements and the spatial resolution of the different conformations was shown using DEER distance determination and subsequent multilateration.

One spin label was attached at the site of interest, in the activation loop near the DFG-motif. With this approach not the whole loop structure can be modeled. But examining this one specific amino acid close to the DFG motif has shown representative behavior to discriminate between the different conformations before (Roser, Weisner, Simard, Rauh, & Drescher, 2018; Simard & Rauh, 2014). To localize the position of this spin label in the activation loop, five reference sites were chosen, distributed in the structurally well-defined domains of the kinase. The reference sites were positioned around the activation loop to receive distance information from all directions, yielding a higher resolved multilateration result. Hence five p38α variants were designed, with the one labeling site in the activation loop and a second reference labeling site, where the latter was different for each variant. After site-directed mutagenesis to cysteine residues, the spin labels were introduced into the protein by labeling with MTSL. The distance distributions were determined via DEER spectroscopy either of p38α without an inhibitor (apo), or when bound to one of two inhibitor types. Type I inhibitors stabilize the activation loop in the DFG-in state, whereas type II inhibitors were shown to stabilize the DFG-out state (Pargellis et al., 2002; Simard et al., 2009).

The set of five distance restraints was used for multilateration, resulting in a sphere that represents the location of the spin label in the activation loop. The mulitlateration was conducted with a Gaussian fit of the distance restraints of the apo kinase and of the kinase bound to type I or type II inhibitors, yielding three different spheres (Fig. 2). Those spheres, shown relative to a structural model of the protein of interest, are now an ideal graphical representation of on the one hand the flexibility of the investigated structural element, represented by the size of the sphere. On the other hand, the position of the loop relative to the other protein domains is displayed for different conditions. Therefore, it is an optimal approach to track conformational changes of flexible structural elements in proteins.

Rotamer simulations combined with a structural model based on a crystal structure of p38α that has bound to the type I inhibitor SB203580 (PDB 1A9U), where also the activation loop was resolved, suggest that the spin labeling did not alter the protein conformations (Fig. 3).

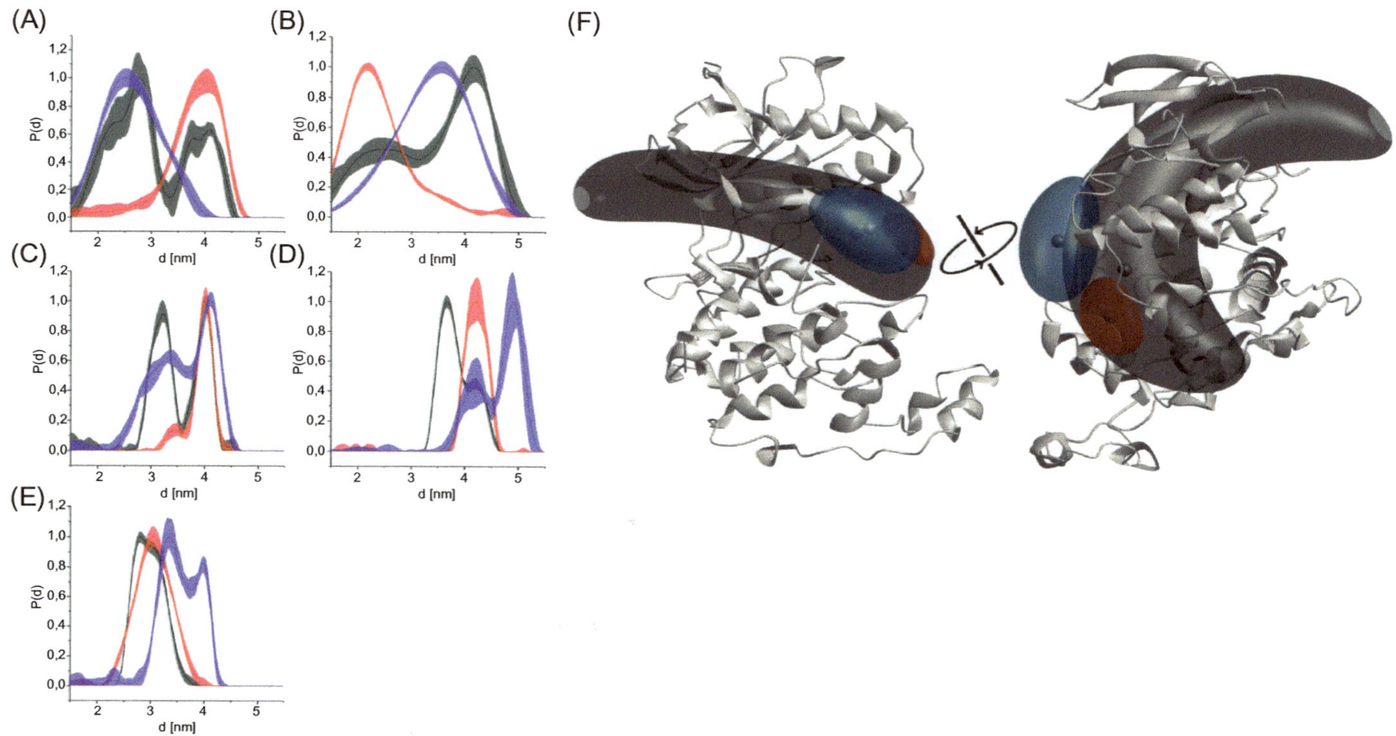

Fig. 2 See figure legend on next page.

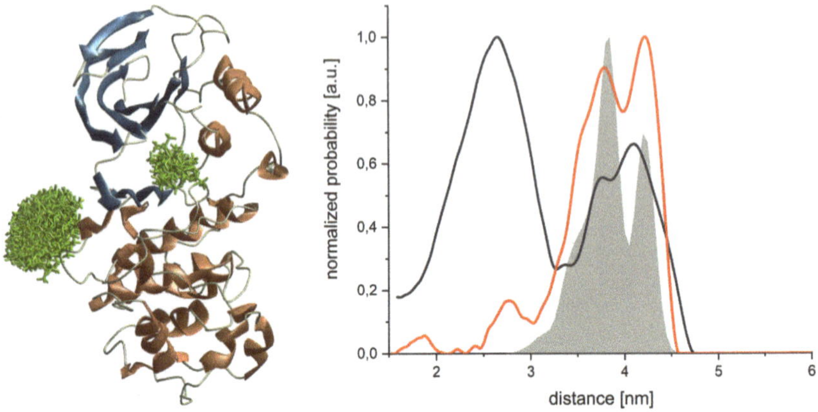

Fig. 3 On the left is shown the crystal structure (PDB 1A9U) of p38α bound to the inhibitor SB203580. In green are shown the spin label rotamers at position 119 and 172. On the right the corresponding distance distributions are displayed. In black is shown the experimental distance distribution without any inhibitor (apo), as red solid line in presence of a type I inhibitor SB203580 and as grey area the simulated distance distribution between the same residues. *Prepared with data provided from Roser, P., Weisner, J., Stehle, J., Rauh, D., & Drescher, M. (2020). Conformational selection: Vs. induced fit: Insights into the binding mechanisms of p38α MAP Kinase inhibitors.* Chemical Communications, 56*(62), 8818–8821. doi:10.1039/d0cc02539a with permission from the Royal Society of Chemistry.*

The comparable width between experimental and simulated distance distribution shows that there should be no additional flexibility from the protein, indicating that this inhibitor locks the activation loop in a quite rigid conformation. Compared to that, the experimental distance distribution of p38α without any inhibitor shows a much broader distribution. This is a hint that the protein itself, here, the activation loop has an inherent flexibility, which adds to the original distribution width of the spin label site chain.

Fig. 2 Activation loop of p38α analyzed by DEER distance determination. Left: Distance distributions of five doubly labeled p38α constructs without inhibitors (black), with a type I inhibitor (red) and a type II inhibitor (blue). Each construct had the same labeling site in the activation loop and a different labeling site distributed in the kinase (A–E). Right: Multilateration result of the distance distributions A–E without inhibitors (black volume), with a type I inhibitor (red volume) and a type II inhibitor (blue volume). The volumes represent 50% of the total probability of the location of the spin label in the activation loop (F). *Reproduced from Roser, P., Weisner, J., Stehle, J., Rauh, D., & Drescher, M. (2020). Conformational selection: Vs. induced fit: Insights into the binding mechanisms of p38α MAP Kinase inhibitors.* Chemical Communications, 56*(62), 8818–8821. doi:10.1039/d0cc02539a with permission from the Royal Society of Chemistry.*

In the apo state of p38α the activation loop gives rise to a broad multimodal distance distribution, for all five protein variants (see Fig. 2A–E, grey distance distributions). This supports the hypothesis, that the apo activation loop lies in a conformational equilibrium between the DFG-in and DFG-out state (Roser et al., 2018; Ruff et al., 2018). The multimodality is not ideally reflected by the grey sphere, representing the multilateration result (Fig. 2F), as they were fitted with a single Gaussian for the multilateration. Nevertheless, it shows a large volume of localization, confirming the great flexibility of the activation loop in the apo state, which was also found for other kinase's activation loops (Burgess, Grazia Concilio, Bayliss, & Fielding, 2016; Peter et al., 2019).

As representative for type I inhibitors, data for SB203580 is shown (Fig. 2A–E, red distance distributions), while all type I inhibitors feature similar distance restraints. Determined by multilateration, the probability volume of the activation loop bound to the type I inhibitor lies within the volume of the apo kinase, but occupies only a small part of it (Fig. 2F, red volume). The binding mechanism, where a ligand restricts the protein in only a part of the already existing conformation, is called a conformational selection mechanism. Considering the localization volumes of the spin label in the activation loop, the binding of the type I inhibitors might therefore be categorized into that type of binding mechanism.

The type II inhibitors induced a broader distance distribution, which is also less consistent between the four inhibitors, than it was the case with the type I inhibitors. One shoulder of the distance distribution overlaps with the distance distribution that corresponds to the DFG-in conformation (Fig. 2A–E, blue distance distributions). This supports the conclusion that the type II inhibitors induce not a complete DFG-out conformation, but an equilibrium between DFG-in and DFG-out. To visualize the pure DFG-out conformation, the distance distributions representing the DFG-in conformation were subtracted from the distances induced by the type II inhibitors, resulting in distances that represent only the DFG-out conformation. Multilateration of those, resulted in a localization volume that is again much smaller than the apo volume and only partially overlaps with it (Fig. 2F, blue volume), an observation that could not have been drawn by looking at the distance distributions alone. As the distance distributions between the four type II inhibitors were not completely homogeneous, the multilateration was repeated for all four, resulting in almost identical volumes. This shows, that even if the distance distributions were not identical, a three-dimensional visualization of the conformation revealed, that they still describe the same conformation. This type of binding could be an induced-fit like mechanism rather than conformational selection, as a new conformation is induced by binding of the type II inhibitor, which was not present in the apo protein.

In summary it was demonstrated that multilateration is an ideal approach to track conformational rearrangements of very flexible structural elements, like the activation loop of the p38α kinase. It was possible to visualize the degree of flexibility of the activation loop, together with the localization of different conformations induced by small molecule ligands. Using that, it was possible to resolve the binding mechanism of those two inhibitor types, an important step in the design of new tailored drugs to regulate out of order kinase activity.

2. Experimental guidelines

2.1 Choice of labeling sites

Choosing the labeling sites needs to be considered carefully, as this might decide over the quality of the multilateration. First, some general factors should be noted. The labeling sites should be at the surface of the protein, best replacing amino acid residues that point outwards, so that the labeling reagent can easily reach the labeling site. The cysteine mutation and also the attached spin label should not alter the protein structure. This could be tested for example with circular dichroism (CD) spectroscopy, comparing the wild-type with the labeled protein. If the protein of interest has a known functionality, which can be tested in a biochemical assay, this also should be considered. Also the positioning of the labels in the protein is substantial. The expected distances between the site of interest and the reference sites must be in the range of DEER, ideally between 2 and 5 nm, if feasible. The reference sites should be evenly distributed around the site of interest to get the best multilateration result. For a good result at least five reference site are recommended. If a conformational change of the site of interest should be investigated, it needs to be ensured that the reference sites stay fixed during that change. This could be checked experimentally by measuring the distances between some reference sites before and after the investigated conformational change. If there are sufficient structural models available, that contain the reference sites and were resolved in comparable conditions like the planned experiment, a structure alignment can also show the structural stability of the reference sites. As an example is shown the

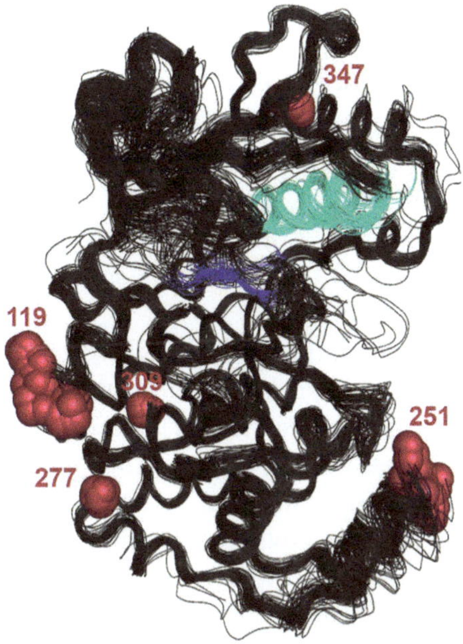

Fig. 4 Alignment of 239 p38α (uniprot ID: Q16539) crystal structures available in the protein data bank. The five used labeling sites are shown as pink spheres and appear to be mainly consistent in their position for all proposed structures. Created using PyMOL (Schrodinger, 2015). *Reproduced from Roser, P., Weisner, J., Stehle, J., Rauh, D., & Drescher, M. (2020). Conformational selection: Vs. induced fit: Insights into the binding mechanisms of p38α MAP Kinase inhibitors.* Chemical Communications, *56(62), 8818–8821. doi:10.1039/d0cc02539a with permission from the Royal Society of Chemistry.*

alignment of 239 crystal structures of p38α in Fig. 4. The pink spheres, representing the reference sites show no large differences in the position of their C-alpha atoms, whereas the DFG motif in the activation loop, colored in dark blue, shows a high structural diversity.

2.2 Site-directed spin labeling

The labeling protocol described here is optimized for the spin label MTSL ((1-oxyl-2,2,5,5-tetramethylpyrroline-3-methyl)methanethiosulfonate) and the protein p38α. By varying the buffer, the labeling time, temperature and the molar equivalents of the label, the protocol can be adapted to most proteins and cysteine based labels.

1. Prepare a stock solution of MTSL with a concentration of 25 mM in dimethyl sulfoxide (DMSO). The stock solution should be stored at −80°C and not longer than several months, as the MTSL label can form dimers, and would not be reactive anymore.
2. Dissolve the protein to a concentration of 50 μM in 20 mM Tris buffer containing 200 mM NaCl, 5% v/v glycerol, pH 7.4.
3. Add six molar equivalents of the label stock solution to the protein. Pay attention that the DMSO concentration stays below 5% v/v during all steps.
4. Incubate the labeling solution at 4 °C overnight while gently shaking.
5. To remove the unreacted spin label after reaction, wash the labeling solution using Vivaspin 4 centrifugal filters (10 kDa MWCO, Sartorius) 6–8x at 4 °C. Check the flow through with continuous wave (CW) EPR spectroscopy and continue washing until no signal of free spin label can be detected anymore. Then recover the protein from the centrifugal filters as instructed by the manufacturer and store it at −80 °C.

To control for successful labeling, mass spectrometry or determination of the labeling yield should be applied. In order to derive the labeling yield, we suggest to determine the spin concentration using an EMXnano X-band spectrometer (Bruker) and the reference-free spin counting module in the Xenon software (Bruker). The protein concentration can be determined with a BioPhotomer® D30 (Eppendorf). Non-quantitative labeling of one or both sites is in general no problem for the DEER experiment, if the signal intensity and the modulation depth of the DEER time trace are sufficient high. Over-stoichiometric labeling might indicate unspecific labeling of a third residue, e.g. some buried native cysteine residues. For unspecific labeling can be checked by labeling a variant of the same protein without the introduced cysteine residues. If exactly the same reaction conditions are used and no spin label signal can be detected after the removal of unreacted spin label, unspecific labeling can be excluded. The presence of more than two spin labels per protein should be absolutely avoided, as it makes a profound data analysis of the resulting distance distribution very complicated.

2.3 The DEER experiment and analysis

The following procedure is described for a Bruker ELEXSYS-E580 Q-band spectrometer equipped with an arbitrary waveform generator (AWG, Bruker Biospin), a QT-II resonator (ER5106QT-2, Bruker Biospin) and a 150 W TWT amplifier (Applied Systems Engineering). If no AWG setup is available, we suggest to use a protocol described previously by Schmidt, Fedoseev, Summerer, & Drescher (2015).

2.3.1 Sample preparation

The optimal spin concentration and therewith the optimal protein concentration for the DEER experiment depends on several factors. Higher concentrations improve the absolute signal intensity, needed to resolve long distances (Jeschke, 2012). A high spin concentration also increases instantaneous diffusion and the intermolecular dipolar coupling contribution, which then again limits the detection of longer distances (de Vera, Blackburn, Galiano, & Fanucci, 2013; Jeschke & Polyhach, 2007). So the optimal concentration also depends on the expected distance. For a start, we recommend using a protein concentration around 50 μM for a comfortable signal intensity, but with this setup also single-digit μM concentrations might be sufficient to achieve an acceptable signal-to-noise ratio. The protein should be dissolved in deuterated buffer containing 20% v/v deuterated glycerol (glycerol-d_8, Sigma-Aldrich). The DMSO concentration should be below 5% at all time during sample preparation. After the individual sample preparation, the sample is filled in a quartz glass tube (inner diameter: 2 mm, outer diameter: 3 mm, length: 12 cm) with a volume of 60 μL using a microliter syringe (Hamilton 700 series). After quick centrifugation with a hand centrifuge to remove air bubbles, the sample is snap frozen in liquid nitrogen and ideally directly transferred into the resonator, which has been cooled to a temperature of 50 K. Alternatively, the sample can be stored at -80 °C, which might lead to a slight decrease in signal intensity, due to loss of the glass characteristic of the water-glycerol mixture. Thawing and re-freezing of the sample directly before the measurement solves that problem, but protein integrity must be ensured in that case.

2.3.2 Optimizing parameters and setting up the 4-pulse DEER experiment

The resonator should be held at a fixed temperature of 50 K. A Hahn echo sequence ($\pi/2 - \tau - \pi - \tau -$ echo) is applied at a frequency of 34 GHz, full microwave power (0 dB attenuation and running TWT) and a magnetic field of around 12100 G to observe the echo of the nitroxide spin. Then the signal phase and the shot repetition time can be optimized to maximize the echo intensity. To determine the magnetic field value that corresponds to the maximum of the nitroxide spectrum at the pump frequency, measure an echo-detected field sweep (EDFS) using the Hahn echo sequence and a magnetic field range of around 12,000–12,200 G (see Fig. 5A). As pump pulse for the DEER experiment, broadband shaped pulses significantly improve the modulation depth compared to rectangular shaped pulses. Due to a broader inversion profile, they excite a larger fraction of the nitroxide spectrum, illustrated in Fig. 5A in dark blue (Doll & Jeschke, 2017; Doll, Pribitzer, Tschaggelar, & Jeschke, 2013; Spindler, Glaser, Skinner, & Prisner, 2013; Spindler, Schöps, Kallies, Glaser, & Prisner, 2017). It has been shown that the parameters for the best performing pulses depend not only on the type of experiment or the spin label, but also on the sample concentration and the dipolar evolution time (Scherer, Tischlik, Weickert, Wittmann, & Drescher, 2020). A symmetric hyperbolic secant pulse HS{1,1} with a length t_p of 100 ns, a β parameter of $8/t_p$ and a frequency sweep width of 110 MHz performs best with spin concentrations around 60 μM or a trace length longer than 5 μs. An asymmetric HS{1,6} pulse could create a higher modulation depth, but also a stronger background decay. This produces more noise with increasing evolution time, which is a disadvantage for longer DEER time traces. So depending on the spin concentration of the sample and the desired evolution time, the best performing broadband-shaped pulse must be chosen individually. As for most protein samples the expected distance range is above 3 nm, we recommend using the symmetric HS{1,1} pulse. For the observer pulse a Gaussian shape with a frequency offset of 90 MHz to the pump frequency performed best in combination with a broadband shaped pump pulse. The pulse length is determined with an inversion recovery pulse sequence ($\pi - \tau$ - $\pi/2 - \tau - \pi - \tau -$ echo) at the observer frequency and lies around 70 ns for the π pulse.

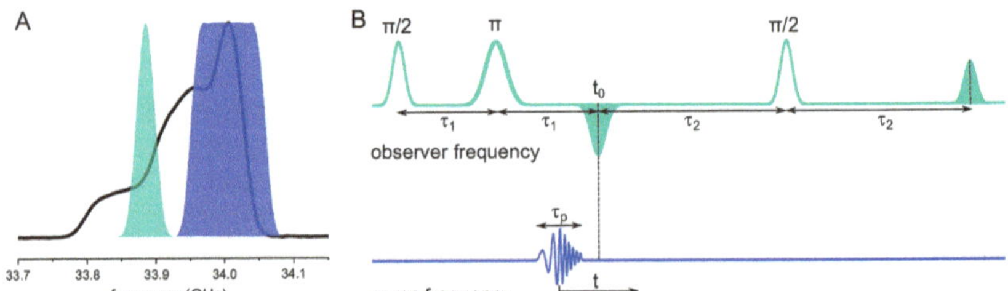

Fig. 5 The 4-pulse DEER experiment. The inversion profiles of a Gaussian observer π-pulse (light blue) and a HS pump pulse (dark blue) are shown in perspective to an EDFS of a nitroxide radical (black) (A). The pulse sequence for the observer frequency (light blue) and the pump frequency (dark blue) of the 4-pulse DEER experiment is represented (B).

After the pulse parameters are optimized, the DEER experiment can be set up. The pulse sequence for the 4-pulse DEER experiment is shown in Fig. 5B. The τ_2 parameter should be determined individually for each sample, depending on the range of the expected distance and the signal intensity. The longer τ_2, the larger is the distance that can be detected. For a τ_2 of 5 μs the shape of the distance distribution is reliable up to 4 nm, the width is reliable up to 5.4 nm and the mean distance up to 6.5 nm. The maximal distance scales with the cubic root of τ_2 (Jeschke, 2012). However, with longer τ_2 the echo intensity decreases due to longitudinal relaxation and hence it should be ensured that the signal-to-noise ratio is still acceptable. For τ_1 we recommend a length of 400 ns and for the time step rate Δt a value between 8 and 20 ns. The shot repetition time should be around 5 ms and is optimized to prevent saturation of the longitudinal relaxation. An 8-step phase cycle according to Tait & Stoll, 2016 is advisable and an 8-step nuclear modulation averaging to suppress nuclear modulation artefacts. The acquisition time is highly dependent on the signal intensity and can lie between 1 h and 72 h. If the phase of the spectrometer is not stable over the measurement times, it might be useful to save each scan separately in a 2D experiment, to be able to correct the phase afterwards for each scan individually.

2.3.3 Analysis of DEER data

1. If the DEER trace was recorded in a two-dimensional experiment (every slice saved separately), first do a phase correction of every slice, then average all slices, e.g. using Matlab (MathWorks®).
2. For DEER data analysis different software is available. Here is described the procedure for DeerAnalysis (Jeschke et al., 2006), which can be downloaded in a Matlab environment following the instructions in the DeerAnalysis user manual (http://www.epr.ethz.ch/software).
3. For extracting the distance distribution follow the instructions in the user manual, completing the following steps:
   a. Determine the zero time (maximum) of the DEER raw data.
   b. Cut the end of the DEER trace, if there is some artefact, e.g. a sudden increase for the last few hundreds nanoseconds.
   c. Do a background correction by choosing a background model (homogeneous, polynomial or experimental) and a starting point of the background fit (background start). The DEER raw data is then divided by the fitted background, resulting in the form factor.
   d. Fit the form factor using a model free approach (Tikhonov regularization), a neural network ensemble (DEERnet) or if already some

information is known about the distance distribution use a model-based method (e.g. different Gaussian models are available).

e. Validate the resulting distance distribution. Vary at least the background start and the white noise, as suggested in the user manual of DeerAnalysis.

f. Save the final distance distribution.

2.4 Multilateration

To perform a multilateration based on DEER distance restraints, a structural model of the studied protein as a starting point is required. This is ideally an experimentally derived structure, for example obtained by X-ray crystallography or NMR. Crucial is, that the position of the reference labeling sites is resolved, as their relative position to each other must be known to yield a reliable multilateration result. We recommend using the localization tool in the software MMM (Multiscale modeling of macromolecular systems, Jeschke, 2018) for the multilateration.

2.4.1 Preparation of the constraint file

To perform the multilateration in MMM, an input file is needed, a so called "Constraint file". In this file the experimental DEER distance data is listed, which will be used for the multilateration. The Constraint file can be prepared following these instructions:

1. The distance distributions can be coded in two ways. One possibility is using parameters of the Gaussian curve. The experimental distributions are fitted with a Gaussian curve to obtain the parameters mean distance and standard deviation. If the distance distribution is multimodal and only one of the peaks indicates the conformation that shall be multilaterated, this might be applied. The fitting can be realized with any data analysis software like Matlab, Microsoft Excel or Origin (OriginLab®). Alternatively, the numerical distance distribution in the DeerAnalysis output format can be used. Then the filename is to be mentioned in the constraint file.
2. A PDB code of a structure of the protein of interest is needed, containing coordinates for all the reference sites.
3. Open an empty .txt file and type the following commands, starting with the hash character #:

   # PDB: This command specifies reference structure for the multilateration.

ENSEMBLE: This command is optional. It defines the number of the small dark spheres, that are sampled to fill the red probability volume and their diameter (in Å). The default value is 1 sphere, with a diameter of 2 Å.

OUTPUT: This command is optional. It defines the name of the output pdb file.

LOCATE: Here the constraints are defined. At least three arguments are needed. (1) The address of the spin-labeled reference site, either given as the residue number or the MMM address. (2) The experimental mean distance in nanometer. (3) The standard deviation of distribution in nanometer. If the file of the numerical distance distribution of DeerAnalysis shall be used, the filename is written as a fourth argument. The file must be in the same folder like the constraint file or the pathname must be added additionally.

END: This command ends the constraint section.

4. Enter in the line below the #LOCATE command the experimental distance constraints, using the three or four arguments mentioned above. An exemplary constraint file can be found in the MMM manual.

The five terms after #LOCATE are used to further describe the multilateration. The first entry tags the distribution volume in the graphical display with that name, the second entry defines the percentage of probability that is displayed as volume, the default value is 0.5, the third entry defines the type of label, the default is MTSL and the fourth entry defines the temperature of the rotamer library, here the default value is 298 K. Except for the type of label, which must be adapted, depending on the used spin label, the default values can be used in most cases. The fifth value defines the display type, either full, constraints, nodes or none. "Full" means the probability volume is shown, the maximal probability of the electron spin of the reference labels, displayed as purple spheres and red lines connecting the purple spheres with the maximal probability of the multilaterated site (see Fig. 1). "Constraints" shows only the distribution volume with the lines, and "nodes" the volume in combination with the purple spheres.

2.4.2 Multilateration in MMM

1. Load the a PDB structure in MMM.
2. Load the prepared Constraint file via the "Display" menu with the method "Localization". The multilateration is now running and the result is shown in the Model window.

3. To verify the multilateration, have a look at the message board after the calculation. There is given a value for the "highest location probability" and the "maximum relative probability at the density cube border." The perfect solution for highest location probability would be 1, as it indicates that all distances are simultaneously fulfilled at any point in space. In reality of course smaller values have to be expected, but it should be above 0.25. The probability at the density cube boarder should not be larger than 10%, otherwise there have not been enough experimental constraint. If the red probability sphere has no closed surface, but is open to one side it is also a sign, that more experimental constraints are needed to improve the result.
4. If different multilateration results, based on the same protein, shall be compared, several multilateration can be conducted successively. After the calculation of the first multilateration, another constraint file can be loaded via the "Localization" method. This can be repeated as often as desired. The probability volumes are displayed transparent over each other in different colors (see Fig. 6).
5. In the "Density" menu each multilateration volume can be addressed via its tag, that was defined in the constraint file. The color, the transparency

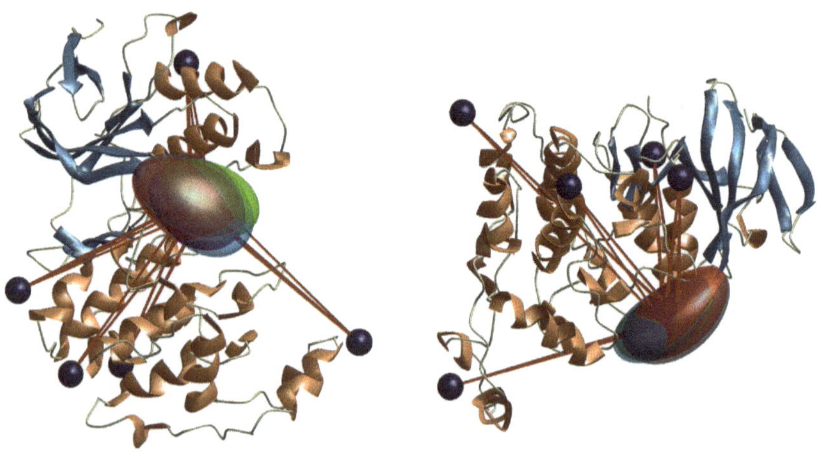

Fig. 6 Multilateration results for p38α with five different type II inhibitors, received after five successive multilaterations with MMM. All localization spheres are placed on top of each other using different colors and are semi-transparent. *Reproduced from Roser, P., Weisner, J., Stehle, J., Rauh, D., & Drescher, M. (2020). Conformational selection: Vs. induced fit: Insights into the binding mechanisms of p38α MAP Kinase inhibitors.* Chemical Communications, *56(62), 8818–8821. doi:10.1039/d0cc02539a with permission from the Royal Society of Chemistry.*

and the contour level, meaning the percentage of the total probability that is represented by the volume, can then be adapted.

6. The multilateration can be saved as mat file to be reloaded with MMM or exported as pdb file to be displayed in any pdb viewer, like Chimera or PyMol. In the detached model window of MMM the visualization can also be directly exported as any typical image format, like jpeg, png, tiff or emf.

References

Abdullin, D., Florin, N., Hagelueken, G., & Schiemann, O. (2015). EPR-based approach for the localization of paramagnetic metal ions in biomolecules. *Angewandte Chemie, 127*(6), 1847–1851. https://doi.org/10.1002/ange.201410396.

Abdullin, D., & Schiemann, O. (2021). Localization of metal ions in biomolecules by means of pulsed dipolar EPR spectroscopy. *Dalton Transactions, 50*(3), 808–815. https://doi.org/10.1039/d0dt03596c.

Altenbach, C., Marti, T., Khorana, H. G., & Hubbell, W. L. (1990). Transmembrane protein structure: Spin labeling of bacteriorhodopsin mutants. *Science, 248*(4959), 1088–1092. https://doi.org/10.1126/science.2160734.

Assafa, T. E., Anders, K., Linne, U., Essen, L. O., & Bordignon, E. (2018). Light-driven domain mechanics of a minimal phytochrome photosensory module studied by EPR. *Structure, 26*(11), 1534–1545.e4. https://doi.org/10.1016/j.str.2018.08.003.

Benjin, X., & Ling, L. (2020). Developments, applications, and prospects of cryo-electron microscopy. *Protein Science, 29*(4), 872–882. https://doi.org/10.1002/pro.3805.

Bleicken, S., Jeschke, G., Stegmueller, C., & Salvador-Gallego, R. (2014). Structural model of active bax at the membrane. *Molecular Cell, 56*(4), 496–505. https://doi.org/10.1016/j.molcel.2014.09.022.Structural.

Bonomi, M., Heller, G. T., Camilloni, C., & Vendruscolo, M. (2017). Principles of protein structural ensemble determination. *Current Opinion in Structural Biology, 42*, 106–116. https://doi.org/10.1016/j.sbi.2016.12.004.

Braun, T. S., Stehle, J., Kacprzak, S., Carl, P., Höfer, P., Subramaniam, V., et al. (2021). Intracellular protein-lipid interactions studied by rapid-scan electron paramagnetic resonance spectroscopy. *Journal of Physical Chemistry Letters, 12*(9), 2471–2475. https://doi.org/10.1021/acs.jpclett.0c03583.

Burgess, S. G., Grazia Concilio, M., Bayliss, R., & Fielding, A. J. (2016). Detection of ligand-induced conformational changes in the activation loop of aurora-A kinase by PELDOR spectroscopy. *ChemistryOpen, 5*(6), 531–534. https://doi.org/10.1002/open.201600101.

Cornish, V. W., Benson, D. R., Altenbach, C. A., Hideg, K., Hubbell, W. L., & Schultz, P. G. (1994). Site-specific incorporation of biophysical probes into proteins. *Proceedings of the National Academy of Sciences of the United States of America, 91*(8), 2910–2914. https://doi.org/10.1073/pnas.91.8.2910.

Cunningham, T. F., Putterman, M. R., Desai, A., Horne, W. S., & Saxena, S. (2015). The double-histidine Cu^{2+}-binding motif: a highly rigid, site-specific spin probe for electron spin resonance distance measurements. *Angewandte Chemie—International Edition, 54*(21), 6330–6334. https://doi.org/10.1002/anie.201501968.

Dalaloyan, A., Martorana, A., Barak, Y., Gataulin, D., Reuveny, E., Howe, A., et al. (2019). Tracking conformational changes in calmodulin in vitro, in cell extract, and in cells by electron paramagnetic resonance distance measurements. *ChemPhysChem, 20*(14), 1860–1868. https://doi.org/10.1002/cphc.201900341.

de Vera, I. M. S., Blackburn, M. E., Galiano, L., & Fanucci, G. E. (2013). Pulsed EPR distance measurements in soluble proteins by site-directed spin labeling (SDSL). *Current Protocols in Protein Science*, *2013*(November), 17.17.1–17.17.29. https://doi.org/10.1002/0471140864.ps1717s74.

Doll, A., & Jeschke, G. (2017). Wideband frequency-swept excitation in pulsed EPR spectroscopy. *Journal of Magnetic Resonance*, *280*, 46–62. https://doi.org/10.1016/j.jmr.2017.01.004.

Doll, A., Pribitzer, S., Tschaggelar, R., & Jeschke, G. (2013). Adiabatic and fast passage ultra-wideband inversion in pulsed EPR. *Journal of Magnetic Resonance*, *230*, 27–39. https://doi.org/10.1016/j.jmr.2013.01.002.

Evans, E. G. G. B., Pushie, M. J., Markham, K. A. A., Lee, H. W., & Millhauser, G. L. L. (2016). Interaction between prion protein's copper-bound octarepeat domain and a charged C-terminal pocket suggests a mechanism for N-terminal regulation. *Structure*, *24*(7), 1057–1067. https://doi.org/10.1016/j.str.2016.04.017.

Fleissner, M. R., Brustad, E. M., Kálai, T., Altenbach, C., Cascio, D., Peters, F. B., et al. (2010). Site-directed spin labeling of a genetically encoded unnatural amino acid. *Proceedings of the National Academy of Sciences of the United States of America*, *107*(12), 5693. https://doi.org/10.1073/pnas.1000908107.

Gaffney, B. J., Bradshaw, M. D., Frausto, S. D., Wu, F., Freed, J. H., & Borbat, P. (2012). Locating a lipid at the portal to the lipoxygenase active site. *Biophysical Journal*, *103*(10), 2134–2144. https://doi.org/10.1016/j.bpj.2012.10.002.

Gamble Jarvi, A., Bogetti, X., Singewald, K., Ghosh, S., & Saxena, S. (2021). Going the dHis-tance: Site-directed Cu2+labeling of proteins and nucleic acids. *Accounts of Chemical Research*. https://doi.org/10.1021/acs.accounts.0c00761.

Gamble Jarvi, A., Cunningham, T. F., & Saxena, S. (2019). Efficient localization of a native metal ion within a protein by Cu^{2+}-based EPR distance measurements. *Physical Chemistry Chemical Physics*, *21*(20), 10238–10243. https://doi.org/10.1039/c8cp07143h.

Griffith, O. H., & McConnell, H. M. (1966). A nitroxide-maleimide spin label. *Proceedings of the National Academy of Sciences of the United States of America*, *55*(1), 8–11. https://doi.org/10.1073/pnas.55.1.8.

Hagelueken, G., Abdullin, D., & Schiemann, O. (2015). MtsslSuite: Probing biomolecular conformation by spin-labeling studies. In *Vol. 563. Methods in enzymology* (1st ed.). Elsevier Inc. https://doi.org/10.1016/bs.mie.2015.06.006.

Hagelueken, G., Abdullin, D., Ward, R., & Schiemann, O. (2013). MtsslSuite: In silico spin labelling, trilateration and distance-constrained rigid body docking in PyMOL. *Molecular Physics*, *111*(18–19), 2757–2766. https://doi.org/10.1080/00268976.2013.809804.

Hett, T., Zbik, T., Mukherjee, S., Matsuoka, H., Wolfgang, B., Klose, D., et al. (2021). *Spatiotemporal resolution of conformational changes in biomolecules by combining pulsed electron—Electron double resonance spectroscopy with microsecond freeze-hyperquenching*. https://doi.org/10.1021/jacs.1c01081.

Hubbell, W. L., Cafiso, D. S., & Altenbach, C. (2000). Identifying conformational changes with site-directed spin labeling. *Nature Structural Biology*, 7(9), 735–739. https://doi.org/10.1038/78956.

Jeschke, G. (2002). Distance measurements in the nanometer range by pulse EPR. *ChemPhysChem*, *3*(11), 927–932. https://doi.org/10.1002/1439-7641(20021115)3:11<927::AID-CPHC927>3.0.CO;2-Q.

Jeschke, G. (2012). DEER distance measurements on proteins. *Annual Review of Physical Chemistry*, *63*(1), 419–446. https://doi.org/10.1146/annurev-physchem-032511-143716.

Jeschke, G. (2018). MMM: A toolbox for integrative structure modeling. *Protein Science*, *27*(1), 76–85. https://doi.org/10.1002/pro.3269.

Jeschke, G., Chechik, V., Ionita, P., Godt, A., Zimmermann, H., Banham, J., et al. (2006). DeerAnalysis2006—A comprehensive software package for analyzing pulsed ELDOR data. *Applied Magnetic Resonance*, *30*(3–4), 473–498. https://doi.org/10.1007/BF03166213.

Jeschke, G., & Polyhach, Y. (2007). Distance measurements on spin-labelled biomacromolecules by pulsed electron paramagnetic resonance. *Physical Chemistry Chemical Physics*, *9*(16), 1895–1910. https://doi.org/10.1039/b614920k.

Kim, M. S., Pinto, S. M., Getnet, D., Nirujogi, R. S., Manda, S. S., Chaerkady, R., et al. (2014). A draft map of the human proteome. *Nature*, *509*(7502), 575–581. https://doi.org/10.1038/nature13302.

Kucher, S., Elsner, C., Safonova, M., Maffini, S., & Bordignon, E. (2021). In-cell double electron–electron resonance at nanomolar protein concentrations. *The Journal of Physical Chemistry Letters*, *12*(14), 3679–3684. https://doi.org/10.1021/acs.jpclett.1c00048.

McHaourab, H. S., Steed, P. R., & Kazmier, K. (2011). Toward the fourth dimension of membrane protein structure: Insight into dynamics from spin-labeling EPR spectroscopy. *Structure*, *19*(11), 1549–1561. https://doi.org/10.1016/j.str.2011.10.009.

Pannier, M., Veit, S., Godt, A., Jeschke, G., & Spiess, H. W. (2000). Dead-time free measurement of dipole-dipole interactions between electron spins. *Journal of Magnetic Resonance*, *142*(2), 331–340. https://doi.org/10.1006/jmre.1999.1944.

Pargellis, C., Tong, L., Churchill, L., Cirillo, P. F., Gilmore, T., Graham, A. G., et al. (2002). Inhibition of p38 MAP kinase by utilizing a novel allosteric binding site. *Nature Structural Biology*, *9*(4), 268–272. https://doi.org/10.1038/nsb770.

Peter, M. F., Tuukkanen, A. T., Heubach, C. A., Selsam, A., Duthie, F. G., Svergun, D. I., et al. (2019). Studying conformational changes of the Yersinia type-III-secretion effector YopO in solution by integrative structural biology. *Structure*, *27*(9), 1416–1426. e3. https://doi.org/10.1016/j.str.2019.06.007.

Polyhach, Y., Bordignon, E., & Jeschke, G. (2011). Rotamer libraries of spin labelled cysteines for protein studies. *Physical Chemistry Chemical Physics*, *13*(6), 2356–2366. https://doi.org/10.1039/c0cp01865a.

Potapov, A., Yagi, H., Huber, T., Jergic, S., Dixon, N. E., Otting, G., & Goldfarb, D. (2010). Nanometer-scale distance measurements in proteins using Gd^{3+} spin labeling. *Journal of the American Chemical Society*, *132*(26), 9040–9048. https://doi.org/10.1021/ja1015662.

Reichenwallner, J., Hauenschild, T., Schmelzer, C. E. H., Hülsmann, M., Godt, A., & Hinderberger, D. (2019). Fatty acid triangulation in albumins using a landmark spin label. *Israel Journal of Chemistry*, *1–17*. https://doi.org/10.1002/ijch.201900073.

Roser, P., Schmidt, M. J., Drescher, M., & Summerer, D. (2016). Site-directed spin labeling of proteins for distance measurements: In vitro and in cells. *Organic and Biomolecular Chemistry*, *14*(24), 5468–5476. https://doi.org/10.1039/c6ob00473c.

Roser, P., Weisner, J., Simard, J. R., Rauh, D., & Drescher, M. (2018). Direct monitoring of the conformational equilibria of the activation loop in the mitogen-activated protein kinase p38α. *Chemical Communications*, *54*(85), 12057–12060. https://doi.org/10.1039/c8cc06128a.

Roser, P., Weisner, J., Stehle, J., Rauh, D., & Drescher, M. (2020). Conformational selection: Vs. induced fit: Insights into the binding mechanisms of p38α MAP Kinase inhibitors. *Chemical Communications*, *56*(62), 8818–8821. https://doi.org/10.1039/d0cc02539a.

Roskoski, R. (2016). Classification of small molecule protein kinase inhibitors based upon the structures of their drug-enzyme complexes. *Pharmacological Research*, *103*, 26–48. https://doi.org/10.1016/j.phrs.2015.10.021.

Ruff, E. F., Muretta, J. M., Thompson, A. R., Lake, E. W., Cyphers, S., Albanese, S. K., et al. (2018). A dynamic mechanism for allosteric activation of aurora kinase a by activation loop phosphorylation. *ELife*, *7*, 1–22. https://doi.org/10.7554/eLife.32766.

Scherer, A., Tischlik, S., Weickert, S., Wittmann, V., & Drescher, M. (2020). Optimising broadband pulses for DEER depends on concentration and distance range of interest. *Magnetic Resonance*, *1*(1), 59–74. https://doi.org/10.5194/mr-1-59-2020.

Schmidt, M. J., Borbas, J., Drescher, M., & Summerer, D. (2014). A genetically encoded spin label for electron paramagnetic resonance distance measurements. *Journal of the American Chemical Society*, *136*(4), 1238–1241. https://doi.org/10.1021/ja411535q.

Schmidt, M. J., Fedoseev, A., Summerer, D., & Drescher, M. (2015). Genetically encoded spin labels for in vitro and in-cell EPR studies of native proteins. In *Vol. 563. Methods in enzymology* (1st ed.). Elsevier Inc. https://doi.org/10.1016/bs.mie.2015.05.023.

Schmidt, T., Wälti, M. A., Baber, J. L., Hustedt, E. J., & Clore, G. M. (2016). Long Distance Measurements up to 160 Å in the GroEL tetradecamer using Q-band DEER EPR spectroscopy. *Angewandte Chemie - International Edition, 55*(51), 15905–15909. https://doi.org/10.1002/anie.201609617.

Schrodinger, L. L. C. (2015). *The PyMOL molecular graphics system, Version 2.1.*

Shah, C., Hegde, B. G., Morén, B., Behrmann, E., Mielke, T., Moenke, G., et al. (2014). Structural insights into membrane interaction and caveolar targeting of dynamin-like EHD2. *Structure, 22*(3), 409–420. https://doi.org/10.1016/j.str.2013.12.015.

Simard, J. R., Getlik, M., Grütter, C., Pawar, V., Wulfert, S., Rabiller, M., et al. (2009). Development of a fluorescent-tagged kinase assay system for the detection and characterization of allosteric kinase inhibitors. *Journal of the American Chemical Society, 131*(37), 13286–13296. https://doi.org/10.1021/ja902010p.

Simard, J. R., & Rauh, D. (2014). FLiK: A direct-binding assay for the identification and kinetic characterization of stabilizers of inactive kinase conformations. In *Vol. 548. Methods in enzymology* (1st ed.). Elsevier Inc. https://doi.org/10.1016/B978-0-12-397918-6.00006-9. Issue C.

Spindler, P. E., Glaser, S. J., Skinner, T. E., & Prisner, T. F. (2013). Broadband inversion PELDOR spectroscopy with partially adiabatic shaped pulses. *Angewandte Chemie - International Edition, 52*, 3425–3429.

Spindler, P. E., Schöps, P., Kallies, W., Glaser, S. J., & Prisner, T. F. (2017). Perspectives of shaped pulses for EPR spectroscopy. *Journal of Magnetic Resonance, 280*, 30–45. https://doi.org/10.1016/j.jmr.2017.02.023.

Steed, P. R., Stein, R. A., Mishra, S., Goodman, M. C., & McHaourab, H. S. (2013). Na +-substrate coupling in the multidrug antiporter NorM probed with a spin-labeled substrate. *Biochemistry, 52*(34), 5790–5799. https://doi.org/10.1021/bi4008935.

Tait, C. E., & Stoll, S. (2016). Coherent pump pulses in double electron electron resonance spectroscopy. *Physical Chemistry Chemical Physics, 18*(27), 18470–18485. https://doi.org/10.1039/c6cp03555h.

Treiber, D. K., & Shah, N. P. (2013). Ins and outs of kinase DFG motifs. *Chemistry and Biology, 20*(6), 745–746. https://doi.org/10.1016/j.chembiol.2013.06.001.

Weickert, S., Cattani, J., & Drescher, M. (2019). Intrinsically disordered proteins (IDPs) studied by EPR and in-cell EPR. *Electron Paramagnetic Resonance, 26*. https://doi.org/10.1039/9781788013888-00001.

Widder, P., Berner, F., Summerer, D., & Drescher, M. (2019). Double nitroxide labeling by copper-catalyzed azide-alkyne cycloadditions with noncanonical amino acids for electron paramagnetic resonance spectroscopy [brief-report]. *ACS Chemical Biology, 14*(5), 839–844. https://doi.org/10.1021/acschembio.8b01111.

Wingler, L. M., Elgeti, M., Hilger, D., Latorraca, N. R., Lerch, M. T., Staus, D. P., et al. (2019). Angiotensin analogs with divergent bias stabilize distinct receptor conformations. *Cell, 176*(3), 468–478.e11. https://doi.org/10.1016/j.cell.2018.12.005.

Yang, Z., Kurpiewski, M. R., Ji, M., Townsend, J. E., Mehta, P., Jen-Jacobson, L., et al. (2012). ESR spectroscopy identifies inhibitory Cu^{2+} sites in a DNA-modifying enzyme to reveal determinants of catalytic specificity. *Proceedings of the National Academy of Sciences of the United States of America, 109*(17), 3–10. https://doi.org/10.1073/pnas.1200733109.

Yang, Z., Liu, Y., Borbat, P., Zweier, J. L., Freed, J. H., & Hubbell, W. L. (2012). Pulsed ESR dipolar spectroscopy for distance measurements in immobilized spin labeled proteins in liquid solution. *Journal of the American Chemical Society, 134*(24), 9950–9952. https://doi.org/10.1021/ja303791p.

CHAPTER SIX

Integrative ensemble modeling of proteins and their complexes with distance distribution restraints

Gunnar Jeschke* and Laura Esteban-Hofer
ETH Zürich, Department of Chemistry and Applied Biosciences, Zürich, Switzerland
*Corresponding author: e-mail address: gunnar.jeschke@phys.chem.ethz.ch

Contents

Abstract

Many proteins and protein complexes exhibit regions that are intrinsically disordered. Whereas an arsenal of techniques exists to characterize structured proteins or protein regions, characterization of the vast conformational space occupied by intrinsically

Methods in Enzymology, Volume 666
ISSN 0076-6879
https://doi.org/10.1016/bs.mie.2022.02.010

disordered regions remains a challenging task due the ensemble-averaging nature of many techniques that provide mean value restraints. More representative information can be gained in the form of distribution restraints, such as EPR-derived distance distributions. Previously we developed the ensemble modeling tool MMM, where we partition the macromolecule into structured and unstructured domains and utilize an integrative structural approach with a focus on EPR-derived distance restraints. Here we present the successor program of MMM: MMMx. All the modeling functionality was ported to MMMx and is now accessed by a uniform script format, allowing to combine the different modules at will to modeling pipelines. During the conception of MMMx many of the tools were improved or updated. We discuss the general functionality of MMMx and its modules, and illustrate some of the modeling tools by application examples.

1. Introduction

Interactions and function of proteins depend on an interplay between precisely structured and flexible parts. Whereas a whole arsenal of techniques exists for determination of precise structures (Curry, 2015), there is still a dearth of techniques for quantitative experimental characterization of the conformer distribution of flexible parts (Shrestha et al., 2019) and of the distribution of relative arrangement of parts. The difficulty in solving the latter two problems arises mostly from the ensemble-averaging nature of most physical measurements. Depending on the measured quantity, information on broadly distributed relative arrangement is either lost completely or reduced to a mean value, implying loss of information on distribution width. Large sets of precise mean-value restraints may be inconsistent with a single conformation, as has been demonstrated for exact nuclear Overhauser enhancement data obtained by NMR spectroscopy (Voegeli, Olsson, Guentert, & Riek, 2016). However, even such data can specify only ensembles consisting of a very small number of conformers, whereas intrinsically disordered domains may approach the limit of a random polymer coil (Hofmann et al., 2012) which needs to be represented by a large number of conformers.

The problem of distributed structure can be solved if a sufficient number of restraints on distances, angles, or torsion angles is available in the form of a probability distribution *P*. At a minimum, such restraints must specify the standard deviation in addition to the mean value, which corresponds to a Gaussian approximation of the distribution and to a harmonic energy potential with respect to deviation from the mean value. Since actual distributions of the geometric parameters can be asymmetric (Peter et al., 2019), bimodal

(Duss et al., 2014), or even multimodal (Yardeni, Mishra, Stein, Bibi, & Mchaourab, 2020), it may be advantageous to renounce parameterization of *P* or to use a multi-Gaussian parametric model (Ritsch et al., 2021).

Such distribution restraints can be obtained by EPR techniques for distances between spin-labeled sites in proteins and protein complexes in a distance range between about 20 and 80 Å (Borbat & Freed, 2017; Jeschke, 2016 a), in favorable cases up to 160 Å (Schmidt, Walti, Baber, Hustedt, & Clore, 2016). This approach comes with two disadvantages that preclude its application to structure determination as a stand-alone technique in most cases. First, spin-label sidechains exhibit conformation distribution of their own, which can be modeled with an accuracy for mean backbone–backbone distances not better than 3 Å (Hagelueken, Ward, Naismith, & Schiemann, 2012; Jeschke, 2013). Second, each site pair requires generation, expression, purification, and labeling of a protein double mutant (Hubbell, Cafiso, & Altenbach, 2000) or involves RNA synthesis (Babaylova et al., 2016; Duss et al., 2014). Thus, the number of distance distribution restraints that can be obtained by the site-directed spin labeling/EPR approach is usually much smaller than the number of free parameters. Further, the combination of the small number of restraints with their moderate accuracy precludes structure determination at atomic resolution, i.e., proper characterization of the precisely structured parts.

Therefore, in order to characterize the distributed structure of biological systems featuring both precisely structured and flexible parts, information from different techniques must be integrated. To this end, we had developed the MATLAB-based open-source software MMM (Jeschke, 2018; Polyhach, Bordignon, & Jeschke, 2011). This software allows for combining information from atomic-resolution structures with information from EPR-derived distance distribution restraints, and, more recently, with small-angle scattering data (Jeschke, 2021). MMM, which was conceived in 2009, is a collection of tools developed since that time. These tools can only be accessed via the graphical user interface of MMM, which makes it difficult to combine them freely into structure modeling and fitting pipelines. From a programmer perspective, internal representation of proteins and nucleic acids in MMM is inconvenient for integrating further approaches. For all these reasons, we have created the successor software MMMx which provides access to all modeling functionality of MMM by a uniform script format. The single exception is the interface provided in MMM to Modeler (Fiser & Šali, 2003). This interface was not ported to MMMx, since MMMx ensemble models are not supposed to depend on homology.

The new format is tailored for generating ensemble modeling pipelines, where the sequence of the tools can be changed at will. During porting them from MMM to MMMx, several tools were improved or streamlined. MMMx is also available at epr.ethz.ch/software.html as a packaged Windows app that does not depend on a MATLAB license, but rather installs the free MATLAB runtime engine during its own installation. In this chapter, we describe the concept of distance-distribution-based ensemble modeling with MMMx as well as the general form of pipelines for modeling and analysis of the obtained ensembles. Further details are found in on-line documentation at mmmx.info.

2. Concept of RigiFlex modeling of ensemble structures

2.1 How to think about heterogeneously distributed structure

According to the broad thermodynamic hypothesis, the conformational distribution of biomolecules and their complexes is governed by the Gibbs free energies G of all possible conformers. Given the distribution of G, i.e., the free energies of all possible conformers c_i, the conformer distribution $P(c_i)$ is given by the Boltzmann distribution

$$P(c_i) = \frac{\exp\left(-G(c_i)/k_\mathrm{B}T\right)}{\Sigma_i \exp\left(-G(c_i)/k_\mathrm{B}T\right)}, \tag{1}$$

where k_B is the Boltzmann constant and T is the temperature. The narrow thermodynamic hypothesis, originally due to Epstein, Goldberger, and Anfinsen (Epstein, Goldberger, & Anfinsen, 1963), assumes dominance of a single conformer. This assumption is good if the free energy difference between the conformer with lowest and second-lowest free energy exceeds $k_\mathrm{B}T$ by a large factor. It is the basis of all those structural biology projects that aim to interpret data in terms of a single structure that is specified at atomic resolution. This approach has been very successful in understanding structure–function relationship for globular proteins and has significantly contributed to such understanding for membrane proteins.

That said, most proteins contain flexible parts at least at the chain termini and some proteins contain extensive intrinsically disordered domains or are intrinsically disordered as a whole in some or all of their known states (Gall, Romero, Cortese, Uversky, & Dunker, 2007; Uversky, 2017; van der Lee et al., 2014). Many aspects of their function depend on shifts in the conformational distribution of the flexible parts. According to the broad thermodynamic hypothesis, such shift arises from a different dependence of the $G(c_i)$ on external parameters, such as temperature, pH, or concentrations of ions, small-molecule ligands, or macromolecular binding partners.

From a structural perspective, conformational distribution is heterogeneous along the chain of proteins and nuclear acids. The energy contribution corresponding to secondary structure and to strong tertiary contacts often exceeds $k_B T$ by a large factor, making sections with such features rigid within experimental uncertainty of structure determination. On the other hand, sections without such features may exist whose energies differ by less than $k_B T$ and whose geometries are governed only by fixed bond length and bond angles, and by the Ramachandran distribution of backbone torsion angles. Such sections adopt the conformation of a Ramachandran coil, which can be approximated by a Flory polymer random coil (Fitzkee & Rose, 2004; Hofmann et al., 2012). In between the extremes of precisely defined structure of rigid parts and random-coil behavior, there exist weakly or moderately structured parts with free energy differences between conformers of the order of $k_B T$.

Returning to the structural perspective, all geometric parameters ξ, such as distances between two sites, angles between three sites, and torsion angles between four sites, are substantially distributed if at least one of the sites resides in a flexible part. This is visualized in a slightly simplified form in Fig. 1. Panel (A) schematically shows the dependence of conformer free energy on the geometric parameter ξ, assuming a small number of conformers that all correspond to different values of ξ. The entropy component of free energy then results from different entropies of solvation. If we define the conformer only by its backbone structure, an entropy component arises from sidechain rotamer distribution. In the situation of distinct ξ_i, $P(\xi)$ can

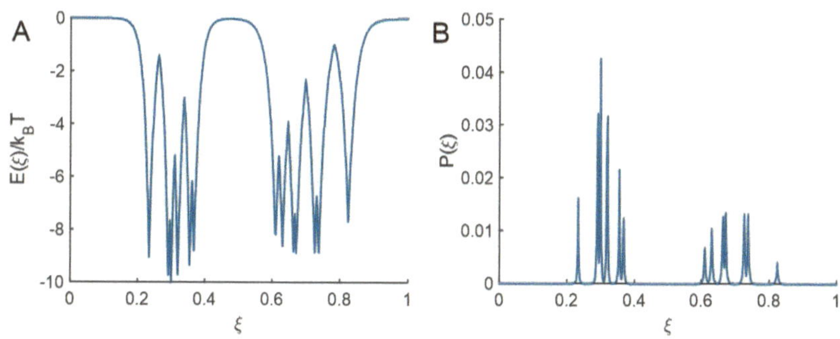

Fig. 1 Relation between the free energy profile with respect to a geometric parameter ξ (A) and the probability density distribution $P(\xi)$ (B).

be computed from $G(\xi)$ in analogy to Eq. (1), as shown in panel (B). Vice versa, given $P(\xi)$, the free energy profile can be computed by Boltzmann inversion,

$$G(\xi) = -k_{\mathrm{B}} T \ln P(\xi). \tag{2}$$

It follows that a distribution $P(\xi)$ defines a force-field term with respect to ξ, which is given by

$$F(\xi) = -\frac{\partial G(\xi)}{\partial \xi}. \tag{3}$$

For the simple case of a single Gaussian distribution $P(\xi)$ with mean value $\langle \xi \rangle$, $G(\xi)$ corresponds to a harmonic potential with its minimum at $\langle \xi \rangle$ and the force F is proportional to the displacement $\Delta\xi = |\xi - \langle \xi \rangle|$. In the following, we restrict ourselves to distance distribution restraints $P(r)$, which are the only type of distribution restraints already implemented in MMMx.

While these considerations provide a thermodynamic foundation for modeling conformer ensembles on the basis of distance, angle, or torsion angle distributions, the actual modeling task remains complex. Even if we consider only backbone conformation for a peptide with R residues, we need to sample $2(R - 1)$ Ramachandran angles. Assuming that a given Ramachandran angle can take only a_j discrete values, exhaustive sampling requires generation of $\prod_{j=1}^{2(R-1)} a_j$ conformers. In most cases of interest, it is impossible to generate and process a raw ensemble of this size. Thus, we rely on a divide-and-conquer approach that roughly corresponds to factorization of the huge parameter space. To this end, we partition the system into b rigid bodies and f flexible sections. Each conformer of the whole system is described by a rigid-body arrangement (RBA), which can be separately determined in a first step that fixes $6(b - 1)$ rotation and translation parameters. Given an RBA, each flexible part can be modeled individually. This RigiFlex approach (Section 2.2 and leftmost pipeline in Fig. 2) is easily extended to an ensemble description.

2.2 The RigiFlex approach

For two or more rigid bodies, we need to generate an RBA ensemble that is consistent with all distance distribution restraints. To this end, the rigid bodies first need to be extracted from atomic-resolution structures, which were usually obtained by X-ray crystallography, NMR, or cryo-EM and are available in protein database (PDB) format. For NMR structures, this

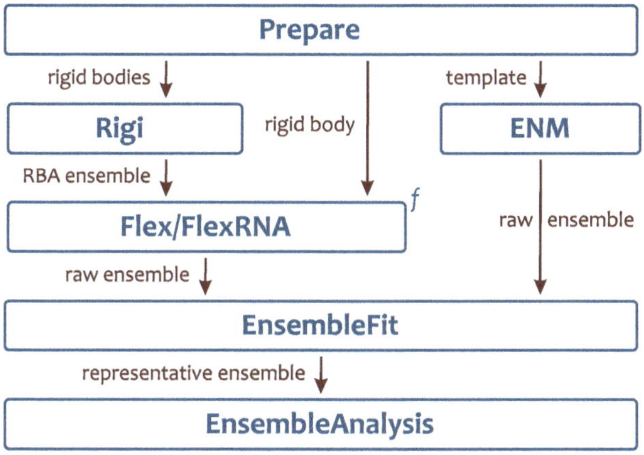

Fig. 2 Examples of the most important modeling pipelines that can be realized in MMMx. Each of the modules (*blue boxes*) can serve as an entry or end point of a pipeline. Rigid bodies, templates, and ensembles are input and output as PDB files, thus allowing for integration with other modeling and analysis software.

may involve determination of the rigid core and in general, it may require removal of flexible sections and occasionally of ligands that are not part of the system under study. The `prepare` module of MMMx can perform these tasks (Section 3.3).

Already before acquisition of experimental distance distributions, three reference sites per rigid body need to be defined. This number guarantees that the number of pair distances between reference sites is equal to ($b = 2$) or exceeds ($b > 2$) the number of translation and rotation degrees of freedom of the RBA (Jeschke, 2018). The `ExperimentDesign` module can be used for this step (Section 3.2).

The `Rigi` module performs exhaustive sampling of RBA space at a resolution requested by the user or derived from a maximum number of trials requested by the user (Section 3.4). RBAs are discarded if clashes between rigid bodies or violation of distance distribution restraints cannot be removed by local optimization. Depending on the system and the restraints, such an exhaustive search may result in an RBA ensemble that is too large for further processing. Therefore, MMMx offers the option to reduce the ensemble to a given number of RBAs by hierarchical clustering. We advise to generate about 2000 RBAs after clustering.

In the next step, the f flexible parts are generated one-by-one (Sections 3.5 and 3.6). We recommend to generate only one conformer per RBA for each flexible part and rather extend the RBA ensemble if a larger raw ensemble is sought. Although `Rigi` allows for specifying a maximum distance between the terminal anchor points of flexible linkers, it may happen that no linker can be generated for some RBAs. If the rigid bodies are densely packed, this may even happen for a large fraction of RBAs. In this case, it is advisable to generate a larger RBA ensemble.

If a system is devoid of rigid bodies, raw ensembles can be generated by the `Flex` approach (Jeschke, 2016 b) or the equivalent `FlexRNA` approach (Jeschke, 2021). This is still true if the system contains a single rigid body. Flexible peptide or nucleic-acid sections are then anchored in this rigid body at one of their termini or at both termini.

Conformers in the RigiFlex raw ensemble may feature weak clashes or other slight violations of canonical geometry of peptide or nucleic-acid chains, as can be seen, for instance, by running MolProbity (Davis et al., 2007) on them. MMMx provides a module that interfaces Yasara structure (Krieger et al., 2009) for conformer refinement. We made this choice because Yasara can solvate and optimize complexes of proteins and nucleic acids without user input and provides refined structures that remain close to the input structure. Since all modules of MMMx accept PDB files as input and provide PDB files as output, users can also perform refinement with other software and return to the MMMx pipeline afterward.

At this point, we have generated a large raw ensemble that is in rough agreement with the distribution restraints that have been used for efficient sampling in the `Rigi`, `Flex`, and `FlexRNA` modules.

2.3 The representative ensemble

The question arises how strongly such a raw ensemble should be contracted before analyzing it. In principle, each individual system copy in the sample—or in the cell—may be a distinct conformer. Even if the problem is discretized at some resolution, a full ensemble may be too large for comprehension or even for some types of analysis. Further, a discretization level is difficult to set, since resolution is heterogeneous across the system and, in general, along each macromolecular chain.

In this situation, we aim for a parsimonious ensemble that is consistent with all restraints within their experimental uncertainty. This request is a bit idealistic since sample conditions (concentration, temperature, state of aggregation, medium) may differ between the measurements of restraints by different techniques. We solve this problem by balancing the loss of merit between approaches (Jeschke, 2021). Briefly, we first partition the total set of restraints into subsets of homogeneous restraints. Restraints are homogeneous if they can be acquired at the same sample conditions and if the deviation between experimental and ensemble-modeled restraint is described by the same type of error function so that a mean of their error functions can be computed. Examples are small-angle scattering curves with the chi-square deviation as an error function, distance distributions with the overlap as an error function (Jeschke, 2021; Ritsch et al., 2021), or NMR paramagnetic relaxation enhancements (PREs) with the root mean-square deviation as an error function. We then fit preliminary ensembles that are based on the same raw ensemble and provide the best agreement with only one of the S homogeneous subsets of restraints. This is done for each subset and provides error functions $\epsilon_{s,\text{ individual}}$ that are lower bounds to the fit deviation for a given restraint subset and raw ensemble. Finally, we fit an ensemble globally to all subsets by minimizing the loss of merit

$$L = \frac{1}{S}\sum_{s=1}^{S}\frac{\epsilon_{s,\text{global}}}{\epsilon_{s,\text{individual}}} - 1. \tag{4}$$

The loss of merit is a measure for the inconsistency between the restraint subsets. The weighting of the individual restraint subsets thus results from normalizing the fit deviation in the integrative fit by the lowest possible fit deviation that is obtained in a subset-only fit.

The problem of contraction of the raw ensemble can be solved by assigning populations to conformers and discarding conformers with very low population. Indeed, some of the populations may approach zero. This is easily seen for the limiting case where a single conformer perfectly fits all restraints. It is unlikely that linear combinations of several conformers provide an equally good fit. Such population fitting and discarding of low-population conformers reduce the amount of information of an ensemble, while still fitting restraints within their experimental uncertainty.

3. Modeling and analysis pipelines

3.1 General MMMx pipeline

The most important ensemble modeling and analysis pipelines of MMMx are visualized in Fig. 2. The additional module `YasaraRefine` can be interspersed at any point for structure refinement of a single conformer or of a whole ensemble. This module has an option for repacking sidechains with SCWRL4 (Krivov, Shapovalov, & Dunbrack Jr., 2009) before refinement, which is recommended, since Yasara may fail for poor initial sidechain geometry. Refinement of conformers can easily be the computationally most costly step in generating a raw ensemble. It is possible to refine only the conformers that are retained in the much smaller representative ensemble. However, this approach runs the risk that the refinement fails for some of these conformers, which invalidates the representative ensemble. In such cases, refinement must be performed before running `EnsembleFit`. The module `Locate` can be used for locating a spin label or paramagnetic metal ion site with respect to a rigid body (Bleicken et al., 2014) as originally introduced by Gaffney et al. (2012). We use an algorithm similar to the one introduced by Hagelueken, Abdullin, Ward, and Schiemann (2013). The `ExperimentDesign` module is not part of the modeling and analysis pipeline. It can help with the choice of spin-labeling sites for projects that use the `ENM`, `Flex`, and `FlexRNA` modules and of reference sites in projects that use the `Rigi` module. Some functionality depends on external software. In particular, MMMx uses MUSCLE (Edgar, 2004) where sequence alignment is required; SCWRL4 (Krivov et al., 2009) where amino acid sidechains have to be added, modified, or repacked; ATSAS (Manalastas-Cantos et al., 2021) for fitting of small-angle scattering curves; Yasara structure (Krieger et al., 2009) for refinement; and MSMS (Sanner, Olson, & Spehner, 1996) for computing solvent-accessible surfaces. Where such dependencies exist, they are mentioned in module descriptions below.

In general, each module can serve as entry point or end of a pipeline, and structure data is communicated between modules via PDB files. Thus, MMMx can be combined at will with other software for modeling, analysis, and visualization. The MMMx ensemble file (`.ens`) format specifies the PDB file name and population of each conformer. By default, log output is written to a file with the same name as the script file, but with extension `log`. This file name can be overwritten by a `logfile` statement at the beginning of a modeling script. Numeric output data is saved to comma-separated value files (`.csv`), which can be imported by most data analysis and visualization software. The only exception is an internal memory-saving format for RBA ensembles that is expanded to atomic coordinates by the next downstream module in the pipeline. Alternatively, RBA ensembles can be output in PDB format. MMMx, unlike its precursor MMM, does not include visualization tools. We plan to add ensemble-specific visualization capabilities later through a ChimeraX (Goddard et al., 2018) interface. Some MMMx modules generate data plots. These plots can be saved in most common graphics formats. In the following, we briefly describe functionality of the modules in plain text. The syntax of MMMx scripts (`.mcx` files) is described in the online documentation, which is also packaged together with MMMx. The package contains some example scripts.

3.2 ExperimentDesign

The `ExperimentDesign` module can perform spin-labeling site scans (Polyhach & Jeschke, 2010), and compute label-to-label distance distributions for individual conformers and ensembles after performing a site scan. Furthermore, the `ExperimentDesign` module can suggest site pairs for elastic network modeling of structure transitions with the `ENM` module and reference site sets within rigid bodies for generating RBA ensembles with the `Rigi` module.

3.3 Prepare

The `Prepare` module is intended for all modifications of an input PDB file that are necessary or convenient before using it as a template of rigid-body input for modeling. Such modifications can also be done with other software. Doing them by a `Prepare` section in an MMMx script helps to ensure reproducibility of the workflow.

`Prepare` can renumber residues in chains, select only certain chains, and remove residues or ligands. This module can also perform frame transformations of coordinates, in particular, centering the structure, transforming it into a symmetry or pseudo-symmetry frame where a rotation axis is the z axis, or superimpose the whole structure or parts of it to corresponding atoms of a template. Such superposition can also be based on sequence alignment, using MUSCLE (Edgar, 2004), which is bundled with MMMx. `Prepare` can compute the center plane and thickness of a best-fitting lipid bilayer by minimizing free energy of immersion of the protein into the bilayer. The coordinates can then be transformed to a frame where the z axis is the bilayer normal and the center plane corresponds to $z = 0$, enabling the use of bilayer depth restraints (Fehr et al., 2015). The bilayer functionality depends on MSMS for computation of the solvent-accessible surface (Sanner et al., 1996). For oligomeric membrane proteins we recommend to associate the bilayer normal with the symmetry axis and optimize only the center plane z coordinate and membrane thickness. It is also possible to substitute a chain in one structure by a corresponding chain from another structure.

Finally, `Prepare` can be used for amino acid sidechain modifications if SCWRL4 (Krivov et al., 2009) is installed. The module can detect and repair incomplete sidechains, mutate residues, or repack all sidechains. The amino acids selenocysteine and selenomethionine, which are sometimes introduced in constructs for X-ray crystallography, can be replaced by their sulfur counterparts cysteine and methionine even if SCWRL4 is not installed.

3.4 Rigi

The `Rigi` module accepts input rigid bodies that may clash with each other and automatically separates them. It allows for specifying the maximum distance between reference sites (default 180 Å). A lower value leads to finer sampling of RBA space if distance distribution restraints do not exist for all pairs of reference points. Note that a too low value may exclude valid solutions.

In the `Rigi` module, distance restraints are specified by mean distances and standard deviations σ_r. Note that as compared to MMM, `Rigi` in MMMx samples a range of $\pm\sqrt{8 \ln 2}\sigma_r$ (two times the full width at half maximum) instead of a range of $\pm\ \sigma_r$ for each distance distribution restraint. Tests have shown that the resulting broader raw ensembles lead to better agreement with distance distribution restraints in later ensemble fitting and contraction. The resolution of sampling has a rather demanding default value of 3 Å. If distributions are broad, this value needs to be increased or a maximum number of trials needs to be specified to complete exhaustive sampling in a reasonable time. In the latter case, the resolution is automatically determined and reported in the log file.

The number of output models is large by default (20,000) and may need to be reduced if computationally expensive steps follow in the pipeline. By default, `Rigi` assumes that up to 500,000 initial RBAs can be reduced to the requested number of models by hierarchical clustering. This number needs to be reduced on computers with small memory. By default, `Rigi` stops after 48 h of computation time. Such a stop by the time limit is reported in the log file and solutions are provided, but they do not correspond to exhaustive sampling. We strongly recommend to achieve exhaustive sampling by reducing resolution or, equivalently, the maximum number of trials or by increasing the maximum computation time.

In addition to distance distribution restraints, `Rigi` accepts link restraints, i.e., maximum distances between sites. We advise to specify peptide links and nucleotide links for all pairs of anchor sites that will be connected by flexible linkers downstream in the pipeline. It is also possible to specify cross-link restraints. In this case, the user can provide the percentage of cross-links that must be possible in a given RBA (default 30%).

By default, RBAs are randomly oriented in space. It is possible to superimpose them at one of the rigid bodies. The RBA ensemble is saved in an internal format. By user request, all RBAs can also be saved as PDB files.

3.5 Flex

For the `Flex` module, distance distribution restraints can be specified by mean distance and standard deviation or as a full distribution. In the latter case, conformers are accepted by a variant of rejection sampling. We do not yet have sufficient data to decide whether one of the approaches is generally more efficient. Rejection sampling has the practical advantage that the acceptance rate can be specified, which allows for an easier control of the compromise between size and fit quality of a raw ensemble that can be computed in a given time.

Distance restraints can be defined between a site in the input model and a site in the flexible section, between two sites in the flexible section, and between symmetry-related sites in flexible sections in an oligomer. In the latter case, the `Prepare` module should be used to transform the input model to the symmetry frame. In addition, `Flex` accepts bilayer immersion depth restraints, and propensities for α structure (helix propensity), β structure (strand propensity), and cis prolines.

If the flexible peptide sections should be attached to existing models, these can be input by PDB file name with wildcard characters allowed, by an MMMx ensemble list (`.ens`), or by expanding a `Rigi` RBA ensemble.

3.6 FlexRNA

The `FlexRNA` module operates in the same way as `Flex`, except that rejection sampling, oligomer and depth restraints and secondary-structure propensities are not implemented.

3.7 ENM

The `ENM` module models a shift between conformer ensembles upon a change in external conditions, for instance, upon ligand binding or release. Generation of the raw ensemble is based on deformation of a simple ball-and-springs model, where the balls represent Cα atoms, the springs effective forces between them, and the deformation force is due to distance restraints (Zheng & Brooks, 2006). Distance distribution restraints are implemented by a Monte Carlo approach assuming a normal distribution with the mean value and standard deviation derived from experiment. Spin labels are implemented by their predicted mean position relative to Cα atoms in their vicinity (Jeschke, 2012 a) and a thermal distribution over deformation modes is assumed (Jeschke, 2012 b).

The `ENM` module allows for removing ligands from an input template structure and to select chains from the template. This obviates the need for running `Prepare` before if these are the only required changes in the template. By default, the output ensemble comprises 100 conformers. The algorithm is sufficiently fast for generating raw ensembles with thousands of conformers. By default, only peptide chains are deformed. The user can specify ligands that should be dragged along with the neighboring Cα atoms. In such dragging, ligands are treated as rigid with only translation and rotation being applied to them. We recommend to refine output conformers of the ENM module, for instance, with the `YasaraRefine` module using the `repack` option for sidechains.

3.8 EnsembleFit

The `EnsembleFit` module accepts raw ensembles as an MMMx ensemble list (`.ens`), by specifying a list of PDB files with wildcard characters in the file name, or by expanding a `Rigi` RBA ensemble. The current version can fit to distance distribution restraints and NMR PRE restraints as well as to small-angle scattering curves, if ATSAS (Manalastas-Cantos et al., 2021) is installed. The ensemble PRE computation follows a Python implementation by Tesei et al. (2021) and allows for specifying restraints as intensity ratios between the paramagnetically and diamagnetically labeled protein or as PRE rates. For small-angle neutron scattering, it is possible to specify deuterium content of the solvent and to provide resolution information.

By default, the output ensemble is saved only as a list of PDB file names and corresponding populations. Assembly of a single PDB file with all conformers included and with the populations reported in a REMARK 400 section is recommended only after checking ensemble size. To this end, `EnsembleFit` can be rerun with the `nofit` and `savepdb` options and by specifying the fitted ensemble as initial ensemble. The `nofit` option can also be used for evaluation of subsets of potential restraints that were not used during fitting, for instance, in jackknife resampling (Ritsch et al., 2021).

3.9 EnsembleAnalysis

The `EnsembleAnalysis` module can process MMMx ensemble list (`.ens`) files with population information and ensembles specified by a single PDB file or by a filename with wildcard characters. In the last case, uniform populations of all conformers are assumed. This also applies to a single PDB file, unless population information is provided in a REMARK 400 section. If in a file list, each individual PDB file corresponds to a frame of an MD trajectory, subsampling can be performed, i.e., only every n^{th} frame is retained. Note that `EnsembleAnalysis` first reads all files so that external subsampling may be preferable if the number of frames is huge.

The module allows for two operations that modify the ensemble, superposition and sorting. Superposition allows for defining a new frame where all conformers are optimally superimposed as a whole or for a certain chain or for a residue range in a chain. Such superposition can be defined with respect to the first conformer, to a template conformer outside the ensemble, or to the central conformer, which has minimal sum of mean square deviations from all other ensemble members. For two input ensembles, superposition is also possible to the common central conformer. Sorting does not change individual conformers but only their order in the ensemble. After such sorting. similar conformers are close in the list.

`EnsembleAnalysis` can compute basic measures of the ensemble, such as its width and density (Jeschke, 2021), radius of gyration and standard deviation of the radii of gyration of all conformers, and the pair rmsd matrix for all conformers. For ensembles that contain parts with high or at least moderate order, the pair correlation matrix (Jeschke, 2021) may reveal structural correlation between residues in different chains or with large sequence distance within a chain. Parts with weak order can be better characterized by the compactness and proximity matrices (Jeschke, 2021) or the matrix of segment-length deviations from a random-coil model (Ritsch et al., 2021). In the latter case, a plot with the distribution of segment lengths is also provided.

Local order can be analyzed by residue-wise flexibility, which is based on circular variance of the Ramachandran angles (MacArthur & Thornton, 1993) and takes the value of 0 for complete order and 1 for random distribution. The order of a residue with respect to the rest of the structure can be characterized by a local order parameter that extends Flory's approach to correlation in freely jointed chains to chains with heterogeneous order. As in Flory's case, we consider the distribution of angles between bond vectors, where we associate a bond with the $C\alpha_i$-$C\alpha_{i+1}$ vector for peptide chains and with the $C4'_i$-$C4'_{i+1}$ vector for nucleic-acid chains. At a given residue i we compute the variance σ_{ij}^2 of $\cos\theta_{ij}$ over all residues j including i. The correlation parameter is then $c_{ij} = 1 - \sqrt{2}\sigma_{ij}^2$ and the local order parameter for residue i is the mean of all c_{ij}. It ranges between 0 for random distribution of all θ_{ij} and 1 for perfect order. Since $c_{ii} = 1$, zero can only be approached but not reached.

For visualization, the `EnsembleAnalysis` module can compute a pseudo-electron density of the ensemble. This computation can be restricted to individual chains or residue ranges in individual chains. The density is stored as an MRC file, which can be read by many visualization programs. Note that this density depends on how the conformers are superimposed and has to be interpreted with care, unless the system contains exactly one rigid body and the ensemble is superimposed at this body.

Pseudo-electron density can also serve for comparing two ensembles A and B, again with the caveat that it must be possible to superimpose conformers of both ensembles in a suitable way. In this case, `EnsembleAnalysis` computes pseudo-electron densities $\rho_{i,\,\mathrm{A}}(x, y, z)$ and $\rho_{i,\,\mathrm{B}}(x, y, z)$ for individual residues i, which are normalized $\left(\int\int\int \rho_{i,\mathrm{A/B}}(x, y, z)\, \mathrm{d}x\, \mathrm{d}y\, \mathrm{d}z = 1\right)$. Residue-wise similarity between the ensembles can then be defined as

$$s_i = \int\int\int \min\left(\rho_{i,\mathrm{A}}, \rho_{i,\mathrm{B}}\right) \mathrm{d}x\, \mathrm{d}y\, \mathrm{d}z. \tag{5}$$

Similarity is 1 for perfect overlap of the pseudo-electron densities at residue i between ensembles A and B, and 0 if the two pseudo-electron densities are distinct. Plots of s_i for all residues can reveal correlations between two ensembles. For two ensembles that were derived from the same restraints, it can provide a measure for the uncertainty of the ensemble description.

4. Application examples

An example for using the RigiFlex approach on an RNA-binding protein and its complexes with two RNA fragments will be published elsewhere. The `Flex` module on its own was used, together with some of the `EnsembleAnalysis` features, in recent work on the ensemble structure of heterogeneous nuclear ribonucleoprotein A1 in its dispersed state in the absence of RNA (Ritsch et al., 2021). Here, we focus on the new capabilities of MMMx for elastic network modeling of state transitions and on new features of the `EnsembleAnalysis` module.

4.1 Elastic network modeling of cAMP release from a HCN channel

Although the initial implementation of spin-label-based elastic network modeling in MMM (Jeschke, 2012 a, 2012 b) was able to generate raw ensembles, it was tested and used in the paradigm of the narrow thermodynamic hypothesis, where a state change occurs between two precisely defined conformers. An early application of the approach to the release of cyclic AMP (cAMP) from a hyperpolarization-activated cyclic nucleotide-gated (HCN) ion channel (Puljung, DeBerg, Zagotta, & Stoll, 2014) was also performed in this spirit, where EPR information is used only in the form of mean distances. Based on the distance distributions measured in this study for the apo and cAMP-bound forms of mouse HCN2 ion channel, which were kindly provided to us by Stefan Stoll, we here illustrate how such modeling is performed in the paradigm of the broad thermodynamic hypothesis with MMMx.

In addition to the data for site pairs 537/624, 563/624, 570/624, 537/635, 563/635, and 570/624 that are shown in Fig. 2 of Puljung et al. (2014) and capture the conformational change, we also received the distributions for site pair 537/570 within a domain with stable conformation. We extracted mean value and standard deviation of the site-to-site distances from the supplied distributions and used this information to generate raw ensembles of 250 conformers each for the apo and cAMP-bound forms using the `ENM` module. By `ENM` directives, we downloaded the crystal structure of the cAMP-bound form (PDB 3etq) from the Protein Database, extracted chain A, and removed the cAMP ligand (residue 1). Most of the 250 trials with Monte Carlo sampling of label-to-label distances from the seven Gaussian distributions converged to mean-distance rmsd between 2 and 3 Å. In a few cases the converged rmsd exceeded 5 Å (largest value 6.26 Å). A small fraction of such poor fits is not unexpected, as the label-to-label distances are correlated and random sampling may pick impossible combinations.

The 250 conformers were then subjected to SCWRL4 sidechain repacking and subsequent refinement by Yasara Structure through the `YasaraRefine` module. We found that the previous sidechain repacking

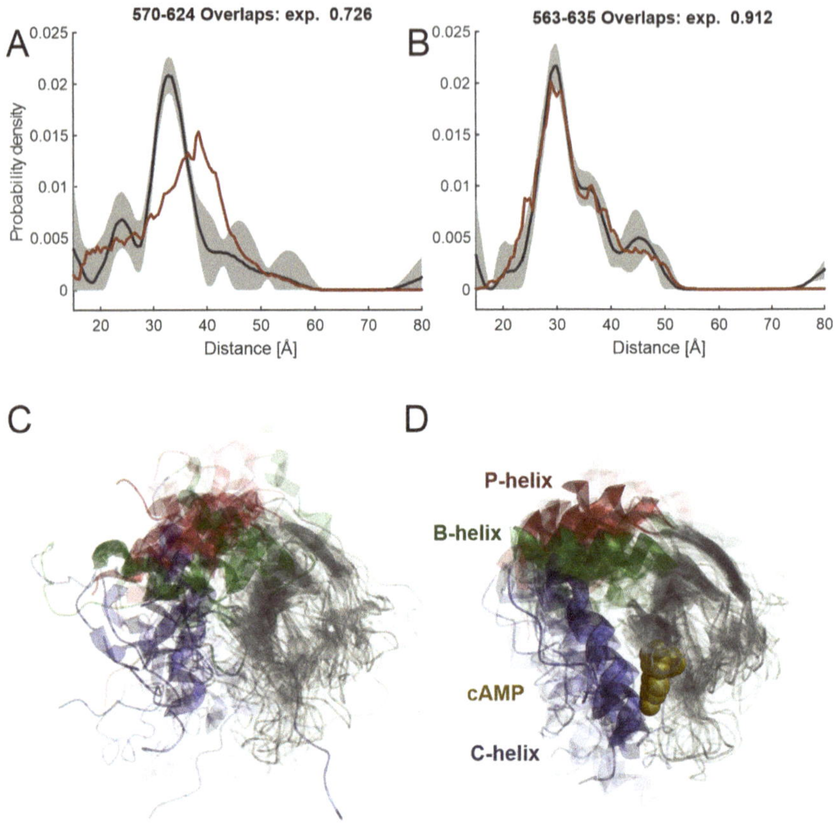

Fig. 3 Elastic network modeling of the release of cAMP from the mouse hyperpolarization-activated cyclic nucleotide-gated ion channel HCN2. Distance distribution data are from Puljung et al. (2014) and were kindly provided by Stefan Stoll. (A) Worst overlap between back-calculated (*red*) and experimental (*black*) distance distributions for the apo form. The *gray* area corresponds to uncertainty estimated by the validation feature of DeerAnalysis2013. (B) Best overlap between back-calculated (*red*) and experimental (*black*) distance distributions for the apo form. (C) Visualization by MMM of the ensemble of the apo form. Conformer populations are encoded by transparency. The P-helix (*crimson*), B-helix (*dark green*), and C-helix (*dark blue*) are emphasized by color. (D) Visualization of the ensemble of the cAMP-bound form. The cAMP molecule is shown as a *gold-colored* space-filling model, based on its position in the crystal structure.

was necessary, as otherwise Yasara runs failed for a few of the conformers. In a final modeling step, the raw ensembles were refined and contracted by maximizing overlap between back-calculated and experimental unparameterized distance distribution. Fig. 3A and B shows the worst- and best-fitting site pairs for the apo state, for which no crystal structure is available. Geometric mean overlap of the seven distance distributions was 0.827 and 0.824 for the refined ensembles for the apo and cAMP-bound states, respectively.

Using the `EnsembleAnalysis` module, we then superimposed all conformers of both ensembles at residues 537–570 by minimizing backbone rmsd. To that end we used the central conformer among all conformers from both ensembles as a template. Ensemble comparison showed that the β-sheet core (residues 537–570) overlaps very well between both states, whereas the positions of the P-, B-, and C-helix (colored in Fig. 3) with respect to this core vary. The general movement toward a more compact arrangement of these helices was already established by the single-conformer modeling used by Puljung et al. (2014). Beyond the conclusions in this work, ensemble modeling reveals more disorder in the apo state as compared to the cAMP-bound state. Inspection of the crystal structure reveals that the packing of the sidechain of R632 against the cAMP base may cause stabilization of the C-helix. This stabilization is in agreement with FRET results reported in Taraska, Puljung, Olivier, Flynn, and Zagotta (2009).

4.2 Analysis and comparison of two ensembles from the protein ensemble database

We have browsed the Protein Ensemble Database (Lazar et al., 2020) for finding a pair of refined ensembles that can serve for showcasing some of the ensemble analysis features of MMMx. In recent work, Lincoff et al. (2020) discussed the Bayesian approach X-EISD to ensemble selection on the basis of a multitude of NMR restraints, a SAXS curve, single-molecule FRET data, and the hydrodynamic radius for the N-terminal SH3 domain of Drk (1–59). From two different raw ensembles, they obtained two refined ensembles with very similar fit quality, but different fractions of conformers devoid of secondary structure. The more structured ensemble (PED00157, ENSEMBLE) was derived from a raw ensemble generated with the ENSEMBLE software (Krzeminski, Marsh, Neale, Choy, & Forman-Kay, 2012), whereas the less structured ensemble (PED00158, MIXED) was derived from a raw ensemble containing 50% of ENSEMBLE-generated conformers and 50% random conformers generated by an unfolding algorithm. Using module `EnsembleAnalysis`, we have superimposed all conformers on the common central conformer of both ensembles. Pseudo-electron density overlap of the two ensembles is 0.867, whereas the overlap for individual residues ranges between 0.40 and 0.71 (Fig. 4A) with minima near the termini.

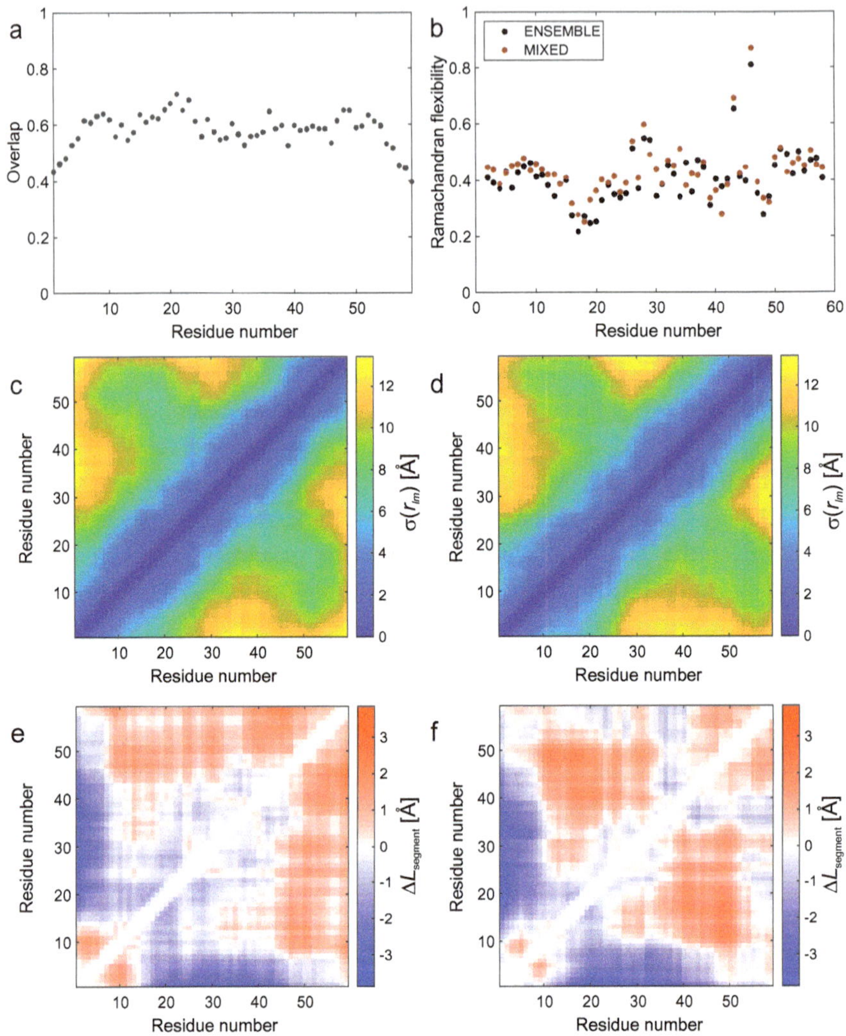

Fig. 4 Comparison of two ensemble models of the N-terminal SH3 domain of Drk (1–59) derived from an ENSEMBLE-based and a MIXED raw ensemble (see Lincoff et al., 2020). (A) Overlap of pseudo-electron density per residue. (B) Variation of flexibility (Ramachandran angle disorder parameter (MacArthur & Thornton, 1993)) for ENSEMBLE (*black*) and MIXED (*red*) raw ensembles. (C, D) Pair correlation matrix with the ENSEMBLE raw ensemble (C) and MIXED raw ensemble (D). (E, F) Segment-length deviation from a random coil with the ENSEMBLE raw ensemble (E) and the MIXED raw ensemble (F).

Despite the higher fraction of conformers devoid of secondary structure in the MIXED ensemble, Ramachandran flexibilities are rather similar for the two ensembles (Fig. 4B). Likewise, only minor differences are seen in the pair correlation plots (Fig. 4C and D), where blue hues correspond to residue pairs whose relative position is strongly correlated. In both ensembles, moderate correlation is seen between a section near the N-terminus (residues 5–22) and a section near the C-terminus (residues 42–56). The differences between the two ensembles are best seen in the plots of segment-length deviation with respect to a random-coil model (Fig. 4E and F), where blue regions correspond to more compact and red regions to more extended segments. The two random-coil models have rather similar Kuhn lengths of 5.37 and 5.33 Å and rather similar stretch exponents of 0.529 and 0.524 for the ENSEMBLE and MIXED ensembles, respectively. Such agreement is expected, as both ensembles agree with the same SAXS curve and hydrodynamic radius. However, in the ensemble derived from the ENSEMBLE-based raw ensemble (panel (E)), the C-terminal section is more uniformly remote from the rest of the protein than in the ensemble derived from the MIXED raw ensemble (panel (F)).

5. Conclusion

Since many proteins and protein complexes feature function-relevant disorder, they ought to be described by ensemble models. With the MMMx package, we introduce an ensemble modeling and analysis toolbox that is based on distance distribution restraints as a unique information source on ensemble width and on heterogeneity of disorder. Distribution information can be used already in generation of raw ensembles, where it improves efficiency of sampling of the conformational space. In a subsequent step of ensemble refinement and contraction, distribution information is integrated with restraints that are available only as ensemble mean values.

The current version of MMMx can fit to small-angle scattering curves and NMR paramagnetic relaxation enhancements in addition to distance distribution restraints. Previous work by Lincoff et al. (2020) suggests that shorter-range NMR restraints, such as J-couplings, chemical shift information, and residual dipolar couplings, are required to properly specify local structure. Ensemble fitting to such restraints, as well as to the hydrodynamic radius and single-molecule FRET restraints can be implemented in the existing framework of MMMx. Work in this direction is now in progress.

Acknowledgments

We thank Ettore Bartalucci for tests of rejection sampling in `Flex`, Daniella Goldfarb for suggesting the drag feature in the `ENM` module, Kresten Lindorff-Larsen for sharing the ensemble implementation of PRE restraints from his group before publication, Stefan Stoll for sharing primary data on cAMP release from HCN channel, and Frédéric Allain for a long-standing collaboration that inspired the `FlexRNA` module and several other features.

References

Babaylova, E. S., Malygin, A. A., Lomzov, A. A., Pyshnyi, D. V., Yulikov, M., Jeschke, G., ... Bagryanskaya, E. G. (2016). Complementary-addressed site-directed spin labeling of long natural RNAs. *Nucleic Acids Research*, *44*(16), 7935–7943. https://doi.org/10.1093/nar/gkw516.

Bleicken, S., Jeschke, G., Stegmueller, C., Salvador-Gallego, R., García-Sáez, A. J., & Bordignon, E. (2014). Structural model of active bax at the membrane. *Molecular Cell*, *56*(4), 496–505. https://doi.org/10.1016/j.molcel.2014.09.022.

Borbat, P. P., & Freed, J. H. (2017). Dipolar spectroscopy—Single-resonance methods. In *eMagRes* (pp. 465–494). American Cancer Society. https://doi.org/10.1002/9780470034590.emrstm1519.

Curry, S. (2015). Structural biology: A century-long journey into an unseen world. *Interdisciplinary Science Reviews*, *40*(3), 308–328. https://doi.org/10.1179/0308018815Z.000000000120.

Davis, I. W., Leaver-Fay, A., Chen, V. B., Block, J. N., Kapral, G. J., Wang, X., ... Richardson, D. C. (2007). Molprobity: All-atom contacts and structure validation for proteins and nucleic acids. *Nucleic Acids Research*, *35*, W375–W383. https://doi.org/10.1093/nar/gkm216.

Duss, O., Michel, E., Yulikov, M., Schubert, M., Jeschke, G., & Allain, F. H. T. (2014). Structural basis of the non-coding RNA RsmZ acting as a protein sponge. *Nature*, *509*(7502), 588–592. https://doi.org/10.1038/nature13271.

Edgar, R. C. (2004). MUSCLE: Multiple sequence alignment with high accuracy and high throughput. *Nucleic Acids Research*, *32*(5), 1792–1797. https://doi.org/10.1093/nar/gkh340.

Epstein, C. J., Goldberger, R. F., & Anfinsen, C. B. (1963). The genetic control of tertiary protein structure: Studies with model systems. *Cold Spring Harbor Symposia on Quantitative Biology*, *28*, 439–449. https://doi.org/10.1101/SQB.1963.028.01.060. http://symposium.cshlp.org/content/28/439.short.

Fehr, N., Dietz, C., Polyhach, Y., von Hagens, T., Jeschke, G., & Paulsen, H. (2015). Modeling of the N-terminal section and the lumenal loop of trimeric light harvesting complex II (LHCII) by using EPR*. *Journal of Biological Chemistry*, *290*(43), 26007–26020. https://doi.org/10.1074/jbc.M115.669804.

Fiser, A., & Šali, A. (2003). Modeller: Generation and refinement of homology-based protein structure models. In *Macromolecular crystallography, Part D: Vol. 374* (pp. 461–491). Academic Press. https://doi.org/10.1016/S0076-6879(03)74020-8.

Fitzkee, N. C., & Rose, G. D. (2004). Reassessing random-coil statistics in unfolded proteins. *Proceedings of the National Academy of Sciences*, *101*(34), 12497–12502. https://doi.org/10.1073/pnas.0404236101. https://www.pnas.org/content/101/34/12497.

Gaffney, B. J., Bradshaw, M. D., Frausto, S. D., Wu, F., Freed, J. H., & Borbat, P. (2012). Locating a lipid at the portal to the lipoxygenase active site. *Biophysical Journal*, *103*(10), 2134–2144. https://doi.org/10.1016/j.bpj.2012.10.002.

Gall, T. L., Romero, P. R., Cortese, M. S., Uversky, V. N., & Dunker, A. K. (2007). Intrinsic disorder in the protein data bank. *Journal of Biomolecular Structure and Dynamics*, *24*(4), 325–341. https://doi.org/10.1080/07391102.2007.10507123.

Goddard, T. D., Huang, C. C., Meng, E. C., Pettersen, E. F., Couch, G. S., Morris, J. H., & Ferrin, T. E. (2018). UCSF chimerax: Meeting modern challenges in visualization and analysis. *Protein Science*, *27*(1), 14–25. https://doi.org/10.1002/pro.3235. https://onlinelibrary.wiley.com/doi/abs/10.1002/pro.3235.

Hagelueken, G., Abdullin, D., Ward, R., & Schiemann, O. (2013). mtsslSuite: In silico spin labelling, trilateration and distance-constrained rigid body docking in PyMOL. *Molecular Physics*, *111*(18-19), 2757–2766. https://doi.org/10.1080/00268976.2013.809804.

Hagelueken, G., Ward, R., Naismith, J. H., & Schiemann, O. (2012). Mtsslwizard: In silico spin-labeling and generation of distance distributions in PyMOL. *Applied Magnetic Resonance*, *42*(3), 377–391. https://doi.org/10.1007/s00723-012-0314-0.

Hofmann, H., Soranno, A., Borgia, A., Gast, K., Nettels, D., & Schuler, B. (2012). Polymer scaling laws of unfolded and intrinsically disordered proteins quantified with single-molecule spectroscopy. *Proceedings of the National Academy of Sciences*, *109*(40), 16155–16160. https://doi.org/10.1073/pnas.1207719109. https://www.pnas.org/content/109/40/16155.

Hubbell, W. L., Cafiso, D. S., & Altenbach, C. (2000). Identifying conformational changes with site-directed spin labeling. *Nature Structural Biology*, 7(9), 735–739. https://doi.org/10.1038/78956.

Jeschke, G. (2012a). Characterization of protein conformational changes with sparse spin-label distance constraints. *Journal of Chemical Theory and Computation*, *8*(10), 3854–3863. https://doi.org/10.1021/ct300113z.

Jeschke, G. (2012b). Optimization of algorithms for modeling protein structural transitions from sparse long-range spin-label distance constraints. *Zeitschrift für Physikalische Chemie*, *226*(11-12), 1395–1414. https://doi.org/10.1524/zpch.2012.0289.

Jeschke, G. (2013). Conformational dynamics and distribution of nitroxide spin labels. *Progress in Nuclear Magnetic Resonance Spectroscopy*, *72*, 42–60. https://doi.org/10.1016/j.pnmrs.2013.03.001.

Jeschke, G. (2016a). Dipolar spectroscopy—Double-resonance methods. In *eMagRes* (pp. 1459–1476). American Cancer Society. https://doi.org/10.1002/9780470034590.emrstm1518. https://onlinelibrary.wiley.com/doi/abs/10.1002/9780470034590.emrstm1518.

Jeschke, G. (2016b). Ensemble models of proteins and protein domains based on distance distribution restraints. *Proteins: Structure, Function, and Bioinformatics*, *84*(4), 544–560. https://doi.org/10.1002/prot.25000. https://onlinelibrary.wiley.com/doi/abs/10.1002/prot.25000.

Jeschke, G. (2018). MMM: A toolbox for integrative structure modeling. *Protein Science*, *27*(1), 76–85. https://doi.org/10.1002/pro.3269. https://onlinelibrary.wiley.com/doi/abs/10.1002/pro.3269.

Jeschke, G. (2021). MMM: Integrative ensemble modeling and ensemble analysis. *Protein Science*, *30*, 125–135.

Krieger, E., Joo, K., Lee, J., Lee, J., Raman, S., Thompson, J., … Karplus, K. (2009). Improving physical realism, stereochemistry, and side-chain accuracy in homology modeling: Four approaches that performed well in CASP8. *Proteins: Structure, Function, and Bioinformatics*, 77(S9), 114–122. https://doi.org/10.1002/prot.22570. https://onlinelibrary.wiley.com/doi/abs/10.1002/prot.22570.

Krivov, G. G., Shapovalov, M. V., & Dunbrack, R. L., Jr. (2009). Improved prediction of protein side-chain conformations with SCWRL4. *Proteins: Structure, Function, and Bioinformatics*, 77(4), 778–795. https://doi.org/10.1002/prot.22488. https://onlinelibrary.wiley.com/doi/abs/10.1002/prot.22488.

Krzeminski, M., Marsh, J. A., Neale, C., Choy, W.-Y., & Forman-Kay, J. D. (2012). Characterization of disordered proteins with ENSEMBLE. *Bioinformatics*, *29*(3), 398–399. https://doi.org/10.1093/bioinformatics/bts701.

Lazar, T., Martínez-Pérez, E., Quaglia, F., Hatos, A., Chemes, L. B., Iserte, J. A., … Piovesan, D. (2020). PED in 2021: A major update of the protein ensemble database for intrinsically disordered proteins. *Nucleic Acids Research*, *49*(D1), D404–D411. https://doi.org/10.1093/nar/gkaa1021.

Lincoff, J., Haghighatlari, M., Krzeminski, M., Teixeira, J. M. C., Gomes, G.-N. W., Gradinaru, C. C., … Head-Gordon, T. (2020). Extended experimental inferential structure determination method in determining the structural ensembles of disordered protein states. *Communications Chemistry*, *3*, 74. https://doi.org/10.1038/s42004-020-0323-0. https://europepmc.org/articles/PMC7409953.

MacArthur, M. W., & Thornton, J. M. (1993). Conformational analysis of protein structures derived from NMR data. *Proteins: Structure, Function, and Bioinformatics*, *17*(3), 232–251. https://doi.org/10.1002/prot.340170303. https://onlinelibrary.wiley.com/doi/abs/10.1002/prot.340170303.

Manalastas-Cantos, K., Konarev, P. V., Hajizadeh, N. R., Kikhney, A. G., Petoukhov, M. V., Molodenskiy, D. S., … Franke, D. (2021). *ATSAS 3.0*: Expanded functionality and new tools for small-angle scattering data analysis. *Journal of Applied Crystallography*, *54*(1), 343–355. https://doi.org/10.1107/S1600576720013412.

Peter, M. F., Tuukkanen, A. T., Heubach, C. A., Selsam, A., Duthie, F. G., Svergun, D., I, … Hagelueken, G. (2019). Studying conformational changes of the yersinia type-III-secretion effector YopO in solution by integrative structural biology. *Structure*, *27*(9), 1416–1426. https://doi.org/10.1016/j.str.2019.06.007.

Polyhach, Y., Bordignon, E., & Jeschke, G. (2011). Rotamer libraries of spin labelled cysteines for protein studies. *Physical Chemistry Chemical Physics*, *13*, 2356–2366. https://doi.org/10.1039/C0CP01865A.

Polyhach, Y., & Jeschke, G. (2010). Prediction of favourable sites for spin labelling of proteins. *Spectroscopy-An International Journal*, *24*(6), 651–659. https://doi.org/10.1155/2010/706498.

Puljung, M. C., DeBerg, H. A., Zagotta, W. N., & Stoll, S. (2014). Double electron–electron resonance reveals cAMP-induced conformational change in HCN channels. *Proceedings of the National Academy of Sciences*, *111*(27), 9816–9821. https://doi.org/10.1073/pnas.1405371111. https://www.pnas.org/content/111/27/9816.

Ritsch, I., Esteban-Hofer, L., Lehmann, E., Emmanouilidis, L., Yulikov, M., Allain, F. H. T., & Jeschke, G. (2021). Characterization of weak protein domain structure by spin-label distance distributions. *Frontiers in Molecular Biosciences*, *8*, 636599.

Sanner, M. F., Olson, A. J., & Spehner, J.-C. (1996). Reduced surface: An efficient way to compute molecular surfaces. *Biopolymers*, *38*(3), 305–320. https://doi.org/10.1002/(SICI)1097-0282(199603)38:3<305::AID-BIP4>3.0.CO;2-Y. https://onlinelibrary.wiley.com/doi/abs/10.1002/%28SICI%291097-0282%28199603%2938%3A3%3C305%3A%3AAID-BIP4%3E3.0.CO%3B2-Y.

Schmidt, T., Walti, M. A., Baber, J. L., Hustedt, E. J., & Clore, G. M. (2016). Long distance measurements up to 160 å in the GroEL tetradecamer using Q-band DEER EPR spectroscopy. *Angewandte Chemie International Edition*, *55*(51), 15905–15909. https://doi.org/10.1002/anie.201609617.

Shrestha, U. R., Juneja, P., Zhang, Q., Gurumoorthy, V., Borreguero, J. M., Urban, V., … Petridis, L. (2019). Generation of the configurational ensemble of an intrinsically disordered protein from unbiased molecular dynamics simulation. *Proceedings of the National Academy of Sciences*, *116*(41), 20446–20452. https://doi.org/10.1073/pnas.1907251116. https://www.pnas.org/content/116/41/20446.

Taraska, J. W., Puljung, M. C., Olivier, N. B., Flynn, G. E., & Zagotta, W. N. (2009). Mapping the structure and conformational movements of proteins with transition metal ion FRET. *Nature Methods*, *6*(7), 532–U94. https://doi.org/10.1038/nmeth.1341.

Tesei, G., Martins, J.a. M., Kunze, M. B. A., Wang, Y., Crehuet, R., & Lindorff-Larsen, K. (2021). DEER-PREdict: Software for efficient calculation of spin-labeling EPR and NMR data from conformational ensembles. *PLOS Computational Biology*, *17*(1), 1–18. https://doi.org/10.1371/journal.pcbi.1008551.

Uversky, V. N. (2017). Intrinsic disorder here, there, and everywhere, and nowhere to escape from it. *Cellular and Molecular Life Sciences*, *74*(17), 3065–3067. https://doi.org/10.1007/s00018-017-2554-5.

van der Lee, R., Buljan, M., Lang, B., Weatheritt, R. J., Daughdrill, G. W., Dunker, A. K., … Babu, M. M. (2014). Classification of intrinsically disordered regions and proteins. *Chemical Reviews*, *114*(13), 6589–6631. https://doi.org/10.1021/cr400525m.

Voegeli, B., Olsson, S., Guentert, P., & Riek, R. (2016). The exact NOE as an alternative in ensemble structure determination. *Biophysical Journal*, *110*(1), 113–126. https://doi.org/10.1016/j.bpj.2015.11.031.

Yardeni, E. H., Mishra, S., Stein, R. A., Bibi, E., & Mchaourab, H. S. (2020). The multidrug transporter MdfA deviates from the canonical model of alternating access of MFS transporters. *Journal of Molecular Biology*, *432*(20), 5665–5680. https://doi.org/10.1016/j.jmb.2020.08.017.

Zheng, W., & Brooks, B. R. (2006). Modeling protein conformational changes by iterative fitting of distance constraints using reoriented normal modes. *Biophysical Journal*, *90*(12), 4327–4336. https://doi.org/10.1529/biophysj.105.076836.

CHAPTER SEVEN

Light-induced pulsed dipolar EPR spectroscopy for distance and orientation analysis

Arnau Bertran[a], Antonio Barbon[b], Alice M. Bowen[a,c,*], and Marilena Di Valentin[b,*]

[a]Centre for Advanced Electron Spin Resonance and Inorganic Chemistry Laboratory, Department of Chemistry, University of Oxford, Oxford, United Kingdom
[b]Department of Chemical Sciences, University of Padova, Padova, Italy
[c]EPSRC National Research Facility for Electron Paramagnetic Resonance Spectroscopy, Department of Chemistry and Photon Science Institute, The University of Manchester, Manchester, United Kingdom
*Corresponding authors: e-mail address: alice.bowen@manchester.ac.uk; marilena.divalentin@unipd.it

Contents

Methods in Enzymology, Volume 666
ISSN 0076-6879
https://doi.org/10.1016/bs.mie.2022.02.012

Abstract

Measuring distances in biology at the molecular level is of great importance for understanding the structure and function of proteins, nucleic acids and other biological molecules and their complexes. Pulsed Dipolar Spectroscopy (PDS) offers advantages with respect to other methods as it is uniquely sensitive and specific to electronic spin centers and allows measurements in near-native conditions, comprising the in-cell environment. PDS methods measure the electron spin–spin dipolar interaction, therefore they require the presence of at least two paramagnetic centers, which are often stable radicals. Recent developments have introduced transient triplet states, photo-activated by a laser pulse, as spin labels and probes, thereby establishing a new family of techniques—Light-induced PDS (LiPDS). In this chapter, an overview of these methods is provided, looking at the chromophores that can be used for LiPDS and some of the technical aspects of the experiments. A guide to the choice of technique that can yield the best results, depending on the type of system studied and the information required, is provided. Examples of previous LiPDS studies of model systems and proteins are given. Characterization data for the chromophores used in these studies is tabulated to help selection of appropriate triplet state probes in future studies.

Abbreviations

AT12	ATTO Thio12
DAF	Delay After Flash
DEER	Double Electron-Electron Resonance
EB	Erythrosine B
EY	Eosin Y
EPR	Electron Paramagnetic Resonance
FC LaserIMD	Frequency-Correlated Laser-Induced Magnetic Dipole Spectroscopy
FRET	Förster Resonance Energy Transfer
ISC	Intersystem Crossing
LaserIMD	Laser-Induced Magnetic Dipole Spectroscopy
LiDEER	Light-induced Double Electron-Electron Resonance
LiPDS	Light-induced Pulsed Dipolar Spectroscopy
LiRIDME	Light-induced Relaxation Induced Dipolar Modulation Enhancement
LITTER	Light-Induced Triplet–Triplet Electron Resonance
MTSSL	Methanethiosulfonate Spin Label
NIT	Nitroxide
NMR	Nuclear Magnetic Resonance
PELDOR	Pulsed Electron Double Resonance
PDS	Pulsed Dipolar Spectroscopy
RB	Rose Bengal
ReLaserIMD	Refocused Laser-Induced Magnetic Dipole Spectroscop
RIDME	Relaxation Induced Dipolar Modulation Enhancement

TAM	Triarylmethyl Radical
TPP	5-(4-carboxyphenyl)-10,15,20-triphenylporphyrin
ZFS	Zero-Field Splitting

1. Introduction

Pulsed Dipolar Spectroscopy (PDS) is a collection of Electron Paramagnetic Resonance (EPR) techniques concerned with measuring the dipolar interaction between pairs of spin centers separated on a length scale of 1.5–8 nm (Banham et al., 2008; Jeschke, 2012), with extension up to 16 nm possible if samples are fully deuterated, to extend the phase memory time, T_m, (Schmidt, Wälti, Baber, Hustedt, & Clore, 2016). Usually PDS is applied to systems containing two (or more) permanent spin-active species, such as organic radicals (e.g., nitroxide and trityl-based spin labels) or metal centers or clusters (e.g., Cu(II), Gd(III), low-spin Fe(III) and iron sulfur clusters) (Abdullin et al., 2015; Cunningham, Putterman, Desai, Horne, & Saxena, 2015; Goldfarb, 2014, 2019; Haugland, Lovett, & Anderson, 2018; Joseph et al., 2016; Roessler et al., 2010; Wort et al., 2019). While PDS techniques were developed on synthetic polymer and organic model systems (Jeschke, 2002; Tsvetkov, Milov, & Maryasov, 2008), they have been widely applied in the field of biophysical studies as the accessible length scale matches the size of biological assemblies (Abdullin & Schiemann, 2020; Astashkin, 2015; Feintuch, Otting, & Goldfarb, 2015; Jeschke & Polyhach, 2007; Kurshev, Raitsimring, & Tsvetkov, 1989; Hagelueken, Abdullin, & Schiemann, 2015). Much of biological PDS relies on the use of site-directed spin labeling with a functionalized 1-oxyl-2,2,5,5-tetra-methylpyrroline-3-methyl methanethiosulfonate, known as methanethiosulfonate spin label (MTSSL), where site-directed mutagenesis is used to incorporate a cysteine at a known position in the primary structure, which is then labeled with this thiol-specific reagent (Altenbach, Flitsch, Khorana, & Hubbell, 1989; Klare & Steinhoff, 2015).

The dipolar interactions measured can be interpreted to provide information on the separation, and in some cases, the relative orientation of the spin centers (Bowen, Tait, Timmel, & Harmer, 2013). PDS EPR is a useful tool for the study of structure (del Alamo et al., 2020) and conformational changes of proteins (del Alamo, Jagessar, Meiler, & McHaourab, 2021), their complexes (Bowen et al., 2018; Yang et al., 2020), RNA and DNA (Domnick et al., 2020; Grytz et al., 2017; Heinz et al., 2020). In biological applications, spin centers may be present innately, e.g., within the active site of an enzyme, or added to a system via site-directed mutagenesis and spin labeling.

PDS EPR measurements are often complementary to other biophysical methods such as X-ray crystallography, Nuclear Magnetic Resonance (NMR), Förster Resonance Energy Transfer (FRET) or cryo-electron microscopy (Wunnicke & Hänelt, 2017). Unlike X-ray crystallography, the system does not have to be crystallized, but rather measured in a solid matrix, such as a frozen solution, which can be disordered. This means that it is possible to study transient interactions between binding partners and regions of structural flexibility, and to study molecules in more native environments and under various conditions. Unlike single molecule techniques, e.g., cryo-electron microscopy or single molecule FRET, PDS EPR is an ensemble measurement, simultaneously recording the full set of dipolar interactions within the sample, enabling different conformational populations to be observed. Additionally, the labels used in EPR are often smaller and less flexible than those required for FRET, providing a more direct readout of the protein structure (Klose et al., 2012). The length scale of PDS EPR is significantly longer than that accessible with most NMR techniques, therefore distance restraints obtained by PDS EPR can be highly complementary to NMR in constraining structural models (Duss, Yulikov, Allain, & Jeschke, 2015). Furthermore, while NMR becomes rapidly more complex as the size of the system increases, PDS EPR does not suffer from this as it is specific to the spin centers only. Finally, the specificity and sensitivity of PDS EPR to spin-active species has allowed the techniques to be used in complex environments, such as inside cells (Azarkh et al., 2019; Kucher, Elsner, Safonova, Maffini, & Bordignon, 2021).

The biological applications of PDS EPR using stable spin centers have previously been detailed within this series (Astashkin, 2015; Feintuch et al., 2015; Hagelueken et al., 2015). In this chapter we focus on a newly implemented adaptation of PDS using light-activated triplet states as spin centers, described by the overarching term of light-induced PDS (LiPDS). The use of light-activated spin centers in PDS experiments has several potential advantages, ranging from the ability to measure in both light and dark conditions, under which the spin center is active or inactive, respectively, to the high sensitivity derived from the non-Boltzmann population of the triplet state sublevels. Additionally, the large orientation resolution of the triplet state spectrum can allow both the distance and orientation between spin centers to be resolved. Many biological molecules, such as photosynthetic proteins (Di Valentin, Buchel, Giacometti, & Carbonera, 2012; Lubitz, Lendzian, & Bittl, 2002), heme-proteins

(Ciuti et al., 2021; Dal Farra et al., 2019; Hoffman, 1975), and flavoproteins (Nohr, Rodriguez, Weber, & Schleicher, 2015), contain moieties that can act as light-activated spin centers.

1.1 The triplet state

In photoexcited systems, the triplet state is populated mostly by intersystem crossing (ISC) from an excited singlet. ISC is not the only process leading to triplet states in chromophores, but there are other mechanisms, such as charge recombination, radical pair ISC (RP-ISC) and spin–orbit charge transfer ISC (SOCT-ISC) (Imran et al., 2021). The quantum yield of the triplet formation (Φ_T) is determined by the competition of these processes with other decay processes of the excited singlet states, namely, fluorescence and internal conversion to lower energy singlet states. In turn, the triplet state can be depopulated via phosphoresce and ISC back to the ground singlet state. These processes are exemplified in the Jablonski diagram reported in Fig. 1A.

The anisotropic magnetic dipole-dipole interaction (10^2 to 10^3 MHz) between the two unpaired electrons in the triplet state is similar in nature to that used for distance measurements in PDS experiments. However, it is orders of magnitude larger, and it is responsible for the features visible in the powder-like triplet EPR spectrum. Triplet state EPR spectra are typically governed by the electronic Zeeman and zero-field splitting (ZFS) interactions (Weil & Bolton, 2006). The ZFS contribution to the Hamiltonian is $\widehat{\boldsymbol{S}}^{\boldsymbol{T}} \cdot \boldsymbol{D} \cdot \widehat{\boldsymbol{S}}$, where $\boldsymbol{D}$ is the traceless dipolar tensor (with principal values X, Y and Z). This term is responsible for lifting the degeneracy of the triplet sublevels also in zero magnetic field and leads to the name of this interaction. The ZFS interaction is also responsible for the presence of non-degenerate EPR transitions under an external field between the high field eigenstates $|T_-\rangle$, $|T_0\rangle$, and $|T_+\rangle$ (see Fig. 1B). The spectral effect is a splitting of the EPR transition bands (typically on the order of 10^2 to 10^3 G, which does not depend on the frequency of the spectrometer) and the appearance of three pairs of features in the powder spectrum, corresponding to resonance conditions with the external magnetic field oriented along the principal direction of the ZFS tensor, the so-called canonical orientations. The position of these turning points depends on the magnitude of the principal values of the ZFS tensor, $\boldsymbol{D}$; because of the zero-trace condition, the principal values X, Y and Z can be expressed as function of two ZFS parameters: D and E. An example is shown in Fig. 1B for a triplet state

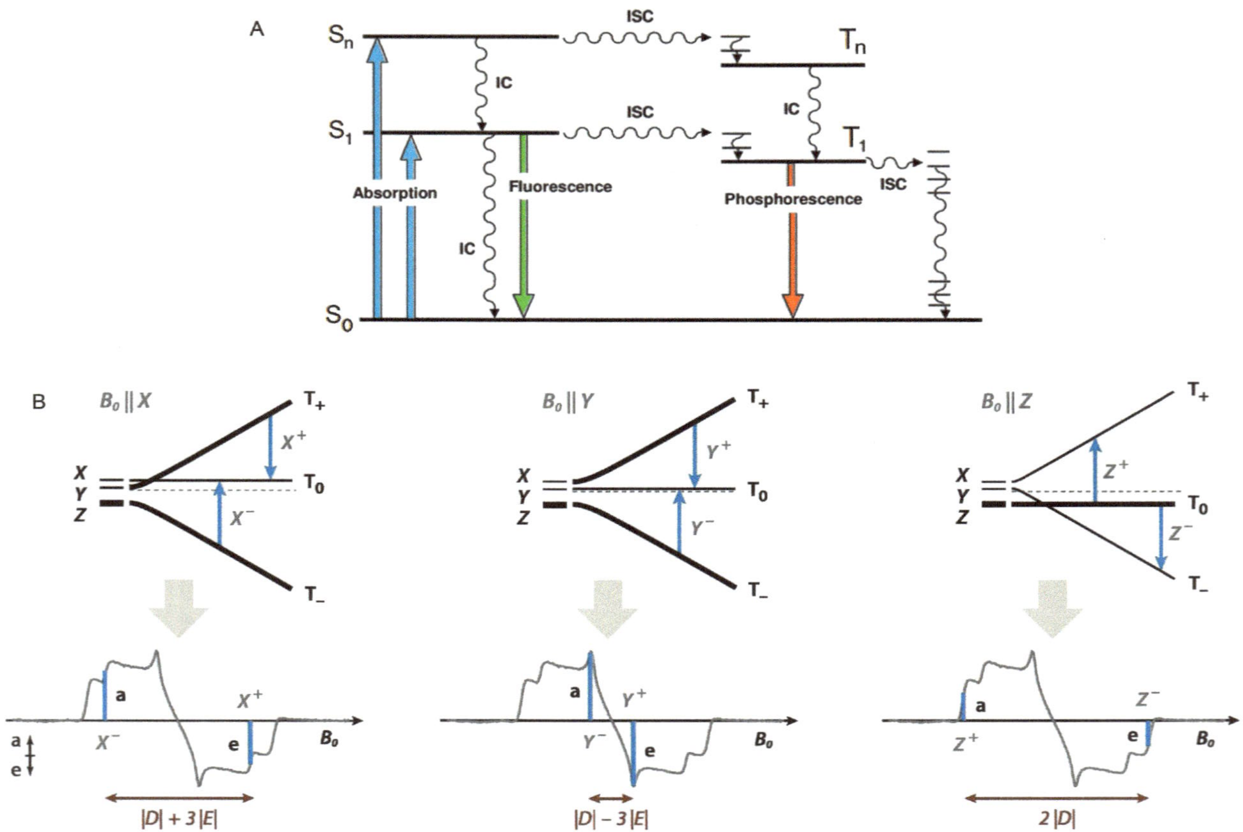

Fig. 1 See figure legend on opposite page.

with $D > 0$, $E < 0$ and $|D|>3|E|$ (Richert, Tait, & Timmel, 2017), which is the case for porphyrin derivatives (Barbon et al., 2020; Bolzonello, Albertini, Collini, & Di Valentin, 2017). Triplet state spectra are typically recorded at cryogenic temperature in frozen solutions using time-resolved EPR techniques.

The enhanced absorption and emission observed in a triplet spectrum arise from another important feature of photoexcited triplet states: the presence of electron spin polarization. This is a result of the mechanisms by which the triplet is formed, leading to significant deviation from the Boltzmann distribution of the spin state populations. In some cases the opposite polarization can occur, having a larger population in the upper energy level than in the lower energy level (Hintze, Steiner, & Drescher, 2017). Both absorptive and emissive transitions are observed and the triplet state spectrum is referred to as being spin polarized.

In order to exploit the triplet state as a paramagnetic probe in PDS experiments, it is important to consider the spin relaxation parameters of the specific chromophore selected. T_1 is the longitudinal spin relaxation, which accounts for the persistence of the spin magnetic component parallel to the magnetic field, when the magnetization is perturbed from the equilibrium value. In the case of the triplet state, this parameter encodes for the relaxation of both the initial inverted spin polarization and any population perturbations that are a result of applying microwave excitation. The T_2 parameter describes the irreversible dephasing of the magnetic spin component perpendicular to the magnetic field. In PDS experiments involving the formation of an echo, it determines the maximum time window that can

Fig. 1 (A) Jablonski diagram for the formation of a triplet state and its decay. (B) Spin energy level diagrams as a function of an external magnetic field with the field vector, $\boldsymbol{B}_0$, aligned with the dipolar axes X, Y and Z, respectively. In the figure, a triplet with $D > 0$ and $E < 0$ and $|D|>3|E|$ is considered. Whenever the energy of the microwave radiation matches the energy gap between $|T_0\rangle$ and either $|T_-\rangle$ or $|T_+\rangle$, a transition can be observed. The relative populations (non-Boltzmann) and the transitions are provided. Therefore, the transitions can be either emissive (e) or enhanced absorptive (a) depending on the relative populations of the high-field sublevels involved, indicated by the thickness of the level bars. In a powder spectrum, the transitions corresponding to the canonical field orientations are characterized by turning points as illustrated by the spectrum shown for a zinc porphyrin monomer triplet state. A single zero-field state, the Z zero-field sublevel, is primarily populated. *Panel (B) reprinted (adapted) from Richert, S., Tait, C. E., & Timmel, C. R. (2017). Delocalisation of photoexcited triplet states probed by transient EPR and hyperfine spectroscopy.* Journal of Magnetic Resonance, 280, *103–116. doi:10.1016/j.jmr.2017.01.005. Copyright 2017 Elsevier.*

be obtained between a refocusing pulse and the echo formation. To be more precise, it is also necessary to consider the contribution to the relaxation induced by the pulses (in general not very large) and this contribution is accounted for in the phase memory time (T_m). Implications for the selection of chromophores and relaxation properties of those chromophores that have been successfully used for LiPDS EPR are provided in Section 7 of this chapter.

2. The theory of pulsed dipolar spectroscopy

PDS EPR methods rely on changing the spin state of one of the spin centers of a pair of coupled spin centers by a value of $\Delta M_S = \pm 1$ at a variable time during a detection sequence. This can be accomplished, for example, by applying a microwave inversion pulse (pump) to one of the two partner species or by formation of the triplet state by a laser pulse (photoexcitation) as it is discussed in this chapter. This perturbation leads to a change in the magnetic environment at the partner spin. Detection of the signal from this partner spin while shifting the perturbation in time results in an oscillating dipolar trace, which can be interpreted to determine the distance distribution and, in cases of rigid spin labels, the relative orientation between the spin centers.

In the case of two permanent spin labels, the secular Hamiltonian of the spin system is described by:

$$\hat{H} = \left[\sum_{i=A,B} \frac{1}{\hbar} \boldsymbol{B}_0^T \cdot \boldsymbol{g}_i \cdot \hat{\boldsymbol{S}}_i \right] + \frac{\mu_0 g_A g_B \beta^2}{4\pi\hbar r^3} \left\{ \hat{S}_{A,z} \hat{S}_{B,z} \left(1 - 3\cos^2(\theta)\right) \right\}$$

The first term relates the interaction of the spins (represented by the spin operator $\widehat{\boldsymbol{S}}_i$) with the external magnetic field vector ($\boldsymbol{B}_0^T$), where $\boldsymbol{g}_i$ is the $\boldsymbol{g}$-tensor, and the second term gives the secular dipolar coupling between the electron spins. The "T" indicates a matrix/vector transpose. From this we can define the dipolar coupling frequency (ω_{dd}), which is the frequency of oscillation of a PDS trace. The dipolar frequency is dependent on the spin–spin separation (r) and angle (θ) between the inter-spin dipolar vector and the applied field (as illustrated in Fig. 2):

$$\omega_{dd} = \frac{\mu_0 g_A g_B \beta^2}{4\pi\hbar r^3} \left(1 - 3\cos^2(\theta)\right)$$

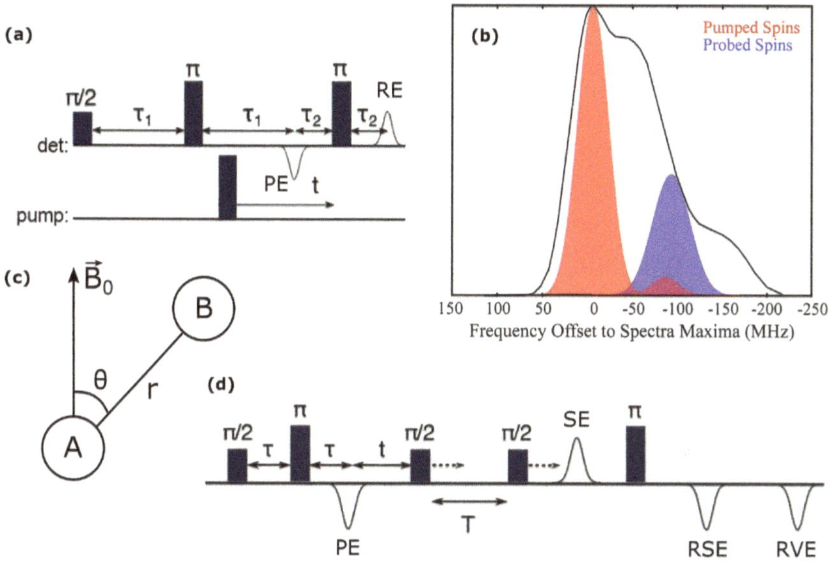

Fig. 2 (A) The 4-pulse DEER/PELDOR sequence. (B) Frequency-domain Q-band (~34 GHz) spectrum of a nitroxide spin label (*black*) plotted with the excitation profiles of 16 ns pump and detection pulses. The pump-detection offset is 100 MHz. (C) Schematic representation of two spins (A and B) connected via a distance vector *r* (dipolar vector). θ is the angle between the dipolar vector and the applied magnetic field $\boldsymbol{B}_0$. (D) The 5-pulse RIDME sequence. *Panel (B) reprinted (adapted) with permission from Schiemann, O., Heubach, C. A., Abdullin, D., Ackermann, K., Azarkh, M., Bagryanskaya, E. G., … Jeschke, G. (2021). Benchmark test and guidelines for DEER/PELDOR experiments on nitroxide-labeled biomolecules.* Journal of the American Chemical Society, 143*(43), 17875–17890. doi:10.1021/jacs.1c07371. Copyright, 2021 American Chemical Society.*

The two most frequently used PDS techniques for measuring the dipolar interaction are Double Electron-Electron Resonance (DEER)/Pulsed Electron Double Resonance (PELDOR) and Relaxation Induced Dipolar Modulation Enhancement (RIDME) (Martin et al., 1998), which are described below. Their pulse sequences are given in Fig. 2.

2.1 DEER

DEER uses two microwave frequencies. The first frequency (blue in Fig. 2B) is on resonance with a set of detection spin packets and is used to generate a refocused echo for detection. The second frequency (red in Fig. 2B) is on resonance with a second set, the pump spins. The pump spins are inverted by a time dependent π pulse, the "pump pulse"; in the presence of an interaction, this causes a perturbation on the detection spins, a condition required to observe the dipolar interaction as an oscillation in the detection spin echo. Where possible, a refocused echo is used, for example in the 4-pulse DEER sequence, Fig. 2A, as it provides a well-defined zero time for the dipolar oscillation when the pump pulse crosses the initial Hahn echo. This is helpful for accurate analysis as described in Section 3.3 in relation to a specific LiPDS sequence based on the refocused echo. While the DEER experiment for measuring distances between nitroxide spin labels is fairly robust, care must be taken during data analysis, particularly in the fitting of backgrounds to the data (Schiemann et al., 2021).

2.2 RIDME

RIDME is a single frequency PDS technique. It replaces the "pump pulse" of the DEER experiment with a "mixing block" sandwiched by two $\pi/2$ pulses at the same frequency as the rest of the pulse sequence (Milikisyants, Scarpelli, Finiguerra, Ubbink, & Huber, 2009). The effect of the $\pi/2$ pulse before the mixing time (T) is to flip the magnetization from the x-y plane to the z-axis, storing it there for the duration of T, as it is no longer subject to T_2 relaxation. During T, some of the coupled spins will naturally invert due to T_1 relaxation. Finally, the detection spins are returned to the x-y plane by application of the second $\pi/2$ pulse. In this experiment the longitudinal relaxation is used to "pump" the spins inducing changes in dipolar interactions. For this reason, RIDME works best when one of the spins relaxes faster than the other, such that the slow relaxing species can be used for detection without significant loss of spin center due to T_1 relaxation of the detection signal during T.

2.3 Orientation selection

In cases where there is resolved anisotropy in the spectrum, and the bandwidth of the pulses is narrow compared to the spectral width, as for frequency-correlated techniques, orientation selection may be observed. This is a result of only a limited number of orientations of the dipolar vector with respect to the applied magnetic field (θ) contributing to the dipolar frequency. It is most pronounced when rigidly linked spin centers are used and the spectra show well-resolved anisotropy, for example, from the g-tensor. Orientation-selective PDS measurements between two permanent spins have been reviewed extensively (Bowen et al., 2013; Giannoulis et al., 2018). Although the additional information contained within the PDS traces can be vital for determining molecular structures, the analysis required to extract this information is more complex and not as straightforward as for orientation-independent datasets. This is exemplified in Section 2.4.

2.4 Analysis of PDS traces

Raw PDS traces are comprised of two contributions to the signal $V(t)$, the oscillating part due to the intra molecular dipolar interactions ($V_{intra}(t)$) and a background component ($V_{Background}(t)$) (Jeschke, Koch, Jonas, & Godt, 2002):

$$V(t) = V_{intra}(t) \cdot V_{Background}(t)$$

The function describing the background component depends on both the system and the PDS experiment used. In the case of a DEER experiment, the background component is due solely to the intermolecular dipolar interactions. In a homogenous sample this will be a three-dimensional monoexponential background; however, the dimensionally can be reduced if the distribution is not homogenous. For example, if the molecules are arranged on a two-dimensional membrane, the dimensionality will reduce towards two, and the background will no longer be a monoexponential. Also, in a RIDME experiment the background is a complex function, as it includes perturbation effects from spin relaxation, therefore it may be reasonable to fit it with a low-order polynomial if the exact form cannot be calculated (Keller et al., 2019).

PDS traces can be analyzed using DeerAnalysis (Jeschke et al., 2006), where the background correction form is chosen on the base of the experiment and of the system. The resultant time trace, also called form factor, has an initial decay followed by oscillations around zero (Fig. 3B). The modulation depth value (Δ) is the initial decay, scaled on the base of the unprocessed echo intensity at time $t=0$; this parameter provides information on the efficiency of the experiment in terms of the number of coupled spins and the efficiency of the pump process. The Fourier transform of the form factor, with the modulation depth normalized to 1, yields a Pake pattern or partial Pake pattern, if only some values of θ contribute to the dipolar trace due to orientation selection (Fig. 3C). In the case of orientation-independent traces, these can be directly analyzed using an orientation-independent kernel such as that implemented in DeerAnalysis (Jeschke et al., 2006), which uses a Tikhonov regularization to extract a distance distribution (Fig. 3D).

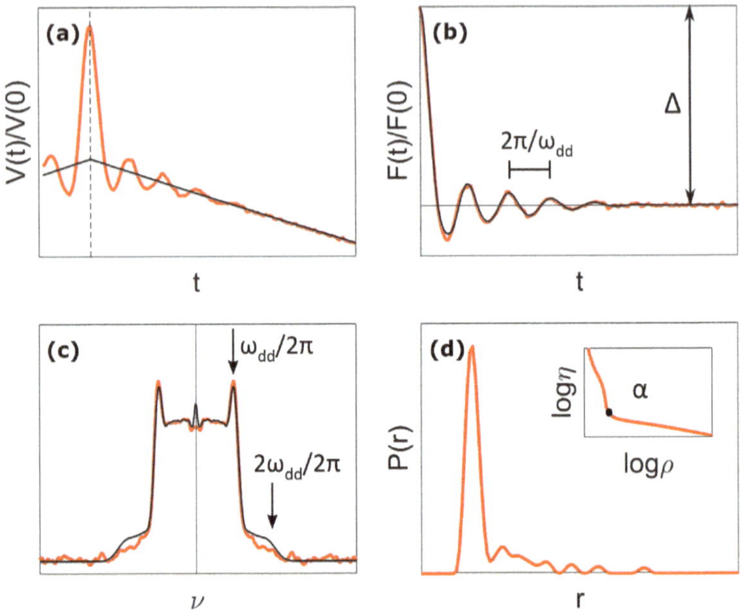

Fig. 3 (A) An example of a PDS time trace as recorded, the background signal is shown as a *thin black line*. (B) Background-corrected time trace with the modulation depth, Δ, and the dipolar frequency, ω_{dd}, indicated. (C) The Fourier transform of the background-corrected time trace with the modulation depth normalized to 1 to remove the zero frequency components yields a Pake pattern as shown. The corresponding fit is also shown. If orientation selection is present, this will yield a partial Pake pattern. (D) Tikhonov regularization using DeerAnalysis of a complete Pake pattern allowing the distance distribution to be extracted. *Inset*: regularization curve showing the optimal regularization parameter (α).

Where orientation selection is present, it may be necessary to simulate a library of data corresponding to possible molecular orientations and then fit this to the experimental data. Alternatively, neural network analysis has been shown to be able to include orientation selection in the datasets (Keeley et al., 2021).

3. LiPDS methods

LiPDS methods are designed to measure the dipolar interaction either between a permanent spin center and a photoexcited triplet state or between two photoexcited triplets.

In the first case, of one permanent spin center (A) and a triplet state (B), the secular Hamiltonian of the spin system is described by:

$$\hat{H} = \left[\sum_{i=A,B} \frac{1}{\hbar} \boldsymbol{B}_0^T \cdot \boldsymbol{g}_i \cdot \hat{\boldsymbol{S}}_i\right] + \hat{\boldsymbol{S}}_B^T \cdot \boldsymbol{D}_B \cdot \hat{\boldsymbol{S}}_B + \frac{\mu_0 g_A g_B \beta^2}{4\pi\hbar r^3}\left\{\hat{S}_{A,z}\hat{S}_{B,z}\left(1 - 3\cos^2(\theta)\right)\right\}$$

This also includes the ZFS term, $\hat{\boldsymbol{S}}_B^T \cdot \boldsymbol{D}_B \cdot \hat{\boldsymbol{S}}_B$. In the case of two triplets with a dipolar interaction between them, the Hamiltonian above is modified to include a ZFS term for the other spin center.

All LiPDS methods rely on using one or more laser pulses to generate a triplet state that can be used for detection and/or using a variable time light pulse to generate the "pump" triplet state. The first LiPDS method published was a modification of the DEER experiment, Light-induced DEER (LiDEER) (Di Valentin, Albertini, Zurlo, Gobbo, & Carbonera, 2014), using a light-activated triplet as the detection spin. Subsequently, several other methods have been proposed, namely Laser-Induced Magnetic Dipole Spectroscopy (LaserIMD) (Hintze, Bücker, Domingo Köhler, Jeschke, & Drescher, 2016) and its refocused form, Refocused Laser-Induced Magnetic Dipole Spectroscopy (ReLaserIMD) (Dal Farra et al., 2019), Light-induced RIDME (LiRIDME) (Dal Farra, Richert, et al., 2019), Frequency-Correlated LaserIMD (FC-LaserIMD) (Bowen et al., 2021) and Light-Induced Triplet–Triplet Electron Resonance (LITTER) (Bertran et al., 2021). The pulse sequences for these techniques are provided in Fig. 4 and the methods are detailed in the following sections.

3.1 LiDEER

LiDEER is a modification of the DEER sequence used to measure the dipolar interaction between one permanent spin center and a photo-activated triplet state. The microwave pulse sequence is identical to that of the DEER experiment but it is preceded by a laser pulse to photoexcite the chromophore, used as the detection spin (Di Valentin, Albertini, et al., 2014). It was initially tested by Di Valentin et al. on a series of peptides with a nitroxide-containing 4-amino-1-oxyl-2,2,6,6-tetramethyl-piperidine-4-carboxylic acid (TOAC) moiety and a 5-(4-carboxyphenyl)-10,15,20-triphenylporphyrin (TPP) attached to the N-terminus (see Fig. 5) (Di Valentin et al., 2016). Here the

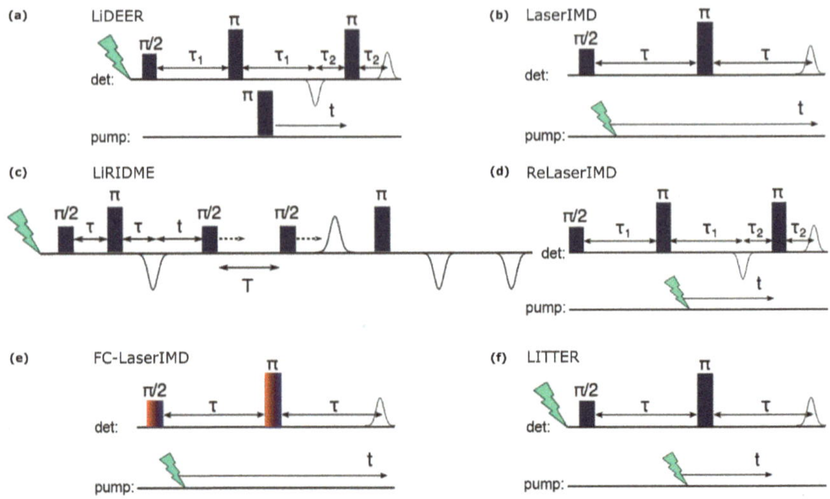

Fig. 4 Pulse sequences for LiPDS techniques: (A) Light-induced Double Electron–Electron Resonance (LiDEER). (B) Laser-Induced Magnetic Dipole (LaserIMD) Spectroscopy. (C) Refocused Laser-Induced Magnetic Dipole (ReLaserIMD) Spectroscopy. (D) Light-induced Relaxation Induced Dipolar Modulation Enhancement (LiRIDME).(E) Frequency-Correlated LaserIMD (FC LaserIMD). (F) Light-Induced Triplet–Triplet Electron Resonance (LITTER).

photo-activated porphyrin triplet state was used as the detection spin center and a microwave pump pulse was applied to the nitroxide radical. A refocused echo detection sequence (4-pulse DEER) was used to yield a dead-time free trace with a clear zero time. The experiments were performed at X-band and the pump pulse was set to be resonant with the center of the nitroxide spectrum. This feature comprises all orientations of the nitrogen $m_I = 0$ level, so orientation selection was negligible in these experiments. Orientation selection could be maximized for these experiments if the pump pulse was set resonant to different positions across the outer nitrogen $m_I = \pm 1$ components.

Using the photo-activated triplet for detection increases the sensitivity and the signal-to-noise ratio of the recorded traces due to the non-Boltzmann polarization of the triplet state. The length scale that can be measured in this experiment is primarily limited by the T_2 or T_m of the triplet species. The modulation depth of the collected data is determined by the inversion efficiency of the microwave pump pulse and this can be improved by using a chirp, or in theory, other adiabatic shaped pulses (Bieber, Bücker, & Drescher, 2018). These pulses can be generated using an Arbitrary Waveform Generator (AWG) and can be designed to have a wide bandwidth and high inversion efficiency. With careful implementation, they can also be designed to compensate for changing microwave power (B_1) accessible at different frequencies due to the bandwidth of the resonator. However, care must be taken to ensure that these pulses are not too long, as evolution of the dipolar coupling within the

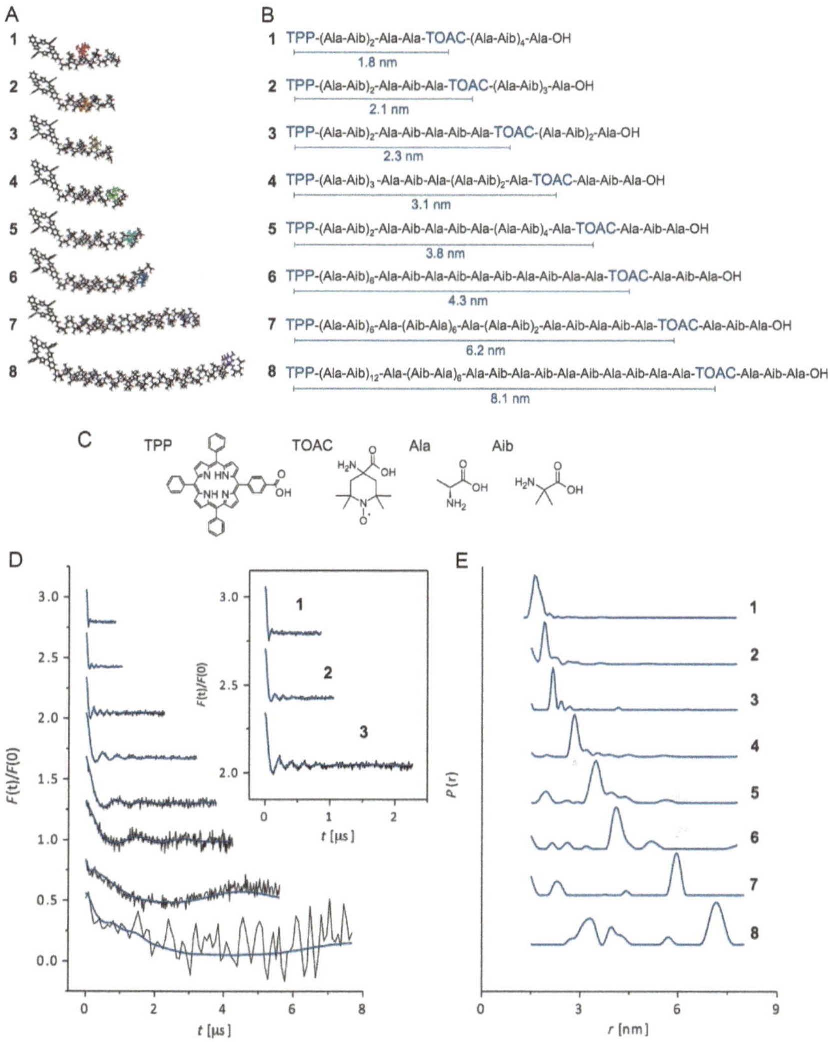

Fig. 5 Density Functional Theory (DFT)-optimized structures (A) and sequences (B) of the model systems, with the inter-spin distances indicated. (C) Building blocks for the model systems. (D) Background-corrected LiDEER traces for the systems. (E) Corresponding distance distributions. *Reprinted (adapted) with permission from Di Valentin, M., Albertini, M., Dal Farra, M. G., Zurlo, E., Orian, L., Polimeno, A., … Carbonera, D. (2016). Light-induced porphyrin-based spectroscopic ruler for nanometer distance measurements.* Chemistry - A European Journal, *22(48), 17204–17214. doi:10.1002/chem.201603666. Copyright 2016 John Wiley and Sons.*

pulse length can lead to distortion of the experimental time trace. Several reviews on the application of shaped pulses to PDS EPR have previously been published (Spindler, Schöps, Bowen, Endeward, & Prisner, 2016; Spindler, Schöps, Kallies, Glaser, & Prisner, 2017).

3.2 LaserIMD

LaserIMD is also used to measure the dipolar interaction between one permanent spin center and a photo-activated triplet state. However, unlike LiDEER, in LaserIMD the permanent spin center is used for detection and a variable-time laser pulse is used to generate the photo-activated triplet during the microwave pulse sequence, this can be thought of as an optical pump pulse (Hintze et al., 2016).

Generation of the photo-activated triplet by the laser pulse changes the spin state of this moiety from $S=0$ in the singlet state to $S=1$ in the triplet state which can lead to $\Delta M_S = \pm 1$ if the triplet adopts $M_S = +1$ or $M_S = -1$ configurations. This change induces a variation in the dipolar interaction and the effect of the laser pulse is equivalent to the microwave pump pulse in the DEER experiment. If a $M_S = 0$ configuration is adopted, then $\Delta M_S = 0$ and there is no change in the magnetic environment at the partner spin and no dipolar oscillation is observed from these spin pairs. The initial experiments were performed by Hinze *et al.* on the system labeled **3** in Fig. 5, the data recorded are shown in Fig. 6 (Hintze et al., 2016).

The detection sequence used in LaserIMD is a Hahn echo sequence, which yields a trace that does not have a well-defined zero time. However, as the same dipolar oscillation is present at both the start and end of the trace, the zero time can be found by knowing the length of the pulses and delays in the pulse sequence and symmetrizing the trace (Hintze et al., 2016). This has the advantage that both sides of the trace can be summed to increase the signal-to-noise ratio. The use of a Hahn echo for detection follows the same principle as for 3-pulse DEER. However, unlike in 3-pulse DEER, where the start of the trace can be obscured by a dead-time artefact as a result of the Traveling-Wave Tube (TWT) amplifier nonlinearity observed when two microwave pulses cross, this is not the case in LaserIMD as the laser pulse does not require TWT amplification. Hence, there is no dead-time in this experiment and accurate determination of the zero time depends solely on how easy it is to accurately symmetrize the trace. This depends on being able to record the complete trace to a good signal-to-noise level and on how symmetric the trace is. Asymmetry in the trace is likely a result of different background contributions being present in each half of the trace (see Fig. 6A).

Care should be taken in applying a background correction to LaserIMD traces. Generation of the triplet state during the evolution time of the permanent spin induces not only a dipolar oscillation but the increase in the number of spin-active species causes a time-variant change in the T_m of

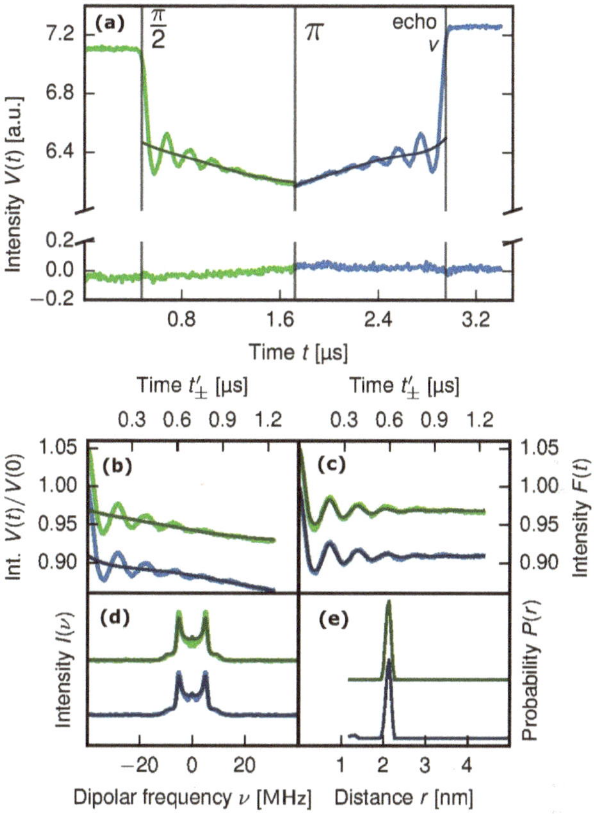

Fig. 6 LaserIMD experiment on a 0.1 mM solution of system **3** in Fig. 5, in methanol-d_4 with 2 vol% D_2O performed in the Q-band spectrometer at 10 K. Light excitation was carried out at 351 nm with pulse energies of 3 mJ. (A) Raw data of the time domain signal with the background contribution determined experimentally. The top and bottom traces correspond to the real and imaginary parts of the quadrature signal, respectively. (B) Comparison of both branches (*blue*, right, and *green*, left) of the LaserIMD raw data. (C) Corresponding form factor after experimental background correction with a model-free fit obtained by Tikhonov regularization. (D) Fourier-transformed dipolar coupling spectra with fit. (E) Corresponding distance distribution with a mean distance of 2.12 nm and a width of 0.06 nm. *Reprinted (adapted) with permission from Hintze C., Bücker D., Domingo Köhler S., Jeschke G., Drescher M. (2016), Laser-induced magnetic dipole spectroscopy.* Journal of Physical Chemistry Letters, *7(12), 2204–2209, doi:10.1021/acs.jpclett.6b00765. Copyright 2016 American Chemical Society.*

the permanent spin. This change causes the background form to differ from the expected three-dimensional exponential for a homogenous sample and also causes the background to differ between the two halves of the trace, which may hamper the symmetrization required to determine the zero time.

3.3 ReLaserIMD

To combat the problem of the ill-defined zero time in LaserIMD, described above, the Hahn echo detection sequence can be replaced with a refocused echo detection sequence, yielding the ReLaserIMD experiment (Dal Farra, Richert, et al., 2019). This borrows the idea of using a refocused echo for detection from 4-pulse DEER, rendering a trace where there is a defined zero time occurring at the time of the initial Hahn echo (see Fig. 7). If the zero time is not accurately defined, then this can lead to artefacts in the distance distribution after analysis. This technique was first applied to the model peptide **3** in Fig. 5.

Unlike 4-pulse DEER, where only the pump microwave pulse is moving, in the implementation of ReLaserIMD (and LaserIMD) it is frequently easier to have the laser pulse static in time and move the microwave pulses relative to this light pulse. As the microwave pulses can usually only be displaced in a positive direction, in effect this leads to the laser pulse moving backwards through the sequence. In this setup it can be easier to have $\tau_1 > \tau_2$, which is opposite to the standard 4-pulse DEER setup, to yield a trace where the zero time occurs at an early time. Alternatively, if $\tau_1 < \tau_2$ is used, this will yield a trace where the zero time occurs at a later time and the time axis must be inverted before analysis with software such as DeerAnalysis (Jeschke et al., 2006).

ReLaserIMD has the advantage over LaserIMD of providing a well-defined zero time. However, it can lead to lower signals than LaserIMD as spin–spin relaxation is more effective during the longer sequence and signal is lost with the application of each pulse due to pulse imperfections. ReLaserIMD therefore may be undesirable in systems where there is a large separation between the spin centers, where small variations in the chosen position of the zero time lead to only very small changes in the distance distribution and a long trace is required to measure sufficient oscillations. Conversely, ReLaserIMD is important for the measurement of systems with small inter-spin separations where only a short trace is required to record the dipolar oscillation frequency and accurate determination of the zero time is vital for correct analysis.

3.4 LiRIDME

LiRIDME is based on the RIDME pulse sequence with an additional laser pulse, to populate the triplet state, before the microwave pulses (Dal Farra, Ciuti, Gobbo, Carbonera, & Di Valentin, 2019). It relies on using the

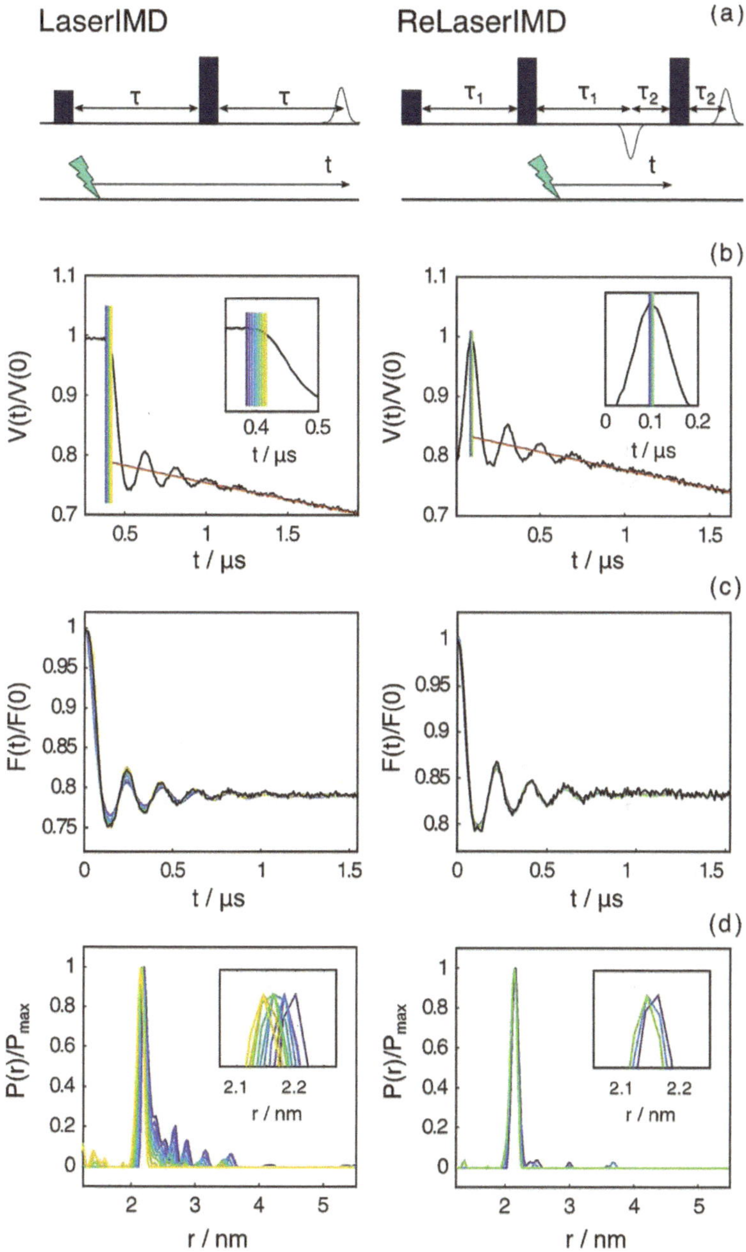

Fig. 7 Influence of the zero-time determination on the distance analysis in LaserIMD and ReLaserIMD measured on system **3** in Fig. 5. For both datasets: pulse sequences (A), raw dipolar time traces with the selected zero-time positions used in the distance analysis (B), form factors with the corresponding fits (C) and distance distributions (D) obtained by DeerAnalysis. *Reproduced (adapted) from Dal Farra, M. G., Ciuti, S., Gobbo, M., Carbonera, D., & Di Valentin, M. (2019). Triplet-state spin labels for highly sensitive pulsed dipolar spectroscopy.* Molecular Physics, 117*(19), 2673–2687. doi:10.1080/00268976.2018.1503749. Copyright 2019 Wiley and Sons.*

longitudinal relaxation of one of the spin species changing the spin state of one of the coupled spin centers during the mixing time (see Fig. 4C). The first example of LiRIDME was measured on a zinc-protoporphyrin IX-substituted and nitroxide-labeled neuroglobin (ZnG19G) (see Fig. 8A). In this case, because the chromophore is a metalled porphyrin, $T_{1,\text{triplet}} < T_{1,\text{nitroxide}}$ under the measurement conditions of 20 K and hence the nitroxide was chosen as the detection species. As the triplet moiety has three energy levels, relaxation is possible to provide $\Delta M_S = \pm 2$ in addition to the desired $\Delta M_S = \pm 1$ change, this leads to the presence of overtones in the data set, seen as a faster oscillation, particularly evident in the first modulation period compared to ReLaserIMD measured on the same system (see Fig. 8C). These overtones must be considered in the analysis and, in this example, a 50 % contribution from the overtone was observed.

3.5 Frequency-correlated LaserIMD and combination with LiDEER for orientation analysis

In the case of systems where the positions of the labels are relatively rigid, it is possible to obtain important information about the relative orientation of the spin centers using orientation selection. This occurs when the spectrum of a species is anisotropic, due either to $\boldsymbol{g}$-tensor (pump pulse on the nitroxide) or $\boldsymbol{D}$-tensor anisotropy (pump pulse on the triplet state), and the dipolar vector has a rigid orientation with respect to the anisotropic tensor. Probing different orientational populations of the spectrum, for example using selective microwave pulses or using a frequency-correlated technique, will allow only certain orientations of the dipolar vector with respect to the applied magnetic field ($\boldsymbol{B}_0$) to be observed. Careful simulation and analysis of the data allows the relative orientation of the two spin centers to be determined.

Application of orientation selection to LiPDS measurements is much less developed, although the principles behind it are the same as for other PDS methods. The first published example for this type of experiment was performed on the model peptide **3** in Fig. 5 (Bowen et al., 2021). ReLaserIMD traces were recorded at different field positions across the nitroxide spectrum at Q-band (see Fig. 9A and B). Since in this experiment it was determined that the laser pulse, working as a pump pulse, excited all orientations of the porphyrin equally, only the spatial position of the porphyrin center relative to the nitroxide $\boldsymbol{g}$-tensor (distance vector) could be determined (see Fig. 9C). To try to resolve the orientation of the porphyrin $\boldsymbol{D}$-tensor, a series of LiDEER experiments were performed (see Fig. 9D-I), where the most intense Y canonical orientations of the triplet state powder spectrum

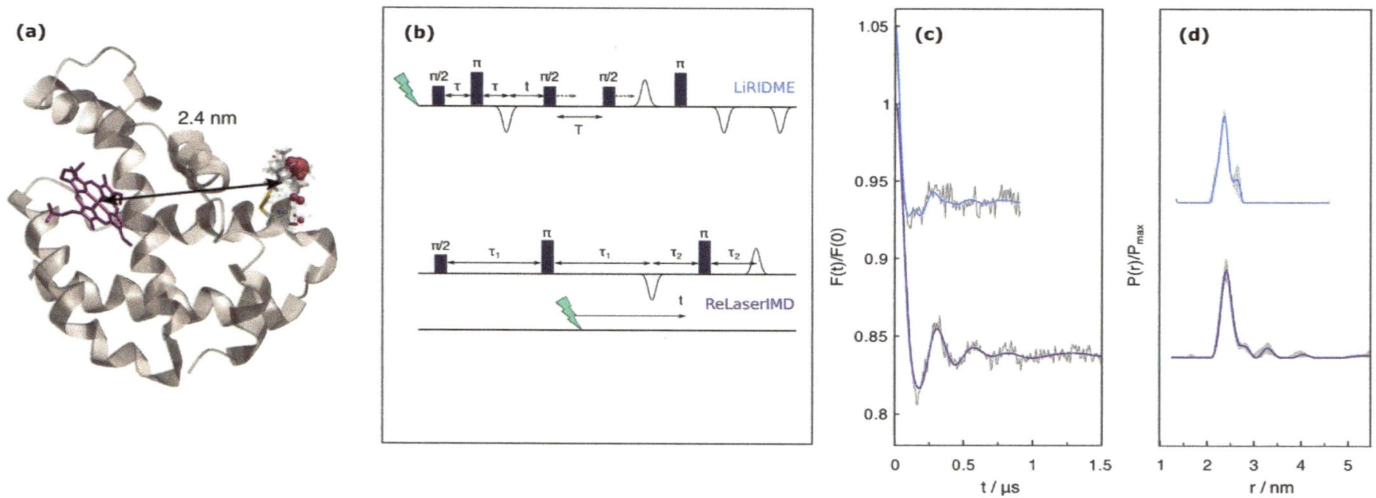

Fig. 8 (A) Ribbon structure of ZnG19, a single-cysteine mutant of human neuroglobin, whcih is nitroxide-labeled and reconstituted with Zn(II) proto-porphyrin IX. The predicted spin density distribution of the nitroxide is shown in magenta. (B) LiRIDME and ReLaserIMD pulse schemes. (C) Form factors (*gray*) and best fits to the LiRIDME (*azure*) and the ReLaserIMD (*violet*) data. (D) Corresponding distances distributions. The distance analyses have been performed with DeerAnalysis, for ReLaserIMD, and with OvertoneAnalysis for LiRIDME (with 50% contribution of the second harmonic overtone). The error bars have been obtained using the validation procedure, implemented in both software, varying the starting point for the background fitting between 300 and 500 ns and adding 50% of the original noise. *Reproduced (adapted) from Dal Farra, M. G., Ciuti, S., Gobbo, M., Carbonera, D., & Di Valentin, M. (2019). Triplet-state spin labels for highly sensitive pulsed dipolar spectroscopy.* Molecular Physics, 117*(19), 2673–2687. doi:10.1080/00268976.2018.1503749. Copyright 2019 Wiley and Sons.*

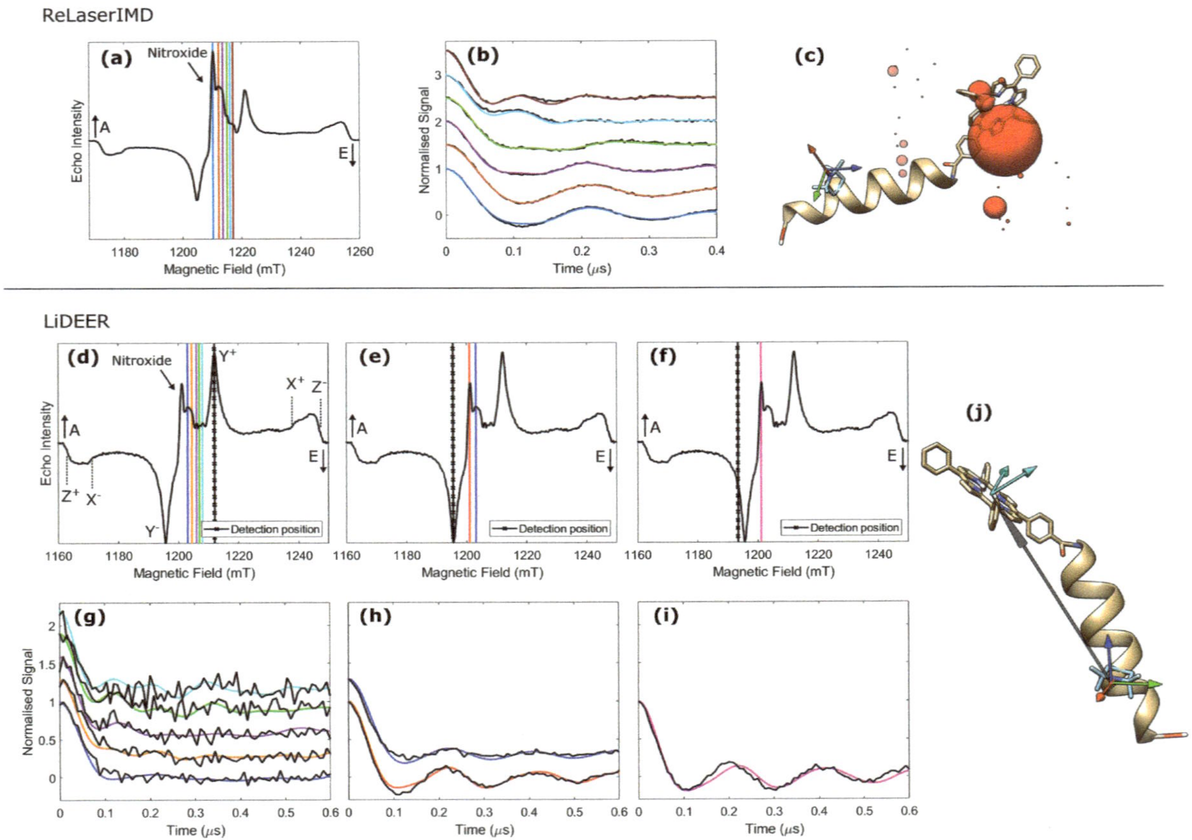

Fig. 9 See figure legend on opposite page.

of the porphyrin chromophore were targeted. While the offset between pump and probe frequencies was limited by the resonator bandwidth, these experiments were sufficient to determine the dominant orientations of the ***D***-tensor (see Fig. 9J). Simulations to determine the relative position and orientation of the two paramagnetic probes were performed using a previously published algorithm (Lovett et al., 2009) and fitted to the experimental data using a library approach similar to that described by Marko et al. (Marko & Prisner, 2013).

Recent developments in PDS applied to systems containing permanent spin centers have primarily revolved around the implementation of shaped pulses generated using an AWG. Pulses can be designed to have a large bandwidth and to be adiabatic, having a turning angle invariant to B_1. If broadband pulses are used in the detection sequence, it is possible to record the complete echo envelope, whose Fourier transform yields the frequency-correlated spectrum. Implementing such pulses in a LaserIMD experiment enables a 2D frequency-correlated experiment to be performed. An example of this experiment, recorded on the model peptide **3**, is presented in Fig. 10 along with a corresponding simulation derived from the orientations fitted to the ReLaserIMD experiments presented in Fig. 9. This is a valuable result as it shows how the orientation dependence changes across the entire

Fig. 9 (A) Echo-detected field-swept spectrum of peptide **3** measured using parameters optimized for the nitroxide radical. The experimental field values used for ReLaserIMD are indicated as vertical lines. (B) Measured ReLaserIMD traces (*black*) shown with the orientation-dependent simulations (*different shades* correspond to field positions in panel A). The modulation depth has been normalized to 1 for ease of analysis. (C) Density Functional Theory (DFT)-optimized structure of peptide **3** showing the different positions of the porphyrin center determined by the fitting procedure as *spheres*, relative to the nitroxide ***g***-tensor frame (arrows: *red* = gx, *green* (lightest) = gy, *blue* (darkest) = gz). The diameter of the sphere is proportional to the number of times a single porphyrin position contributes to the complete fit shown in panel (B). (D–F) Electron spin echo field swept spectra showing the detection (*black crossed vertical lines*) and pump (*gray/colored vertical lines*) positions used for LiDEER. (G–I) Experimental (*black*) and simulated (*gray/colored*) LiDEER traces. The colors of the simulated traces correspond to the pump positions shown in panels (D–F). (J) DFT-optimized structure of peptide **3** showing the positions and z axes of the ZFS tensors of the porphyrin centers (*small arrows*) most contributing to the best fitting set of simulated LiDEER traces, relative to the nitroxide ***g***-tensor frame (arrows: *red* = gx, *green* (lightest) = gy, *blue* (darkest) = gz) and the spin–spin vector (*gray arrow*). *Reprinted (adapted) with permission from Bowen, A. M., Bertran, A., Henbest, K. B., Gobbo, M., Timmel, C. R., & Valentin, M. Di. (2021). Orientation-selective and frequency-correlated light-induced pulsed dipolar spectroscopy.* The Journal of Physical Chemistry Letters, 12, *3819–3826. doi:10.1021/acs.jpclett.1c00595. Copyright 2021 American Chemical Society.*

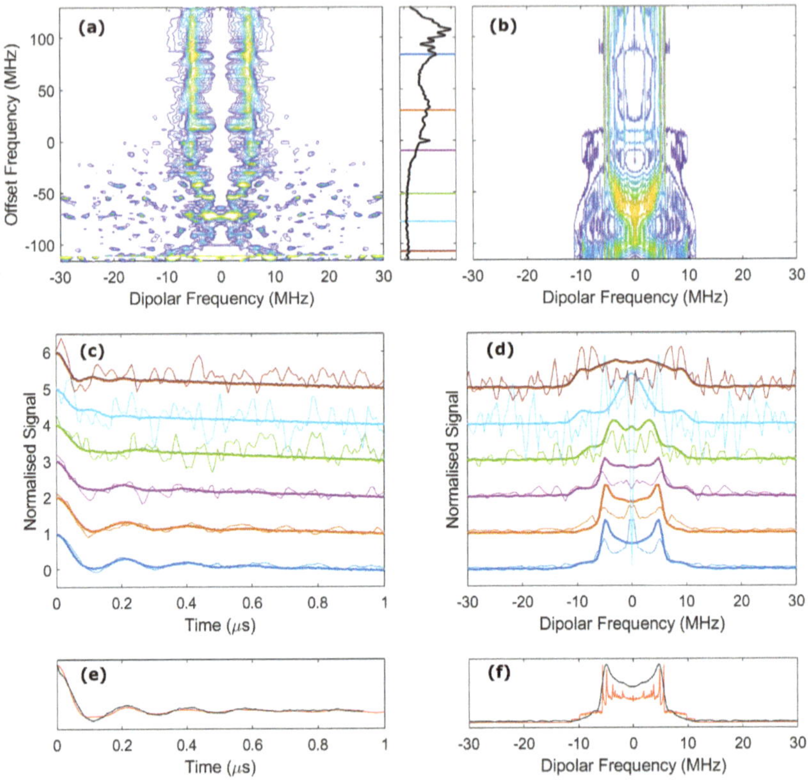

Fig. 10 (A) Complete normalized experimental FC ReLaserIMD spectrum and (B) corresponding simulation with the panel between panels (A and B) displaying the Fourier transform of the FC ReLaserIMD echo-recorded at the zero time of the experiment, with the experimental frequency values used for the ReLaserIMD indicated as horizontal lines. (C) Time domain slices of the FC ReLaserIMD (*thin lines*) plotted with the orientation-selective ReLaserIMD traces recorded at equivalent fields (*thick lines*). (D) Frequency domain slices of the FC ReLaserIMD (*thin lines*) plotted with the orientation-selective ReLaserIMD results recorded at equivalent fields (*thick lines*). The *colors* in panels (C and D) correspond to the frequency and field positions indicated in Fig. 9A and the panel between panels (A and B). (E) Sum over the offset frequency dimension of the FC ReLaserIMD experimental data shown in black and simulation shown in red. (F) Fourier transform of panel (E). *Reprinted (adapted) with permission from Bowen, A. M., Bertran, A., Henbest, K. B., Gobbo, M., Timmel, C. R., & Di Valentin, M. (2021). Orientation-selective and frequency-correlated light-induced pulsed dipolar spectroscopy.* The Journal of Physical Chemistry Letters, 12, *3819–3826. doi:10.1021/acs.jpclett.1c00595. Copyright 2021 American Chemical Society.*

spectrum performing a single experiment and, therefore, may be useful to aid selection of spectral positions where one-dimensional traces with optimal signal-to-noise ratios can be recorded afterwards. This 2D experiment can also yield orientationally-independent data by summing the offset frequency domain across the width of the EPR spectrum. This has the advantage that the summed trace can be analyzed using an orientationally independent algorithm such as that implemented in DeerAnalysis (Jeschke et al., 2006), reducing the likelihood of artefacts in the analysis.

3.6 LITTER

LITTER can be thought of as a combination of both LiDEER and LaserIMD (Bertran et al., 2021). It makes use of two lasers and it is designed to both detect and optically pump two triplet states, thereby probing the dipolar interaction between them. The second laser flash switches on the dipolar interaction and gives rise to a time-dependent modulation of the primary echo intensity. LITTER combines the advantages of both LiDEER and LaserIMD: the detection of a spin polarized triplet state formed by an initial laser flash, the use of the short Hahn echo detection sequence, and the unlimited pump bandwidth afforded by using a second variable time laser flash to form the second triplet.

This method is unique among the LiPDS techniques as it begins with a system that is completely EPR silent in the ground state. It was pioneered on a model system containing two porphyrin moieties, molecule [**A**] in Fig. 11B, and validated using a system containing only one porphyrin moiety, molecule [**B**] in Fig. 11B, for which no oscillation was observed. The large spectral anisotropy present in the triplet state powder EPR spectra due to the ***D***-tensor, including those of the porphyrin triplet, means that when LITTER is recorded at different canonical positions significant orientation selection can be observed in the dipolar traces, resulting in different oscillation frequencies (see Fig. 11C–E). When these traces are analyzed using an orientation independent algorithm, this results in significantly different distance distributions (red and blue lines in Fig. 11F). As a result, it is necessary to analyze LITTER data considering orientation selection. In this procedure, the pump laser pulse was considered to excite all orientations equally and the orientations contributing to the orientationally-dependent fits are presented in Fig. 11G, showing good agreement with the structure of the peptide system modelled using Density Functional Theory.

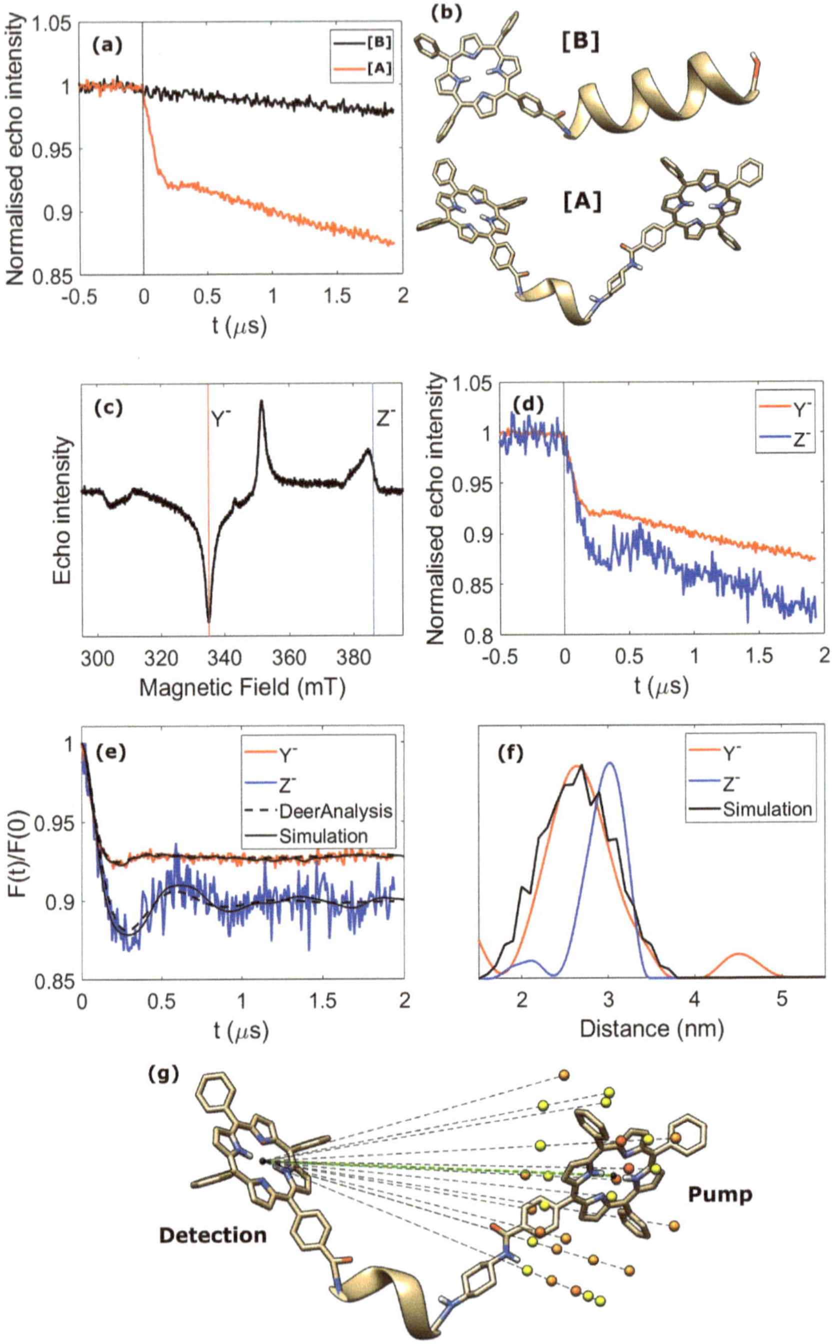

Fig. 11 See figure legend on opposite page.

LITTER is an exciting new technique with many potential applications to biological systems containing chromophores; there is the potential for it to be used in cellular environments where the reducing conditions make using conventional nitroxide spin labels difficult, and careful selection of spin labels may allow direct comparison with FRET distance measurements.

4. Examples of biological applications for LiPDS

As already mentioned in the previous section, implementation of the arsenal of pulse sequences, now available to perform LiPDS, is based on the employment of geometrically well-defined model compounds. The porphyrin-based molecular ruler, introduced in Fig. 5, covers the complete range of applicability of PDS spectroscopy and provides the required constrains to test the performance of different techniques (Di Valentin, Albertini, et al., 2016). The choice of a porphyrin chromophore as photoactive probe is based on its favorable spectroscopic properties, which will be discussed in detail in Section 7, on the large amount of EPR studies reported in the literature and on the presence of a porphyrinoid cofactor in

Fig. 11 (A) LITTER traces for the bis-porphyrin peptide model system TPP-Ala-[(αMe) Val]$_4$-NH-C_6H_{10}-NH-TPP **[A]** (*red*) and the single-porphyrin peptide TPP-(Ala-Aib)$_6$-Ala-OH **[B]** used as control (*black*). The external magnetic field was set to the emissive maximum of the photoexcited triplet spectrum, and only the first half of the LITTER trace was recorded, using a τ of 2004 ns and a t_{pp} of 6.82 μs. (B) Lowest-energy geometries of molecules **[A]** and **[B]** optimized by Density Functional Theory (DFT), with an expected distance between the centers of the two porphyrins in **[A]** of 2.6 nm. (C) Echo-detected field-swept spectrum after laser flash for molecule **[A]**, showing the Y^- (*red*) and Z^- (*blue*) turning points. (D) Raw LITTER traces. (E) Form factors with the corresponding fits obtained with DeerAnalysis and simulation fits, showing modulation depths of 8% and 13%. (F) Resulting spin–spin distance distributions, with maxima at 2.65 nm (*red*) and 3.02 nm (*blue*) determined by DeerAnalysis and the distance distribution from the simulated fits (black). (G) DFT-optimized geometry with the different positions of the pump spin center contributing to the simulated form factor shown as spheres representing the centers of the corresponding porphyrins. The relative weight of each pump position in the simulated form factor is represented by the color intensity of the sphere (*yellow*, 1; *orange*, 2; *dark orange*, 3; *red*, 4). The dipolar vector between the centers of the two porphyrins of the DFT-optimized geometry is highlighted in *green*. Hydrogen atoms have been omitted for the sake of clarity. *Reprinted (adapted) with permission from Bertran, A., Henbest, K. B., De Zotti, M., Gobbo, M., Timmel, C. R., Di Valentin, M., & Bowen, A. M. (2021). Light-Induced Triplet–Triplet Electron Resonance Spectroscopy.* The Journal of Physical Chemistry Letters, 12, *80–85. doi:10.1021/acs.jpclett.0c02884. Copyright 2021 American Chemical Society.*

relevant classes of proteins. Applying the design criteria of the distance ruler presupposes the introduction of the pair of spin label functionalities in highly rigid structures with a defined conformational preference. These properties are collectively possessed by the series of α-helical peptides synthesized for this purpose: they consist of alternating L-alanine (Ala) and α-aminoisobutyric acid (Aib) residues and are covalently linked at the N-terminus with TPP. The nitroxide label is inserted at various sequence positions, spanning distances in the range of 1.8–8.0 nm, replacing the original Aib amino acid with the structurally comparable nitroxide-containing 4-amino-1-oxyl-2,2,6,6-tetramethyl-piperidine-4-carboxylic acid (TOAC) residue, in which the nitroxyl radical is fixed on a rigid heterocyclic structure. Aib residues are strong promoters of helical conformations and ensure a rigid linker between the spin probes with only a slight bending for the longer members of the series, as established by quantum mechanical calculations of the optimized structures. All sorts of LiPDS sequences have been performed on the porphyrin-based spectroscopic ruler, the main results are summarized in Fig. 5. LiDEER experiments have demonstrated that distance measurements can be performed with high sensitivity and accuracy in the overall distance range (Di Valentin, Albertini, et al., 2016), while the potential of orientation selection has been explored in 1D and 2D LaserIMD or ReLaserIMD experiments, providing both distance and mutual orientation information (Bowen et al., 2021).

The feasibility of the distance measurement between two porphyrin chromophores was demonstrated applying the LITTER sequence to a bis-porphyrin model peptide (see Fig. 11) (Bertran et al., 2021). The homo-oligopeptide spacer consists in a repeating sequence of α-methyl-valine ((αMe)-Val), with a strong propensity to form helical structures, like the Aib amino acid, due to C_{α}-tetrasubstitution. For synthetic reasons, the N-terminal is an Ala, which is a less sterically bulky residue but at the same time is not as conformationally restricted as (αMe)-Val, while at the C-terminus the trans-1,4-diaminocyclohexane semirigid moiety is inserted in order to link identical chromophores at either end. LITTER measurements on the bis-porphyrin peptide provided clear-cut results, demonstrating *in primis* the feasibility of the experiments based on two light-induced probes. From LITTER traces, distance distributions and relative orientations of the porphyrin moieties were extracted, allowing the dominant conformation of the peptide in frozen solution to be identified.

Among model systems, not only porphyrin-based peptides but also covalent dyads have been considered. The chromophore is in this case a C_{60}

moiety and the stable paramagnetic probes are an a spirocyclohexyl-substituted nitroxide (NIT) or a triarylmethyl radical (TAM) (Krumkacheva et al., 2019; Timofeev et al., 2022). LiDEER experiments were performed on these molecular dyads to validate the use of fullerene as promising alternatives to porphyrin in the role of photoswitchable probes as shown in Fig. 12. Further comparison of different triplet chromophores for use in LiPDS is discussed in Section 7.

At present, LiPDS applications to proteins are restricted to distance measurements between the nitroxide probe, introduced by site-directed spin labeling, and the photoexcitable center (Dal Farra, Ciuti, et al., 2019; Di Valentin et al., 2016; Hintze et al., 2016). When present, the photoactive prosthetic group can be utilized as endogenous probe, while for proteins that lack the intrinsic photoexcitable center, site-directed spin labeling procedures, based on the use of fluorophores as spin labels, have been developed to introduce the triplet state probe. Alternatively, the non-covalent binding of the chromophore to specific sites in proteins acting as supramolecular hosts was also exploited (Sannikova et al., 2020).

Hemeproteins, together with photosynthetic proteins and flavoproteins, contain photoactive cofactors and represent three important classes of proteins involved in fundamental biological processes. Tentatively, the hemeprotein cytochrome C, spin labeled with MTSSL at the free cysteine position, was investigated in order to demonstrate that LaserIMD could be employed for distance measurements between the endogenous prosthetic group and a nitroxide label (Hintze et al., 2016). However, no triplet state was observed by EPR spectroscopy, as expected for a low-spin ferric heme. The modulations appearing in the corresponding LaserIMD trace can probably be ascribed to higher spin states populated by photoexcitation. Unambiguous results on the feasibility of the LiPDS experiment were instead obtained with human neuroglobin. A mutant of this benchmark hemeprotein, containing a single cysteine, was expressed to avoid multiple labeling with MTSSL and reconstituted with Zn(II) proto-porphyrin IX to introduce a photoactivated triplet state probe in the heme pocket (Dal Farra, Richert, et al., 2019). Distance analysis of the corresponding ReLaserIMD and LiRIDME traces provided accurate determination of distance distributions, in perfect agreement with high-resolution X-ray structural data (see Fig. 8). In the future, the next goal is to apply the orientation-resolved techniques on proteins containing bound optically-active cofactors to gain further structural information beyond distance distribution, enabling more accurate reporting of protein structure changes. Rigidly embedded endogenous spin centers fulfil the requirement for orientation selection since they

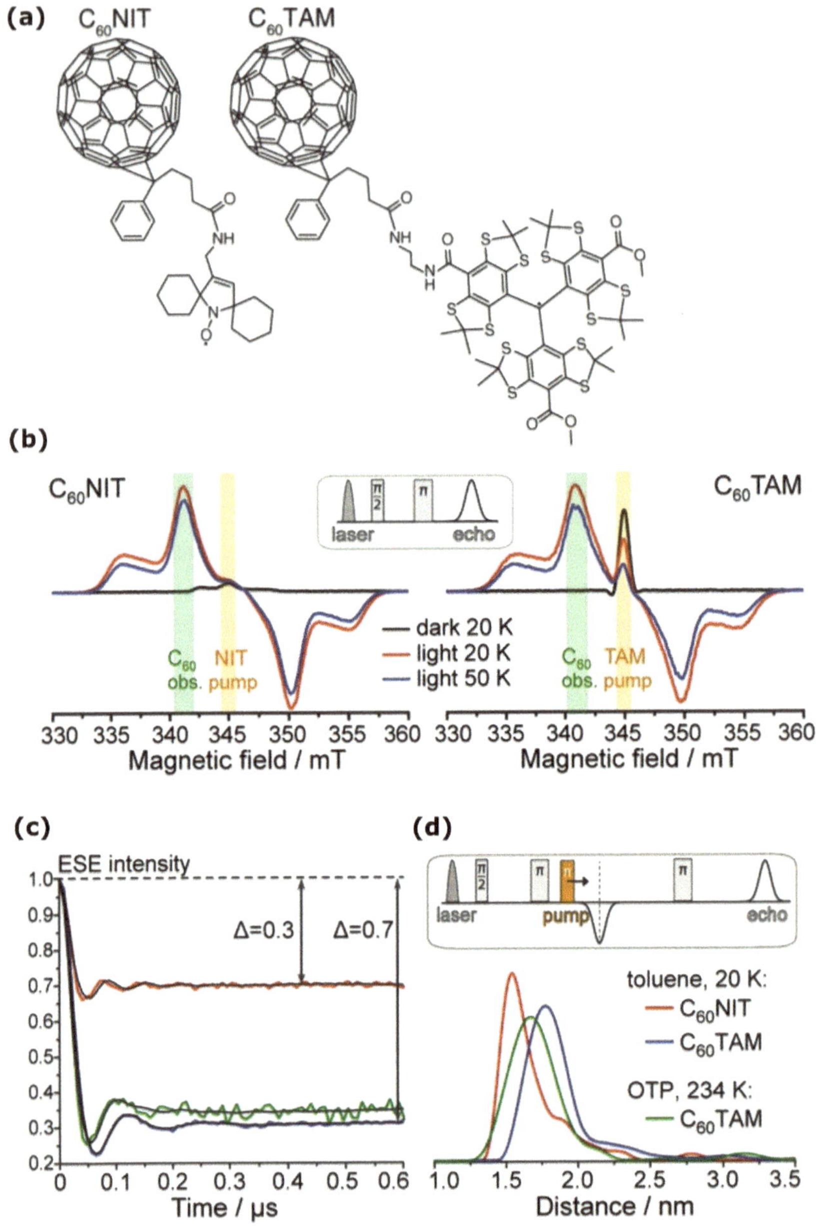

Fig. 12 (A) Chemical structures of the fullerene-based covalent dyads. (B) X-band echo-detected field-swept spectra of the dyads in the dark (*black*), and after laser excitation at 20 K (*red*) and 50 K (*blue*). The observer (*green*) and pump (*orange*) positions for the LiDEER experiments are indicated. (C) Background-corrected and normalized LiDEER traces for C_{60}NIT in toluene at 20 K (*red*), C_{60}TAM in toluene at 20 K (*blue*) and C_{60}TAM in ortho-terphenyl at 234 K (*green*). Fits using DeerAnalysis are overlaid (*black*). (D) Corresponding distance distributions obtained by Tikhonov regularization. The spin-echo and LiDEER pulse sequences are shown as insets. *Reprinted (adapted) with permission from Krumkacheva, O. A., Timofeev, I. O., Politanskaya, L. V., Polienko, Y. F., Tretyakov, E. V., Rogozhnikova, O. Y., … Fedin, M. V. (2019). Triplet fullerenes as prospective spin labels for nanoscale distance measurements by pulsed dipolar EPR spectroscopy.* Angewandte Chemie, 131*(38), 13405–13409. doi:10.1002/ange.201904152. Copyright 2019 John Wiley and Sons.*

are firmly anchored in the protein, while spin labels, typically used as markers in proteins, are inherently flexible and orientation selection cannot be easily achieved.

Not only porphyrin derivatives but other photoactive cofactors have been considered as endogenous probes. Investigation on the photosynthetic light-harvesting protein Peridinin Chlorophyll Protein (PCP) from *Amphidinium cartarae* is based on a triplet state carotenoid probe (a peridinin molecule) populated by triplet–triplet energy transfer from a chlorophyll pigment located in Van-der-Waals contact (Di Valentin, Dal Farra, et al., 2016). In general, the Φ_T for carotenoid molecules is negligible and a photosensitizer is required in order to use this chromophore as a photo-switchable probe. In the specific case of PCP, the LiDEER experiments was performed with the precise aim of identifying the specific pair of pigments involved in the photoprotection mechanism based on triplet–triplet energy transfer. The structure-function relationship could be unambiguously derived in the LiPDS experiments, exploiting a photoactive endogenous probe (see Fig. 13), by measuring the distance between the carotenoid chromophore and MTSSL, introduced in the refolded protein by site-specific spin labeling, and comparing all sets of distances between the radical and each of the peridinins in PCP derived from the X-ray structure.

Porphyrin and C_{60} moieties are recognized by specific proteins, such as lysozyme and human serum albumin (HSA), that can be considered as supramolecular hosts for these chromophores. The recognition is well defined and the chromophore should bind rather selectively in the protein substrate. Exploiting this property, LiDEER measurements were performed on these systems to provide direct information on the location of the binding sites (Krumkacheva et al., 2019; Sannikova et al., 2020). Site-directed spin labeling of HSA with a nitroxide (NIT) radical derivative ensured the dipolar coupling between the chromophore ligand and the radical species inserted on a specific cysteine site (Sannikova et al., 2020). Besides C_{60}, two distinct porphyrins with different properties were used: amphiphilic meso-tetrakis (4-hydroxyphenyl)porphyrin (TMPyP4) and water soluble cationic meso-tetrakis(N-methyl-4-pyridyl)porphyrin (mTHPP), which are potential photosensitizers for photodynamic therapy of cancer. While modulation depths are significant especially for the supramolecular complex with C_{60}, providing clear evidence of the complex formation, the absence of defined oscillations in the LiDEER traces and the consequent broad distance distribution in the protein samples are indicative of unexpected lack of specificity in chromophore binding. These experiments are reported in Fig. 14.

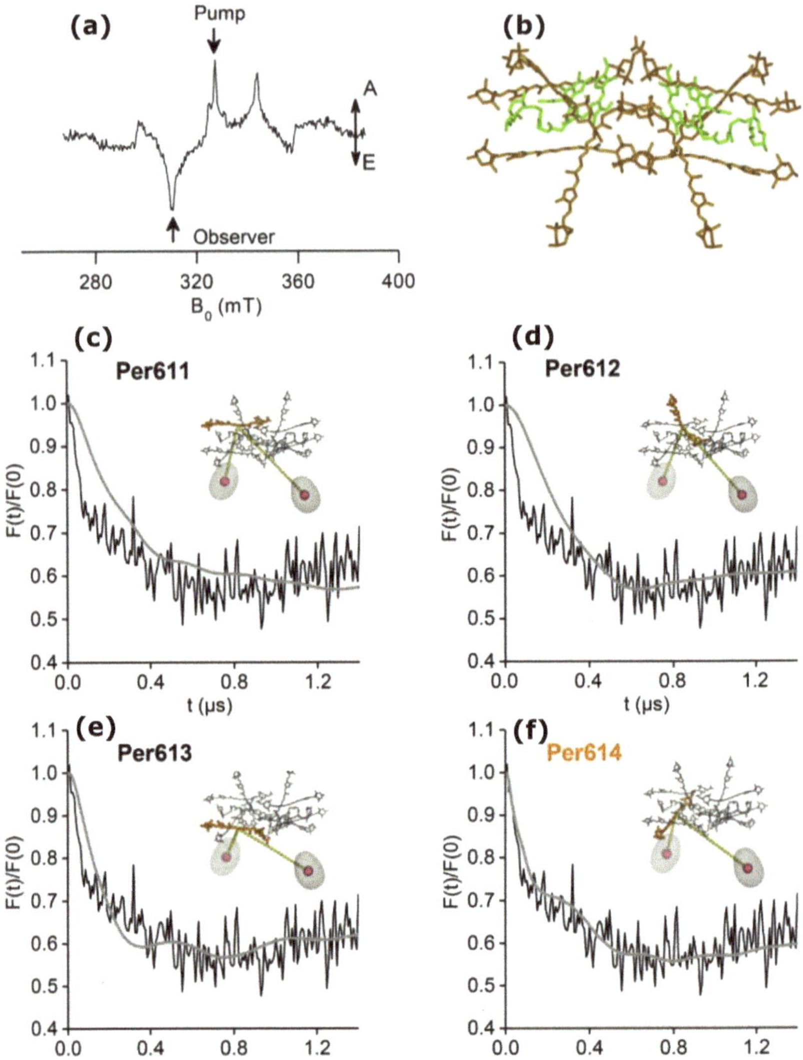

Fig. 13 (A) X-band echo-detected field-swept spectrum recorded under photoexcitation for spin-labeled refolded Peridinin Chlorophyll Protein (PCP). The arrows indicate the positions of the pump (nitroxide) and the detection spin (peridinin triplet state). (B) Structure of the pigments associated with the monomeric basic unit of the PCP complex from *Amphidinium cartarae*. (C–F) Comparison of background-corrected experimental and calculated LiDEER time traces for the triplet state localized on each of the four peridinin molecules. Peridinin 614 is highlighted as it corresponds to the peridinin involved in triplet-triplet energy transfer. *Reprinted (adapted) with permission from Dal Farra, M. G., Ciuti, S., Gobbo, M., Carbonera, D., & Di Valentin, M. (2019). Triplet-state spin labels for highly sensitive pulsed dipolar spectroscopy.* Molecular Physics, 117*(19), 2673–2687. doi:10.1080/00268976.2018.1503749. Copyright 2019 Taylor & Francis Ltd., https://www.tandfonline.com.*

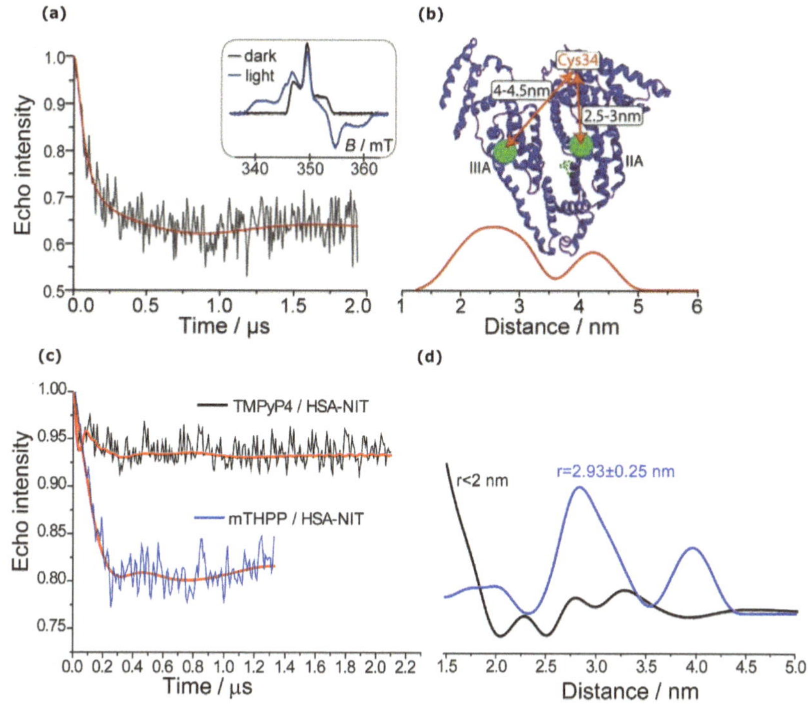

Fig. 14 (A) Background-corrected and normalized X-band LiDEER trace for the HSA–NIT: PCBM complex at 50 K (*black*), and fit with DeerAnalysis (*red*). *Inset*: echo-detected field-swept EPR spectra at 50 K in the dark (*black*) where only a signal due to NIT is observed and after laser excitation (*blue*) where signals from both NIT and the C_{60} triplet state are seen. (B) Corresponding distance distribution obtained by Tikhonov regularization. *Inset*: favorable binding sites of C_{60} (*green*) in HSA. (C) Background-corrected and normalized X-band LiDEER traces for TMPyP4/HSA-NIT (*black*) and mTHPP/HSA-NIT (*blue*) complexes at 10 K, with DeerAnalysis fits (*red*). (D) Corresponding distance distributions obtained by Tikhonov regularization. *Panels (A and B) reprinted (adapted) with permission from Krumkacheva, O. A., Timofeev, I. O., Politanskaya, L. V., Polienko, Y. F., Tretyakov, E. V., Rogozhnikova, O. Y., … Fedin, M. V. (2019). Triplet fullerenes as prospective spin labels for nanoscale distance measurements by pulsed dipolar EPR spectroscopy.* Angewandte Chemie, 131*(38), 13405–13409. doi:10.1002/ange.201904152. Copyright 2019 John Wiley and Sons. Panels (C and D) reprinted (adapted) with permission from Sannikova, N. E., Timofeev, I. O., Chubarov, A. S., Lebedeva, N. S., Semeikin, A. S., Kirilyuk, I. A., … Krumkacheva, O. A. (2020). Application of EPR to porphyrin-protein agents for photodynamic therapy.* Journal of Photochemistry & Photobiology, B: Biology, 211, *112008. doi:10.1016/j.jphotobiol.2020.112008. Copyright 2020 Elsevier.*

An important requisite to broaden the scope of the triplet spin labels in biological macromolecules is the availability of a labeling procedure for photoexcitable spin labels which allows the extension of LiPDS to classes of proteins that lack an intrinsic photoactive center. A successful procedure was proposed by Williams et al., selecting Rose Bengal (RB), Eosin Y (EO) and ATTO Thio12 (AT12) as label candidates (Williams, Tischlik, Scherer, Fischer, & Drescher, 2020). All these chromophores bear heavy atom substitutions that promote ISC to the triplet state through increased spin–orbit coupling (details on the spectroscopic properties are discussed in Section 7.1). For orthogonal labeling, the conventional maleimide–cysteine conjugation chemistry was combined with the copper-catalyzed azido–alkyne cycloaddition reaction between azide-functionalized labels and an alkyne-bearing non-canonical amino acid. The photoexcitable spin labels were functionalized with a linker containing either a maleimide or an azide moiety with a complementary functionalization of the nitroxide derivative (some of these compounds are commercially available). LaserIMD spectroscopy was performed on the spin labeled protein named oxidoreductase thioredoxin. The latter is a well-established model system in site-directed spin labeling owing to its rigid three-dimensional structure, the availability of X-ray structural data and the possibility to remove the native cysteines by site-directed mutagenesis without perturbing the structural integrity. Although the results, as shown in Fig. 15, cannot compare to the narrow inter-spin distance distributions obtained for biomolecules with an intrinsic rigidly anchored chromophore, the achieved distance width is satisfactory and the interspin distance between the labeling sites is in agreement with the value derived from the X-ray structure. The bulk of the extended aromatic system of the photoexcitable labels, which might lead to steric clash with the protein backbone, as well as the flexibility of the appended linkers, have not compromised the distance determination by broadening the distribution.

5. Choice of LiPDS technique

5.1 General considerations

Several aspects must be taken into account when selecting which LiPDS technique to use. The most important aspect is the characteristics of the molecular system of study. If it already contains an endogenous chromophore that can be used for LiPDS, then the properties of its photo-activated triplet state will set the main experimental constraints. Triplets with very short T_m can only be used as a pump spin and the study will be limited to LaserIMD techniques, while triplets with long T_m can also be used as detection spin and allow for both LiDEER and LaserIMD techniques.

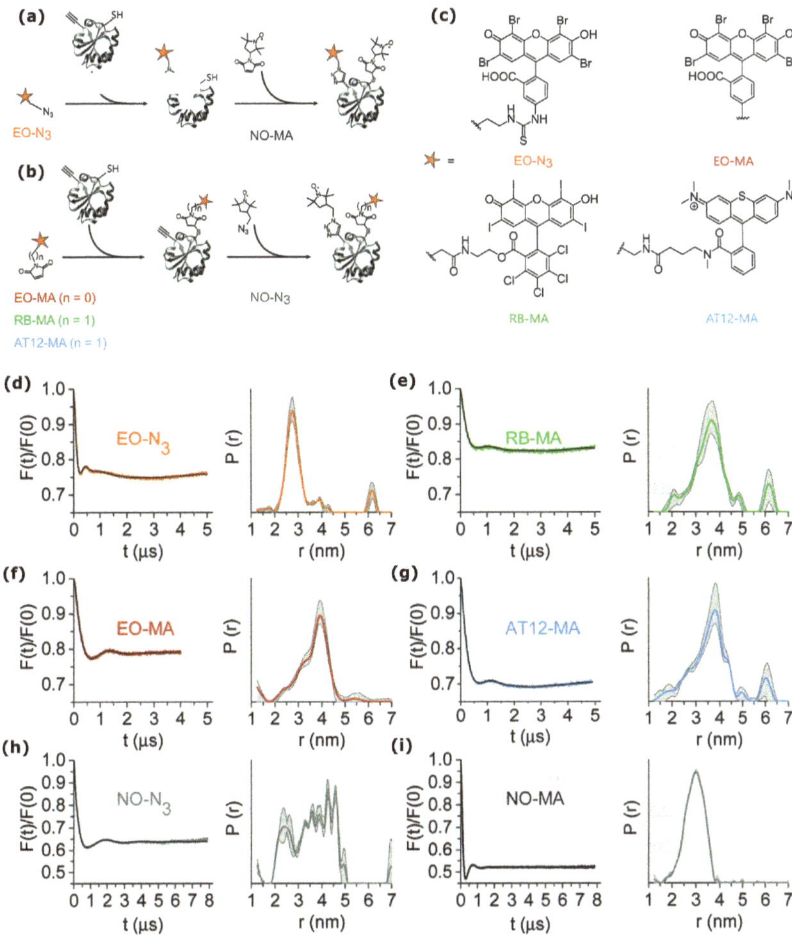

Fig. 15 Orthogonal labeling of TRX at the ncAA pPrF (residue 14) and at the cysteine (residue 74). (A) EO-N_3 is attached via CuAAC, followed by nitroxide labeling with NO-MA via cysteine–maleimide coupling. (B) EO-MA, RB-MA and AT12-MA are attached via cysteine–maleimide coupling, followed by nitroxide labeling with NO-N_3 via CuAAC. (C) Chemical structure of the dyes. (D–G) Background-corrected and normalized LaserIMD traces (*colored*) and fits (*black*) to obtain the corresponding distance distributions (*right*). (H and I) Bis-nitroxide control samples. *Reprinted (adapted) from Williams, L., Tischlik, S., Scherer, A., Fischer, J. W. A., & Drescher, M. (2020). Site-directed attachment of photoexcitable spin labels for light-induced pulsed dipolar spectroscopy.* Chemical Communications, 56, *14669–14672. doi:10.1039/d0cc03101a. Copyright 2020 Royal Society of Chemistry.*

If T_1 of the triplet is short, much shorter than T_1 of the other spin label introduced in the molecule, then LiRIDME would be a suitable technique.

The range of distances to measure will determine the detection sequence to use. For the determination of short distances, a well-defined zero time in the dipolar trace is essential, and a refocused-echo detection sequence, such as those employed in 4-pulse LiDEER and ReLaserIMD, is necessary. For long distances, however, the long-time traces necessary to capture the dipolar oscillations might lead to very low signal intensities if long 4-pulse sequences are used. The effects of the zero-time uncertainty on the determination of long distances are often small and the use of the Hahn-echo version of the 3-pulse LiDEER and LaserIMD techniques is acceptable. Alternatively, short refocused-echo traces can be combined with long Hahn-echo traces and "stitched" together as described by Lovett et al. for DEER (Lovett, Lovett, & Harmer, 2012), simultaneously providing an accurate determination of the zero time and a higher signal intensity in the long trace.

The type of information desired should also orient toward the use of LiPDS technique. If, in addition to the distance between the two spin centers, information on their relative orientations is required, then the 1D dipolar experiment must be repeated at different parts of the detection spin EPR spectrum. Prior knowledge on the expected orientational dependence of the dipolar interaction is useful to guide the selection of these spectral positions in order to obtain the maximum orientational information. When this prior knowledge is not available, a 2D frequency-correlated experiment, using shaped pulses, can be carried out to measure the complete orientational spectrum.

Finally, for molecular systems with more than one chromophore LITTER is the best suited technique, as it allows for the direct study of such systems without any further modification. The use of endogenous chromophores is desirable for their structural rigidity, leading to more accurate distance and orientational information. Even when the molecular system of study needs to be modified prior to LiPDS, LITTER can be still advantageous for cases when the use of permanent spin labels is not possible. A significant example is the experiment in the cellular environment where nitroxides are quickly degraded by reduction.

Once the most suitable LiPDS technique for the study has been identified, adequate experimental parameters must be selected and optimized to obtain the best results. The choice of microwave frequency is often limited by instrumentation, but if multiple frequencies are available, detection on permanent radicals should be done at high frequencies, such as Q-band. Signal intensity in these cases strongly benefits form the enhanced population difference between the two radical spin sublevels provided by a larger Zeeman interaction. In addition, higher frequencies often render higher orientational resolution. Detection on the photo-activated triplet, on the other hand, does not benefit from high microwave frequencies, as their enhanced signal intensity comes from the non-Boltzmann population of the triplet spin sublevels via ISC. Orientation resolution from triplets is frequency-insensitive too, because it relies on the ZFS, which is often much larger than the g-tensor anisotropy.

Temperature can have a strong effect on spin relaxation and must be optimized for the detection spin center. The temperature chosen must come from a trade-off between T_1 and T_m for the detection spin; T_1 must be shorter than the repetition rate of the experiment, which is determined by the repetition rate of the laser, and T_m must be suffciently long to allow a pulse sequence that yields a dipolar trace of a suitable length to observe the dipolar evolution. Some photo-activated triplet states require very low temperatures (e.g., $\leq$20 K) to show reasonable phase-memory times, while in the case of permanent radicals, such as nitroxide, higher temperatures may be beneficial for an optimal balance between T_1 and T_m (e.g., ca. 60 K for nitroxides at Q-band). The length of the pulse sequence must be kept as short as possible in order to maximize signal intensity, and it is limited by the length of trace necessary to capture the dipolar oscillations from the molecular system of study, which should match the inter-pulse delay for dipolar evolution (often τ for Hahn-echo detection and τ_2 for refocused-echo detection).

5.2 LiDEER vs. LaserIMD

Both LiDEER and (Re)LaserIMD can potentially be applied to the same triplet–radical systems and complement each other. For the particular case of porphyrin–nitroxide systems, LiDEER is better suited for the measurement of long inter-spin distances thanks to the longer phase-memory time of the porphyrin triplet compared to that of the nitroxide (Dal Farra, Richert, et al., 2019). With increasing microwave frequency, the inversion efficiency of the pump pulse decreases as the spectral width of the nitroxide increases, leading to a degradation of the modulation depth in LiDEER. This can be compensated by using a broadband chirp pump pulse, that

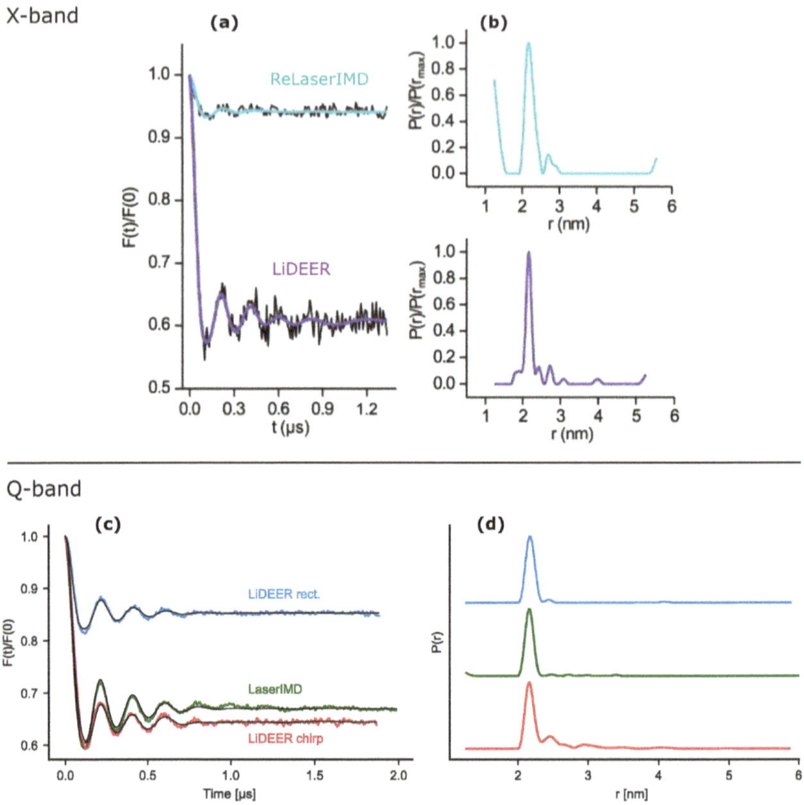

Fig. 16 Comparison between LiDEER and ReLaserIMD at X-band (*top*) and Q-band (*bottom*) for peptide 3 in Fig. 5. Experimental form factors with fits (A and C) and corresponding distance distributions obtained by Tikhonov regularization (B and D). The *red* dataset was acquired with a chirp pump pulse. *Panels (A and B) reprinted (adapted) with permission from Dal Farra, M. G., Ciuti, S., Gobbo, M., Carbonera, D., & Di Valentin, M. (2019). Triplet-state spin labels for highly sensitive pulsed dipolar spectroscopy.* Molecular Physics, *117(19), 2673–2687. doi:10.1080/00268976.2018.1503749. Copyright 2019 Taylor & Francis Ltd., https://www.tandfonline.com. Panels (C and D) reprinted (adapted) with permission from Bieber, A., Bücker, D., & Drescher, M. (2018). Light-induced dipolar spectroscopy—A quantitative comparison between LiDEER and LaserIMD.* Journal of Magnetic Resonance, *296, 29–35. doi:10.1016/j.jmr.2018.08.006. Copyright 2018 Elsevier.*

excites the whole bandwidth of the nitroxide spectrum (Fig. 16C and D), red). However, as with DEER using shaped pump pulses, care must be taken to ensure that the length of the chirp pulse is sufficiently short as to not "wash out" the dipolar oscillation (Doll, Qi, Godt, & Jeschke, 2016). This can be explained by the fact that during the frequency sweep, spins with different resonance frequencies are inverted at different times resulting in an offset-dependent dispersion of the effective dipolar evolution times, since the zero time varies for spins at different spectral positions.

LaserIMD techniques do not suffer from modulation depth reduction at higher microwave frequencies because the depolarized laser pulse excites all orientations of the chromophore (Bieber et al., 2018). Despite not taking advantage of the triplet spin polarization, sensitivity can be enhanced by using a critically-coupled resonator with a high Q-factor, which is not possible in two-frequency techniques such as LiDEER.

For LiDEER, the detection may be conducted either on the triplet or on the permanent paramagnetic species, and the pumping on the other species, in analogy to normal DEER experiments. The implication is that pumping on the nitroxide with unselective pulses allows to observe triplet states which are certainly affected by radical spin inversion while taking advantage of the triplet spin polarization. In the other case, because of the limited photoexcitation capability, a fraction of radicals is bound to chromophores in the ground state, unaffected by the pumping at the triplet. Consequently, a less sensitive detection of the modulation is expected (Spindler et al., 2017; Spindler, Glaser, Skinner, & Prisner, 2013).

Another aspect to consider is the cavity bandwidth. Indeed, the requirements are different for LiDEER and LaserIMD, and the other single frequency experiments, as both frequencies need to be accommodated within the cavity bandwidth (up to 250 MHz separation) (Dal Farra, Richert, et al., 2019). This implies that a strongly overcoupled cavity must be used for LiDEER experiments. For other experiments a less overcoupled cavity can be used, especially at Q-band, where even near critically coupled conditions can be used (Bieber et al., 2018). This yields beneficial effects for the echo intensity, as shorter pulses can be used, exciting a larger spin ensemble, and for the enhanced sensitivity connected to the narrower bandwidth.

6. Methods of illumination

So far, pulsed sources have been used to excite the sample. This is naturally demanding for LaserIMD experiments, where dephasing is obtained by switching on the triplet states at the desired timing within the microwave pulse sequence (Hintze et al., 2016), but it is in general convenient because it can produce a transient concentration of triplet species comparable to that of persistent radicals (10^{-4} to 10^{-5} M). These values cannot be achieved with continuous-wave (cw) illumination because lamps would need to be some orders of magnitude more intense than commercially available options.

Pulsed sources provide ideal conditions for running PDS experiments with the maximum sensitivity. These conditions are reached in all techniques (except from LITTER) when all the systems are excited, so that all permanent paramagnetic states, observed by pulse detection sequences, are affected by the generation of the triplet state. Consequently, the best set up is that in which all the sample is activated by light. LITTER represents an exception, as in this case the best results are obtained when about 50 % of chromophores are excited with each of the two pulses. In dealing with this aspect, one has to consider the number of photoexcited molecules together with the triplet formation yield, and the fraction of sample inside the cavity that is reached by light.

To give an estimate of the fraction of excited molecules that can form from excitation by a 3 mm diameter laser beam hitting a 1×10^{-5} M sample of a strongly absorptive chromophore ($\varepsilon = 1 \times 10^{5}\ M^{-1}\ cm^{-1}$ for porphyrins), it can be shown that under these conditions saturation in the absorption of the sample is reached with a pulse energy of 0.1 mJ at the typical wavelenghth of 500 nm. For a 3 mm optical path, the solution has an absorption of 0.3, corresponding to 50 % of energy being absorbed and the number of photons that have photoexcited the chromophores present in solution corresponding to $\approx 1 \times 10^{-5}$ M. This means that all the chromophores are photoexcited and saturation is reached under this condition. Photoexcited (singlet state) chromophores are converted into triplet states according to the ISC efficiency, corresponding to the Φ_T. Hence a complete conversion of the chromophores into triplet states is rarely reached.

An instrumental limiting factor preventing the complete photoactivation of the sample is represented by the efficiency of the energy transmission between the laser optical bench and the sample tube. As the sample is located inside the EPR cavity, the method of irradiation is primarily determined by the aperture of the EPR cavity used. Most cavities have a dedicated optical access for direct excitation of the sample. Bruker flex-line pulsed X-band resonators, such as split-ring MS3 or dielectric MD5, have relatively large optical access compared to the cavity size, e.g., 3 mm for the MD5 resonator. In the case of MD5 there is a thin grid that partially limits the direct excitation of the entire sample. However, in the case of the MS3 resonator there is no impediment, and the laser beam can be used directly (Fig. 17A).

As for Q-band resonators, due to the reduced dimensions of the cavities, the optical access is narrow. Both standard Bruker Q-band pulsed cavities D2 and TII offer 2 mm optical access, therefore the laser beam needs to be focused (Fig. 17B) in order to pass into the resonator. Fig. 17B highlights

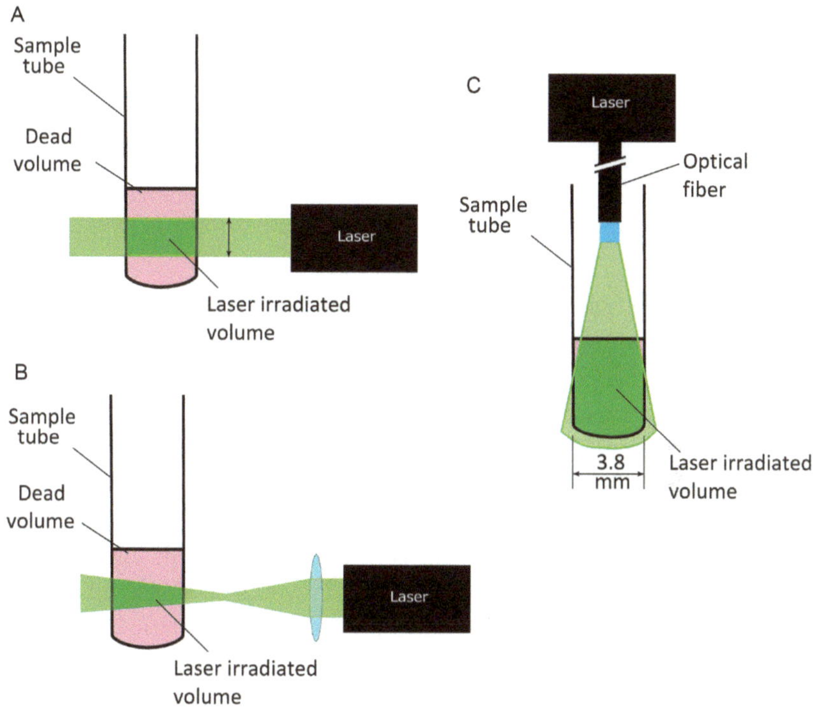

Fig. 17 Irradiation of the sample. (A) Direct excitation. (B) Direct excitation with focusing lens to vary the beams width. (C) Irradiation with the optical fiber from the top of the spectrometer. The dead volumes (sample inside the cavity not reached by light) and the laser irradiated volumes are represented by *different colors*.

the fact that using a focusing lens can have consequences on the fraction of volume that is illuminated with respect to the volume contained in the cavity. In order to minimize the dead volume, a low volume of sample must be used. The implications that a limited sample irradiation has on the measurements are discussed in the next section.

In the case of direct excitation of the sample through the resonator optical window, the laser must also pass through the optical window of the cryostat. To prevent damage of this window it is advisable to keep the laser power below ca. 3 mJ. The beam should also not be focused directly onto the cryostat window to avoid burning the window. Precise tolerances will depend on the laser wavelength used and the window material, and these should be known prior to the start of an experiment. The beam should also be directed at the center of the cryostat window, because if it strikes the edge it could vaporize the seal or grease holding the window in place, which may then get deposited on its inner surface reducing its optical transparency.

Irradiation of the sample can also be obtained by exploiting the sample positioning rod, that can allocate an optical fiber, reaching the sample from the top, and perpendicular to the direct optical excitation, as represented in Fig. 17C. When the optical fiber is inserted inside the tube, both the sample geometry and the multiple reflexions of light from the internal walls of the tube allow an efficient irradiation of the sample. The limitation of the method is the ability to convey a high energy density inside the fiber without damages at the fiber end facing the laser.

Another advantage of the irradiation with the fiber is the loss of the laser light polarization, which might induce undesired magnetophotoselection (MPS) effects (Toffoletti, Wang, Zhao, Tommasini, & Barbon, 2018). In transparent glasses irradiated with polarized laser light, a non-uniform orientational distribution of excited chromophores is obtained, and this might generate a partially oriented powder-like distribution of the interactions between the chromophore and the persistent paramagnetic center. On the other hand, MPS effects could be exploited to obtain more detailed information about the relative orientation of the chromophore with respect to the persistent paramagnetic center. This possibility is currently under investigation.

For LITTER experiments, two laser beams need to be simultaneously delivered to the EPR resonator (Fig. 18A) (Bertran et al., 2021). This is achieved using two lasers and combining the different illumination methods described in Fig. 17: direct beam path, either with or without focusing, and optical fiber (Fig. 18B and C). The optimal settings for LITTER use a moderate laser flash energy for the detection laser and a higher energy for the pump laser. In this way, chromophore excitation is not saturated with the first laser flash and there are enough ground state chromophores left for excitation with the second laser flash, which does saturate the sample excitation. Due to the limited laser power that can be directed to the optical window of the spectrometer cryostat, because of its damage threshold, it is preferable to deliver the detection laser (Laser 1) via direct beam path through the optical window of the cryostat, and the pump laser (Laser 2) to the top of the sample via optical fiber.

The dipolar modulation in LITTER originates only from the fraction of the sample that is irradiated by both lasers (overlap volume; Fig. 18B and C).

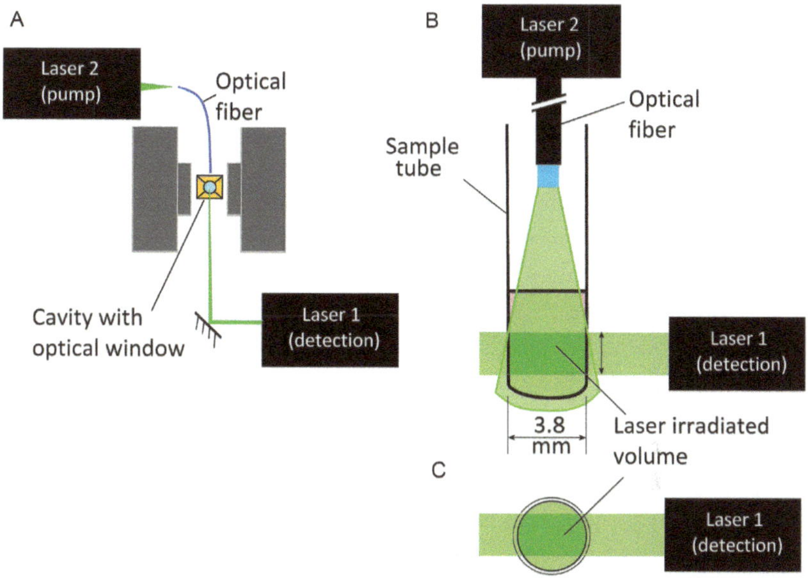

Fig. 18 Two-laser setup for LITTER. (A) Schematic view of the delivery of the two laser beams to the EPR spectrometer resonator. (B) Side and (C) top depictions of the overlap between the two beams at the sample.

It is therefore important to maximize this volume in order to obtain the best signal. Sample height needs to be matched to the diameter of the Laser 1 beam at the sample, in order to minimize the dead volume above the overlap region, which attenuates the Laser 2 beam without contributing to the dipolar signal. A good optical solvent glass, free of cracks, is also essential for efficient laser overlap. With this respect, the presence of dissolved gasses can be detrimental for the formation of the glass. This can be reduced by either preparing the sample in an anaerobic environment, such as within a glovebox using solvents that have previously been degassed, or the sample can be degassed using several freeze-pump-thaw cycles. Here the sample is frozen, the space above the sample pumped to a vacuum, the vacuum pump removed but the vacuum not released and the sample, thawed. Thawing the sample in a vacuum causes the dissolved gasses in the sample to bubble out of solution. The procedure is usually repeated until no bubbling is seen on thawing. The disadvantage of the freeze-pump-thaw procedure is that a small amount of the solvent/sample is also lost on each cycle and therefore concentrations may vary.

6.1 Triggering the laser with respect to the microwave pulses

Triggering the microwave and the laser time bases concurrently, so that they run correctly with respect to the desired experimental scheme, is a fundamental issue for LiPDS. For Q-switched lasers, it is the flashlamp that must be triggered. As the delay between the flashlamp and the opening of the Q-switch determines the laser energy output, this parameter needs to be kept fixed, while the microwave sequence can be delayed separately. In order to prevent jittering problems in some lasers, it is convenient to trigger the Q-switch with one channel of the spectrometer pulse-forming unit. It is then possible with a simple echo-detected transient experiment to determine the minimum delay between the laser trigger and the microwave two-pulse echo sequence at which an echo is observed. The microwave sequence is translated in time until the echo from the triplet species is observed, see Fig. 19. The time delay between the first microwave pulse position of the two-pulse echo sequence and the Q-switch trigger is the offset between the microwave and laser time bases. Once this "zero" time has been determined, the proper experiment pulse sequence can be translated in time. A variable delay before the microwave pulse sequence and the trigger pulse allows the position of the laser pulse to be changed with respect to the detection sequence, although the absolute time of the laser pulse in each

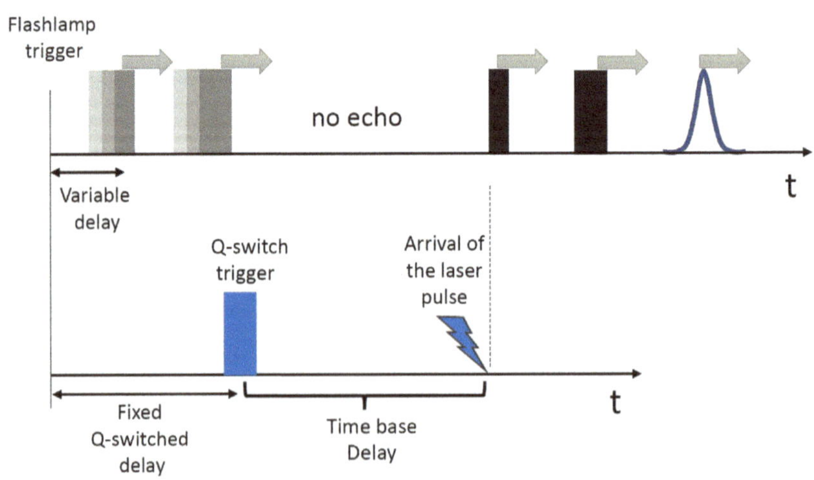

Fig. 19 The pulse sequence for an experiment to determine the time base delay between the optical and the microwave time scale. A Hahn echo pulse sequence is delayed by a variable delay time, relative to the Q-switch trigger, until an echo is observed.

experiment is kept constant and it is the time of the microwave pulses that varies between each point on the acquired time trace.

In the case of the LITTER experiment, it can be more convenient to clock the complete experiment with respect to one of the lasers. A delay pulse generator is used to trigger both the second laser and the spectrometer and for ease of running the experiment the time delay between the two laser pulses is kept constant. The position of the microwave pulse sequence with respect to the two lasers is changed throughout the experiment. The relative positions of both laser pulses with respect to the microwave pulses can be determined using a time-variant Hahn echo detection sequence as described above and noting where the echo intensity increases, which occurs when the $\pi/2$ microwave pulse crosses the laser pulse. The delay of the second laser pulse with respect to the clocking laser pulse can be changed using the pulse delay generator.

7. Choice of chromophore

A chromophore must have certain spectroscopic properties to perform well in LiPDS. First of all, it must have a high Φ_T. For organic chromophores, Φ_T can be improved by addition of heavy atoms, such as covalently linked halogens or coordinated transition metals, which enhance ISC (Marian, 2020). A high extinction coefficient at the wavelength of excitation, ideally within the visible range, is also beneficial but not essential, as it can be compensated by increasing the laser power. Photoexcitation in the UV is often detrimental due to the formation of higher excited states that lead to faster photodegradation of the chromophore.

The resulting triplet state after photoexcitation must have strong and long-lived spin polarization, together with a long T_1 compared to the duration of the experiment. Short triplet optical lifetime is rarely an issue at the cryogenic temperatures of LiPDS, as typical values are well above the duration of any LiPDS pulse sequence. However, extremely long-lived triplets could limit the maximum repetition frequency of the experiment or could reduce signal intensity due to ground-state depletion if very high-repetition frequencies were used. When the spin-echo detection is performed on the triplet, a long T_m relative to the length of the pulse sequence is also necessary. In these cases, a narrow EPR spectrum (small ZFS parameters) is also beneficial to improve the signal intensity, although this usually comes at expenses of orientational resolution if multiple experiments at different spectral positions are carried out.

In addition to the aforementioned spectroscopic properties, chromophores for LiPDS must show good photostability under sustained high-power laser irradiation in degassed frozen solutions, typically for several days. Removing oxygen, according to the procedures described above, is important in order to improve the photostability.

When used for structural studies of biomolecules, chromophores should be small, in order not to affect the structure and function of the system under investigation. In addition, smaller chromophores can label less exposed sites of the biomolecule, which might be inaccessible with bulky labels. Solubility in water is often a requirement for the labeling process too. Extensive electron spin delocalization in the triplet state causes additional broadening of the distance distributions obtained by LiPDS; however, this can be taken into account in the analysis and is not necessarily detrimental for the final results.

Finally, it is preferable to use a chromophore commercially available in a functionalized form suitable for labeling the biomolecule of interest at the desired site. Functionalized chromophores for biolabeling are often more expensive than their corresponding non-functionalized forms, but given the small amounts required for this type of experiments, price is not critical.

7.1 Chromophore characterization

Different types of chromophores have been successfully used as spin labels for LiPDS so far, namely porphyrins, xanthene and thioxanthene dyes, and fullerenes (Di Valentin, Albertini, et al., 2014; Krumkacheva et al., 2019; Williams et al., 2020). In addition, naturally-occurring chromophores such as heme (after Fe exchange by Zn) and carotenoids have been used as endogenous light-activated spin centers too (Dal Farra, Richert, et al., 2019; Di Valentin, Dal Farra, et al., 2016). This section discusses and compares these different chromophores and shows how to proceed for the characterization of a new chromophore for LiPDS.

Porphyrins, in particular TPP, have been the most recurrently used chromophores for LiPDS, especially in model peptides, thanks to their high Φ_T, strong spin polarization, long lifetimes and phase-memory times, and good photostability (Di Valentin, Albertini, et al., 2014). They have high extinction coefficient features in the visible region, corresponding to $S_0 \rightarrow S_1$ vibronic transitions (the so-called Q-bands), which are four in the free-base case. The complexation of a diamagnetic metal ion, such as Zn(II) or Mg(II), by the tetrapyrrolic ring of porphyrins leads to the collapse of the four Q-bands into two due to the increase from 2-fold to 4-fold symmetry of the chromophore. These are naturally-occurring chromophores, such as chlorophylls. For other endogenous porphyrin-based chromophores such as heme, substitution of the paramagnetic metal ion by Zn(II) or Mg(II) allows to convert them into diamagnetic ground-state chromophores suitable for LiPDS. The presence of the metal further increases Φ_T via heavy-atom enhanced ISC, but it makes triplet spin relaxation and decoherence faster. The very favorable spectroscopic properties of both free-base and metalated porphyrins allow them to be used both as detection and/or pump spin center in LiPDS.

More recently, several xanthene- and thioxanthene-based dyes have been proposed and successfully used as pump spin centers for LaserIMD (Serrer et al., 2019; Williams et al., 2020). These chromophores, namely EO, Erythrosine B (EB), RB and AT12, are smaller than porphyrins and therefore more suitable for biolabeling, and some of them are already widely used as dyes in biological assays and can be purchased in suitable functionalized forms (Fig. 20). These chromophores display very high extinction coefficients in the visible range, and high Φ_T thanks to the heavy-atom enhanced ISC. However, the presence of heavy atoms significantly decreases their spin relaxation times and triplet lifetime, restricting their use to only pump spin center. Their smaller size and shorter spatial spin delocalization lead to stronger dipolar interactions between the two unpaired electrons in the triplet state, resulting in larger ZFS parameters compared to those of porphyrins. This is another reason to use these chromophores as pump and not detection spin centers in LiPDS.

Fullerenes, namely C_{60}, have also been successfully used for LiDEER in chemical model systems and proteins (Krumkacheva et al., 2019). Their spectroscopic properties are excellent: $\Phi_T = 1$, narrow triplet EPR spectrum, slow spin relaxation and decoherence, and very good photostability. However, they are very bulky labels with bad biocompatibility, and they are not commercially available in functionalized forms suitable for biolabeling. Therefore, they are not a desirable choice for biological applications of LiPDS, unless the non-covalent binding of this chromophore to specific sites in proteins acting as supramolecular hosts is exploited, as already demonstrated (Krumkacheva et al., 2019)

For chromophores with low Φ_T, where the triplet cannot be efficiently populated by direct photoexcitation, sensitization through triplet-triplet energy transfer from a high-Φ_T chromophore has proven to be a viable option to enable LiPDS. Carotenoids have negligible Φ_T and occur naturally in close

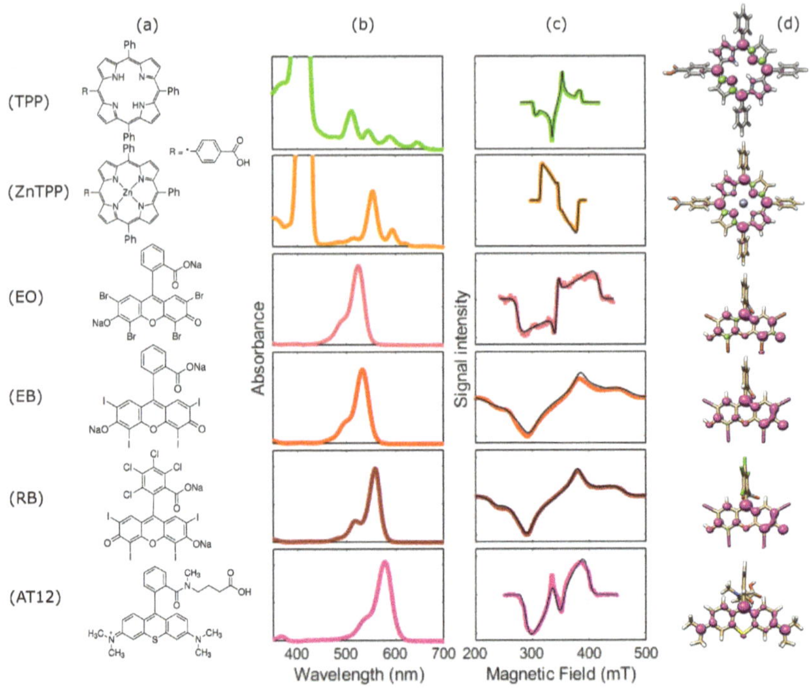

Fig. 20 Some suitable chromophores for biological LiPDS: TPP, zinc(II) 5-(4′-carboxyphenyl)-10,15,20-triphenylporphyrin (ZnTPP), EO, EB, RB and AT12. (A) Chemical structures of the non-functionalized chromophores. (B) UV–Vis absorption spectra at room temperature in ethanol (Methyltetrahydrofuran for ZnTPP). (C) X-band cw transient EPR spectra at 20 K in d_6-ethanol (Methyltetrahydrofuran for ZnTPP) after laser excitation (*color*) and simulation to extract triplet state parameters (*black*). (D) Triplet spin density distribution calculated at the Density Functional Theory level.

proximity to high- Φ_T chlorophylls in light-harvesting proteins (Di Valentin et al., 2013); they have been used for LiDEER via chlorophyl-carotenoid sensitization (Di Valentin, Dal Farra, et al., 2016).

To assess the suitability of a chromophore for LiPDS and to decide on the best pulse sequence and experimental parameters to use, it is useful to perform optical and EPR spectroscopic characterization beforehand. The UV-Vis absorption spectrum of the chromophore in the ground state will determine the laser excitation wavelength to use in the LiPDS experiments, ideally close to the absorption maximum, and the optimal concentration of chromophore in the sample, which should allow for efficient irradiation of the whole sample volume (sample absorbance < 0.7 at X-band). It is important to note that the shape and position of the UV-Vis absorption features of the chromophore might be solvent-and temperature-dependent. Therefore, a low-temperature UV-Vis absorption measurement in the solvent and at the same temperature that will be used for the LiPDS experiments, or at least below the freezing point of the solvent, is recommended whenever possible. For example, for ZnTPP, EO and AT12, narrowing and shifts of the absorption bands occur upon cooling (Fig. 21), while only narrowing occurs for TPP, EB and RB. The triplet state lifetime in those conditions can be determined by transient absorption spectroscopy and needs to be taken into account if very high-repetition frequencies are used for the LiPDS experiment. This must allow enough time between two consecutive repeats of the experiment for the chromophore to return to the ground state to a reasonable fraction.

To characterize the EPR spectrum of the photoexcited triplet state, field-swept cw transient EPR after laser excitation and without field modulation should be used (Fig. 20C). Simulation and fitting of this spectrum renders the triplet state parameters often needed for the analysis of LiPDS data: g-values, ZFS parameters, triplet sublevel populations, linewidths and strains (see Table 1). When LiPDS with detection on the triplet state is intended, it is also useful to record an echo-detected field-swept spectrum of such triplet state, using a light-induced Hahn-echo pulse sequence with the same Delay After Flash (DAF) and pulse lengths of the dipolar experiment (flash–DAF–$\pi/2$–τ–π–τ–acquisition), to aid in the choice of field positions for LiPDS (Fig. 22A). The different relative intensities of the spectral features in the echo-detected spectrum with respect to the cw one reflect the different relaxation rates of these features and will determine the intensities of the LiPDS traces.

Characterization of the spin relaxation of the triplet state is key for choosing the optimal experimental parameters in LiPDS. The type of experiment, temperature, DAF and inter-pulse delays need to be chosen accordingly. The longitudinal spin relaxation and the decay of spin polarization due to thermalization of the non-Boltzmann triplet sublevel populations originated via ISC can be characterized by a light-induced Hahn-echo experiment where the DAF is incremented stepwise (Fig. 22B). A biexponential fit to the echo intensity trace obtained renders a short lifetime dominated by T_1 and thermalization, and a longer lifetime dominated by triplet decay to the ground state. Similarly, spin decoherence can be characterized by an analogous experiment with fixed DAF and variable τ, and a monoexponential fit to the resulting echo intensity trace renders T_m (Fig. 22C). In general, in order to use a certain chromophore for LiPDS, the lifetimes obtained from the

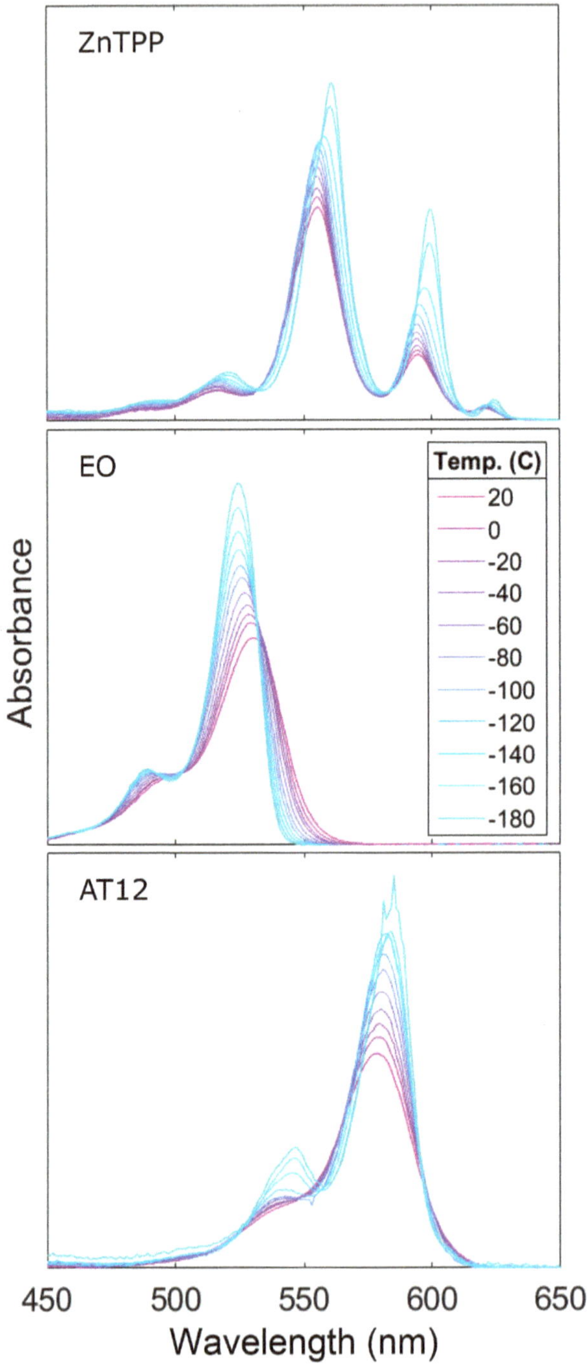

Fig. 21 Low-temperature UV–Vis spectra of ZnTPP in Methyltetrahydrofuran (*left*), EO in ethanol (*center*) and AT12 in ethanol (*right*).

Table 1 Triplet state parameters for reported LiPDS chromophores.

	Φ_T	Triplet lifetime (ms)	DAF lifetime (ms)	T_m (μs)	|D| (MHz)	|E| (MHz)	px:py:pz
TPP	0.88[a]	3.4	0.27; 1.6	1.87	1183	228	0.30:0.53:0.17
ZnTPP	0.90[a]		0.17; 1.4	0.84	933	278	0.29:0.30:0.41
EO	0.61[a]	3.7	0.36	1.16	2003	621	0.32:0.62:0.06
EB	0.97[a]	0.69	0.01; 0.10	0.35	3486	328	0.40:0.60:0.00
RB	0.93[a]	0.43	0.07; 0.89	0.38	3336	278	0.37:0.61:0.02
AT12	0.2[b]	5.3	0.14; 0.84	0.56	1715	451	0.75:0.00:0.25
C_{60}	1[a]	0.4[c]		3.5[c]	342[d]	14[d]	
Zn-heme			0.08; 2.0[e]	0.52[e]	1078[e]	108[e]	
Peridinin			0.01[f]		1348[f]	132[f]	

[a]Murov, Carmichael, and Hug (1993).
[b]ATTO-TECH (2021), https://www.atto-tec.com.
[c]Krumkacheva et al. (2019).
[d]Bennati, Grupp, and Mehring (1995).
[e]Dal Farra, Ciuti, et al. (2019).
[f]Di Valentin, Dal Farra, et al. (2016).
Triplet lifetimes and T_m values have been obtained by monoexponential fits. Delay After Flash (DAF) lifetimes have been obtained by biexponential fits (unless in the cases where a single lifetime is reported).

variable DAF experiment need to be significantly longer than the duration of the LiPDS pulse sequence. An exceptional case is LiRIDME, where a fast-relaxing triplet state is required to induce dipolar modulations within the mixing time, T (see Sections 2.2 and 3.4). In the particular experiments where detection is performed on the triplet (i.e. LiDEER and LITTER), T_m also needs to be longer than the LiPDS pulse sequence. Therefore, detection on a triplet with fast spin decoherence will be limited to short LiPDS time traces and correspondingly short attainable interspin distances. Nevertheless, given that the echo intensity does not always decay monotonously with increasing τ due to Electron Spin Echo Envelope Modulation (ESEEM) (Fig. 22C), there might be cases where using a longer τ is beneficial for signal intensity.

In addition to spectroscopic characterization, calculation of the electron spin density distribution in the triplet state prior to experiment is useful to both assess the suitability of the chromophore and aid in the posterior analysis of the LiPDS data (Fig. 20D). This can be easily done at Density

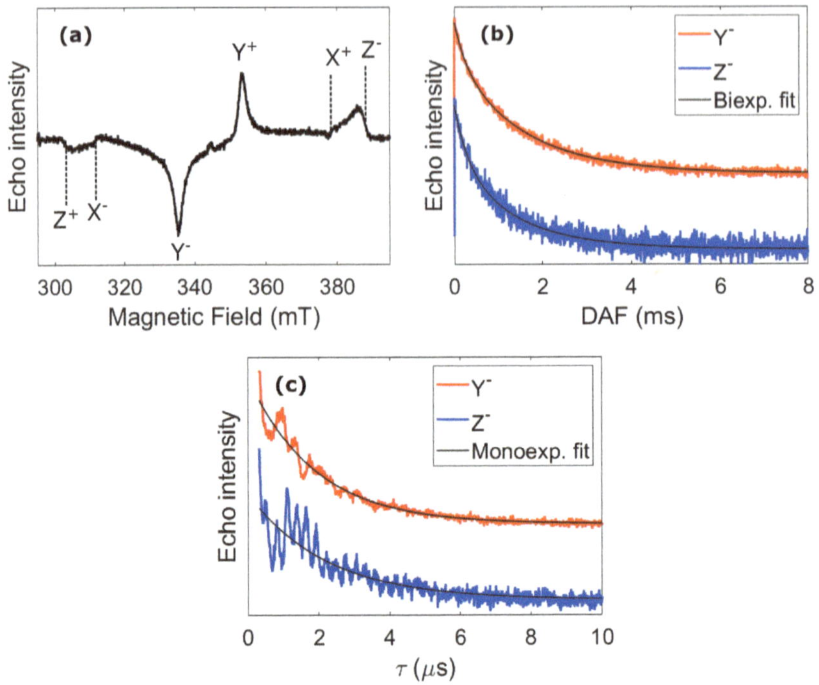

Fig. 22 Pulsed EPR characterization of the photoexcited triplet states of TPP in d_6-ethanol, at X-band and 20K. (A) Echo-detected field-swept spectrum indicating the six turning points. (B) DAF traces recorded at the Y- (*red*) and Z- (*blue*) turning points, and biexponential fits to obtain lifetimes (*black*). (C) Phase memory time traces recorded at the same positions, and monoexponential fit to obtain T_m.

Functional Theory level using standard quantum chemistry software such as Gaussian® or the freely available ORCA (Frisch et al., 2009; Neese, 2012). Triplet states with extensive spatial spin delocalization will typically lead to broader distributions of dipolar frequency components and might be less suited for the measurement of short inter-spin distances. Electron-Nuclear Double Resonance spectroscopy (ENDOR) on the triplet state can be used to confirm experimentally the mapping of the electron spin density distribution.

7.2 Labeling strategies

The biological application of chromophores in LiPDS requires attachment of such molecules to specific sites of the biomolecular systems of interest, from which to measure distances and orientations. A myriad of different strategies has been developed to attach chromophores to biomolecules, and the topic is outside the scope of this book. Only some aspects and examples directly related to the field of LiPDS are covered in this section. For systematic reviews on biolabeling, the reader is directed to the literature (Braun, Drescher, & Summerer, 2019; Fielding, Concilio, Heaven, & Hollas, 2014; Klare & Steinhoff, 2009).

The initial work on LiPDS of model peptide systems used TPP labels attached to the N-terminus of the peptides via amide bond with a carboxyl derivatization of one of the phenyl rings (Di Valentin, Albertini, et al., 2014). If suitable for the system of study, this direct labeling strategy is very desirable because it only introduces a single-carbon linker between the biomolecule and the chromophore, minimizing the distance distribution broadening and uncertainty due to label flexibility. However, this strategy is problematic for xanthene and thioxanthene dyes, as secondary amide formation with the ortho-carboxyl group enables an intramolecular ring closing reaction to render a spirolactam form (Fig. 23A). The electronic conjugation over the chromophore is disrupted in the closed form, leading to the disappearance of the absorption bands in the visible region and therefore preventing photoexcitation. For rhodamine and thiorhodamine chromophore derivatives, the spirolactam formation process is pH dependent and can be reverted to the colored open form by acidification (Fig. 23B). This approach was not successful for fluorescein derivatives such as EO, EB and RB.

Further modification of these xanthene and thioxanthene chromophores is necessary for their practical use in biological LiPDS. Williams et al. recently reported biolabeling strategies for EO, RB and AT12 which do not involve forming a secondary amide on the ortho-carboxyl group (Fig. 15 in Section 4)

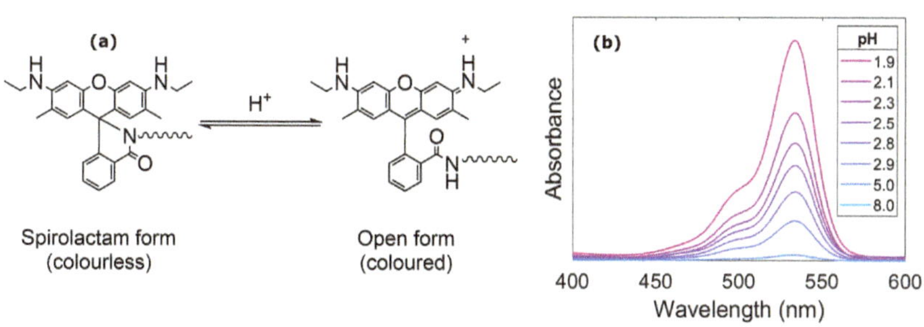

Fig. 23 (A) Spirolactam formation equilibrium for Rhodamine 6G (R6G) attached to the N-terminus of a peptide via direct amide bond. (B) Corresponding spectrometric pH titration for the peptide R6G-(Ala-Aib)$_7$-Ala in ethanol.

(Williams et al., 2020). Additional modifications of the carboxyphenyl ring were used to introduce azide or maleimide moieties to EO, for CuAAC and cysteine–maleimide coupling, respectively. Alternatively, modification of the ortho-carboxy group to form esters or tertiary amides allowed successful introduction of maleimide moieties in RB and AT12 without compromising the electronic conjugation in the chromophore.

8. Conclusions

LiPDS represents a new set of tools for obtaining structural information regarding biomacromolecules, based on the determination of the dipolar interaction between a stable radical and a transient triplet or between two triplet states. As with conventional PDS methods, LiPDS is suitable for determining distances in the range 1.5–8 nm. This is likely extendable to longer distances in the case of complete deuteration, as is the case for conventional PDS methods (Schmidt et al., 2016), as the maximum distance observable is dependent on the phase memory time of the spin used for detection. In order to perform LiPDS it is necessary that the system under investigation contains either a suitable chromophore and a stable spin center (typically a nitroxide) or two chromophores. It is required that chromophores can produce triplet states in high yield. Site-directed mutagenesis and spin labeling procedures have already been developed as methods to introduce suitable labels/probes.

Using a triplet state as a spin probe introduces some additional experimental factors that can affect the data quality. Consideration must be given to sample bleaching due to extensive irradiation time, irradiation efficiency and glass formation. Hence LiPDS experiments are typically more challenging to optimize than conventional PDS methods. However, the light-induced methods also come with certain advantages: triplet states can be strongly polarized leading to enhanced signal-to-noise ratio in the experimental data when the triplet state is used for detection. From a technical point of view, all of the LiPDS variants, except from LiDEER and frequency-correlated LaserIMD, require only a single microwave frequency to conduct the experiment; therefore it is possible to use a cavity that is not strongly overcoupled. This can help improve the experimental sensitivity, due to an enhanced B_1 field and the option to use shorter pulses, compared to multi-frequency experiments which require an overcoupled cavity.

Once the LiPDS experiments are optimized, the information can be extracted from the acquired data in a straightforward manner using available PDS analysis software tools such as DeerAnalysis. In cases where orientation selection is present this can provide additional information about the relative positions of the spin centers but the data must be carefully analyzed to extract this information and knowledge of the spin distribution may be required for a complete analysis.

The experiments LITTER is at X-band require only a relatively small volume of sample (ca. 100 μL) at a concentration which allows sufficient penetration of the incident light throughout the sample; exact values depend on the chromophore and the wavelength used for irradiation. For porphyrins it is typical to use a chromophore concentration of ≈100–200 μM, which may reduced further for LITTER experiments where it is important that the sample has good optical transparency in the laser overlap volume. The combination of the small sample volumes and low concentrations make these techniques suitable to study macromolecule of biological interest.

Future applications of LiPDS could include combinations of standard PDS methods with LiPDS as chromophores are EPR silent until they are optically excited. In this way it may be possible to measure systems containing combinations of stable spin centers and chromophores in the presence and absence of light irradiation to specifically measure individual pairwise dipolar interactions in a complex spin systems containing multiple spin centers.

The stability of triplet states to the in-cell environment has seen them utilized for phosphorescence studies and FRET. This therefore raises the option of using LITTER to interrogate dipolar interactions between two triplet state chromophores in this environment, which is difficult to access using conventional spin labels such as nitroxides, as they are reduced in cells.

Acknowledgments

The authors acknowledge the Centre for Advanced Electron Spin Resonance at Oxford University, funded by the UK EPSRC (EP/L011972/1), for use of facilities. A.M.B. is grateful to the Royal Society and the EPSRC for their support of a Dorothy Hodgkin fellowship (DH160004). A.M.B. and A. Bertran thank the Royal Society for their financial support in the form of a research grant for research fellows (RGF/R1/180099) and enhancement award (RGF/EA/201050). A.M.B. is also grateful to the Royal Society of Chemistry, the analytical chemistry Trust Fund, and the Community for Analytical and Measurement Science fellowship (CAMS Fellowship 2020 ACTF ref. 600310/09). M.D.V. is grateful to the University of Padova (Project 08SIDID2017) for funding. Molecular graphics were plotted with UCSF Chimera, developed by the Resource for Biocomputing, Visualization, and Informatics at the University of California, San Francisco, with support from National Institutes of Health Grant P41-GM103311. M.D.V. is grateful to the University of Padova (Biomolecular DSSCs Project from Interdepartmental Center Giorgio Levi Cases for Energy Economics and Technology).

References

Abdullin, D., Duthie, F., Meyer, A., Müller, E. S., Hagelueken, G., & Schiemann, O. (2015). Comparison of PELDOR and RIDME for distance measurements between nitroxides and low-spin Fe(III) ions. *Journal of Physical Chemistry B, 119*(43), 13539–13542. https://doi.org/10.1021/acs.jpcb.5b02118.

Abdullin, D., & Schiemann, O. (2020). Pulsed dipolar EPR spectroscopy and metal Ions: Methodology and biological applications. *ChemPlusChem, 85*(2), 353–372. https://doi.org/10.1002/cplu.201900705.

Altenbach, C., Flitsch, S. L., Khorana, H. G., & Hubbell, W. L. (1989). Structural studies on transmembrane proteins. 2. Spin labeling of bacteriorhodopsin mutants at unique cysteines. *Biochemistry, 28*(19), 7806–7812. https://doi.org/10.1021/bi00445a042.

Astashkin, A. V. (2015). Mapping the structure of metalloproteins with RIDME. In P. Z. Qin, & K. Warncke (Eds.), *Electron paramagnetic resonance investigations of biological systems by using spin labels, spin probes, and intrinsic metal ions, Part A* (1st ed., pp. 251–284). Waltham, San Diego, Oxford, London: Elsevier. https://doi.org/10.1016/bs.mie.2015.06.031.

ATTO-TECH. (2021). https://www.atto-tec.com.

Azarkh, M., Bieber, A., Qi, M., Fischer, J. W. A., Yulikov, M., Godt, A., et al. (2019). Gd(III)-Gd(III) relaxation-induced dipolar modulation enhancement for in-cell electron paramagnetic resonance distance determination. *Journal of Physical Chemistry Letters, 10*(7), 1477–1481. https://doi.org/10.1021/acs.jpclett.9b00340.

Banham, J. E., Baker, C. M., Ceola, S., Day, I. J., Grant, G. H., Groenen, E. J. J., et al. (2008). Distance measurements in the borderline region of applicability of CW EPR and DEER: A model study on a homologous series of spin-labelled peptides. *Journal of Magnetic Resonance, 191*(2), 202–218. https://doi.org/10.1016/j.jmr.2007.11.023.

Barbon, A., Dal Farra, M. G., Ciuti, S., Albertini, M., Bolzonello, L., Orian, L., et al. (2020). Comprehensive investigation of the triplet state electronic structure of free-base 5,10,15,20-tetrakis(4-sulfonatophenyl)porphyrin by a combined advanced EPR and theoretical approach. *Journal of Chemical Physics, 152*(3). https://doi.org/10.1063/1.5131753.

Bennati, M., Grupp, A., & Mehring, M. (1995). Electron paramagnetic resonance lineshape analysis of the photoexcited triplet state of C_{60} in frozen solution. Exchange narrowing and dynamic Jahn-Teller effect. *The Journal of Chemical Physics, 102*(24), 9457–9464. https://doi.org/10.1063/1.468814.

Bertran, A., Henbest, K. B., De Zotti, M., Gobbo, M., Timmel, C. R., Di Valentin, M., et al. (2021). Light-induced triplet−triplet electron resonance spectroscopy. *The Journal of Physical Chemistry Letters, 12*, 80–85. https://doi.org/10.1021/acs.jpclett.0c02884.

Bieber, A., Bücker, D., & Drescher, M. (2018). Light-induced dipolar spectroscopy—A quantitative comparison between LiDEER and LaserIMD. *Journal of Magnetic Resonance, 296*, 29–35. https://doi.org/10.1016/j.jmr.2018.08.006.

Bolzonello, L., Albertini, M., Collini, E., & Di Valentin, M. (2017). Delocalized triplet state in porphyrin J-aggregates revealed by EPR spectroscopy. *Physical Chemistry Chemical Physics, 19*, 27173–27177. https://doi.org/10.1039/C7CP02968C.

Bowen, A. M., Bertran, A., Henbest, K. B., Gobbo, M., Timmel, C. R., & Di Valentin, M. (2021). Orientation-selective and frequency-correlated light-induced pulsed dipolar spectroscopy. *The Journal of Physical Chemistry Letters, 12*, 3819–3826. https://doi.org/10.1021/acs.jpclett.1c00595.

Bowen, A. M., Johnson, E. O. D., Mercuri, F., Hoskins, N. J., Qiao, R., McCullagh, J. S. O., et al. (2018). A structural model of a P450-ferredoxin complex from orientation-selective double electron-electron resonance spectroscopy. *Journal of the American Chemical Society, 140*(7), 2514–2527. https://doi.org/10.1021/jacs.7b11056.

Bowen, A. M., Tait, C. E., Timmel, C. R., & Harmer, J. R. (2013). Orientation-selective DEER using rigid spin labels, cofactors, metals, and clusters. In C. R. Timmel, & J. R. Harmer (Eds.), *Structural information from spin-labels and intrinsic paramagnetic centres in the biosciences* (1st ed., pp. 283–328). Berlin, Heidelberg: Springer. https://doi.org/10.1007/430_2013_115.

Braun, T., Drescher, M., & Summerer, D. (2019). Expanding the genetic code for site-directed spin-labeling. *International Journal of Molecular Sciences*, *20*(2). https://doi.org/10.3390/ijms20020373.

Ciuti, S., Barbon, A., Bortolus, M., Agostini, A., Bergantino, E., Martin, C., et al. (2021). Neuroglobin provides a convenient scaffold to investigate the triplet-state properties of porphyrins by time-resolved epr spectroscopy and magnetophotoselection. *Applied Magnetic Resonance*, *0123456789*. https://doi.org/10.1007/s00723-021-01421-3.

Cunningham, T. F., Putterman, M. R., Desai, A., Horne, W. S., & Saxena, S. (2015). The double-histidine Cu^{2+}-binding motif: A highly rigid, site-specific spin probe for electron spin resonance distance measurements. *Angewandte Chemie, International Edition*, *54*(21), 6330–6334. https://doi.org/10.1002/anie.201501968.

Dal Farra, M. G., Ciuti, S., Gobbo, M., Carbonera, D., & Di Valentin, M. (2019). Triplet-state spin labels for highly sensitive pulsed dipolar spectroscopy. *Molecular Physics*, *117*(19), 2673–2687. https://doi.org/10.1080/00268976.2018.1503749.

Dal Farra, M. G., Richert, S., Martin, C., Larminie, C., Gobbo, M., Bergantino, E., et al. (2019). Light-induced pulsed EPR dipolar spectroscopy on a paradigmatic hemeprotein. *ChemPhysChem*, *20*(7), 931–935. https://doi.org/10.1002/cphc.201900139.

del Alamo, D., Jagessar, K. L., Meiler, J., & McHaourab, H. S. (2021). Methodology for rigorous modeling of protein conformational changes by Rosetta using DEER distance restraints. *PLoS Computational Biology*, *17*(6), 1–18. https://doi.org/10.1371/journal.pcbi.1009107.

del Alamo, D., Tessmer, M. H., Stein, R. A., Feix, J. B., Mchaourab, H. S., & Meiler, J. (2020). Rapid simulation of unprocessed DEER decay data for protein fold prediction. *Biophysical Journal*, *118*(2), 366–375. https://doi.org/10.1016/j.bpj.2019.12.011.

Di Valentin, M., Albertini, M., Dal Farra, M. G., Zurlo, E., Orian, L., Polimeno, A., et al. (2016). Light-induced porphyrin-based spectroscopic ruler for nanometer distance measurements. *Chemistry - A European Journal*, *22*(48), 17204–17214. https://doi.org/10.1002/chem.201603666.

Di Valentin, M., Albertini, M., Zurlo, E., Gobbo, M., & Carbonera, D. (2014). Porphyrin triplet state as a potential spin label for nanometer distance measurements by PELDOR spectroscopy. *Journal of the American Chemical Society*, *136*(18), 6582–6585. https://doi.org/10.1021/ja502615n.

Di Valentin, M., Buchel, C., Giacometti, G. M., & Carbonera, D. (2012). Chlorophyll triplet quenching by fucoxanthin in the fucoxanthin-chlorophyll protein from the diatom Cyclotella meneghiniana. *Biochemical and Biophysical Research Communications*, *427*(3), 637–641. https://doi.org/10.1016/j.bbrc.2012.09.113.

Di Valentin, M., Dal Farra, M. G., Galazzo, L., Albertini, M., Schulte, T., Hofmann, E., et al. (2016). Distance measurements in peridinin-chlorophyll *a*-protein by light-induced PELDOR spectroscopy. Analysis of triplet state localization. *Biochimica et Biophysica Acta - Bioenergetics*, *1857*(12), 1909–1916. https://doi.org/10.1016/j.bbabio.2016.09.008.

Di Valentin, M., Meneghin, E., Orian, L., Polimeno, A., Büchel, C., Salvadori, E., et al. (2013). Triplet-triplet energy transfer in fucoxanthin-chlorophyll protein from diatom *Cyclotella meneghiniana*: Insights into the structure of the complex. *Biochimica et Biophysica Acta - Bioenergetics*, *1827*(10), 1226–1234. https://doi.org/10.1016/j.bbabio.2013.07.003.

Doll, A., Qi, M., Godt, A., & Jeschke, G. (2016). CIDME: Short distances measured with long chirp pulses. *Journal of Magnetic Resonance*, *273*, 73–82. https://doi.org/10.1016/j.jmr.2016.10.011.

Domnick, C., Eggert, F., Wuebben, C., Bornewasser, L., Hagelueken, G., Schiemann, O., et al. (2020). EPR distance measurements on long non-coding RNAs empowered by genetic alphabet expansion transcription. *Angewandte Chemie, International Edition, 59*(20), 7891–7896. https://doi.org/10.1002/anie.201916447.

Duss, O., Yulikov, M., Allain, F. H. T., & Jeschke, G. (2015). Combining NMR and EPR to determine structures of large RNAs and protein-RNA complexes in solution. In S. A. Woodson, & F. H. T. Allain (Eds.), *Structures of large RNA molecules and their complexes* (1st ed., pp. 279–331). Waltham, San Diego, Oxford, London: Elsevier. https://doi.org/10.1016/bs.mie.2015.02.005.

Feintuch, A., Otting, G., & Goldfarb, D. (2015). Gd^{3+} Spin Labeling for Measuring Distances in Biomacromolecules: Why and How? In P. Z. Qin, & K. Warncke (Eds.), *Electron paramagnetic resonance investigations of biological systems by using spin labels, spin probes, and intrinsic metal ions, Part A* (1st ed., pp. 415–457). Waltham, San Diego, Oxford, London: Elsevier. https://doi.org/10.1016/bs.mie.2015.07.006.

Fielding, A. J., Concilio, M. G., Heaven, G., & Hollas, M. A. (2014). New developments in spin labels for pulsed dipolar EPR. *Molecules, 19*(10), 16998–17025. https://doi.org/10.3390/molecules191016998.

Frisch, M. J., Trucks, G. W., Schlegel, H. B., Scuseria, G. E., Robb, M. A., Cheeseman, J. R., et al. (2009). *Gaussian 09*. Wallingford, CT: Gaussian, Inc.

Giannoulis, A., Motion, C. L., Oranges, M., Bühl, M., Smith, G. M., & Bode, B. E. (2018). Orientation selection in high-field RIDME and PELDOR experiments involving low-spin Co II ions. *Physical Chemistry Chemical Physics, 20*(4), 2151–2154. https://doi.org/10.1039/c7cp07248a.

Goldfarb, D. (2014). Gd 3+ spin labeling for distance measurements by pulse EPR spectroscopy. *Physical Chemistry Chemical Physics, 16*(21), 9685–9699. https://doi.org/10.1039/c3cp53822b.

Goldfarb, D. (2019). Pulse EPR in biological systems—Beyond the expert's courtyard. *Journal of Magnetic Resonance, 306*, 102–108. https://doi.org/10.1016/j.jmr.2019.07.038.

Grytz, C. M., Kazemi, S., Marko, A., Cekan, P., Güntert, P., Sigurdsson, S. T., et al. (2017). Determination of helix orientations in a flexible DNA by multi-frequency EPR spectroscopy. *Physical Chemistry Chemical Physics, 19*(44), 29801–29811. https://doi.org/10.1039/c7cp04997h.

Hagelueken, G., Abdullin, D., & Schiemann, O. (2015). mtsslSuite: Probing Biomolecular Conformation by Spin-Labeling Studies. In P. Z. Qin, & K. Warncke (Eds.), *Electron paramagnetic resonance investigations of biological systems by using spin labels, spin probes, and intrinsic metal ions, Part A* ((1st ed., pp. 595–622). Waltham, San Diego, Oxford, London: Elsevier. https://doi.org/10.1016/bs.mie.2015.06.006.

Haugland, M. M., Lovett, J. E., & Anderson, E. A. (2018). Advances in the synthesis of nitroxide radicals for use in biomolecule spin labelling. *Chemical Society Reviews, 47*(3), 668–680. https://doi.org/10.1039/c6cs00550k.

Heinz, M., Erlenbach, N., Stelzl, L. S., Thierolf, G., Kamble, N. R., Sigurdsson, S. T., et al. (2020). High-resolution EPR distance measurements on RNA and DNA with the non-covalent G′ spin label. *Nucleic Acids Research, 48*(2), 924–933. https://doi.org/10.1093/nar/gkz1096.

Hintze, C., Bücker, D., Domingo Köhler, S., Jeschke, G., & Drescher, M. (2016). Laser-induced magnetic dipole spectroscopy. *Journal of Physical Chemistry Letters*, 7(12), 2204–2209. https://doi.org/10.1021/acs.jpclett.6b00765.

Hintze, C., Steiner, U. E., & Drescher, M. (2017). Photoexcited triplet state kinetics studied by electron paramagnetic resonance spectroscopy. *ChemPhysChem, 18*(1), 6–16. https://doi.org/10.1002/cphc.201600868.

Hoffman, B. M. (1975). Triplet state electron paramagnetic resonance studies of zinc porphyrins and zinc-substituted hemoglobins and myoglobins. *Journal of the American Chemical Society, 97*(7), 1688–1694. https://doi.org/10.1021/ja00840a011.

Imran, M., Zhang, X., Wang, Z., Chen, X., Zhao, J., Barbon, A., et al. (2021). Electron spin dynamics in excited state photochemistry: Recent development in the study of intersystem crossing and charge transfer in organic compounds. *Physical Chemistry Chemical Physics, 23*(30), 15835–15868. https://doi.org/10.1039/d1cp01937f.

Jeschke, G. (2002). Distance measurements in the nanometer range by pulse EPR. *ChemPhysChem*, *3*(11), 927–932. https://doi.org/10.1002/1439-7641(20021115)3:11<927::AID-CPHC927>3.0.CO;2-Q.

Jeschke, G. (2012). DEER distance measurements on proteins. *Annual Review of Physical Chemistry*, *63*(1), 419–446. https://doi.org/10.1146/annurev-physchem-032511-143716.

Jeschke, G., Chechik, V., Ionita, P., Godt, A., Zimmermann, H., Banham, J., et al. (2006). DeerAnalysis2006—A comprehensive software package for analyzing pulsed ELDOR data. *Applied Magnetic Resonance*, *30*(3–4), 473–498. https://doi.org/10.1007/BF03166213.

Jeschke, G., Koch, A., Jonas, U., & Godt, A. (2002). Direct conversion of EPR dipolar time evolution data to distance distributions. *Journal of Magnetic Resonance*, *155*(1), 72–82. https://doi.org/10.1006/jmre.2001.2498.

Jeschke, G., & Polyhach, Y. (2007). Distance measurements on spin-labelled biomacromolecules by pulsed electron paramagnetic resonance. *Physical Chemistry Chemical Physics*, *9*(16), 1895–1910. https://doi.org/10.1039/b614920k.

Joseph, B., Tormyshev, V. M., Rogozhnikova, O. Y., Akhmetzyanov, D., Bagryanskaya, E. G., & Prisner, T. F. (2016). Selective high-resolution detection of membrane protein–ligand interaction in native membranes using trityl–nitroxide PELDOR. *Angewandte Chemie, International Edition*, *55*(38), 11538–11542. https://doi.org/10.1002/anie.201606335.

Keeley, J., Choudhury, T., Galazzo, L., Bordignon, E., Feintuch, A., Goldfarb, D., et al. (2021). Neural networks in double electron-electron resonance: A practical guide. *ArXiv*, 2106.07465.

Keller, K., Qi, M., Gmeiner, C., Ritsch, I., Godt, A., Jeschke, G., et al. (2019). Intermolecular background decay in RIDME experiments. *Physical Chemistry Chemical Physics*, *21*(16), 8228–8245. https://doi.org/10.1039/c8cp07815g.

Klare, J. P., & Steinhoff, H. J. (2009). Spin labeling EPR. *Photosynthesis Research*, *102*(2), 377–390. https://doi.org/10.1007/s11120-009-9490-7.

Klare, J. P., & Steinhoff, H.-J. (2015). Spin labeling studies of transmembrane signaling and transport: Applications to phototaxis, ABC transporters and symporters. In P. Z. Qin, & K. Warncke (Eds.), *Electron paramagnetic resonance investigations of biological systems by using spin labels, spin probes, and intrinsic metal ions, Part B* (1st ed., pp. 315–347). Waltham, San Diego, Oxford, London: Elsevier. https://doi.org/10.1016/bs.mie.2015.05.025.

Klose, D., Klare, J. P., Grohmann, D., Kay, C. W. M., Werner, F., & Steinhoff, H. J. (2012). Simulation vs. reality: A comparison of in silico distance predictions with DEER and FRET measurements. *PLoS One*, 7(6). https://doi.org/10.1371/journal.pone.0039492.

Krumkacheva, O. A., Timofeev, I. O., Politanskaya, L. V., Polienko, Y. F., Tretyakov, E. V., Rogozhnikova, O. Y., et al. (2019). Triplet fullerenes as prospective spin labels for nanoscale distance measurements by pulsed dipolar EPR spectroscopy. *Angewandte Chemie*, *131*(38), 13405–13409. https://doi.org/10.1002/ange.201904152.

Kucher, S., Elsner, C., Safonova, M., Maffini, S., & Bordignon, E. (2021). In-cell double electron-electron resonance at nanomolar protein concentrations. *Journal of Physical Chemistry Letters*, *12*(14), 3679–3684. https://doi.org/10.1021/acs.jpclett.1c00048.

Kurshev, V. V., Raitsimring, A. M., & Tsvetkov, Y. D. (1989). Selection of dipolar interaction by the "2+1" pulse train ESE. *Journal of Magnetic Resonance (1969)*, *81*(3), 441–454. https://doi.org/10.1016/0022-2364(89)90080-2.

Lovett, J. E., Bowen, A. M., Timmel, C. R., Jones, M. W., Dilworth, J. R., Caprotti, D., et al. (2009). Structural information from orientationally selective DEER spectroscopy. *Physical Chemistry Chemical Physics*, *11*, 6840–6848. https://doi.org/10.1039/b913085n.

Lovett, J. E., Lovett, B. W., & Harmer, J. (2012). DEER-Stitch: Combining three- and four-pulse DEER measurements for high sensitivity, deadtime free data. *Journal of Magnetic Resonance*, *223*, 98–106. https://doi.org/10.1016/j.jmr.2012.08.011.

Lubitz, W., Lendzian, F., & Bittl, R. (2002). Radicals, radical pairs and triplet states in photosynthesis. *Accounts of Chemical Research*, *35*(5), 313–320. https://doi.org/10.1021/ar000084g.

Marian, C. M. (2020). Understanding and controlling intersystem crossing in molecules. *Annual Review of Physical Chemistry*, *72*, 617–640. https://doi.org/10.1146/annurev-physchem-061020-053433.

Marko, A., & Prisner, T. F. (2013). An algorithm to analyze PELDOR data of rigid spin label pairs. *Physical Chemistry Chemical Physics*, *15*(2), 619–627. https://doi.org/10.1039/c2cp42942j.

Martin, R. E., Pannier, M., Diederich, F., Gramlich, V., Hubrich, M., & Spiess, H. W. (1998). Determination of end-to-end distances in a series of TEMPO diradicals of up to 2.8 nm length with a new four-pulse double electron electron resonance experiment. *Angewandte Chemie, International Edition*, *37*(20), 2833–2837. https://doi.org/10.1002/(SICI)1521-3773(19981102)37:20<2833::AID-ANIE2833>3.0.CO;2-7.

Milikisyants, S., Scarpelli, F., Finiguerra, M. G., Ubbink, M., & Huber, M. (2009). A pulsed EPR method to determine distances between paramagnetic centers with strong spectral anisotropy and radicals: The dead-time free RIDME sequence. *Journal of Magnetic Resonance*, *201*(1), 48–56. https://doi.org/10.1016/j.jmr.2009.08.008.

Murov, S. L., Carmichael, I., & Hug, G. L. (1993). *Handbook of photochemistry (second)*. New York: Marcel Dekker, Inc.

Neese, F. (2012). The ORCA program system. *Wiley Interdisciplinary Reviews: Computational Molecular Science*, *2*(1), 73–78. https://doi.org/10.1002/wcms.81.

Nohr, D., Rodriguez, R., Weber, S., & Schleicher, E. (2015). How can EPR spectroscopy help to unravel molecular mechanisms of flavin-dependent photoreceptors? *Frontiers in Molecular Biosciences*, *2*(Sep), 1–16. https://doi.org/10.3389/fmolb.2015.00049.

Richert, S., Tait, C. E., & Timmel, C. R. (2017). Delocalisation of photoexcited triplet states probed by transient EPR and hyperfine spectroscopy. *Journal of Magnetic Resonance*, *280*, 103–116. https://doi.org/10.1016/j.jmr.2017.01.005.

Roessler, M. M., King, M. S., Robinson, A. J., Armstrong, F. A., Harmer, J., & Hirst, J. (2010). Direct assignment of EPR spectra to structurally defined iron-sulfur clusters in complex I by double electron-electron resonance. *Proceedings of the National Academy of Sciences of the United States of America*, *107*(5), 1930–1935. https://doi.org/10.1073/pnas.0908050107.

Sannikova, N. E., Timofeev, I. O., Chubarov, A. S., Lebedeva, N. S., Semeikin, A. S., Kirilyuk, I. A., et al. (2020). Application of EPR to porphyrin-protein agents for photodynamic therapy. *Journal of Photochemistry and Photobiology, B: Biology*, *211*, 112008. https://doi.org/10.1016/j.jphotobiol.2020.112008.

Schiemann, O., Heubach, C. A., Abdullin, D., Ackermann, K., Azarkh, M., Bagryanskaya, E. G., et al. (2021). Benchmark test and guidelines for DEER/PELDOR experiments on nitroxide-labeled biomolecules. *Journal of the American Chemical Society*, *143*(43), 17875–17890. https://doi.org/10.1021/jacs.1c07371.

Schmidt, T., Wälti, M. A., Baber, J. L., Hustedt, E. J., & Clore, G. M. (2016). Long distance measurements up to 160 Å in the GroEL tetradecamer using Q-Band DEER EPR spectroscopy. *Angewandte Chemie, International Edition*, *55*(51), 15905–15909. https://doi.org/10.1002/anie.201609617.

Serrer, K., Matt, C., Sokolov, M., Kacprzak, S., Schleicher, E., & Weber, S. (2019). Application of commercially available fluorophores as triplet spin probes in EPR spectroscopy. *Molecular Physics*, *117*(19), 2688–2699. https://doi.org/10.1080/00268976.2019.1608379.

Spindler, P. E., Glaser, S. J., Skinner, T. E., & Prisner, T. F. (2013). Broadband inversion PELDOR spectroscopy with partially adiabatic shaped pulses. *Angewandte Chemie, International Edition*, *52*(12), 3425–3429. https://doi.org/10.1002/anie.201207777.

Spindler, P. E., Schöps, P., Bowen, A. M., Endeward, B., & Prisner, T. F. (2016). Shaped pulses in EPR. *EMagRes*, *5*(4), 1477–1492. https://doi.org/10.1002/9780470034590.emrstm1520.

Spindler, P. E., Schöps, P., Kallies, W., Glaser, S. J., & Prisner, T. F. (2017). Perspectives of shaped pulses for EPR spectroscopy. *Journal of Magnetic Resonance*, *280*, 30–45. https://doi.org/10.1016/j.jmr.2017.02.023.

Timofeev, I. O., Politanskaya, L. V., Tretyakov, E. V., Polienko, Y. F., Tormyshev, V. M., Bagryanskaya, E. G., et al. (2022). Fullerene-based triplet spin labels: Methodology aspects for pulsed dipolar EPR spectroscopy. *Physical Chemistry Chemical Physics*, *24*, 4475–4484. https://doi.org/10.1039/D1CP05545C.

Toffoletti, A., Wang, Z., Zhao, J., Tommasini, M., & Barbon, A. (2018). Precise determination of the orientation of the transition dipole moment in a Bodipy derivative by analysis of the magnetophotoselection effect. *Physical Chemistry Chemical Physics*, *20*(31), 20497–20503. https://doi.org/10.1039/c8cp01984c.

Tsvetkov, Y. D., Milov, A. D., & Maryasov, A. G. (2008). Pulsed electron-electron double resonance (PELDOR) as EPR spectroscopy in nanometre range. *Russian Chemical Reviews*, 77(6), 487–520. https://doi.org/10.1070/RC2008v077n06ABEH003782.

Weil, J. A., & Bolton, J. R. (2006). *Electron paramagnetic resonance: Elementary theory and practical applications (second)*. Hoboken: John Wiley & Sons, Inc.

Williams, L., Tischlik, S., Scherer, A., Fischer, J. W. A., & Drescher, M. (2020). Site-directed attachment of photoexcitable spin labels for light-induced pulsed dipolar spectroscopy. *Chemical Communications*, *56*, 14669–14672. https://doi.org/10.1039/d0cc03101a.

Wort, J. L., Ackermann, K., Giannoulis, A., Stewart, A. J., Norman, D. G., & Bode, B. E. (2019). Sub-micromolar pulse dipolar EPR spectroscopy reveals increasing Cu^{II}-labelling of double-histidine motifs with lower temperature. *Angewandte Chemie, International Edition*, *58*(34), 11681–11685. https://doi.org/10.1002/anie.201904848.

Wunnicke, D., & Hänelt, I. (2017). The synergetic effects of combining structural biology and EPR spectroscopy on membrane proteins. *Crystals*, 7(4), 117. https://doi.org/10.3390/cryst7040117.

Yang, Y., Chen, S. N., Yang, F., Li, X. Y., Feintuch, A., Su, X. C., et al. (2020). In-cell destabilization of a homodimeric protein complex detected by DEER spectroscopy. *Proceedings of the National Academy of Sciences of the United States of America*, *117*(34), 20566–20575. https://doi.org/10.1073/pnas.2005779117.

CHAPTER EIGHT

Controlling and exploiting intrinsic unpaired electrons in metalloproteins

Katherine H. Richardson, Maryam Seif-Eddine, Adam Sills, and Maxie M. Roessler*
Imperial College London, Molecular Sciences Research Hub, London, United Kingdom
*Corresponding author: e-mail address: m.roessler@imperial.ac.uk

Contents

Abstract

Electron paramagnetic resonance spectroscopy encompasses a versatile set of techniques that allow detailed insight into intrinsically occurring paramagnetic centers in metalloproteins and enzymes that undergo oxidation-reduction reactions. In this chapter, we discuss the process from isolating the protein to acquiring and analyzing pulse EPR spectra, adopting a practical perspective. We start with considerations when preparing the protein sample, explain techniques and procedures available for determining the reduction potential of the redox-active center of interest and provide details on methodologies to trap a given paramagnetic state for detailed pulse EPR studies, with an emphasis on biochemical and spectroscopic tools available when multiple EPR-active species are present. We elaborate on some of the most commonly used

Methods in Enzymology, Volume 666
ISSN 0076-6879
https://doi.org/10.1016/bs.mie.2022.02.014

pulse EPR techniques and the choices the user has to make, considering advantages and disadvantages and how to avoid pitfalls. Examples are provided throughout.

Abbreviations

CE	counter electrode
CW	continuous wave
DEER	double electron-electron resonance
EC-EPR	electrochemical EPR
EDFS	echo detected field sweep
ENDOR	electron-nuclear double resonance
ESEEM	electron spin echo envelope modulation
HEMT	high electron mobility transistor
HYSCORE	hyperfine sublevel correlation spectroscopy
hfi	hyperfine interaction
ITO	indium tin oxide
nqi	nuclear quadrupole interaction
PC	paramagnetic center
PFE	protein film electrochemistry
PFE-EPR	protein film electrochemical EPR
RE	reference electrode
SNR	signal to noise ratio
SRT	shot repetition time
WE	working electrode

1. Introduction

It is estimated that nearly half of all enzymes require a metal to function (Andreini, Bertini, Cavallaro, Holliday, & Thornton, 2008). In oxidoreductases the vast majority contain Fe, Cu or Mo that act as redox centers, whereas transferases, hydrolases and lyases make use of typically Zn or Mg as electrostatic stabilizers or non-redox "activators" (Andreini et al., 2008). Metal centers play a crucial role in modulating activity—and the protein environment modulates the properties of the metal center. It is thus no surprise that the study of metal centers in metalloenzymes and metalloproteins has a long history, attracting interest from biochemists and chemists alike. In this journey of characterizing metal centers, electron paramagnetic resonance (EPR) spectroscopy has made and continues to make an essential contribution. Here, we focus on the story that intrinsically occurring unpaired electrons in redox-active proteins can tell.

Proteins that contain metals with open shells, most commonly oxidoreductases, are particularly amenable to EPR studies, although substitution with non-native metals (e.g. Mn^{2+} for Mg^{2+} or Co^{2+} for Zn^{2+}) provides a route to investigating other classes of enzymes. EPR often predicts the presence of redox-active centers and precedes structural information, e.g. the correct prediction of the structure of a [2Fe—2S] cluster in spinach ferredoxin in 1966 (Gibson, Hall, Thornley, & Whatley, 1966). Even today, at a time when structural techniques such as cryo-electron microscopy have enabled major structural advances and insights into very complex membrane-bound oxidoreductases, the information obtainable from EPR spectroscopy is far from obsolete.

The contributions of EPR spectroscopy to studying metal-containing oxidoreductases is most and foremost in understanding their mechanism. The presence of unpaired electrons in oxidoreductases is not usually haphazard and is often tightly controlled given that any unwanted redox chemistry, resulting in for example the production of reactive oxygen species, can be dangerous to the host organism. Thus, trapping and studying states with unpaired electrons in these enzymes often provides information on catalysis—without convolution from the remainder of the protein structure. Indeed, unlike many other techniques, the size of the protein is almost immaterial for EPR. Sensitivity and trapping mechanistically relevant paramagnetic intermediates can however be a challenge, and we address both of these points in detail.

Although continuous-wave (CW) EPR has an important role to play, especially in the initial characterization of a new paramagnetic state, it is through pulse techniques that EPR can make detailed contributions to understanding coordination, bonding and hence reactivity in metalloenzymes (Roessler & Salvadori, 2018). Structural studies with advanced EPR should always be correlated with biochemical studies, such as enzyme activity, to validate the mechanistic role of the trapped paramagnetic states. Moreover, a major challenge and yet often a necessity for such pulse EPR studies, especially when multiple redox-active centers are present, is to control where the unpaired electrons go. Potential adjustment may enable this control and EPR can then provide detailed information on the redox properties of metal centers in proteins, in conjunction with electrochemical methods. The assignment of redox properties to a particular center can be a challenge, especially when multiple centers are present, and we will show how pulse EPR techniques can be useful for this purpose. There is currently a lack of direct correlation between EPR and reactivity studies—a challenge that we address through the development of film electrochemical EPR.

In this chapter we aim to provide a toolkit to investigate redox-active proteins using EPR-based techniques. We discuss what to consider when preparing the protein sample, outline strategies to control the location of unpaired electrons and trap paramagnetic intermediates, and elaborate on how the wide range of pulse EPR techniques can give detailed insight into the role of unpaired electrons in redox-active proteins and oxidoreductases.

2. First things first: The protein

Protein expression and isolation underpin our ability to understand and manipulate metalloproteins. However, the presence of metals in proteins can make purification more challenging. Non-specific metal-ion interactions, high concentrations of redox reagents, avoiding chelating agents and air sensitivity all contribute to this challenge. A plethora of expression systems (in *E coli* (Rosano & Ceccarelli, 2014), in yeast (Baghban et al., 2019), mammalian (Khan, 2013) and in cell free (Gregorio, Levine, & Oza, 2019)) and purification methods can be used, but no one system is appropriate as proteins have very different characteristics. This section will focus on the general considerations and common problems encountered when purifying metalloproteins (Scheme 1). We highlight the options available to control or investigate the cofactor of interest, as well as factors which are often not compatible with downstream characterization of metalloproteins.

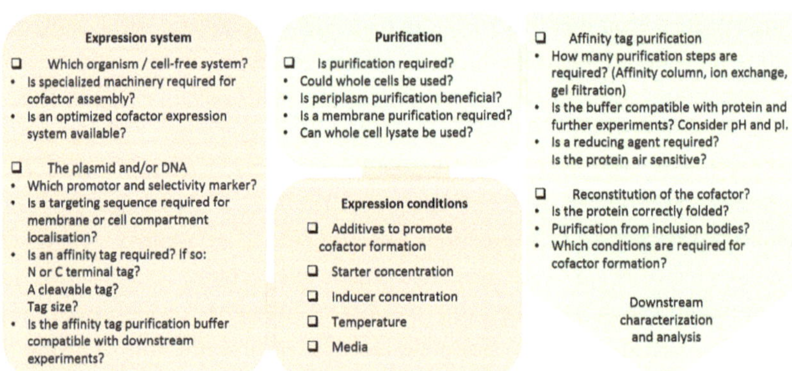

Scheme 1 Considerations when preparing metalloproteins for EPR investigations.

2.1 Expression considerations

Studying a protein from its native host is considered the ideal scenario as it greatly reduces artefacts. Because of low native expression levels, this approach frequently requires large volumes of starting material.

The multistep purification processes necessary are lengthy and many cofactor-containing proteins are unstable. An affinity tag is typically required and although some proteins have native affinity tags, this is rare. Genetically engineered systems and recombinant expression systems are therefore more commonly used than native expression. While *E. coli* is the most frequently employed, models closer to the native system can be beneficial if studying protein complexes which require specific assembly machinery, post translational modification, or secretory expression. *E. coli* strains have been developed to improve expression of Mo (Palmer et al., 1996), heme (Fiege, Querebillo, Hildebrandt, & Frankenberg-Dinkel, 2018; Varnado & Goodwin, 2004) or iron-sulfur cluster (Chaudhuri, Farr, Fenton, Rospert, & Horwich, 2001; Corless, Mettert, Kiley, & Antony, 2020; dos Santos, Johnson, Ragle, Unciuleac, & Dean, 2007) containing proteins, although these strains are not commercially obtainable.

The availability of the metal cofactor often becomes limiting when over-expressing a given metalloprotein. Therefore, growth medias are supplemented with the metal to improve specific binding e.g. $CoCl_2$, $FeCl_3$, $FeSO_4$ or $NaMoO_4$. Care should be taken if the cofactor of interest is unknown as metal ions in the +2 oxidation state are often interchangeable if concentrations are limited, with a bias towards the incorporation of Zn^{2+} (Foster, Osman, & Robinson, 2014; Kocabas, 2013). Elemental analysis or absorption spectroscopy can complement EPR in cofactor identification.

Alteration to metal homeostasis due to over-expression results in metalloproteins having a predisposition to aggregate and to being incorporated into lipid compartments, termed inclusion body formation (the protein of interest is then observed in the post-lysis cell pellet). To tackle aggregation, gentle expression systems such as pCold vectors, autoinduction media or low concentration of the inducer (most commonly IPTG, isopropyl β-D-1-thiogalactopyranoside) are used, sometimes combined with a reduction in growth temperature and increased growth periods.

Because the cytoplasm is a reducing environment, for some cofactors (e.g. Cu^{2+}) or proteins containing many disulfide bonds it can be beneficial to direct them to the periplasm using a targeting sequence (Cantu-Bustos et al., 2016; Santos et al., 2019). Other cofactors, e.g. iron-sulfur or [NiFe] clusters, are highly sensitive to oxygen. For many of these proteins anaerobic expression and purification is thus required (Kim, Hoon Jo, & Joon Cha, 2010). In vitro protein expression using biomolecular translation machinery extracted from cells ("cell-free" systems) can offer a controlled and specialized environment to produce metalloproteins, as well as selective incorporation of isotopic labels, which can be useful for studying protein mechanisms (Abdelkader et al., 2015; Britt, Rao, & Tao, 2020). However, commercially available systems are limited, and yields are low compared to other expression systems (Boyer, Stapleton, Kuchenreuther, Wang, & Swartz, 2008; Kuchenreuther, Shiigi, & Swartz, 2014; Li et al., 2016).

2.2 Purification considerations

Affinity tags are most commonly used as they offer fast, usually two-step, purifications. Many affinity tags of different sizes and with different buffer compatibility exist (Young, Britton, & Robinson, 2012). Poly-histidine tags are frequently used but interactions with Ni^{2+}, Cu^{2+}, Co^{2+} or Zn^{2+} may not be compatible with proteins containing cofactors that also interact or exchange with these metals, e.g. deacetylase MshB (Huang, Kocabas, & Hernick, 2011).

The introduction of affinity tags affects the primary sequence and can therefore have an impact on the folding and catalytic activity of the enzyme. N-terminal tags are most common, however depending on the structure and function of the protein, the terminus of the tag may affect catalytic activity (Aslantas & Surmeli, 2019; Kutyshenko et al., 2019; Sabaty et al., 2013). Alternatively, cleavage sites can be added between the affinity tag and protein sequence of interest. The affinity tag can then be enzymatically removed from the protein after it has been purified, resulting in a more native product. This introduces an additional cleavage and possible purification step post-affinity purification but will produce a protein closer to the native enzyme.

Purification column steps and buffer exchanges lead to reduced yield and therefore it is preferable to choose a purification method which is compatible with the buffer required for downstream experiments. pH and ionic strength should be chosen to aid stability of the protein. Tris buffers should be avoided if divalent cations are required for protein function as they will be chelated (Fischer, Haring, Tribolet, & Sigel, 1979). Dithiothreitol (DTT) and other reducing agents stabilize cysteine interactions and can improve stability of some cofactors, for example [4Fe—4S] and [2Fe—2S] clusters. Detergents can also be used to stabilize membrane proteins, although they may affect protein-protein interactions (Kotov et al., 2019; Yang et al., 2014).

Given the complexities of expressing and purifying cofactor-containing proteins, intentionally purifying the halo-enzyme and reconstituting the metal center (e.g. iron-sulfur clusters (Gervason et al., 2019), heme (Curatti et al., 2007) and FeMo (Hagen, 2018)) is another technique to produce active enzymes. If the protein is small and refolds easily such reconstitution can even be achieved from inclusion bodies.

Paramagnetic species intrinsic to the protein are observed only under certain conditions and depend on the redox state. It may thus be possible to use non- or semi-purified systems to investigate metalloproteins by EPR, taking advantage of the fact that EPR is independent of size and blind to paired electrons. This approach has been used in whole cells (Hagen, 2018), in submitochondrial particles aided by inhibitors to infer specificity of EPR signals (Ingledew & Ohnishi, 1980; Wright et al., 2020) and in thylakoid membranes utilizing light vs dark spectra to assign EPR signals (Anderson & Malkin, 1982; Cammack, Luijk, Maguire, Fry, Packer, 1979; Malkin & Bearden, 1973). This relatively "crude" approach has the advantage of conferring stability given the more native environment of the protein, but isolating and assigning EPR signals definitively can be much more complex compared to studying the "pure" isolated protein.

3. Determining reduction potentials and "directing" electrons biochemically

3.1 Protein film electrochemistry

Protein film electrochemistry (PFE) (Blanford, 2013) can provide detailed insight into protein redox centers and their reactions. In PFE, the redox protein is immobilized directly onto the working electrode (WE) surface such that, in contrast to solution-based electrochemistry, direct potential control is achieved and very little protein (typically picomoles) is required. This section focusses on using PFE to determine reduction potentials of protein redox centers, exploiting the technique's ability for direct potential control. However, PFE has much more to offer as a technique, most notably for studying enzymes under catalysis (Armstrong, Evans, & Megarity, 2018; Armstrong, Heering, & Hirst, 1997; Fourmond & Léger, 2020; Léger et al., 2003).

3.1.1 A key requirement for PFE: The protein "film"

PFE relies on creating a redox-active protein film on the WE. There are two major requirements:

- The protein must be stably immobilized on the WE surface for the duration of the experiment.

- Electron exchange between the WE and the protein's electroactive cofactors must be possible and should not be rate-limiting.

Depending on the protein, making a film can range from being easy to very difficult. However, ensuring the best possible film is obtained will be rewarded by the large amount of information obtained by PFE which is difficult if not impossible to obtain with "classical," solution-based, enzymology (Armstrong et al., 2016).

There are many strategies to make a protein film. The specifics of an approach are affected, amongst other factors, by the wide choice of WE materials, ranging from carbon-based materials such as pyrolytic graphite "edge" to gold (Yates et al., 2018). Here, we outline the porous Indium Tin Oxide (ITO) electrode and its associated film making strategies.

3.1.2 Choice of working electrode: Porous ITO electrodes

The semiconducting mixed metal oxide ITO was used as the WE material in one of the first studies using PFE (Yeh & Kuwana, 1977). Since then, ITO has been used for combined spectroelectrochemical studies on account of its compatibility with spectroscopic techniques (Reuillard et al., 2017), including EPR (Abdiaziz et al., 2019).

ITO electrodes can be made to have highly porous structures by relying on gaps between particles (Hoertz et al., 2010; Kato et al., 2012; Siritanaratkul et al., 2017) or by using polystyrene beads to template larger pore sizes that are then removed by combustion (Mersch et al., 2015). The 3D porous network allows much more sample to be loaded for a given geometric "2D" electrode surface area and hence generally results in increased sensitivity for PFE and spectroscopic measurements. The large surface area of porous electrodes is however coupled to a major increase in non-faradaic "background" current relative to the desirable faradaic current (Box 1).

Protein films can be formed on porous ITO electrodes by electrostatic immobilization (Siritanaratkul et al., 2017). These films rely on the ITO being sufficiently past its isoelectric point of ca. 6.0 (Sun et al., 2004) to give a net negative surface charge that attracts proteins with a degree of net positive surface charge. Porosity can aid electrostatic film stability as any protein material that dissociates from the surface is still entrapped in the pore (Megarity et al., 2019); this increases the likelihood that it can re-adsorb onto the pore wall (Siritanaratkul et al., 2017). Another strategy for forming a PFE film on ITO is to covalently attach the protein to the ITO surface with a chemical linker. Such linkers usually exploit the high affinity of phosphonate groups for ITO (Kato et al., 2013).

BOX 1 Basic information that can be deduced from a non-turnover CV.

Faradaic current is caused by electrochemical reactions in which the electrode oxidizes or reduces redox active species in the cell. Electrons cross between the electrode-cell interface. In PFE, the redox active species of interest is the protein film.

Non-Faradaic current originates from processes that are not electrochemical reactions. No transfer of electrons occurs at the electrode-cell interface. For PFE, the most important non-Faradaic process to consider is double-layer capacitance at the WE surface.

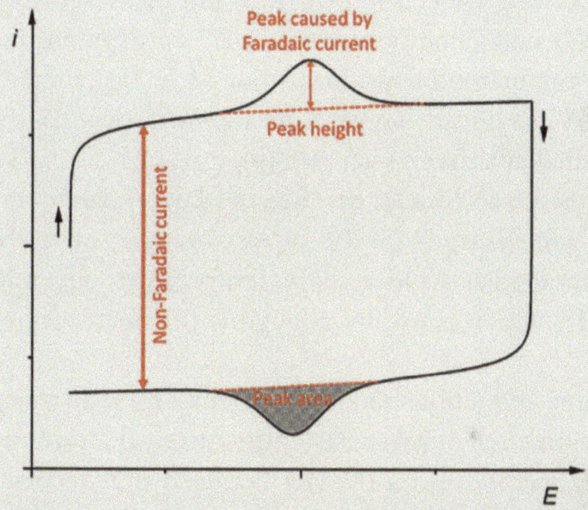

The peak area in the absence of substrate (non-turnover conditions) is proportional to the amount of electroactive protein in the film: amount of protein = $\frac{\text{peak area}}{nF\nu}$ where n is the stoichiometric number of electrons transferred, F is the Faraday constant and ν is the scan rate. For a stable film the amount of protein remains constant and assuming that all redox centers are equally electrochemically accessible, n can therefore be determined. The number of electrons transferred can also be determined from the peak width at half height, which is 90.6/n mV at 25 °C (Saveant, 2006).

3.1.3 Measuring potentials of protein redox centers with PFE

The most common electrochemical technique for determining reduction potentials is cyclic voltammetry (CV). A CV experiment uses the three-electrode setup which consist of:

- A working electrode (WE), where the electrochemical reaction of interest is occurring.
- A reference electrode (RE), a redox half-cell with a constant and well-known potential, to measure the potential of the WE against.
- A counter electrode (CE) that completes the circuit with the WE; current is measured between the WE and CE. The CE should have a surface area much larger (ideally >10 times) than the WE.

The three electrodes are placed in buffer contained in an electrochemical cell, which is usually a custom glass vessel with a water jacket to control the temperature during the measurement, and connected to a potentiostat. Experimentals can be viewed in real time and measurement parameters as well as experimental conditions (e.g., pH, concentrations of substrates and inhibitors) can be modified ad hoc. For a CV experiment, the potential applied to the WE is linearly swept back and forth over time from an initial to a final potential at a given scan rate (v).

It is good practice to convert the measured potentials to be relative to the standard hydrogen electrode (SHE) for easy comparison of data. Reference system redox potentials have a slight temperature dependence (Bard & Faulkner, 2001), which is usually negligible for fluctuations around room temperature.

To determine reduction potentials of enzyme redox centers, CVs are acquired in the absence of substrate ("non-turnover" conditions), resulting in peaks arising from electron transfer between the electrode and protein redox center (Scheme 2) (Armstrong et al., 1997, 2018; Léger et al., 2003). Such "non-turnover" peaks range from being obvious, e.g. the heme center in cytochrome *c* (Eddowes & Hill, 1979; Yeh & Kuwana, 1977), to hardly noticeable, e.g. the Mo center in Methionine Sulfoxide reductase P (Adamson et al., 2015). Unfortunately, it is not uncommon for non-turnover peaks to be invisible, especially for larger proteins, even if a protein film has been made successfully. However, in these cases excellent catalytic

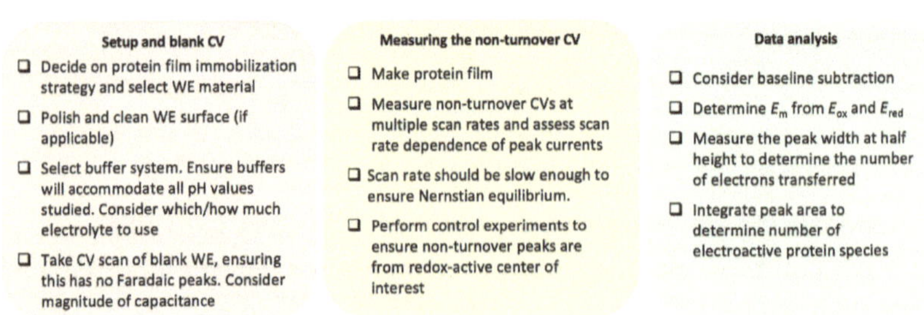

Scheme 2 Flowchart for non-turnover PFE.

signals for kinetic studies may still be obtainable. This is because the magnitude of "catalytic" current (proportional to turnover rates, which can be very high) is much larger compared to the "non-turnover" current (proportional to the number of electrochemically active redox centers).

Control experiments are necessary to ensure that the observed non-turnover signals arise from the protein redox center of interest. CVs with the blank WE should be recorded and no peaks should be seen. If the capacitance is reproducible, such a "blank," can also be subtracted from the non-turnover CV to make the redox peaks clearer. Scan-rate experiments should then be conducted to ascertain that the protein is surface bound. For an immobilized electroactive species, peak heights should vary linearly with scan rate below the critical scan rate (discussed in Section 3.1.4) (Saveant, 2006). In contrast, for electroactive species diffusing from solution, peak heights are proportional to the square root of the scan rate (Saveant, 2006).

To ascertain that there are no contributions to the CV from solution-based electroactive species, the cell solution should be stirred, or the WE rotated. No change is expected in the non-turnover peaks, although "film loss" (often accelerated by rotation) may lead to a reduction in the peak intensity over time. Moreover, at least in an ideal-case scenario, the oxidative and reductive peak potential should be the same, and invariant with scan rate (Section 3.1.4). The film stability should be tested by investigating whether any decrease in Faradaic current is observed over time.

3.1.4 Analysis of non-turnover CVs

3.1.4.1 Reduction potentials

The reduction potential (also called mid-point or redox potential) is obtained by averaging the potential values for the oxidation and reduction peak maxima. Theoretically, the peak separation ($E_{ox} - E_{red}$) should be zero for an immobilized species under Nernstian equilibrium (Saveant, 2006), i.e. when electron transfer between the electrode and immobilized species is sufficiently fast on the timescale of reaction, such that an equilibrium can be reached throughout the sweep of potential. However, above a critical scan rate, the change in potential is too fast for a Nernstian equilibrium to be established. Under these conditions the peak separation rapidly increases with scan rate. A "trumpet plot" of oxidation and reduction peak divergence from midpoint potential vs scan rate can be used to determine the critical scan rate and establish the electron transfer rate constant (Armstrong, Evans, et al., 2018; Laviron, 1979).

3.1.4.2 Electron to redox center stoichiometry

To determine the amount of electroactive protein adsorbed, the average area of the non-turnover peaks is calculated. The number of electrons transferred must be known; this can be determined from the shape of the peak (Box 1). Knowing the amount of protein adsorbed on the WE can be challenging, especially for porous electrodes for which it can be difficult to obtain the "true" (not geometric) surface area. Spectroscopic methods (e.g. UV-vis) may then be useful in conjunction with CVs (Abdiaziz et al., 2019).

3.1.5 Practical considerations when performing PFE measurements

3.1.5.1 Cell solution

As with all electrochemical reagents, ultrapure water should be used, which is commonly available in biochemical laboratories. While the buffers themselves can act as electrolyte, extra electrolyte (commonly KCl or NaCl) is typically added. However, sometimes electrolytes must be minimized or not used at all as the increase in ionic strength can cause instability in electrostatically immobilized protein films. To counteract this lowering of ionic strength, a Luggin capillary can be employed to minimize the ohmic drop between the WE and RE (Evans & Armstrong, 2014).

3.1.5.2 Oxygen

Most PFE measurements need to be carried out anaerobically as the reduction of oxygen can interfere with measurements. Although it is possible to remove oxygen by purging a closed cell with N_2 or Ar on the open bench, an anerobic glovebox is the preferred option. A glovebox setup for PFE has some unusual specifications, namely the high moisture content. This is non-trivial as most gloveboxes used for chemistry require dry atmospheres. An essential specification (not normally found on gloveboxes) is the ability to take samples into the glovebox under nitrogen purge as well as under vacuum, because many proteins can be unstable when subjected to vacuum. A small fridge or freezer can also be a very useful accessory for the storage of protein samples.

3.1.5.3 Electrical interference and "noise"

Interference from electrical appliances can present a particular source of difficulty. The characteristic identifier is a regular repeating waveform superimposed over the voltammogram. Switching off unnecessary electrical appliances in the vicinity of the setup may resolve the issue. However, it is not always possible to identify the source of electrical interference, or the source of interference cannot be turned off. For electrochemistry performed on the bench, placing the electrochemical setup inside an earthed metal Faraday cage can prevent interference. Interference can also be reduced by gloveboxes as they act as a faraday cage. However, it is important to ensure that electrical equipment (magnetic stirrers etc.) inside the glovebox does not cause interference. Loose connections inevitably lead to noisy data. Bubbles in electrochemical cells have a similar effect by insulating key components, such as electrodes (often the RE) and Luggin capillaries.

3.1.6 Limitations of PFE and solutions

Despite the many advantages of PFE there are also limitations:

- Direct structural information cannot be determined.
- It is not always possible or it may require much effort to create a protein film that enables electron transfer between the WE and protein redox centers.

The former has led to the development of spectroelectrochemical techniques (Ash & Vincent, 2012), e.g. PFE combined with UV-Vis and Raman spectroscopies to investigate heme centers in *Shewanella oneidensis* decaheme cytochrome MtrC (Reuillard et al., 2017). Although EPR can be a powerful method for investigating redox-active centers in proteins, for technical reasons its combination with PFE is challenging (Ash & Vincent, 2012). We recently demonstrated the proof-of-concept to realize PFE-EPR (Abdiaziz et al., 2019), discussed further in Section 3.3.

Besides the lack of structural information, sometimes it is not possible to determine reduction potentials with PFE. It may not be possible to make a useable film or non-turnover signals may not be visible. In these cases, other techniques must be used, such potentiometric EPR titrations.

3.2 Potentiometric titrations

Potentiometric titrations monitored by EPR spectroscopy are a powerful tool to identify and quantify the properties of cofactors within proteins, allowing accurate determination of their midpoint potentials. Unlike in most electrochemical techniques, including PFE, the potential is not controlled at the WE by the potentiostat. Instead, the open-circuit potential of the cell solution is measured, which is chemically poised by the addition of oxidizing and reducing agents, typically potassium ferricyanide and sodium dithionite, respectively. Upon reaching the desired potential, an EPR sample of the cell solution is taken and flash frozen. Samples are taken at a range of potentials to study the appearance or disappearance of a radical species as the protein cofactor is oxidized or reduced. Potentiometric titrations rely on establishing an equilibrium; when used to investigate enzymes, the absence of substrate (i.e. non-turnover conditions) is essential.

Although more time- and sample-consuming than PFE, potentiometric titrations can be used to study multiple cofactors at the same time, as well as complex cofactor systems within proteins. They also have the advantage of identifying which species are observed. One limitation of potentiometric EPR titrations has been the quantity of protein sample required (>2 mL, often >5 mL), a large barrier for using the technique with proteins that are difficult to purify in large quantities. However, our recently developed small-volume setup only requires a volume of ca. 200 μL, with as little as 140 μL possible, for an entire titration resulting in at least ten EPR samples (Wright et al., 2016).

3.2.1 Preparing the sample

Because samples are flash frozen, adding a cryoprotectant can help to prevent aggregation upon freezing. Glycerol (>10%) is most commonly used but sucrose, lactose, mannitol or (poly)ethylene glycol are also frequently employed. Indeed, the choice of cryoprotectant can be important when investigating metalloproteins with low-temperature EPR (Serra et al., 2021). The choice of the cryoprotectant can have an effect on metal binding and biological activity levels (Morrissey et al., 2000), hence several options may have to be considered. Ethylene glycol is often a good choice.

The majority of biological buffers shift in pH upon freezing (Orii & Morjta, 1977). pH often not only affects the activity or stability of proteins (Heering et al., 1995), but also clearly matters in the case of pH-dependent reduction potentials, where the value measured could shift by as much as +120 mV (for a two-proton, one-electron redox couple, see Box 2) with a decrease in just one pH unit (which is the case for e.g. phosphate buffers upon freezing (Okada et al., 2019)). Temperature-independent pH buffers, for which we have reported a "recipe" from pH 6–9 (le Breton et al., 2017), should be used when the activity of the protein, and the thermodynamics of its cofactors, are known or suspected to be pH dependent. These buffers have relatively high ionic strengths and contain phosphate which can affect protein stability and solubility, so may not be appropriate in all cases. It is possible to quantify the shift in pH upon freezing using fluorescent pH-dependent dyes (le Breton et al., 2017; Orii & Morjta, 1977), so that other temperature-independent buffer recipes can be found.

BOX 2 Expression for fitting "Nernst plots" from potentiometric titrations, including potential dependence on pH.

The Nernst equation for a species A in equilibrium in solution is given by

$$E = E^0 + \frac{RT}{nF} \ln \frac{[A_{ox}]}{[A_{red}]},$$

where n is the number electrons transferred in the reaction, F is the Faraday constant (96,500 C mol^{-1}), T is temperature, R is the gas constant (8.314 J mol^{-1} K^{-1}) and E is the cell potential.

For a redox titration E is the measured potential and $[A_{ox}]$ is the relative EPR signal intensity at each E, enabling determination of E^0 by fitting the following equation to the data (with $[A_{red}] = 1 - [A_{ox}]$):

$$[A_{ox}] = \frac{1}{1 + e^{-\frac{nF}{RT}(E - E^0)}}$$

For a pH dependent redox couple

$$A_{ox} + mH^+ + ne^- \rightarrow A_{red}H_m,$$

using the Nernst equation, at 298 K, we find:

$$E = E^0 + \frac{2.3RT}{nF} \log \frac{[A_{ox}]}{[A_{red}H_m]} - 0.059 \frac{m}{n} \mathrm{pH}.$$

Hence the reduction potential will decrease by 59 mV per increase in pH unit for a 1H/1e$^-$ couple.

Sodium dithionite is less stable in acidic buffers (Lem & Wayman, 1970), which may prevent reaching very low potentials during the titration (−450 mV vs SHE) at low pH values.

Sample volume is a key consideration and is dictated by the quantity of protein obtainable and the frequency at which EPR measurements will be performed (typically X-band, though Q and W-band have been used), as well as EPR resonator sensitivity. With small-volume potentiometric titrations, the lower limit is ~5 μM 10 μL samples and a ~150–250 μL starting volume of protein sample in the cell. As titration samples for EPR have to be prepared in an anaerobic chamber or an argon/vacuum manifold to prevent interference of oxygen, going below this volume leads to significant evaporation of buffer in the dry atmosphere of the chamber. This will alter the concentration over time and the relative EPR signal intensities, which are used to determine reduction potentials. The addition of cryoprotectants will mitigate against this (e.g. glycerol is hygroscopic) and evaporation can also be prevented by the addition of mineral oil (Hagedoorn et al., 2014). To ensure that evaporation does not introduce a significant error, a control experiment should be performed with just buffer and cryoprotectant (but no protein), monitoring the cell volume for the duration of a typical titration (~1–2 h) and ensuring that it remains unchanged.

3.2.2 Preparing the equipment

Only a WE and RE are required given that the open circuit potential is measured. The electrochemical cell must be of adequate dimensions to enable both electrodes to be submerged in the sample once all 10 or so EPR samples have been removed. For small-volume titrations suggested cell dimensions are shown in Fig. 1. Multiple (tiny) stirrer bars close to the WE can be useful to ensure fast and complete mixing, which can be assessed by watching the potential read-out following the addition of e.g. oxidant.

Redox mediators (typically small redox-active organic molecules, see Fig. 2) need to be added to the buffer to "communicate" the potential of the protein redox center to the WE, at approximately equimolar concentration to the protein. The mediator solution should be added in tiny amounts (<1% of the cell solution, often in sub-μL volumes) at a higher concentration to avoid interference with the buffering capacity of the cell solution (not all mediators are soluble in water and many are dissolved in DMSO). Mediators are chosen to cover the potential of range of interest (Fig. 2), often 6–13 different mediators are combined in one titration. Light sensitive mediators (Fig. 2) should be protected from light, although carrying out the entire titration in the dark is not necessary. Some mediators, e.g. methyl and benzyl viologen, result in an EPR signal at $g=2$ when reduced. Mediators that closely resemble the native substrate of the enzyme (e.g. quinones) are best avoided.

The cell solution (with mediators and protein) should be equilibrated by stirring in the glovebox to remove all oxygen and until the potential of the solution is stable (ca. 30 min). To improve protein stability, slow down evaporation and prevent temperature-dependent fluctuations, the titration cell solution is kept cold (4 °C) using a water bath attached to the electrochemical cell (Fig. 1).

3.2.3 Preforming the titration

Once the potential of the solution is steady, small volumes of sodium dithionite (SDT) or ferricyanide ($K_3[Fe(CN)_6]$) are added to reduce or increase the potential to the desired value. Typically, 0.1–0.2 μL of 5–20 mM SDT is used for small volume titrations or 1–2 μL 2–20 mM SDT or $K_3[Fe(CN)_6]$ for 2 mL titrations. The volume of added solution should be noted in order to correct for the sample dilution during the titration (which should ideally be negligible).

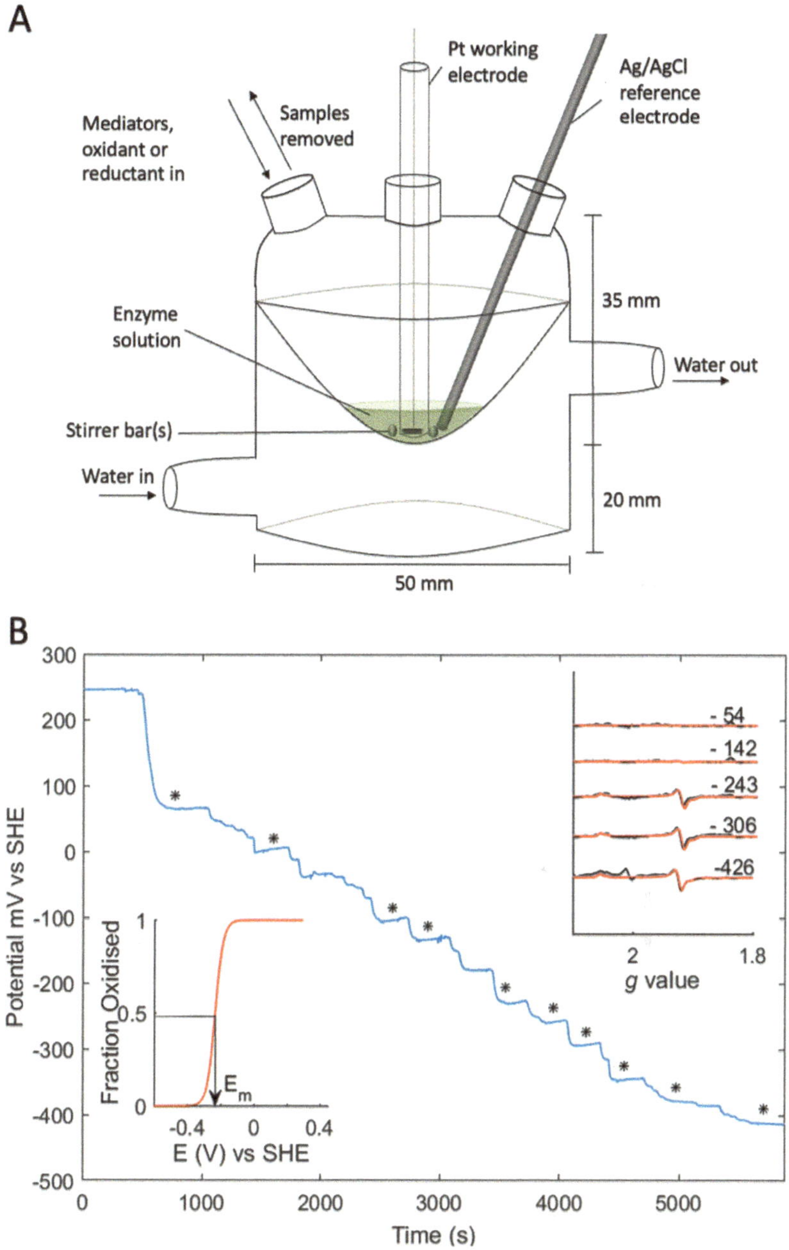

Fig. 1 Set up for EPR monitored small-volume potentiometric titrations. (A) Electrochemical cell with dimensions. Stirrer bars should be placed as close as possible to the WE and RE without touching these. Samples are removed from the cell solution using a glass syringe to avoid moving the electrodes or stirrer bars. (B) Example titration of the reduction of FeS centers in photosynthetic complex I (unpublished data). The potential of the cell solution is monitored throughout the titration and should be stable when a sample is removed (indicated with *). Top right inset: FeS cluster CW EPR spectra with corresponding sample potential. Bottom inset: Nernst curve with reduction potential. *Part A: Adapted from Wright, J. J., Salvadori, E., Bridges, H. R., Hirst, J., & Roessler, M. M. (2016). Small-volume potentiometric titrations: EPR investigations of Fe-S cluster N2 in mitochondrial complex I.* Journal of Inorganic Biochemistry, 162, *201–206. doi: 10.1016/j.jinorgbio.2016.04.025.*

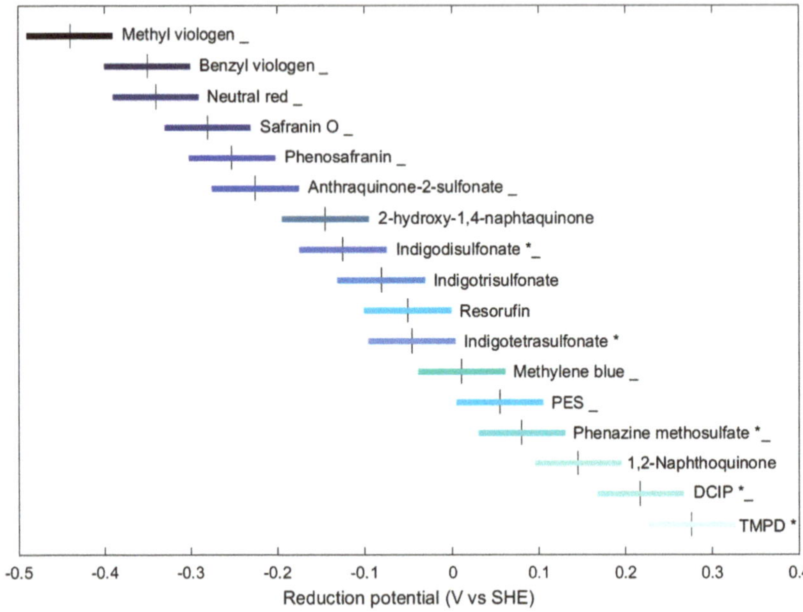

Fig. 2 Reduction potentials of titration mediators. Light-sensitive and water-soluble mediators are indicated by * and _ respectively. CAS numbers: *N,N,N′,N′*-tetramethyl-p--phenylendediamine (TMPD) (CAS 637-01-4), 2,6-Dichlorophenol indophenol (DCIP) (CAS 620-45-1), 1,2-Naphthoquinone (CAS 524-42-5), Phenazine methosulfate (CAS 299-11-6), Phenazine ethosulfate (PES) (CAS 10510-77-7), Methylene blue (CAS 122965-43-9), Inigotetrasulfonate (CAS 67627-19-4), Resorufin (CAS 34994-50-8), Inigotrisulfonate (CAS 67627-18-3), Inigodisulfonate (CAS 860-22-0), 2-Hydroxy-1, 4-naphtaquinone (CAS 83-72-7), Anthraquinone-2-sulfonate (CAS 153277-35-1), Phenosafranin (CAS 81-93-6), Safranin O (CAS 477-73-6), Neutral red (CAS 553-24-2), Benzyl viologen (CAS 1102-19-8) and Methyl viologen (CAS 1910-42-5). Midpoint reduction potential indicated by vertical line, bar width represents the useful range of each mediator.

It is beneficial to have a general idea of the potentials at which to take EPR samples. However, achieving a specific potential value can be difficult given that some potential ranges are better covered by the mediators employed than others. Therefore, recording the potentials at which the samples are taken, immediately prior and post freezing is important. Samples should only be taken once the potential has stabilized as any drift in potential while the sample is removed and flash frozen should be minimal (ideally no more than 1–2 mV). The volume of all EPR samples taken should be identical and introducing bubbles as the sample is injected into the EPR tube needs to be avoided. The inner diameter of the EPR tubes must be invariant to quantify the signal and hence calibrated tubes are required. To assess and ensure reversibility of the titration, it is advisable to take a re-oxidized sample following reduction (or vice versa).

Small-volume samples can be flash-frozen in liquid nitrogen directly. However, this needs to be done slowly and carefully from the bottom of the tubes upwards, so that the tubes do not crack. If possible, and this becomes a necessity for larger samples where freezing in liquid N_2 is not fast enough, it is beneficial to freeze EPR samples in ethanol cooled by a dry ice/acetone bath (this also works well for a cold-finger set-up in a glovebox), prior to transferring them to liquid N_2. Any excess ethanol needs to be wiped off from the outside of the tube before transferring to liquid nitrogen to prevent enlarging their effective outer diameter, which can cause problems for sample loading into the corresponding EPR resonator. Other cooling mixtures, such as cold isopentane are also used.

3.2.4 Determining reduction potentials

To determine the midpoint potentials the titration samples are measured by CW EPR. The modulation frequency and amplitude need to be optimized to resolve all species and prevent distortion of the EPR signals. The microwave power needs to be such that the signals are not saturated (Cammack, 1993). For most resonators, which exhibit a non-uniform field distribution (usually with a maximum at the center), all samples need to be placed in the same location. The loaded resonator quality factor Q can be used as a guide to position each sample. When working with low sample concentrations, the resonator background, which is sensitive to frequency and temperature, can make clean background subtractions challenging. In some resonators, e.g. split-ring resonators, orientation of the sample should also be considered.

The double integral of the EPR signals obtained is used to quantify species with characteristic g values. If there are multiple species present which overlap, or if signal intensity is low, integration over the g values associated with the species or the peak-to-peak height may be used as an approximation. This quantification can be done either from raw data, or from simulation of individual species fitted to the data. The intensity of the maximum EPR signal is typically normalized to 1 or converted to a percentage. Signal intensities as a function of potential (Nernst plots) can then be fitted using the Nernst equation (for most redox centers the number of electrons = 1) at the temperature at which the experiment was performed, normally 278–293 K. The reduction potential is then obtained from least-squares fitting (Box 2).

3.2.5 Limitations and outlook

Potentiometric titrations are a robust method to determine the reduction potentials of redox-active cofactors, provided that their oxidized or reduced form is EPR active (Scheme 3). These cofactors often determine the thermodynamic driving force for electron transfer, and therefore activity of proteins or whole processes such as respiration and photosynthesis (Adamson et al., 2017; Ayikpoe et al., 2019; Shleev et al., 2005; Wright et al., 2016). There are however several drawbacks to potentiometric titrations. CW EPR signals may overlap and can be very broad, making their quantification difficult (Richardson et al., 2021). Although potentiometric titrations are typically monitored using X-band CW EPR, a recent report suggests that Q-band pulse EPR could be useful in circumventing some of the issues (Heghmanns et al., 2021); indeed, it may become possible to separate overlapping species based on their different relaxation properties and the sometimes increased resolution at the higher microwave frequency.

Although EPR-based potentiometric titrations remain a powerful technique in the study of metalloproteins, especially for paramagnetic species which require low-temperature EPR to be observed, they can only be carried out on steady-state radicals and are therefore not able to detect any fleeting paramagnetic intermediates formed during enzyme catalysis.

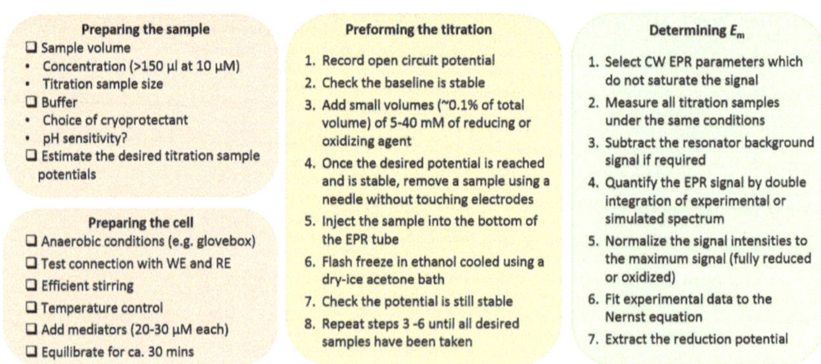

Scheme 3 Flowchart for EPR-monitored potentiometric titrations.

Moreover, the presence of mediators may interfere with the signal of interest (many protein-intrinsic paramagnets have signals at $g \sim 2$, such as semiquinones).

3.3 Protein film electrochemical EPR

The combination of PFE and EPR was born out of the need to avoid mediators and go beyond non-turnover conditions. Although prior to PFE-EPR (Abdiaziz et al., 2019) many electrochemical EPR set-ups have been reported, there was only one application to an enzyme, or rather an enzyme domain from nitric oxide synthase reductase (Murray et al., 2011); mediators were still required and there was no attempt to achieve accurate potential control. This showcases a major problem for electrochemical EPR (EC-EPR): slow diffusion of the enzyme to the WE. The direct attachment of proteins to the working-electrode surface (i.e. PFE) eliminates the "diffusion problem" and we have demonstrated that PFE-EPR allows direct and accurate potential control of buried redox-active centers in proteins, without the need for any mediators (Abdiaziz et al., 2019).

PFE-EPR may be divided into two main categories:

(1) Ex-situ PFE-EPR: PFE followed by low-temperature EPR measurements. Chronopotentiometry is used to control the potential and the sample is then flash frozen to investigate the redox-active species at a given redox state. The procedure may be repeated at different potentials. This method eliminates the need for mediators and requires very little protein sample, but does not otherwise offer significant advantages over classical EPR-based potentiometric titrations.

(2) In-situ PFE-EPR: PFE is performed while the cell containing the WE is inserted into the EPR cavity. This yet unpublished method enables the real-time observation of radicals during a redox reaction, and may be applied under catalytic conditions. However, it is currently only applicable to PCs that are EPR visible at non-cryogenic temperatures.

In this section, we focus on (1). However, in both cases performing PFE-EPR requires the insertion of the WE into the EPR resonator. Hence, the properties of both the WE and the cell suitable for PFE-EPR need to be considered carefully. We discuss practical considerations of PFE-EPR and provide an outlook of this technique that is still under development.

3.3.1 Working electrode and cell

For PFE-EPR the WE must be:

- *Biocompatible*: immobilizing the protein should not disrupt its structure and function.
- *Functionalizable*: the surface should be adaptable to anchor molecules with different functional groups and surface charges.
- *Conductive*: the material should be conductive enough to allow electrochemical measurements but not too conductive to affect EPR measurements.
- *Tunable surface*: the material should be tunable in terms of its structure, roughness and surface area to accommodate proteins of different sizes and properties.
- *Microwave transparent*: to allow EPR measurements and microwave power control in the resonator, the material of the PFE-EPR WE should be microwave transparent.
- *EPR silent*: the material should not give rise to any EPR signal.

3D ITO structures fulfill the majority of the criteria listed above and were used successfully as WE for PFE-EPR (Abdiaziz et al., 2019). However, electrodes prepared from commercially available ITO nanoparticles give rise to an EPR signal at $g = 1.89$.

The PFE-EPR cell must also fulfill several requirements:

- The cell material should be electrochemically inert, microwave transparent and EPR silent.
- The distance between the WE and the RE should be minimized.
- Efficient exchange between bulk solution and electrode surface must occur.
- Dimensions and geometry must be suitable for insertion into the desired EPR resonator (for in-situ PFE-EPR).

To perform ex-situ PFE-EPR, a quartz EPR tube (O.D. 4 mm, I.D. 3 mm) fulfils all the criteria listed above. The primary challenge is the fiddly assembly of electrodes inside such a small "cell."

3.3.2 Working electrode assembly and PFE-EPR measurements

ITO electrodes for PFE measurements are assembled from ITO nanoparticles into a mesoporous or inverse-opal mesoporous structures on flat glass slides (Bajada et al., 2020; Kato et al., 2013, 2014; Reuillard et al., 2017). The procedure for preparing cylindrical mesoITO electrodes, which can be placed inside an EPR tube and are compatible with commercial resonators, is detailed in Fig. 3. Following preparation of the WE, the three-electrode PFE-EPR cell is assembled with teflon-coated platinum and silver wire as CE and pseudo-RE, respectively (Fig. 4).

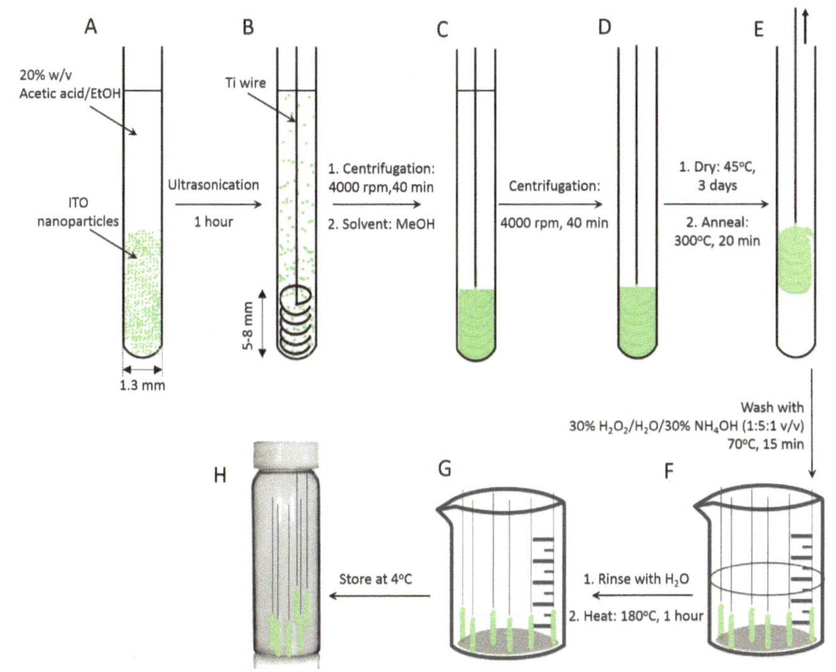

Fig. 3 Preparation procedure of mesoITO WE for PFE-EPR. (A) ITO nanoparticles (Sigma-Aldrich, CAS number: 544876) are suspended in 20% w/v acetic acid/EtOH mixture, (B) ultrasonic treatment of ITO suspension for 1h in the presence of the coiled Ti wire, (C) centrifugation of ITO suspension with Ti wire in the glass tube followed by the removal of excess acetic acid:EtOH mixture replaced with methanol, (D) centrifugation of ITO sample in methanol and removal of the excess MeOH, (E) ITO electrode is dried at 45 °C for 3 days, followed by annealing at 300 °C for 20 min, (F) once cooled down, electrodes are washed with 30% $H_2O_2/H_2O/30\%NH_4OH$ for 15 min at 70 °C, (G) rinsing with deionized water and heating at 180 °C for 1 h, (H) electrodes are stored in a closed vial at 4 °C. It is important to clean the Ti wire thoroughly with isopropanol and ethanol using an ultrasonic bath to remove the titanium oxide "barrier" which otherwise prevents the ITO nanoparticles from attaching to the wire.

In ex-situ PFE-PFE, a current is applied until the desired potential is reached (potentiometry) and the sample is then flash-frozen for EPR measurements, as described for potentiometric titrations (Section 3.2.3). Because ITO WEs do not withstand freeze-thaw cycles, a new WE is currently required for each data point and hence to construct the Nernst plot (if desired). We demonstrated proof-of-concept of PFE-EPR and the possibility to control the reduction potential of a buried redox center in a protein without mediators with Ni/Cu superoxide dismutase: the reduction potential of the $Cu^{2+/+}$ couple deduced from non-turnover peaks obtained with PFE matched that obtained from the Nernst plot perfectly (Abdiaziz et al., 2019).

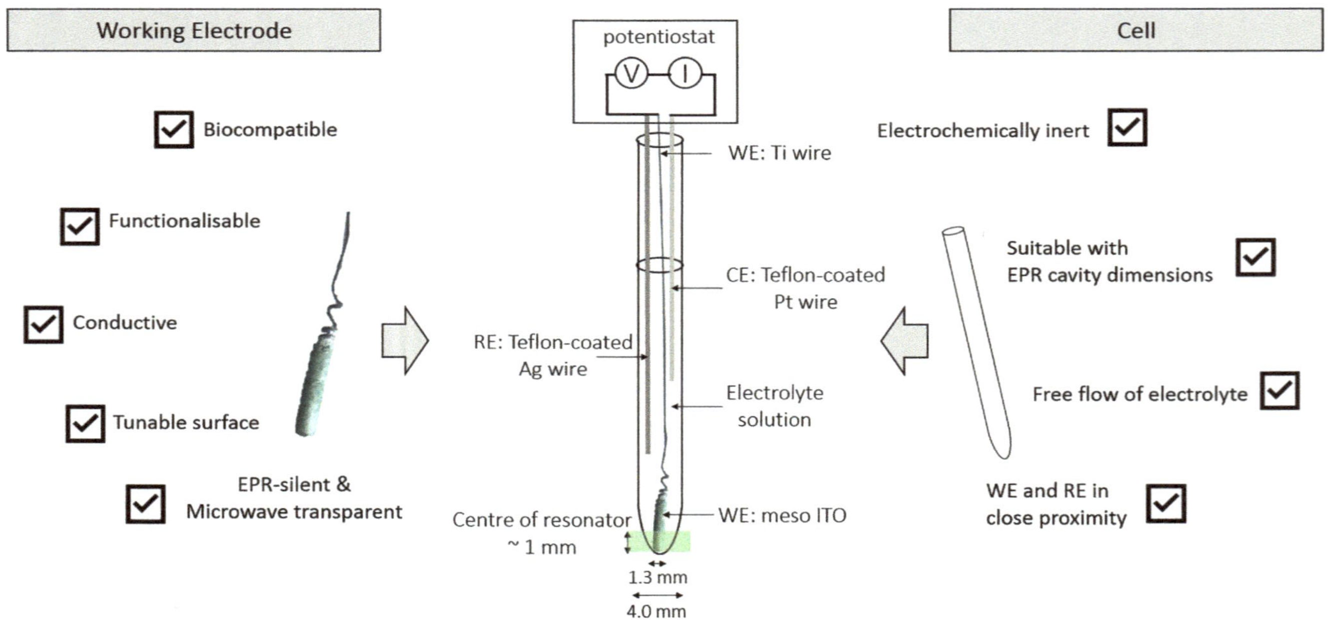

Fig. 4 Essential criteria for PFE-EPR. Conditions needed to obtain functional WE and cell with assembly into a PFE-EPR setup (Abdiaziz et al., 2019).

3.3.3 Limitations and outlook

Compared to the well-established solution-based potentiometric titrations methods and PFE, PFE-EPR is in its infancy. The experimental challenge in assembling the WE and electrochemical cell to conduct ex-situ PFE-EPR measurements may not outweigh the relative convenience of carrying out CW EPR-monitored potentiometric titrations, but it is worthwhile when mediator signals cause serious interference with the paramagnetic species of interest. However, PFE-EPR opens the door for studying the direct effects of potential change under turnover conditions on EPR-active centers in enzymes (and indeed other catalysts), for which there is currently no other methodology available. PFE-EPR thus promises to enhance our understanding of the structure-function relationships of redox-active proteins and enzymes.

4. Using pulsed EPR to gain detailed insight into unpaired electrons

Our discussion so far has focused on CW EPR spectroscopy. While it is the "to-go-to" method for investigating almost any new spin system, including complex metalloproteins, there are limits in the information it can provide. Pulse EPR on the other hand offers the possibility to gain much more detailed understanding of the paramagnetic species of interest, as well as its surroundings (Roessler & Salvadori, 2018). Here, we will discuss some of the most commonly employed techniques, namely hyperfine methods and distances measurements, with examples and practical considerations.

In metalloproteins the presence of multiple and redox-active centers can make it a challenge to obtain samples that are sufficiently simple for detailed pulse EPR investigations. For instance, when carrying out hyperfine EPR investigations (i.e. probing the nuclear spins in the vicinity of the unpaired electron), having more than one PC can make the interpretation of the data extremely difficult and ambiguous. In some cases, potential control may offer the possibility to "isolate" a given center. But in cases where it cannot (e.g. when the reduction potential of two redox-active centers is very similar), pulse EPR may offer a solution, and we discuss possible approaches.

4.1 Considerations prior to pulse EPR experiments

4.1.1 Protein concentration

The concentration of metalloproteins necessary to perform hyperfine experiments is typically 100–200 μM. However, lower (Richardson et al., 2021) and higher (García-Rubio et al., 2003) concentrations (μM–1 mM) can also be measured. Highly concentrated samples are required when the CW EPR signals of the protein are broad, since the number of spins on resonance can change substantially over the field range covering the signal (Section 4.1.3). More concentrated samples (>1 mM) should be avoided since they can lead to faster relaxation processes and broader signals caused by the interactions between adjacent PCs. The relevant concentration for pulse EPR studies is the local rather than the bulk concentration, hence the need to use cryoprotectants to prevent aggregation (Section 3.2.1).

The intensity of the EPR-active redox species of interest ought to be monitored by CWEPR combined with potentiometric titration or PFE, prior to pulse hyperfine experiments. The sample corresponding to the highest intensity of redox active species should be considered for pulse measurements.

4.1.2 Type of EPR tube and sample size

Pulse EPR techniques at multiple frequencies might be necessary. Somewhat frustratingly, this often necessitates using EPR tubes of different sizes, so this is an important consideration when the protein sample is very precious. Prior to pulse measurements it is thus worth contemplating the choice of the resonator(s)—and hence correct EPR tube. It is always possible to measure small tubes (capillaries) in resonators with wider sample access (Table 1), e.g., it is possible to load the sample in a 1.6 O.D. capillary and use it to conduct measurements in the MD5 or T2 resonators, but not vice versa. Luckily, some choices of resonators (e.g. X-band MS2 and Q-band D2 or X-band MS3 and Q-band T2; see Table 1) require tubes with the same outer diameter so that measurements at multiple microwave frequencies on the same sample can be performed. This is useful (Richardson et al., 2021) when adjustment of the reduction potential is required and highlights a further advantage of small-volume potentiometric titrations (Section 3.2) which allows both CW and pulse X- and Q-band measurements to be conducted on the same <10 μL sample.

Preparing samples in smaller-diameter EPR tubes can be more challenging, especially with relatively viscous solutions. Centrifugation at low speeds can "force" the sample to the bottom of the tube. The required sample height to fill the active region of the resonator is also resonator-dependent (Table 1) and therefore the sample volume has to be chosen carefully to achieve maximum sensitivity. For more information about the pulse EPR instrumentation, see (Reijerse & Savitsky, 2017; Stoll, 2017).

Table 1 List of commonly used and commercially available pulse EPR resonators from Bruker with their main characteristics.

Resonator name Resonator type Frequency	Resonator bandwidth when fully overcoupled [MHz]	Active height [mm]	Distance: sample center to sample holder [mm]	Sample access tube/capillary: (I.D., O.D.) [mm]	Required volume [μL]	Applications (+): encouraged (−): discouraged
ER 4118 X - MD5 Dielectric resonator **X-band**	~300	13	39	5.0 (3.0, 4.0)	92	Pulse EPR, big sample volume (+) DEER (−) CW EPR (+)
EN 4118 X -MD4 Dielectric resonator **X-band**	~300	13	55	4.0 (2.8, 3.8) **'Standard' 4-mm tubes are not suitable**	80	Pulse ENDOR(+) CW EPR (−) Pulse EPR (−)
ER 4118 X -MS2/3/5 Split ring resonators **X-band**	MS5: ~320 MS2: ~350	6	39	MS5: 5.0 (3.0, 4.0) MS3: 3.0 (2.0, 2.8) MS2: 2.0 (1.1, 1.6)	MS5: 42 MS3: 19 MS2: 6	Pulse EPR (+) DEER (+) CW EPR (−) Shaped pulses (+)
EN 5107 Q D2 Dielectric resonator **Q-band**	~100	8	25	2.0 (1.1, 1.6)	8	Pulse EPR, small sample volume (+) Pulse ENDOR(+) DEER, small sample volume (+) CW EPR(−)
ER 5106 Q T2 * Dielectric resonator **Q-band**	~150	8 (point sample) 16.5 (line sample)	30 (point sample) 38 (line sample)	3.0 (2.0, 2.8)	52	Pulse EPR, big sample volume (+) DEER, big sample volume (+) CW-EPR (+)

X-band refers to microwave frequency of ~ 9.5 GHz and Q-band refers to ~34 GHz. * The T2 resonator has two maxima of the microwave field B_1; for limited-volume (point) samples only the upper half of the resonator is used, whereas for large-volume (line) samples the whole resonator mode is used.

4.1.3 Orientation selection

EPR spectra of metalloproteins are often very broad, e.g. even axial [4Fe—4S] clusters can easily be 250 G/700 MHz wide (Fig. 5). Theoretically, a pulse with infinitesimal length can excite very broad spectra. However, practically the shortest experimentally achievable pulse on a commercial spectrometer is around 4 ns, with an excitation bandwidth of ~200 MHz (Fig. 5, yellow bar). This is much narrower than the spectral width of metal centers in metalloproteins (Fig. 5), therefore only selected orientations—which depend on the field—will be excited by the pulse sequence (Fig. 5); this phenomenon is known as *orientation selection.* Orientation selection can be overcome if multiple experiments are conducted at different magnetic fields, covering the full width of the field-sweep spectrum. However, in practice this can be unfeasible due to the long acquisition time required. On the other hand, orientation selection can be very useful if it can be incorporated into data analysis (as discussed for pulsed distance measurements in Section 4.4).

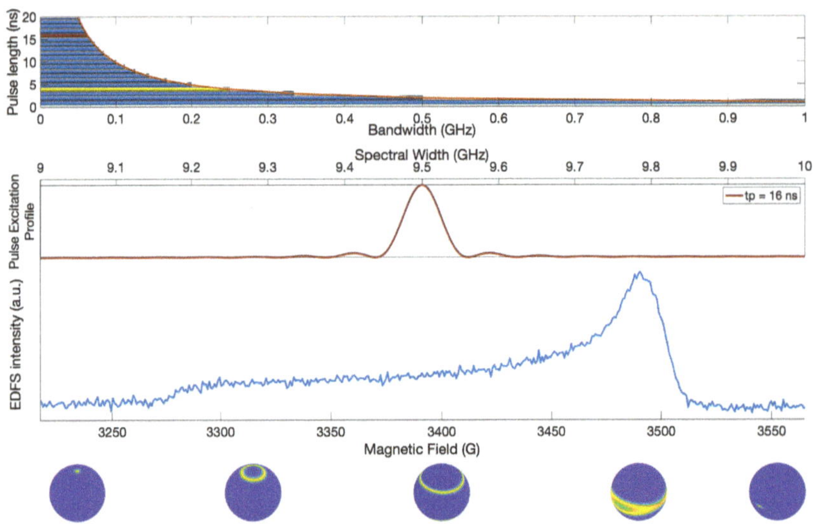

Fig. 5 Illustration of orientation selection with an FeS cluster spectrum. Excitation bandwidth of pulses with different lengths (top); the yellow bar at 4 ns refers to the shortest experimentally achievable pulse length, the red bar at 16 ns refers to the pulse length used to obtain the two-pulse echo-detected field sweep (in blue) of FeS cluster N2 of respiratory complex I ($\pi/2 = 16$ ns, $\pi = 32$ ns, microwave frequency 9.44 GHz). The red trace shows the excitation profile of the $\pi/2 = 16$ ns rectangular pulse. Blue spheres indicate the orientation of the excited spins at different field positions (3230 G, 3312.5 G, 3400 G, 3490 G, 3550 G).

4.2 "Selecting" spins in metalloproteins using pulsed EPR

4.2.1 Relaxation pathways and nomenclature

Relaxation is the recovery to equilibrium from a non-equilibrium state. Relaxation processes contribute to all pulsed EPR techniques, as they are sensitive to the relative electron spin population states. Relaxation times or rates are indicative of the electron's environment, including the electronic structure of the PC, molecular motions, local concentrations, hyperfine and Zeeman anisotropies. A detailed account of relaxation processes in EPR can be found in (Eaton & Eaton, 2002).

Many related yet distinct processes contribute to the relatively fast relaxation times of electron spins in metalloproteins, frequently requiring cryogenic temperatures (<100 K) to observe them. Understanding which processes are contributing to any given experiment to allow quantification of true relaxation times is very challenging for metalloproteins (see e.g. (Gayda et al., 1979) for T_1 of [2Fe—2S] proteins). Fortunately, this is rarely practically useful when setting up pulsed experiments. The complexities of understanding relaxation have resulted in terminology for both theoretically defined and experimentally measured parameters, unhelpfully often used interchangeably (Scheme 4). Broadly speaking, relaxation is divided into spin-spin relaxation (T_2) pathways (or more appropriately for frozen solutions the phase-memory time T_m), and spin-lattice relaxation (T_1) pathways, with $T_1 >> T_2$ at cryogenic temperatures. In our discussion below we shall adhere to this over-simplification for illustration purposes.

Definitions

T_1 or longitudinal relaxation:
The time dependence of magnetization along the z axis
T_1 is also termed **spin-lattice relaxation** in which lattice refers to all the surrounding atoms.

Spectral or spin diffusion:
Anisotropic motion, electron-electron exchange, nuclear-electron cross relaxation and nuclei which take spins off-resonance without changing the net population of Zeeman levels.

Spin-lattice relaxation + spectral diffusion often collectively called **effective T_1**.

T_2 or transverse relaxation:
1. The rate of mutual spin flips and the time constant of dephasing in the xy plane (Atherton 1993)
2. The contribution of rotational motion of hyperfine and Zeeman anisotropies (Goldman 1972); may not result in an electron spin flip

This type of relaxation is referred to as **spin-spin relaxation**.

Experimentally measured parameters

The decay constant of the 2-pulse echo decay trace is known as T_m (phase memory time).
T_m can be used to estimate T_2 but as the relatively dilute electron spins are surrounded by magnetically concentrated nuclear spins, nuclear spin diffusion can be a large contributor; this leads to stretched exponential echo decays (Milov 1973).

Exponential fitting of the inversion recovery traces yields T_1. Because of spectral diffusion a biexponential recovery is normally observed; this issue can be alleviated by using the picket-fence saturation recovery pulse sequence (Dalton 1972).

Quantifying T_1 and T_2

Bloch equations can be used to fit single exponentials to M_z or M_{xy} to extract T_1 and T_2. Experimental data rarely fit a simple exponential as any rate overserved is the sum of competing relaxation pathways, the fastest of which will dominate.
If reasonable assumptions can be made about the relaxation processes at work, modified Bloch equations can be used to quantify relaxation times (Maryasov 2013, Roscher 1996, Klug 1998, Zhou 2000).

Scheme 4 Overview of relaxation parameters and definitions (Atherton, 1993; Dalton et al., 1972; Goldman et al., 1973; Klug et al., 1998; Maryasov & Bowman, 2013; Roscher et al., 1996; Zhou et al., 2000).

4.2.2 Exploiting relaxation to differentiate between paramagnets

Differences in the relative relaxation rates of paramagnetic species can be a useful way to disentangle EPR signals. As we illustrate below, this approach can work even for multiple cofactors of the same type, with overlapping signals in field swept spectra, by varying parameters such as temperature, shot repetition time (SRT) and inter-pulse delays.

4.2.2.1 Temperature

At high temperatures species may relax too fast to be observable; at low temperature species may not relax sufficiently fast, leading to saturation and therefore also evading detection. This results in a temperature "window" that can be used to investigate a particular species (e.g. <40 K for $[4Fe—4S]^+$ clusters).

4.2.2.2 Shot repetition time

The short repetition time (SRT), which is the time between consecutive pulse sequences, can be used to differentiate species based on their T_1 relaxation time (while keeping the temperature constant). To avoid saturation of slow-relaxing species, the SRT has to be long enough for the net magnetization to be fully "restored." By using a shorter SRT a given species may hence be deliberately saturated to remove its contribution to the EPR spectrum (Fig. 6A).

4.2.2.3 Inter-pulse delays

To differentiate PCs on the basis of their T_m, the inter-pulse delay (τ) can be adjusted in the basic Hahn echo sequence ($\pi/2$–τ–π–τ–echo). Spins which de-phase over the time τ, i.e. those that have short T_m, are not detected (Fig. 6B).

4.2.2.4 Inversion recovery

This is the most common pulse experiment used to filter signals based on relaxation (Fig. 7A). A π pulse inverts the net magnetization, which is allowed to recover to varying extent in the subsequent variable filtration time (T_f), with the electron spin echo then detected through, in the simplest case, the Hahn echo sequence. Because the magnetization associated with different spin packets (e.g. different paramagnetic species in the sample) will recover to a varying extent as dictated by T_1, this experiment provides a means of "filtering" out species. Although T_1 can in principle be determined, spectral diffusion can introduce a significant error and may need to be suppressed by using shaped pulses (Doll et al., 2013). An inversion

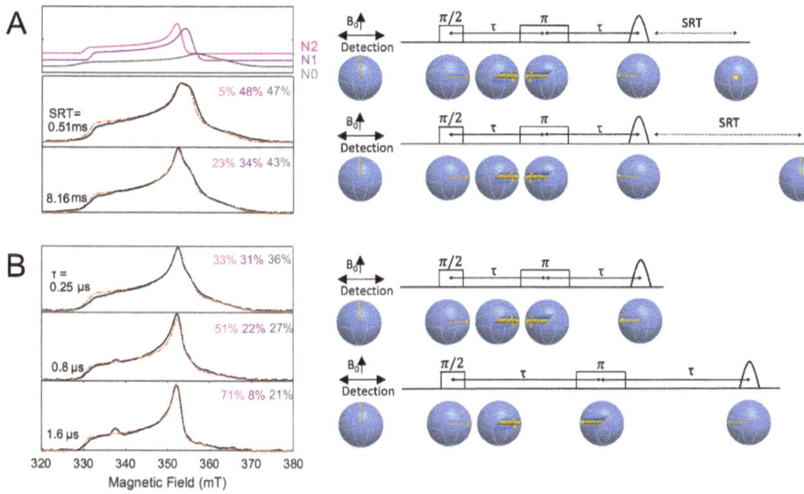

Fig. 6 Untangling the three [4Fe—4S] clusters of photosynthetic complex I (N0, N1 and N2) through relaxation filtering. (A) Simulations of clusters N0, N1 and N2 with experimental (black) EDFS dependence on SRT (ms). Partial recovery of the magnetization to the z-axis (yellow "spin packets") with a shorter SRT is illustrated schematically with the Hahn echo sequence. (B) EDFS dependence on τ (μs), with schematic illustration. % contribution of N0, N1 and N2 to the simulated EDFS (red) are indicated. On most commercial devices τ is set between the rising edges of the pulses.

pulse added prior to any pulse sequence, e.g. HYSCORE or DEER (Maly et al., 2004, 2009), can filter spins contributing to the observed echo; in the context of hyperfine spectroscopy, this family of sequences is known as REFINE (Maly & Prisner, 2004).

4.2.3 Limitations and outlook

Due to the complexity and orientation selection of relaxation processes in metalloproteins, few have been systematically determined; a table of examples can be found in (Eaton & Eaton, 2002). Even with the strategies discussed above, it may not be possible (or necessary) to determine absolute relaxation times. Relative relaxation rates can still inform on the spin environment, including protein-protein interactions (Lyubenova et al., 2010), or through relaxation-filtered spectroscopy assist in assigning basic magnetic parameters such as g values in multi-component systems (Richardson et al., 2021).

If relaxation rates are very fast, it can become difficult to conduct pulsed EPR experiments. Strategies to extend the lifetime of the echo include

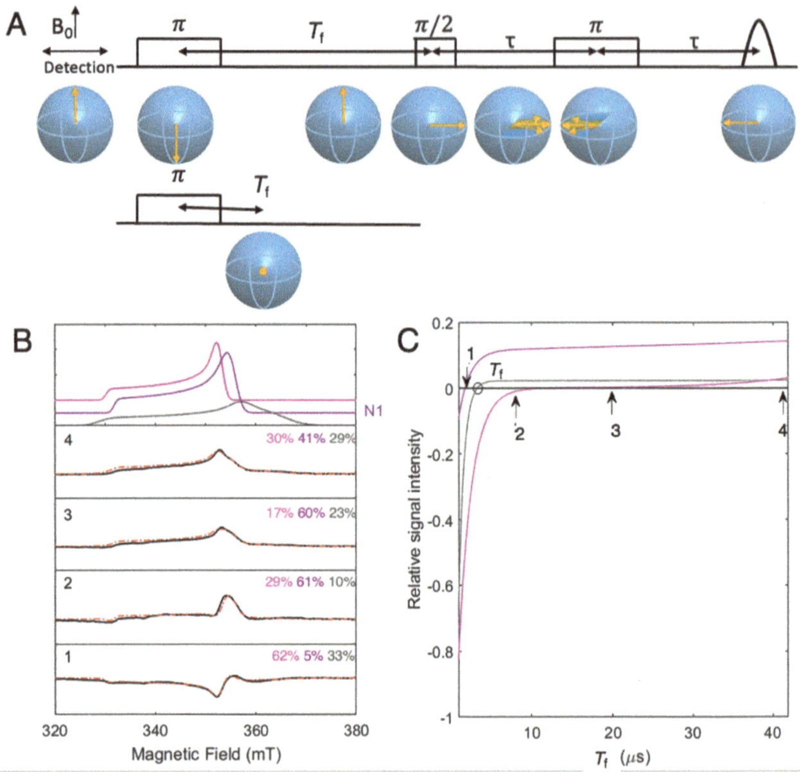

Fig. 7 Untangling the three [4Fe—4S] clusters of photosynthetic complex I (N0, N1 and N2) through inversion recovery experiments. (A) Inversion recovery pulse sequence and schematic diagram of spin relaxation. The length of T_f dictates how much of the magnetization recovers. (B) EDFS at different values of T_f, corresponding to the arrows in (C), with raw data (black), total simulations (red) and individually simulated spectra of N0, N1 and N2. (C) Exponential fit of inversion recovery EDFS signal intensity used to estimate T_f. As a word of caution, this kind of analysis can be complicated by the variation of T_1 across the EPR spectrum of the individual components.

changing measurement frequency (Ward et al., 2010), deuterating the sample, reducing the local spin concentration (by forming a glass) and measuring at extremely low temperatures (Berliner, Eaton, & Eaton, 2000).

4.3 Hyperfine techniques

4.3.1 The parameters to be determined and why they are useful

Hyperfine spectroscopy allows the determination of hyperfine and quadrupolar coupling parameters of $I \neq 0$ nuclei interacting with a $S \neq 0$ PC. This enables the deduction of structure around the PC and its mode of interaction with the environment. For the discussions that follow, it is helpful to have some understanding of energy level diagrams and the scenarios that are typically encountered when studying some of the most common interactions in proteins of unpaired electrons with ^{1}H ($S = ½$) and ^{14}N ($I = 1$). From the frequencies of the transitions observed, the isotropic (a_{iso}) and anisotropic (T) components of the hyperfine interaction are determined (Box 3). These inform on the amount of electron spin density delocalized onto the nucleus (or nuclei) of interest (a_{iso}), and hence bonding interactions, as well as the spatial orientation of the nucleus with respect to the PC (T), and hence geometry. Although energy level diagrams become significantly more complex with $I > ½$ nuclei, the information from the quadrupole coupling parameters κ and η can be extremely useful in providing additional insight into e.g. hybridisation and symmetry (axial or rhombic).

Hyperfine interaction and quadrupolar coupling parameters thus hold valuable information about the environment of the PC, namely: (i) the electronic structure of the PC, (ii) the spin distribution on it, (iii) the distance between nuclei and unpaired electrons. Consequently, hyperfine techniques allow insight into the structure and interactions within active sites or metal centers in metalloproteins.

4.3.2 Most typical hyperfine interactions encountered: ^{1}H and ^{14}N

4.3.2.1 Interactions with ^{1}H

Protons in metalloproteins can be either: (1) matrix protons resulting from ^{1}H nuclei beyond the second coordination sphere of the PC, (2) non-exchangeable protons from amino acids or cofactors in the first and second coordination sphere (covalently linked or interacting through space), (3) exchangeable protons from the first coordination sphere that are H-bonded to the PC, e.g. from amino acids, cofactors or substrates in the active site. To distinguish between exchangeable and non-exchangeable protons, buffer exchange (H_2O to $^{2}H_2O$) is required. The peaks of (only) the exchangeable protons will then become centered around the Larmor frequency of ^{2}H (~ 2.2 MHz at X-band) instead of ^{1}H (~14.5 MHz at X-band), as illustrated in Fig. 8. In line with their gyromagnetic ratios, the hyperfine interaction constants of ^{1}H should be about 6.5-fold larger than those of ^{2}H (le Breton et al., 2017; Grimaldi et al., 2012; Seif-Eddine et al., 2020).

4.3.2.2 Interactions with ^{14}N

Nitrogen interactions can unravel important information about the structure of active sites (Colaneri & Peisach, 2005). These nitrogen nuclei originate

BOX 3 Energy level diagrams.

Diagrams are shown for a $S=\frac{1}{2}$ paramagnetic center interacting with $I=\frac{1}{2}$ and $I=1$ nuclei, with expressions for the hyperfine and quadrupolar coupling constants.

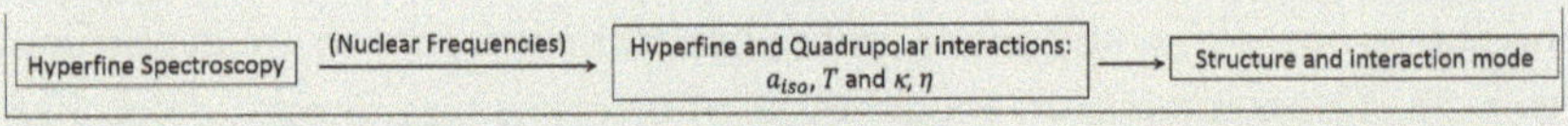

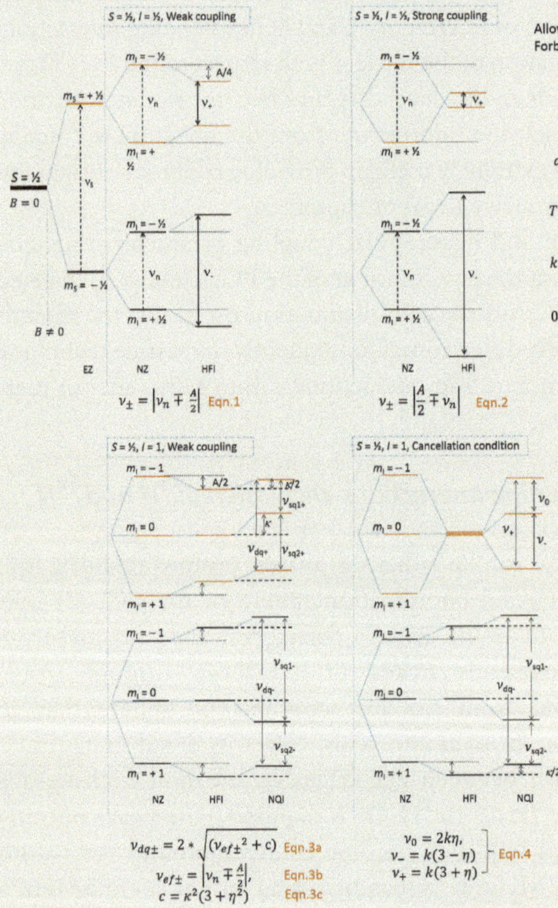

Allowed EPR transitions: $\Delta m_S = \pm 1, \Delta m_I = 0$
Forbidden EPR transitions: $\Delta m_S = \pm 1, \Delta m_I = \pm 1$

$$A = a_{iso} + T,$$

$$a_{iso} = \frac{2\mu_0}{3\hbar} g_e \beta_e g_n \beta_n |\psi_0(0)|^2 \quad \text{Eqn.5}$$

$$T = \frac{\mu_0}{4\pi\hbar} \frac{g_e \beta_e g_n \beta_n}{r^3} \quad \text{Eqn.6}$$

$$k = \frac{e^2 qQ}{4I(2I-1)\hbar} \quad \text{Eqn.7}$$

$$0 \le \eta \le 1, \qquad k = a - \frac{b}{r^3_{(H\ldots A)}} \quad \text{Eqn.8}$$

a_{iso} and T in MHz, κ is measured in Hz.
μ_0 : magnetic permeability of vacuum.
$\hbar$: reduced Planck constant.
$g_{e,n}$: electronic and nuclear g factors.
$\beta_{e,n}$: electronic and nuclear Bohr magnetons.
$|\psi_0(0)|^2$: the probability to have the electron spin density on the nucleus.
r : the distance between the interacting nucleus and the centre of the atomic orbital containing the unpaired electron.
eq : contribution of the electric field gradient.
eQ : nuclear quadrupole moment.
a = 319 kHz, b = 607 kHz· Å^3 (Flores et al., 2007).

EZ: electron Zeeman interaction
NZ: nuclear Zeeman interaction
HFI: hyperfine interaction
NQI: nuclear quadrupolar interaction
S: electron spin, I: nuclear spin
B: external magnetic field
dq: double quantum transition
sq: single quantum transition

from: (i) remote and directly-coupled histidine-imidazole (N_δ and N_ε) (Seif-Eddine et al., 2017), (ii) peptides (Taguchi et al., 2014), (iii) non-imidazole side chains (Seif-Eddine et al., 2017) and (iv) an intermediate or substrate (Singh et al., 2000). Nuclear quadrupole parameters can reveal the type of nitrogen interacting with the PC (Grimaldi et al., 2010; Wright et al., 2020). Because three nuclear transitions can occur in each electron spin manifold (Box 3), analysis of ^{14}N hyperfine spectra can be quite complex, with up to 9/18 peaks observed in one/two dimensional spectra, respectively. Luckily, the "cancellation condition" (where the electron Zeeman and hyperfine interactions "cancel" in one of the electron spin manifolds, Box 3) is not uncommon, or can be achieved by changing the measurement frequency. This greatly simplifies the interpretation of spectra because nuclear quadrupole parameters can be determined directly from the observed nuclear frequencies (ν_0, ν_+, ν_-). When the cancellation condition cannot be achieved, ^{15}N ($I = ½$) isotope labeling can simplify analysis, decreasing the number of resulting peaks to a doublet centered around the ^{15}N Larmor frequency (~1.5 MHz at X-band).

4.3.2.3 $^{14}N—^{1}H\cdots A$

The point-dipole approximation (Box 3, eqn. 6) can be used to calculate the length (r) of the hydrogen bonding when $r > 1.3$–1.4 Å, see e.g. (Roessler et al., 2012). Below this length, the point-dipole approximation is no longer suitable due to the increased covalent character of the ^{1}H-bond (Sinnecker et al., 2004, Sinnecker, Flores, & Lubitz, 2006). In this case, instead of T, κ of the exchanged deuterium (le Breton et al., 2017; Grimaldi et al., 2012; Seif-Eddine et al., 2020) can be used to determine r (Box 3, eqn. 8).

We now introduce pulse EPR techniques used to determine the hyperfine and nuclear quadrupole parameters. We discuss how to choose the most suitable experiment, important parameters to consider while measuring, and analysis of the data obtained.

4.3.3 Techniques to detect hyperfine interactions

Hyperfine and nuclear quadrupole interactions are usually too weak to be resolved by CW EPR spectroscopy. In this chapter we focus on applications of ESEEM (electron-spin-echo-envelope-modulation); as will become apparent below 2D ESEEM techniques are particularly useful when studying a metalloprotein PC interacting with multiple different nuclei. For details on other hyperfine techniques, such as ENDOR (electron nuclear double resonance) or ELDOR detected NMR, see (Goldfarb, 2017; Harmer et al., 2009; Harmer, 2016).

In ESEEM pulse sequences only microwaves are used, and nuclear frequencies are detected indirectly by exciting forbidden transitions in anisotropic systems. ESEEM records the modulation of the spin echo amplitude, observed due to the anisotropy of the hyperfine coupling, as a function of

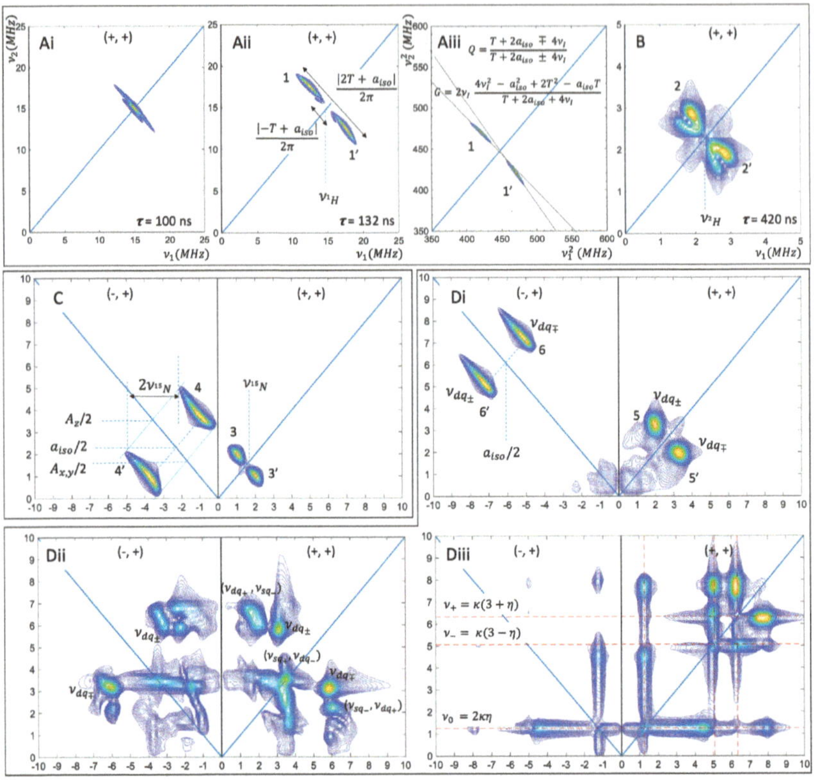

Cross peaks	(1, 1′)	(2, 2′)	(3, 3′)	(4, 4′)	(5, 5′)	(6, 6′)	sq, dq
Nucleus	^{1}H	^{2}H	^{15}N		^{14}N		
HFI [a_{iso}, T] [A_{xx}, A_{yy}, A_{zz}]	[0.9, 3.9] [-3.0 -3.0 8.7]	[0.14, 0.6] [-0.46 -0.46 1.34]	[0.29 0.41 1.68]	[-0.04, +3.69] [3.65 3.65 7.43]	[0.7, 0.4] [0.5 0.5 1.1]	[13, 1.5] [11.5 11.5 16]	[2.8 4.6 3.5]
NQI [κ, η]	N.A.	[0.19, 0]	N.A.	N.A.	[0.42, 0.5]	- -	[1.0, 0]
References	le Breton et al., 2017		Fugate et al., 2012		Seif Eddine et al., 2017	Roessler et al., 2012	

Fig. 8 Simulated HYSCORE spectra based on example literature, showing couplings, with different strength, between $S = ½$ and $I = ½$, 1. (Ai-ii) (1, 1′) hfi between ^{1}H nucleus and FeS cluster in respiratory complex I. (Aiii) quadrature analysis of peaks (1, 1′). (B) (2, 2′) hfi between FeS cluster in respiratory complex I and buffer-exchanged ^{2}H. (C) (3, 3′) weak and (4, 4′) strong hfi between isotopically labeled ^{15}N and reduced FeS cluster in Biotin synthase from *E. coli*. (Di) (5, 5′) weak hfi between ^{14}N of histidine and the electron of menasemiquinone in the membrane nitrate reductase from *E. coli*. (6, 6′) strong hfi between cysteine ^{14}N and [4Fe—3S] cluster in O_2-tolerant [NiFe]-hydrogenase from *E. coli*. (Dii) (+, +) and (-, +) quadrants: intermediate hfi of ^{14}N interacting with [4Fe—3S] cluster in O_2-tolerant [NiFe]-hydrogenase from *E. coli*. (Diii) hyperfine coupling of same ^{14}N as in (Dii) but near the cancellation condition. The hfi and nqi constants used in the simulations are listed in the table with the references from which the couplings were taken. Other parameters used in the simulations: $\pi/2$ pulse: 8 ns (A), 16 ns (B), 16 ns (C), 12 ns (Di (+, +)), 16 ns (Di (−, +)), 16 ns (Dii), 16 ns (Diii), τ: 100 ns (Ai), 132 ns (Aii), 420 ns (B), 132 ns (C), 204 ns (Di (+, +)), 108 ns (Di (−, +)), 108 ns (Dii), 108 ns (Diii), magnetic field: 348.2 mT (A), 345.68 mT (B), 347.5 mT (C), 344.8 mT (Di (+, +)), 343.7 mT (Di (−, +)), 343.7 mT (Dii), 346 mT (Diii), microwave frequency: 9.41 GHz ns (A), 9.70 GHz ns (B), 9.75 GHz ns (C), 9.68 GHz (Di (+, +)), 9.74 GHz ns (Di (−, +)), 9.74 GHz (Dii), 9.74 GHz (Diii).

time delays between pulses. Therefore, the output of an ESEEM experiment requires Fourier transform to obtain frequency-based information, while this is not required for ENDOR. Here, we take a practical approach to help identify the most appropriate ESEEM experiment for a given hyperfine interaction. For further theoretical and practical details of ESEEM, see (van Doorslaer, 2017).

4.3.3.1 How large is the coupling?

A good starting point is to estimate the magnitude of the hyperfine coupling expected. If the nuclear spin of interest is in the first coordination sphere, it is likely that it will be "strongly coupled" and ENDOR techniques will likely be useful. If the coupling is anticipated to be "weak," e.g. the nucleus of interest is hydrogen-bonded or in the 2nd or 3rd coordination sphere, then ESEEM techniques should be attempted in the first instance. When multiple nuclei are expected to interact with the same PC, 2D techniques are better to disentangle the different contributions. In some cases, e.g. when the coupling is small but completely isotropic, nothing will be observed with ESEEM techniques (which rely on anisotropy). Some trial-and-error with different techniques is therefore somewhat inevitable and often ESEEM and ENDOR, as well as ELDOR-detected NMR at Q-band and higher frequencies, can provide complementary information. For instance, Mims-ENDOR has been reported to resolve electron-nuclear couplings as small as 10 kHz (Meyer et al., 2020) that would be challenging to obtain with ESEEM due to the very long period of the oscillations (100 μs).

4.3.3.2 Which microwave frequency?

X-band is commonly used to detect interactions with ^{1}H since the Larmor frequency (~14.5 MHz) is sufficiently high to enable separation from peaks belonging to other interacting nuclei, e.g. ^{2}H (~2.3 MHz) and ^{14}N (~1.1 MHz) or ^{15}N (~1.5 MHz). For weak couplings and nuclei with low nuclear frequency, e.g. ^{2}H, ^{14}N, ^{15}N (Table 2), going to Q-band can help to increase spectral resolution and to differentiate peaks from multiple nuclei interacting with the same PC. High-frequency hyperfine spectroscopy is also very useful to (i) resolve peaks arising from nuclear quadrupolar coupling (Grimaldi et al., 2012), (ii) force strongly coupled ^{14}N into the cancellation condition and hence allow easy determination

Table 2 Nuclear spins, natural abundance and Larmor frequencies of some biologically relevant nuclei at X- and Q-band frequencies.

Nucleus	Nuclear spin	Natural abundance (%)	ν_I at X-band (MHz)	ν_I at Q-band (MHz)
^{1}H	1/2	99.99	14.9	51.1
^{2}H	1	0.01	2.3	7.8
^{14}N	1	99.63	1.1	3.7
^{15}N	1/2	0.37	1.5	5.2
^{13}C	1/2	1.07	3.7	12.9
^{31}P	1/2	100	6.0	20.7

of nuclear quadrupole parameters (Stich et al., 2010; Yeagle et al., 2008), (iii) "escape" the cancellation condition of ^{15}N at X-band that may complicate spectra (Taguchi et al., 2014).

Due to the dependence of the ESEEM modulation depth parameter on the microwave frequency, care should be exerted not to suppress the desired information. For example, high microwave frequencies might be disadvantageous for detecting couplings with nuclei that possess a high gyromagnetic ratio (van Doorslaer, 2017). In contract, the ENDOR efficiency is frequency independent, so that nuclei with high gyromagnetic ratios can be studied at high frequency (Schweiger & Jeschke, 2001).

4.3.3.3 Which type of ESEEM experiment?

2-pulse ESEEM is simple, fast and efficient to get information about the interacting nuclei. However, because of the dependency on (short) T_m (Scheme 4), frequency peaks are broadened, leading to reduced spectral resolution. Analysis is also challenging because of peaks at the sum and difference of the basic nuclear frequencies (van Doorslaer, 2017), especially if multiple nuclei are interacting with the same PC. 2-pulse ESEEM is therefore useful with simple spin systems (one nucleus coupled to an electron spin) and is usually only performed to detect the ESEEM effect and optimize parameters before conducting other (more time consuming) experiments.

3-pulse ESEEM, a T_1-dependent experiment, generally gives rise to sharper peaks compared to 2-pulse ESEEM since $T_1 >> T_m$. Spectra are usually simpler to interpret given that no sum and difference peaks are observed. However, "phase cycling" (see Section 4.3.4.) is necessary in order to select the "right" (stimulated) echo, as a three-pulse sequence gives rise to five echoes. Moreover, experiments must be conducted at several τ to ensure that blind spots (Box 4) are not suppressing the peaks of interest. 1D ESEEM experiments are fast to acquire, so that the need for data with several τ values is normally feasible.

4-pulse ESEEM (HYSCORE) is powerful when multiple hyperfine couplings, from different nuclei and of different magnitude are present, owing to the "spatial resolution" possible with this 2D experiment. HYSCORE is therefore one of the most popular experiments applied to metalloproteins. The main downside is that HYSCORE and its variants (van Doorslaer, 2017) are much more time consuming (extensive phase cycling is required to remove unwanted echoes) and it is not uncommon that experiments need to be acquired overnight or even over multiple days. Combined with the need for measurements at multiple τ values to mitigate against blind spots (since HYSCORE is also a stimulated echo-based experiment) and at multiple field positions to address orientation selection (Section 4.1.3), this time factor can be detrimental. However, technology advancements provide an exciting outlook in this respect, with cryogenic preamplifiers enabling a 100-fold reduction in measurement time (Šimėnas et al., 2021). Non-uniform sampling (Nakka et al., 2014) has also been shown to allow successful reconstruction of HYSCORE spectra in a fraction of the measurement time.

When EPR spectra overlap, complementing HYSCORE with a DONUT-HYSCORE measurement (Goldfarb et al., 1998) can assist with assignment because it is possible to distinguish between nuclear spins coupled to the same or to different electron spins. If the evolution times

BOX 4 Blind spots in ESEEM and Mims ENDOR.

3- and 4-pulse ESEEM and Mims ENDOR are based on stimulated echo sequences and suffer from blind spots.

ESEEM: the modulation of the echo envelope of 3-/4-pulse ESEEM is affected by the value of the first evolution time (τ). This leads to a suppression effect, where peaks at $v = \frac{n}{\tau}$ completely disappear.

Mims-ENDOR: the ENDOR efficiency is a factor of $(1 - \cos A\tau)$ where A is the hyperfine constant, which means that hyperfine coupling peaks appearing at $A - \frac{2\pi n}{\tau}, (n - 1,\ 2,\ \ldots)$ are not detectable.

t_1 and t_2 (see Scheme 4) are not varied independently of each other, 1D versions of the HYSCORE sequence ('4p-ESEEM' experiments) can be obtained yielding projections of the HYSCORE spectrum onto the υ_2 axis ($t_1 = t_{1,0} =$constant) or onto the main diagonal $\upsilon_2 = \upsilon_1$ ($t_1 = t_2 = T/2$); the two experiments are named DEFENCE (deadtime free ESEEM by nuclear coherence-transfer echoes) and CP (combination peak), respectively (Ponti & Schweiger, 1995).

4.3.4 Important parameters to consider when carrying out hyperfine experiments

4.3.4.1 Pulse length

Short pulses, also called hard pulses, are non-selective because the excited frequency bandwidth is large. Hence the shorter the pulse, the more spins with different orientations are excited. Due to technical limitations, the shortest pulse length achievable on commercial spectrometers is around 4 ns, which (for rectangular pulses) corresponds to an excitation bandwidth of only ca. 200 MHz (Fig. 5); although shorter pulses with sub-optimal turning angles can be in principle generated, the resonator bandwidth limits the effective pulse excitation bandwidth.

In all pulse sequences for which hard pulses are required, such as ESEEM and Mims-ENDOR, to enhance the detected spin-echo signal and avoid narrowing the excitation bandwidth of the sequence, π and $\pi/2$ pulses are typically of the same duration; this is achieved by adjusting the individual pulse amplitudes such that the amplitude of the π pulse is the double of that of $\pi/2$ pulse. In HYSCORE the π pulse should ideally be less long than the $\pi/2$ pulses to avoid strong diagonal artifact peaks (Schweiger & Jeschke, 2001).

4.3.4.2 Time delay between pulses

As shown in the Scheme 5 the time delays between two consecutive pulses fall into two categories: incremented time and constant time. Keeping the time between the first and second pulses, τ, constant (e.g. in 3p and 4p ESEEM) leads to "blind spots" (see Box 4). Blind spots can be used to intentionally suppress undesirable high-intensity "matrix" peaks appearing at the Larmor frequency of nuclei such as ^1H (~14.5 MHz at 9.5 GHz, suppressed by setting $\tau = 138$ ns). Acquiring spectra at different τ values overcomes blind spots.

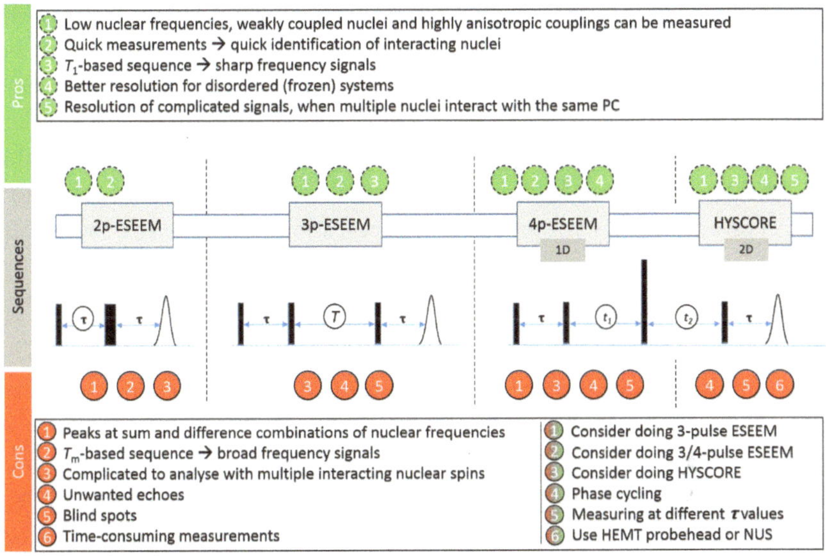

Scheme 5 ESEEM basic sequences with advantages and disadvantages. Troubleshooting methods are proposed for the disadvantages. The times, T, t_1 and t_2 between pulses refer to the incremented time in each sequence.

4.3.4.3 Number of averages

Sensitivity increases with the square root of the number of averages. Spectral averaging is usually required to improve the signal-to-noise ratio (SNR) and is achieved by increasing the number of scans and/or the number of averages. The number of averages is the product of the number of scans (n) multiplied by the shots per point (h). Note that h does not include phase cycling. For example, let's say that h = 100 and n = 1 is equivalent to 10-min spectral acquisition; the sensitivity will be enhanced however the result (after averaging) will appear only after 10 min. In contrast, with h = 25 and n = 4, the displayed SNR will be updated once every 2.5 min but the same overall result will be obtained (assuming a 1D experiment). Such "time economy" is useful when optimizing parameters during test measurements.

4.3.4.4 Shot repetition time

The SRT is the time delay between two consecutive pulse sequences. Generally, it should be long enough to allow the magnetization to recover before the next experiment, which makes it dependent on T_1. However, a long SRT increases the overall acquisition time, especially for long measurements such as HYSCORE. A rule of thumb is that the SRT should be approximately equal to 5 × T_1. Shorter SRT (~1.5 × T_1) should only be considered when "filtering" of different contributing paramagnetic species is desirable (Section 4.2.2). As T_1—and hence the optimal SRT—depends on the temperature, finding the optimal experimental conditions to perform the pulse EPR experiments leads to a considerable gain in sensitivity; generally, this involves finding a compromise between a long T_m (strong echo intensity) and a short T_1 (fast repetition rate).

4.3.4.5 Phase cycling

This is used to eliminate artefacts from unwanted echoes, e.g. 4 out of 5 echoes in the 3p stimulated echo sequence. 4 and 8 step phase cycling are usually required to eliminate unwanted echoes in multi-pulse experiments such as 3-pulse ESEEM and HYSCORE, respectively (Stoll & Kasumaj, 2008). Strictly, HYSCORE requires a 16 step phase cycle (Schweiger & Jeschke, 2001).

4.3.5 Data processing, analysis and simulation: HYSCORE case study

ESEEM raw data are subject to several signal processing steps before Fourier transform (Scheme 6). Signal processing can be carried out using spectrometer (Bruker) software, or in MATLAB using home-written or open-source programs such as HYSCOREAN for HYSCORE data (Fábregas Ibáñez et al., 2019). Simulation is then usually required to obtain a_{iso}, T, and κ and η (where appliable). Easyspin (in conjunction with MATLAB) is the most versatile and widely used open-source simulation program (Stoll & Britt, 2009; Stoll & Schweiger, 2006) and excellent tutorial material is available online. Simulations should then enable assignment of the interacting nuclei. Isotopic labeling (le Breton et al., 2017; Seif-Eddine et al., 2017, 2020), site-directed mutagenesis (le Breton et al., 2017) and DFT calculations (Seif-Eddine et al., 2017, 2020) can all aid with assignment. However, first estimates can and should be obtained by inspecting the spectra; we discuss example X-band HYSCORE spectra below. Once a first set of parameters is obtained from the spectra, numerical simulation should be performed to fine tune the estimated parameters and determine any missing parameters.

In HYSCORE cross-peaks appear in four quadrants. Usually only the (+, +) and (−, +) quadrants are shown (Fig. 8) since they are equivalent to the (−, −) and (+, −) quadrants, respectively.

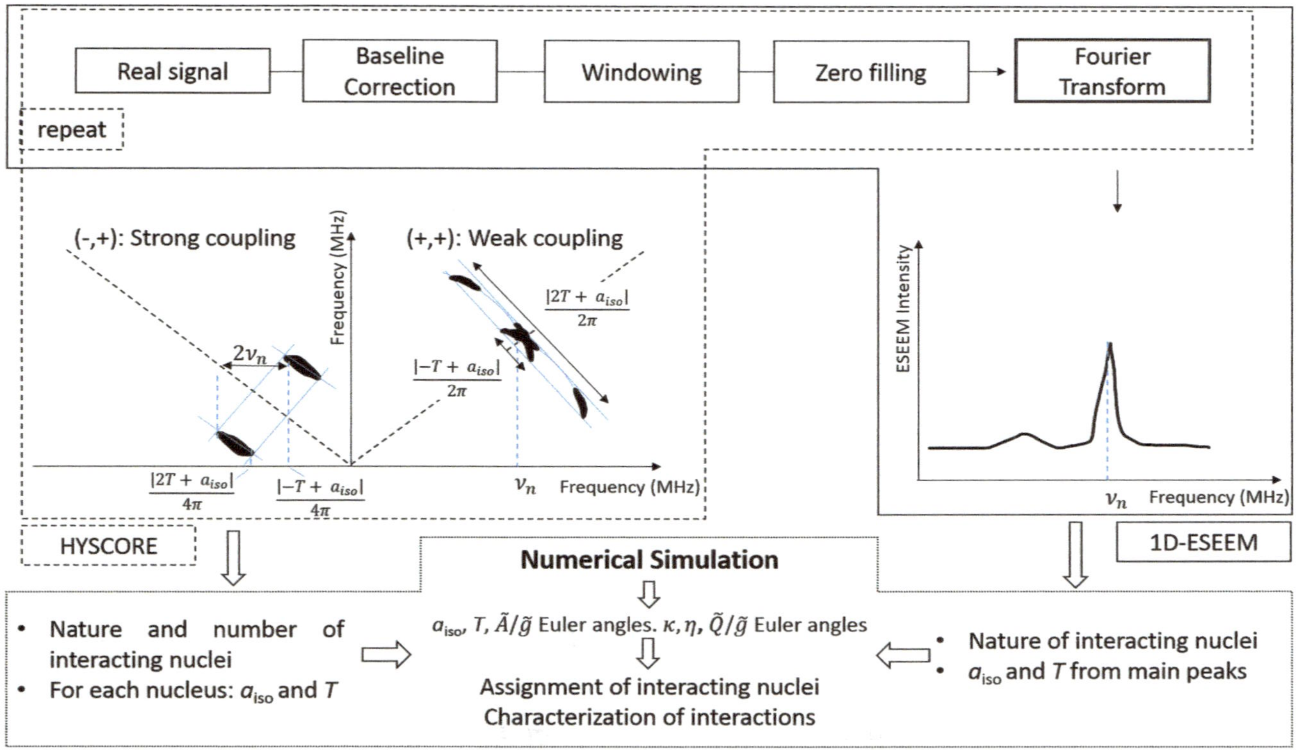

Scheme 6 Signal processing and spectral analysis of 1D and 2D ESEEM spectra. Example spectra are given for an electron spin coupling to a strongly and a weakly coupled nucleus with $I = ½$. A single peak at the Larmor frequency of the interacting nucleus is obtained in a 1D ESEEM spectrum and two cross peaks centered around the Larmor frequency are obtained in HYSCORE. The order of the baseline polynomial correction, type of window function and amount zero filling need to be adjusted for optimal results. For HYSCORE the signal processing needs be performed twice, on the two sets of time-domain data obtained from incrementing two time constants.

4.3.5.1 ^{1}H ($S = 1/2$, $I = 1/2$; weak coupling)

Cross-peaks are centered around the ^{1}H Larmor frequency and the effect of the τ value in suppressing parts of peaks is illustrated in Fig. 8Ai–ii. Peak shape informs on the magnitude of T; circular and stretched peaks (or cross ridges) indicate isotropic and anisotropic coupling, respectively. The separation between the anti-diagonal cross-ridges depends on a_{iso} and T. For example, the cross ridges of ^{1}H interacting with the electron on reduced [4Fe—4S] cluster N2 of respiratory complex I (le Breton et al., 2017) are centered around the ^{1}H Larmor frequency (Fig. 8Ai–ii). To deduce a_{iso} and T of protons, "quadrature analysis" is commonly used, where peaks appear as straight lines in the square frequency axes. Their slope (Q) and y-intercept (G) are directly related to a_{iso} and T (Fig. 8Aiii). Although the quadratic equations in principle result in four combinations of a_{iso} and T, only the "right" combination allows the experimental cross-ridges to be reproduced by numerical simulation.

4.3.5.2 ^{2}H ($S = ½$, $I = 1$)

Because of the low natural abundance of ^{2}H, interactions with ^{2}H are seen only after isotope labelling. In the example shown in Fig. 8B, the ^{2}H nucleus corresponds to the exchanged ^{1}H in Fig. 8Aii after buffer exchange. Cross-peaks appear almost centered around the ^{2}H Larmor frequency, indicating that κ is low and not affecting its nuclear transitions (Box 3). This leads to correlation peaks between nuclear frequencies of single quantum transitions in a similar way to the interaction with a $I = ½$ nucleus; double quantum transitions are very broad and not observed. As expected based on the change in gyromagnetic ratio, $a_{iso}(^{2}H) = a_{iso}(^{1}H)/6.5$ and $T(^{2}H) = T(^{1}H)/6.5$. Complete suppression of the ^{2}H Larmor frequency peak results in a splitting within each peak caused by the quadrupolar interaction and thus allows determination of κ (Seif-Eddine et al., 2020).

4.3.5.3 ^{15}N ($S = 1/2$, $I = 1/2$)

As for protons, ^{15}N cross-peaks are centered around the Larmor frequency in the weak coupling regime (3–3' in Fig. 8C). In the strong coupling regime, peaks appear in the (−, +) quadrant, centered around A/2 and separated by $2\nu_{15_N}$. The ridges appearing in the (−, +) quadrant should theoretically be linear. However due to structural conformers of PCs in frozen solution, the line shape is distorted, leading to a "boomerang" (Dikanov & Taguchi, 2020).

4.3.5.4 ^{14}N ($S = 1/2$, $I = 1$)

Weakly-coupled ^{14}N are observed in the (+, +) quadrant (5 – 5', Fig. 8Di). Peaks arising from double quantum nuclear transitions appear at ($\nu_{dq\pm}$, $\nu_{dq\mp}$ hyperfine and quadrupolar parameters can be estimated by solving equations 3a–c (Box 3) (Arias-Cartin et al., 2010; Seif-Eddine et al., 2017). Strong-coupled ^{14}N give rise to double-quantum transitions appearing in the (−, +) quadrant. In this case, the quadrupolar parameters are negligible compared to the hyperfine parameters, so that $\nu_{dq\pm} = 2\nu_{eff\pm} = |2\nu_{14N} \mp a_{iso}|$ (Roessler et al., 2012). For intermediate coupling regimes, peaks can be seen in both (−, +) and (+, +) quadrants and arise from both single and double quantum transitions, leading to up to 18 peaks whose position can be calculated (Dikanov et al., 1996). However not all 18 peaks are experimentally observable, and their intensities can vary between the two quadrants. Changing the frequency to reach the cancellation condition greatly simplifies the spectra (Fig. 8Diii). Alternatively, ^{15}N labeling would lead to just a doublet centered around ν_{15N} (as in Fig. 8C) and the hyperfine parameters of ^{14}N can then be deduced by multiplying $a_{iso}(^{15}N)$ and $T(^{15}N)$ by the gyromagnetic ratio $\frac{g_{14N}}{g_{15N}} = 0.7$ (Seif-Eddine et al., 2017); κ and η can then be deduced from simulation.

4.4 Measuring distances

Pulsed EPR is a useful tool to measure distances between PCs. Compared with structural techniques, it can be performed on rigid or flexible proteins of any size, and in different physiological states. Most pulsed dipolar techniques are based on detection of the dipole interaction from oscillations of the electron spin-echo envelope as a function of inter-pulse interval length (Berliner et al., 2000; Jeschke et al., 2004). By far the most common method to measure distances by EPR is double electron-electron resonance (DEER, also known as PELDOR). This technique is widely applied to spin-probes (most commonly nitroxides) and therefore there is a wealth of investment in its analysis (Schiemann et al., 2021).

Distances from 15 to 60 Å can typically be measured with nitroxide spin labels (Jeschke, 2012), but with intrinsic PCs in metalloproteins the upper limit is usually much lower owing to the fast T_m (Scheme 4) of metal centers. The lower limit is determined by the need for an excitation bandwidth which exceeds the electron-electron coupling. Below ~15 Å other effects such as exchange coupling and electron-proton forbidden transitions can occur. These processes may not prohibit measurements but can greatly complicate, or even prevent, robust analysis (Jeschke, 2012). Furthermore, the width of the dipolar spectrum can exceed the excitation bandwidth of the pulse sequence (pump pulse), leading to distortions of the detected dipolar pattern.

4.4.1 Double electron-electron resonance (DEER) with metalloproteins

The four pulse DEER sequence is most frequently used as it has no deadtime (Box 5) (Milov et al., 1984; Pannier et al., 2000). Practical guides to setting up a DEER experiment and analysis for nitroxide samples are readily available (Pliotas, 2017). We discuss DEER, also referred to as PELDOR, in the context of metalloproteins.

4.4.1.1 Sample considerations

Given the difficulties in purifying metalloproteins in high quantities, intermolecular artefacts as a result of high local spin concentration are rarely an issue. In any case these can be avoided by using glassy frozen samples. Reconstitution into lipid vesicles can increase local spin concentration and these samples should be treated cautiously (Dastvan et al., 2010). Sample concentrations of 30–50 μM already require several hours of measurement time at X-band. A probe with a cryogenic preamplifier (Šimėnas, O'Sullivan, Zollitsch, Kennedy, Seif-Eddine, Ritsch, Hülsmann, Qi, Godt, Roessler, Jeschke, & Morton, 2021), previously mentioned in Section 4.3.3, made measurements on low concentration (15 μM) and highly anisotropic samples feasible (Richardson et al., 2021). Despite the typically short T_m of metalloprotein PCs, deuteration has enabled access to distances as long as 80 or 115 Å (Ward et al., 2010; Ward, Bowman, El-Mkami, Owen-Hughes, & Norman, 2009).

4.4.2 Setting up the DEER experiment

4.4.2.1 Choosing frequencies

The DEER pulse sequence uses two microwave ("pump" and "detection") frequencies which should excite two separate populations of spins. Therefore, the excitation bandwidths should not overlap, which for a typical set up results in a minimum separation of ~65 MHz between the pump and detection frequency; this issue can be mitigated by the use of Gaussian shaped pulses (Teucher & Bordignon, 2018). The difference may however need to be much larger for anisotropic PCs, with the maximum achievable separation depending on the resonator bandwidth–typically less than 300 MHz (Table 1).

BOX 5 Pulse sequence and description of the DEER experiment.

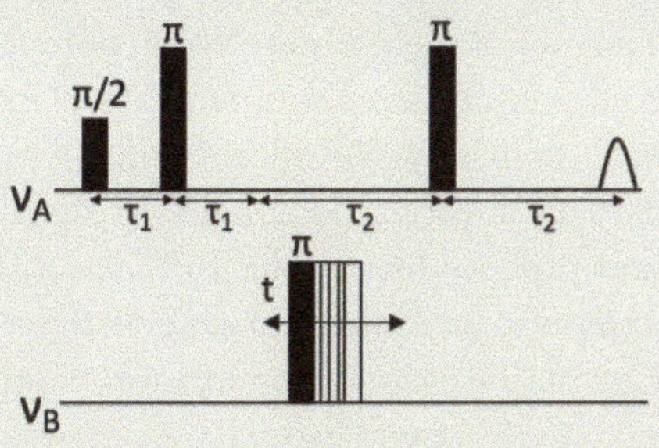

The detection spins (A spins) are refocused using a pulse sequence applied at a given microwave frequency (ν_A). An additional π pulse whose position is varied in time is applied at a different fixed frequency (ν_B) and excites the pump spins (B spins). In the absence of any B spins, the A spins will refocus (assuming T_m is long enough). If B spins are excited, the A spins experience a change of the local magnetic field which can prevent refocusing. The frequency with which the echo intensity is modulated is inversely proportional to the distance cubed between the A and B spins.

For a pair m of A and B spins, inversion of spin B changes frequency of spin A by the electron-electron coupling ω_{dd}, which leads to a phase gain of $\omega_{dd}t$ a fraction λ of the A spin magnetization. The echo amplitude as a function of time t is given by:

$$v(t) = \{1 - \lambda[1 - \cos(\omega_{dd}t)]\},$$

where λ describes the signal modulation, which is proportional to the magnetization of the A spin and to the flip probability of the B spin. If exchange coupling can be neglected and both spins are assumed to be quantized along the external magnetic field, the dipole-dipole coupling simplifies to:

$$\omega_{\mathrm{dd}}^{m} = C_m \frac{1 - 3\cos^2\psi^m}{r_m^3},$$

where C_m is proportional to the product of the g values of the spins and ψ^m is the angle between the spin-spin vector and the external magnetic field.

For FeS clusters, using a local spin model for spin delocalization the dipolar coupling frequency can be approximated to:

$$\omega_{\mathrm{dd}}^{m} = \frac{\mu_0\beta^2}{2h} g_A^m g_B^m \sum_i \sum_j k_{A_i} k_{B_j} \frac{3\cos^2\psi_{ij}^m - 1}{r_{ij}^3},$$

where g_A^m and g_B^m are the experimental g values observed at the orientation of the unit magnetic field vector $B_0^m(\theta, \phi)$, k are the spin projection factors (see text) and ψ is the angle between $B_0^m(\theta, \phi)$ and n_{ij}, the unit vectors between two FeS cluster Fe subsites (red) as so:

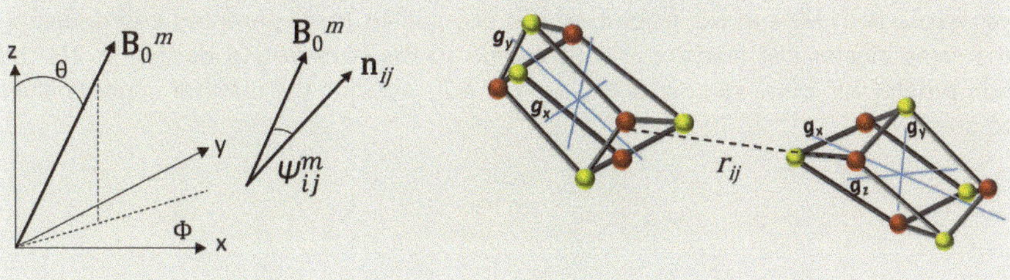

To maximize inversion of the spins the pump pulse is typically placed in the center of the resonance, however symmetrically placing the pulses about the center can increase sensitivity (Salvadori et al., 2015). As metalloprotein cofactors typically have anisotropic and broad EPR signals, the optimal position of the pump and probe pulses in the EPR spectrum depends on the choice of the components to be excited. In general, more than one measurement is required (Fig. 9). The largest modulation depth is observed if the

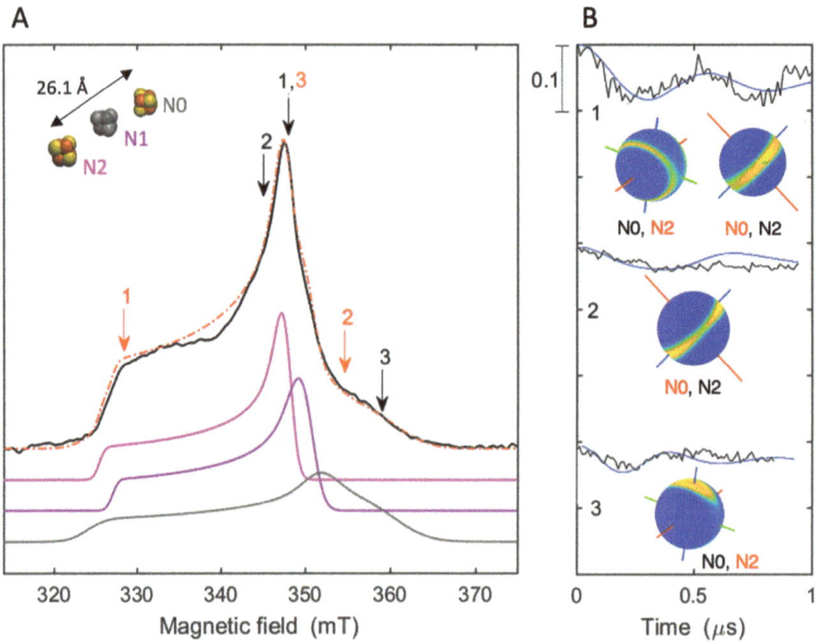

Fig. 9 Orientation selective DEER spectra of the three FeS clusters of photosynthetic complex I. (A) Pump positions in red, detection positions in black. EDFS of the reduced FeS clusters (black), with simulation (red) and individual simulations of N2, N1 and N0. (B) Orientation selective DEER spectra (black) with simulation (blue), obtained at the pump and detection positions as indicated in A and with a distance of 26 Å between N2 and N0. Orientation of the excited spins for each field position and each pair, for a chosen *g* tensor (g_z red, g_y green and g_x blue). In 1 the pump and detection pulses can excite *both* N0 and N2, leading to *two* possibilities (note pumping and detecting the same cluster just leads to a contribution to the background decay). In 2/3 the pump/detection pulse can excite N0 (but not N2), reducing the number of possibilities to one.

pump frequency is at the signal maximum. Although higher sensitivity at Q-band (34 GHz) with respect to X-band (9.5 GHz) has been reported for nitroxide spin labels (Ghimire et al., 2009; Polyhach et al., 2012), the larger anisotropy of the electron Zeeman interaction for metal centers in metalloproteins can offset this gain.

4.4.2.2 Choosing the pulse lengths

Nutation experiments can be used to determine the optimum pulse length for a particular species at a given frequency (Torrey, 1949). To optimize the excitation bandwidth of the detection sequence, the same length for both the π and $\pi/2$ observer pulses should be used.

4.4.2.3 Suppression of proton/deuterium modulation

Excitation of detection spins by the pump pulse can take place via forbidden transitions leading to nuclear modulation. In metalloproteins this is dominated by protons of side chains and the matrix (or deuterons if applicable). This effect can be suppressed by varying systematically the τ_1 delay (see Box 5) and summing the obtained traces (Jeschke et al., 2009, 2004); typical incrementation schemes are 8×8 ns for protons at X-band, 8×56 ns for deuterons at X-band and 8×16 ns for deuterons at Q-band (Jeschke, 2012).

4.4.2.4 Phase cycling

On modern pulse EPR spectrometers the detection and pump pulses originate from the same microwave source, whereby the frequency offset of the pump pulse is generated by an arbitrary waveform generator (AWG); this experimental setup requires at least an 8 step phase cycle to get rid of all unwanted echoes (Tait & Stoll, 2016). On spectrometers using a different—hence incoherent—microwave source for the generation of the pump pulse, a 2 step phase cycle on the detection $\pi/2$ pulse is sufficient.

4.4.3 Analysis of metalloprotein DEER spectra

Since the EPR spectra of metalloproteins are often much broader than the excitation bandwidth that can be achieved, orientation selection is common (Section 4.1.3) and needs to be addressed experientially. Field averaging using one pump and multiple observer frequencies can reduce orientation selection (Jeschke et al., 2010), or it may be possible to average over different pump and observer frequencies (Marko et al., 2009). An alternative approach has been proposed which makes use of fast field jumps (Kaminker et al., 2011). If orientation selection can be reasonably ignored through these means, analysis software such as DeerAnalysis can be used (Ghosh et al., 2020).

Orientation selection can also be addressed by accounting for it using analysis programs such as PeldorFit (Abdullin et al., 2015). This type of analysis requires an appropriate theoretical model for the paramagnetic species of interest which can vary greatly between metal centers. An example workflow for FeS clusters is shown in Scheme 7. For this reason, specific models for Cu^{2+} and FeS based cofactors have been developed (Lovett et al., 2009; Richardson et al., 2021; Roessler et al., 2010; Yang et al., 2010) (discussed in Section 4.4.5). Despite the complexities of orientation selection, DEER has successful been used to study protein conformational changes (Smirnova et al., 2007; Vileno et al., 2011), complex formation (Hilger et al., 2005) and to assign EPR signals to structural positions in metalloproteins (Richardson et al., 2021; Roessler et al., 2010). In most analyses, exchange couplings are ignored as they should not be detected by DEER. However, if two radicals are connected by a fully conjugated pathway it is possible that DEER will detect exchange interactions (Riplinger et al., 2009).

4.4.4 Other pulse EPR techniques to measure distances

Single-frequency techniques, most commonly relaxation-induced dipolar modulation enhancement (RIDME), can be used to measure distances in metalloproteins (Abdullin & Schiemann, 2021) in particular, RIDME is advantageous when two PCs have widely different T_1 relaxation times. Only the excitation of the detection spins is orientation selective so optimal echo conditions can be selected, improving sensitivity. These techniques can offer better SNR and potentially access shorter distances than DEER. However, strong ESEEM effects over a wide range of experimental conditions require higher field strengths than X-band to suppress them. Moreover, finding conditions in which the dipolar interaction dominates the echo decay can be complicated as nuclear dipolar interactions, spectral diffusion and phase memory all contribute to the echo intensity (Keller et al., 2019). This often requires multiple experiments or different samples to definitively attribute the observed modulation to dipolar coupling. DEER therefore still allows the best "isolation" of the dipole-dipole interaction between two paramagnetic species.

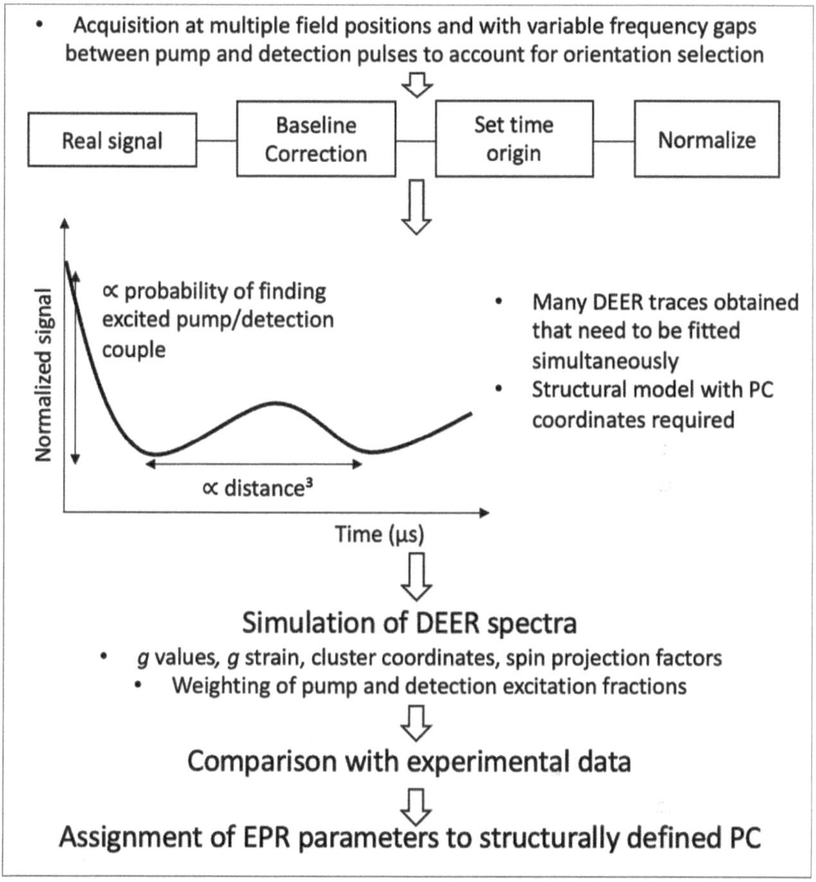

Scheme 7 Signal processing and spectral analysis of orientation selective DEER spectra. The signal is phased to the real domain, a baseline is then subtracted, followed by shifting of the time origin if required and normalization of the signal intensity to 1. Simulation of the spectra is then preformed followed by comparison with experimental data, often using least squares fitting.

5- and 7-pulse versions of the DEER pulse sequence (Borbat et al., 2013; Spindler et al., 2015) make use of the Carr–Purcell sequence (Carr & Purcell, 1954; Meiboom & Gill, 1958) to extend the length of the measured time trace. Longer distances can thus be obtained compared to the 4-pulse sequence; especially if the echo decay is dominated by nuclear spin diffusion (Scheme 4). While 5-pulse DEER can be performed using rectangular pulses (postprocessing may be necessary to get rid of the partial excitation artefact occurring in the middle of the measured trace) (Breitgoff et al., 2017), the 7-pulse DEER requires shaped adiabatic pump pulses for more uniform control of the excitation bandwidth. For further information about shaped pulses, see (Spindler et al., 2016).

4.4.5 DEER with metalloprotein FeS clusters: Case study

By exploiting the orientation selective DEER spectra of the three [4Fe—4S] clusters of photosynthetic complex I (PS-CI), EPR signals can be assigned to their structural coordinates (Richardson et al., 2021). The EPR parameters (g values and g strains) of clusters N0, N1 and N2 (determined by exploiting differences in their relaxation time, Section 4.2) were used to select pump and probe DEER positions which are dominated by interaction between a particular cluster pair (Fig. 9A). Using the EPR parameters, the structural coordinates from the cyro-EM structure, and a model of the spin delocalization over subsites in FeS clusters (spin projection factors), orientation selective DEER spectra can be simulated (Fig. 9B) (Richardson et al., 2021; Roessler et al., 2010). The distance between the two centers correlates with the modulation frequency. The modulation depth of a given DEER trace is determined by the probability of finding the excited pair in the sample (Scheme 7 and Fig. 9). Assignment of the EPR signature of each cluster to its structural counterpart is then made possible by comparing the simulations resulting from different possible assignments to the experimentally determined data. Importantly, this then allows the properties of each of these FeS clusters (namely their reduction potentials obtained from potentiometric titrations) to be assigned to their structural location (Richardson et al., 2021).

5. Conclusions and outlook

In this chapter we hope to have provided a toolkit for investigating redox-active centers in metalloproteins using EPR spectroscopy. Although we have placed our focus on the EPR investigations and analyses, we discussed the importance of sample preparation, from making the protein and downstream implications for EPR experiments, to the need for careful potential control. The latter is especially important when multiple redox-active centers are present, to enable detailed pulse EPR investigations. Potential control requires knowledge of the reduction potential of the redox center and in this regard PFE can be a very useful complementary tool, particularly for smaller proteins that form a "film" relatively easily. Potentiometric titrations, now possible with very little protein, open the door to investigating even precious metalloproteins, of any size, in detail with pulse EPR at X- and Q-band frequencies and potential control. Our development of PFE-EPR gives the prospect of investigating the evolution of radicals during catalysis.

Using examples, alongside general considerations, we have discussed how hyperfine EPR can give detailed insight into bonding and hence reactivity, with HYSCORE often a useful technique when multiple different nuclei surround the PC. Although DEER is mainly known as a technique to measure unknown distances, we have shown how it can be used for assigning the properties of redox centers in combination with known distances from structures. Although the analysis of pulse EPR spectra can be complex, one should not be deterred from investigating proteins and enzymes with numerous multinuclear (e.g. FeS) redox-active centers that are highly anisotropic and even when they exhibit similar potentials, as a combination of biochemical and spectroscopic "tricks" almost always offers a solution.

Acknowledgments

MMR is grateful to the UK Engineering and Physical Sciences Research Council (EP/T031425/1 for PEPR, the Centre for Pulse EPR at Imperial College) and the Leverhulme Trust (Research grant RPG-2018-183) for supporting some of the work described in this chapter. We thank the many co-authors of the papers cited, who have made invaluable contributions to the field and continue to be an inspiration for our work. Dr Alberto Collauto and Dr Clare Megarity are gratefully acknowledged for very helpful comments on the manuscript.

References

Abdelkader, E. H., Feintuch, A., Yao, X., Adams, L. A., Aurelio, L., Graham, B., et al. (2015). Protein conformation by EPR spectroscopy using gadolinium tags clicked to genetically encoded p-azido-l-phenylalanine. *Chemical Communications*, *51*(88), 15898–15901. https://doi.org/10.1039/C5CC07121F.

Abdiaziz, K., Salvadori, E., Sokol, K. P., Reisner, E., & Roessler, M. M. (2019). Protein film electrochemical EPR spectroscopy as a technique to investigate redox reactions in biomolecules. *Chemical Communications*, *55*(60), 8840–8843. https://doi.org/10.1039/C9CC03212F.

Abdullin, D., Hagelueken, G., Hunter, R. I., Smith, G. M., & Schiemann, O. (2015). Geometric model-based fitting algorithm for orientation-selective PELDOR data. *Molecular Physics*, *113*(6), 544–560. https://doi.org/10.1080/00268976.2014.960494.

Abdullin, D., & Schiemann, O. (2021). Localization of metal ions in biomolecules by means of pulsed dipolar EPR spectroscopy. *Dalton Transactions*, *50*(3), 808–815. https://doi.org/10.1039/D0DT03596C.

Adamson, H., Robinson, M., Wright, J. J., Flanagan, L. A., Walton, J., Elton, D., et al. (2017). Retuning the catalytic bias and overpotential of a [NiFe]-hydrogenase via a single amino acid exchange at the electron entry/exit site. *Journal of the American Chemical Society, 139*(31), 10677–10686. https://doi.org/10.1021/jacs.7b03611.

Adamson, H., Simonov, A. N., Kierzek, M., Rothery, R. A., Weiner, J. H., Bond, A. M., et al. (2015). Electrochemical evidence that pyranopterin redox chemistry controls the catalysis of YedY, a mononuclear Mo enzyme. *Proceedings of the National Academy of Sciences of the United States of America, 112*(47), 14506–14511. https://doi.org/10.1073/pnas.1516869112.

Anderson, J. M., & Malkin, R. (1982). An EPR study of the lateral organization of electron carriers in chloroplast thylakoids. *FEBS Letters, 148*(2), 293–296. https://doi.org/10.1016/0014-5793(82)80827-2.

Andreini, C., Bertini, I., Cavallaro, G., Holliday, G. L., & Thornton, J. M. (2008). Metal ions in biological catalysis: From enzyme databases to general principles. *JBIC Journal of Biological Inorganic Chemistry, 13*(8), 1205–1218. https://doi.org/10.1007/S00775-008-0404-5.

Arias-Cartin, R., Lyubenova, S., Ceccaldi, P., Prisner, T., Magalon, A., Guigliarelli, B., et al. (2010). HYSCORE evidence that endogenous mena- and ubisemiquinone bind at the same Q site (QD) of Escherichia coli nitrate reductase A. *Journal of the American Chemical Society, 132*(17), 5942–5943. https://doi.org/10.1021/ja1009234.

Armstrong, F. A., Evans, R. M., Hexter, S. V., Murphy, B. J., Roessler, M. M., & Wulff, P. (2016). Guiding principles of hydrogenase catalysis instigated and clarified by protein film electrochemistry. *Accounts of Chemical Research, 49*(5), 884–892. https://doi.org/10.1021/acs.accounts.6b00027.

Armstrong, F. A., Evans, R. M., & Megarity, C. F. (2018). Protein film electrochemistry of iron–sulfur enzymes. *Methods in Enzymology, 599*, 387–407. https://doi.org/10.1016/BS.MIE.2017.11.001.

Armstrong, F. A., Heering, H. A., & Hirst, J. (1997). Reaction of complex metalloproteins studied by protein-film voltammetry. *Chemical Society Reviews, 26*(3), 169–179. https://doi.org/10.1039/cs9972600169.

Ash, P. A., & Vincent, K. A. (2012). Spectroscopic analysis of immobilised redox enzymes under direct electrochemical control. *Chemical Communications, 48*(10), 1400–1409. https://doi.org/10.1039/c1cc15871f.

Aslantas, Y., & Surmeli, N. B. (2019). Effects of N-terminal and C-terminal polyhistidine tag on the stability and function of the thermophilic P450 CYP119. *Bioinorganic Chemistry and Applications, 2019*. https://doi.org/10.1155/2019/8080697.

Atherton, N. M. (1993). *Principles of electron spin resonance*. Ellis Horwood Limited.

Ayikpoe, R., Ngendahimana, T., Langton, M., Bonitatibus, S., Walker, L. M., Eaton, S. S., et al. (2019). Spectroscopic and electrochemical characterization of the mycofactocin biosynthetic protein, MftC, provides insight into its redox flipping mechanism. *Biochemistry, 58*(7), 940–950. https://doi.org/10.1021/acs.biochem.8b01082.

Baghban, R., Farajnia, S., Rajabibazl, M., Ghasemi, Y., Mafi, A. A., Hoseinpoor, R., et al. (2019). Yeast expression systems: Overview and recent advances. *Molecular Biotechnology, 61*(5), 365–384. https://doi.org/10.1007/s12033-019-00164-8.

Bajada, M. A., Roy, S., Warnan, J., Abdiaziz, K., Wagner, A., Roessler, M. M., et al. (2020). A precious-metal-free hybrid electrolyzer for alcohol oxidation coupled to CO_2-to-syngas conversion. *Angewandte Chemie (International Ed. in English), 59*(36), 15633–15641. https://doi.org/10.1002/anie.202002680.

Bard, A. J., & Faulkner, L. R. (2001). *Electrochemical methods: Fundamentals and applications* (2nd ed.). John Wiley & Sons, Inc.

Berliner, L. J., Eaton, S. S., & Eaton, G. R. (Eds.). (2000). *Distance measurements in biological systems by EPR*. In Biological magnetic resonance (Vol. 19). Springer

Blanford, C. F. (2013). The birth of protein electrochemistry. *Chemical Communications, 49*(95), 11130–11132. https://doi.org/10.1039/C3CC46060F.

Borbat, P. P., Georgieva, E. R., & Freed, J. H. (2013). Improved sensitivity for long-distance measurements in biomolecules: Five-pulse double electron–electron resonance. *The Journal of Physical Chemistry Letters, 4*(1), 170–175. https://doi.org/10.1021/jz301788n.

Boyer, M. E., Stapleton, J. A., Kuchenreuther, J. M., Wang, C. W., & Swartz, J. R. (2008). Cell-free synthesis and maturation of [FeFe] hydrogenases. *Biotechnology and Bioengineering, 99*(1), 59–67. https://doi.org/10.1002/bit.21511.

Breitgoff, F. D., Polyhach, Y. O., & Jeschke, G. (2017). Reliable nanometre-range distance distributions from 5-pulse double electron electron resonance. *Physical Chemistry Chemical Physics, 19*(24), 15754–15765. https://doi.org/10.1039/C7CP01487B.

Britt, R. D., Rao, G., & Tao, L. (2020). Biosynthesis of the catalytic H-cluster of [FeFe] hydrogenase: The roles of the Fe-S maturase proteins HydE, HydF, and HydG. *Chemical Science, 11*(38), 10313–10323. https://doi.org/10.1039/d0sc04216a.

Cammack, R. (1993). Electron paramagnetic resonance spectroscopy of metalloproteins. In C. Jones, B. Mulloy, & A. H. Thomas (Eds.), *Spectroscopic methods and analyses: NMR, mass spectrometry, and metalloprotein techniques* (pp. 327–344). Humana Press. https://doi.org/10.1385/0-89603-215-9:327.

Cammack, R., Luijk, L. J., Maguire, J. J., Fry, I. V., & Packer, L. (1979). EPR spectra of photosystem I and other iron protein components in intact cells of cyanobacteria. *Biochimica et Biophysica Acta, 548*, 267–275.

Cantu-Bustos, J. E., Vargas-Cortez, T., Morones-Ramirez, J. R., Balderas-Renteria, I., Galbraith, D. W., McEvoy, M. M., et al. (2016). Expression and purification of recombinant proteins in Escherichia coli tagged with the metal-binding protein CusF. *Protein Expression and Purification, 121*, 61–65. https://doi.org/10.1016/j.pep.2016.01.007.

Carr, H. Y., & Purcell, E. M. (1954). Effects of diffusion on free precession in nuclear magnetic resonance experiments. *Physical Review, 94*(3), 630–638. https://doi.org/10.1103/PhysRev.94.630.

Chaudhuri, T. K., Farr, G. W., Fenton, W. A., Rospert, S., & Horwich, A. L. (2001). GroEL/GroES-mediated folding of a protein too large to be encapsulated. *Cell, 107*(2), 235–246. https://doi.org/10.1016/S0092-8674(01)00523-2.

Colaneri, M. J., & Peisach, J. (2005). Electron spin-echo envelope modulation studies Of 14N in biological systems. In S. R. Eaton, G. R. Eaton, & L. J. Berliner (Eds.), *Biomedical EPR, Part A: Free radicals, metals, medicine, and physiology* (pp. 455–491). USA: Springer. https://doi.org/10.1007/0-387-26741-7_15.

Corless, E. I., Mettert, E. L., Kiley, P. J., & Antony, E. (2020). Elevated expression of a functional Suf pathway in Escherichia coli BL21(DE3) enhances recombinant production of an iron-sulfur cluster-containing protein. *Journal of Bacteriology, 202*(3). https://doi.org/10.1128/JB.00496-19.

Curatti, L., Hernandez, J. A., Igarashi, R. Y., Soboh, B., Zhao, D., & Rubio, L. M. (2007). *In vitro synthesis of the iron-molybdenum cofactor of nitrogenase from iron, sulfur, molybdenum, and homocitrate using purified proteins*. www.pnas.org/cgi/content/full/.

Dalton, L. R., Kwiram, A. L., & Cowen, J. A. (1972). Electron spin-lattice and cross relaxation in irradiated malonic acid. *Chemical Physics Letters, 14*(1), 77–81. https://doi.org/10.1016/0009-2614(72)87146-X.

Dastvan, R., Bode, B. E., Karuppiah, M. P. R., Marko, A., Lyubenova, S., Schwalbe, H., et al. (2010). Optimization of transversal relaxation of nitroxides for pulsed electron-electron double resonance spectroscopy in phospholipid membranes. *Journal of Physical Chemistry B, 114*(42), 13507–13516. https://doi.org/10.1021/jp1060039.

Dikanov, S. A., & Taguchi, A. T. (2020). Influence of hyperfine coupling strain on two-dimensional ESEEM spectra from I = 1/2 nuclei. *Applied Magnetic Resonance, 51*(9), 1177–1200. https://doi.org/10.1007/s00723-020-01246-6.

Dikanov, S. A., Xun, L., Karpiel, A. B., Tyryshkin, A. M., & Bowman, M. K. (1996). Orientationally-selected two-dimensional ESEEM spectroscopy of the Rieske-type iron – sulfur cluster in 2,4,5-trichlorophenoxyacetate monooxygenase from Burkholderia cepacia AC1100. *Journal of the American Chemical Society, 118*(35), 8408–8416. https://doi.org/10.1021/ja960781x.

Doll, A., Pribitzer, S., Tschaggelar, R., & Jeschke, G. (2013). Adiabatic and fast passage ultra-wideband inversion in pulsed EPR. *Journal of Magnetic Resonance, 230*, 27–39. https://doi.org/10.1016/j.jmr.2013.01.002.

dos Santos, P. C., Johnson, D. C., Ragle, B. E., Unciuleac, M. C., & Dean, D. R. (2007). Controlled expression of nif and isc iron-sulfur protein maturation components reveals target specificity and limited functional replacement between the two systems. *Journal of Bacteriology, 189*(7), 2854–2862. https://doi.org/10.1128/JB.01734-06.

Eaton, S. S., & Eaton, G. R. (2002). *Relaxation times of organic radicals and transition metal ions* (pp. 29–154). https://doi.org/10.1007/0-306-47109-4_2.

Eddowes, M. J., & Hill, H. A. O. (1979). Electrochemistry of horse heart cytochrome c. *Journal of the American Chemical Society, 101*(16), 4461–4464. https://doi.org/10.1021/JA00510A003.

Evans, R. M., & Armstrong, F. (2014). Electrochemistry of metalloproteins: Protein film electrochemistry for the study of E. coli [NiFe]-hydrogenase-1. *Methods in Molecular Biology, 1122*, 73–94. https://doi.org/10.1007/978-1-62703-794-5_6.

Fábregas Ibáñez, L., Soetbeer, J., Klose, D., Tinzl, M., Hilvert, D., & Jeschke, G. (2019). Non-uniform HYSCORE: Measurement, processing and analysis with Hyscorean. *Journal of Magnetic Resonance, 307*. https://doi.org/10.1016/j.jmr.2019.106576.

Fiege, K., Querebillo, C. J., Hildebrandt, P., & Frankenberg-Dinkel, N. (2018). Improved method for the incorporation of heme cofactors into recombinant proteins using Escherichia coli Nissle 1917. *Biochemistry, 57*(19), 2747–2755. https://doi.org/10.1021/acs.biochem.8b00242.

Fischer, B. E., Haring, U. K., Tribolet, R., & Sigel, H. (1979). Metal ion/buffer interactions stability of binary and ternary complexes containing 2-amino-2(hydroxymethyl)-~l, 3-propanediol (Tris) and adenosine 5'-triphosphate (ATP). *European Journal of Biochemistry, 94*(2), 523–530. https://doi.org/10.1111/j.1432-1033.1979.tb12921.x.

Foster, A. W., Osman, D., & Robinson, N. J. (2014). Metal preferences and metallation. *Journal of Biological Chemistry, 289*(41), 28095–28103. https://doi.org/10.1074/jbc.R114.588145.

Fourmond, V., & Léger, C. (2020). An introduction to electrochemical methods for the functional analysis of metalloproteins. *Practical Approaches to Biological Inorganic Chemistry*, 325–373. https://doi.org/10.1016/B978-0-444-64225-7.00009-2.

García-Rubio, I., Martínez, J. I., Picorel, R., Yruela, I., & Alonso, P. J. (2003). HYSCORE spectroscopy in the cytochrome b559 of the photosystem II reaction center. *Journal of the American Chemical Society, 125*(51), 15846–15854. https://doi.org/10.1021/ja035364g.

Gayda, J.-P., Bertrand, P., Deville, A., More, C., Roger, G., Gibson, J. F., et al. (1979). Temperature dependence of the electronic spin-lattice relaxation time in a 2-iron-2-sulfur protein. *Biochimica et Biophysica Acta (BBA) - Protein Structure, 581*(1), 15–26. https://doi.org/10.1016/0005-2795(79)90216-2.

Gervason, S., Larkem, D., Mansour, A.b., Botzanowski, T., Müller, C. S., Pecqueur, L., et al. (2019). Physiologically relevant reconstitution of iron-sulfur cluster biosynthesis uncovers persulfide-processing functions of ferredoxin-2 and frataxin. *Nature Communications, 10*(1). https://doi.org/10.1038/s41467-019-11470-9.

Ghimire, H., McCarrick, R. M., Budil, D. E., & Lorigan, G. A. (2009). Significantly improved sensitivity of Q-band PELDOR/DEER experiments relative to X-band is observed in measuring the intercoil distance of a Leucine zipper motif peptide (GCN4-LZ). *Biochemistry, 48*(25), 5782–5784. https://doi.org/10.1021/bi900781u.

Ghosh, S., Lawless, M. J., Brubaker, H. J., Singewald, K., Kurpiewski, M. R., Jen-Jacobson, L., et al. (2020). Cu2+-based distance measurements by pulsed EPR provide distance constraints for DNA backbone conformations in solution. *Nucleic Acids Research, 48*(9), e49. https://doi.org/10.1093/nar/gkaa133.

Gibson, J. F., Hall, D. O., Thornley, J. H., & Whatley, F. R. (1966). The iron complex in spinach ferredoxin. *Proceedings of the National Academy of Sciences of the United States of America, 56*(3), 987–990.

Goldfarb, D. (2017). ELDOR-detected NMR. In *eMagRes* John Wiley & Sons, Ltd. https://doi.org/10.1002/9780470034590.emrstm1516.

Goldfarb, D., Kofman, V., Libman, J., Shanzer, A., Rahmatouline, R., van Doorslaer, S., et al. (1998). Double nuclear coherence transfer (DONUT)-HYSCORE: A new tool for the assignment of nuclear frequencies in pulsed EPR experiments. *Journal of the American Chemical Society, 120*(28), 7020–7029. https://doi.org/10.1021/ja973271r.

Goldman, S. A., Bruno, G., & v., & Freed, J. H. (1973). ESR studies of anisotropic rotational reorientation and slow tumbling in liquid and frozen media. II. Saturation and nonsecular effects. *The Journal of Chemical Physics, 3071*(1973), 3071–3091. https://doi.org/10.1063/1.1680446.

Gregorio, N. E., Levine, M. Z., & Oza, J. P. (2019). A user's guide to cell-free protein synthesis. *Methods and Protocols, 2*(1), 1–34. https://doi.org/10.3390/mps2010024.

Grimaldi, S., Arias-Cartin, R., Lanciano, P., Lyubenova, S., Endeward, B., Prisner, T. F., et al. (2010). Direct evidence for nitrogen ligation to the high stability semiquinone intermediate in Escherichia coli nitrate reductase A. *The Journal of Biological Chemistry, 285*(1), 179–187. https://doi.org/10.1074/jbc.M109.060251.

Grimaldi, S., Arias-Cartin, R., Lanciano, P., Lyubenova, S., Szenes, R., Endeward, B., et al. (2012). Determination of the proton environment of high stability menasemiquinone intermediate in Escherichia coli nitrate reductase A by pulsed EPR. *The Journal of Biological Chemistry, 287*(7), 4662–4670. https://doi.org/10.1074/jbc.M111.325100.

Hagedoorn, P.-L., van der Weel, L., & Hagen, W. R. (2014). EPR monitored redox titration of the cofactors of Saccharomyces cerevisiae Nar1. *Journal of Visualized Experiments: JoVE, 93*, e51611. https://doi.org/10.3791/51611.

Hagen, W. R. (2018). EPR spectroscopy of complex biological iron–sulfur systems. *Journal of Biological Inorganic Chemistry, 23*(4), 623–634. https://doi.org/10.1007/s00775-018-1543-y.

Harmer, J. R. (2016). Hyperfine spectroscopy—ENDOR. In *eMagRes* John Wiley & Sons, Ltd. https://doi.org/10.1002/9780470034590.emrstm1515.

Harmer, J., Mitrikas, G., & Schweiger, A. (2009). Advanced pulse EPR methods for the characterization of metalloproteins. In L. Berliner, & G. Hanson (Eds.), *High resolution EPR: Applications to metalloenzymes and metals in medicine* (pp. 13–61). New York: Springer. https://doi.org/10.1007/978-0-387-84856-3_2.

Heering, H. A., Bulsink, Y. B. M., Hagen, W. R., & Meyer, T. E. (1995). Influence of charge and polarity on the redox potentials of high-potential iron-sulfur proteins: Evidence for the existence of two groups. *Biochemistry, 34*(45), 14675–14686. https://doi.org/10.1021/bi00045a008.

Heghmanns, M., Günzel, A., Brandis, D., Kutin, Y., Engelbrecht, V., Winkler, M., et al. (2021). Fine-tuning of FeS proteins monitored via pulsed EPR redox potentiometry at Q-band. *Biophysical Reports, 1*(2), 100016. https://doi.org/10.1016/j.bpr.2021.100016.

Hilger, D., Jung, H., Padan, E., Wegener, C., Vogel, K. P., Steinhoff, H. J., et al. (2005). Assessing oligomerization of membrane proteins by four-pulse DEER: pH-dependent dimerization of NhaA Na+/H+ antiporter of E. coli. *Biophysical Journal, 89*(2), 1328–1338. https://doi.org/10.1529/biophysj.105.062232.

Hoertz, P. G., Chen, Z., Kent, C. A., & Meyer, T. J. (2010). Application of high surface area tin-doped indium oxide nanoparticle films as transparent conducting electrodes. *Inorganic Chemistry, 49*(18), 8179–8181. https://doi.org/10.1021/IC100719R.

Huang, X., Kocabas, E., & Hernick, M. (2011). The activity and cofactor preferences of N-acetyl-1-D-myoinosityl-2-amino- 2-deoxy-α-D-glucopyranoside deacetylase (MshB) change depending on environmental conditions. *Journal of Biological Chemistry*, *286*(23), 20275–20282. https://doi.org/10.1074/jbc.M111.234229.

Ingledew, W. J., & Ohnishi, T. (1980). An analysis of some thermodynamic properties of iron-sulphur centres in site I of mitochondria. *The Biochemical Journal*, *186*(1), 111–117. https://doi.org/10.1042/bj1860111.

Jeschke, G. (2012). DEER distance measurements on proteins. *Annual Review of Physical Chemistry*, *63*, 419–446. https://doi.org/10.1146/annurev-physchem-032511-143716.

Jeschke, G., Bender, A., Paulsen, H., Zimmermann, H., & Godt, A. (2004). Sensitivity enhancement in pulse EPR distance measurements. *Journal of Magnetic Resonance*, *169*(1), 1–12. https://doi.org/10.1016/j.jmr.2004.03.024.

Jeschke, G., Sajid, M., Schulte, M., & Godt, A. (2009). Three-spin correlations in double electron–electron resonance. *Physical Chemistry Chemical Physics*, *11*(31), 6553–6554. https://doi.org/10.1039/b913085n.

Jeschke, G., Sajid, M., Schulte, M., Ramezanian, N., Volkov, A., Zimmermann, H., et al. (2010). Flexibility of shape-persistent molecular building blocks composed of *p*-phenylene and ethynylene units. *Journal of the American Chemical Society*, *132*(29). https://doi.org/10.1021/ja102983b.

Kaminker, I., Florent, M., Epel, B., & Goldfarb, D. (2011). Simultaneous acquisition of pulse EPR orientation selective spectra. *Journal of Magnetic Resonance*, *208*(1), 95–102. https://doi.org/10.1016/j.jmr.2010.10.010.

Kato, M., Cardona, T., Rutherford, A. W., & Reisner, E. (2012). Photoelectrochemical water oxidation with photosystem II integrated in a mesoporous indium–tin oxide electrode. *Journal of the American Chemical Society*, *134*(20), 8332–8335. https://doi.org/10.1021/JA301488D.

Kato, M., Cardona, T., Rutherford, A. W., & Reisner, E. (2013). Covalent immobilization of oriented photosystem II on a nanostructured electrode for solar water oxidation. *Journal of the American Chemical Society*, *135*(29), 10610–10613. https://doi.org/10.1021/ja404699h.

Kato, M., Zhang, J. Z., Paul, N., & Reisner, E. (2014). Protein film photoelectrochemistry of the water oxidation enzyme photosystem II. *Chemical Society Reviews*, *43*(18), 6485–6497. https://doi.org/10.1039/C4CS00031E.

Keller, K., Qi, M., Gmeiner, C., Ritsch, I., Godt, A., Jeschke, G., et al. (2019). Intermolecular background decay in RIDME experiments. *Physical Chemistry Chemical Physics*, *21*(16), 8228–8245. https://doi.org/10.1039/C8CP07815G.

Khan, K. H. (2013). Gene expression in mammalian cells and its applications. *Advanced Pharmaceutical Bulletin*, *3*(2), 257–263. https://doi.org/10.5681/apb.2013.042.

Kim, J. Y., Hoon Jo, B., & Joon Cha, H. (2010). Production of biohydrogen by recombinant expression of [NiFe]-hydrogenase 1 in Escherichia coli. *Microbial Cell Factories*, *9*-(54). https://doi.org/10.1186/1475-2859-9-54.

Klug, C. S., Eaton, S. S., Eaton, G. R., & Feix, J. B. (1998). Ligand-induced conformational change in the ferric Enterobactin receptor FepA as studied by site-directed spin labeling and time-domain ESR. *Biochemistry*, *37*(25). https://doi.org/10.1021/bi980144e.

Kocabas, E. (2013). Metalloenzymes: Use of recombinant protein expression and affinity tags to aid identification of native metal ion cofactors. *Biochemistry & Analytical Biochemistry*, *02*(02). https://doi.org/10.4172/2161-1009.1000132.

Kotov, V., Bartels, K., Veith, K., Josts, I., Subhramanyam, U. K. T., Günther, C., et al. (2019). High-throughput stability screening for detergent-solubilized membrane proteins. *Scientific Reports*, *9*(1), 10379. https://doi.org/10.1038/s41598-019-46686-8.

Kuchenreuther, J. M., Shiigi, S. A., & Swartz, J. R. (2014). Cell-free synthesis of the h-cluster: A model for the in vitro assembly of metalloprotein metal centers. *Methods in Molecular Biology*, *1122*, 49–72. https://doi.org/10.1007/978-1-62703-794-5_5.

Kutyshenko, V. P., Mikoulinskaia, G. V., Chernyshov, S. V., Yegorov, A. Y., Prokhorov, D. A., & Uversky, V. N. (2019). Effect of C-terminal His-tag and purification routine on the activity and structure of the metalloenzyme, L-alanyl-D-glutamate peptidase of the bacteriophage T5. *International Journal of Biological Macromolecules*, *124*, 810–818. https://doi.org/10.1016/j.ijbiomac.2018.11.219.

Laviron, E. (1979). General expression of the linear potential sweep voltammogram in the case of diffusionless electrochemical systems. *Journal of Electroanalytical Chemistry and Interfacial Electrochemistry*, *101*(1), 19–28. https://doi.org/10.1016/S0022-0728(79)80075-3.

le Breton, N., Wright, J. J., Jones, A. J. Y., Salvadori, E., Bridges, H. R., Hirst, J., et al. (2017). Using hyperfine electron paramagnetic resonance spectroscopy to define the proton-coupled electron transfer reaction at Fe-S cluster N2 in respiratory complex I. *Journal of the American Chemical Society*, *139*(45), 16319–16326. https://doi.org/10.1021/jacs.7b09261.

Léger, C., Elliott, S. J., Hoke, K. R., Jeuken, L. J. C., Jones, A. K., & Armstrong, F. A. (2003). Enzyme electrokinetics: Using protein film voltammetry to investigate redox enzymes and their mechanisms. *Biochemistry*, *42*(29), 8653–8662. https://doi.org/10.1021/bi034789c.

Lem, W. J., & Wayman, M. (1970). Decomposition of aqueous dithionite. Part I. Kinetics of decomposition of aqueous sodium dithionite. *Canadian Journal of Chemistry*, *48*(5), 776–781. https://doi.org/10.1139/v70-126.

Li, J., Lawton, T. J., Kostecki, J. S., Nisthal, A., Fang, J., Mayo, S. L., et al. (2016). Cell-free protein synthesis enables high yielding synthesis of an active multicopper oxidase. *Biotechnology Journal*, *11*(2), 212–218. https://doi.org/10.1002/biot.201500030.

Lovett, J. E., Bowen, A. M., Timmel, C. R., Jones, M. W., Dilworth, J. R., Caprotti, D., et al. (2009). Structural information from orientationally selective DEER spectroscopy. *Physical Chemistry Chemical Physics*, *11*(31), 6840–6848. https://doi.org/10.1039/b907010a.

Lyubenova, S., Maly, T., Zwicker, K., Brandt, U., Ludwig, B., & Prisner, T. (2010). Multifrequency pulsed electron paramagnetic resonance on metalloproteins. *Accounts of Chemical Research*, *43*(2), 181–189. https://doi.org/10.1021/ar900050d.

Malkin, R., & Bearden, A. J. (1973). Detection of a free radical in the primary reaction of chloroplast photosystem II. *Proceedings of the National Academy of Sciences of the United States of America*, *70*(2), 294–297. https://doi.org/10.1073/pnas.70.2.294.

Maly, T., MacMillan, F., Zwicker, K., Kashani-Poor, N., Brandt, U., & Prisner, T. F. (2004). Relaxation filtered hyperfine (REFINE) spectroscopy: A novel tool for studying overlapping biological electron paramagnetic resonance signals applied to mitochondrial complex I. *Biochemistry*, *43*(13), 3969–3978. https://doi.org/10.1021/bi035865e.

Maly, T., & Prisner, T. F. (2004). Relaxation filtered hyperfine spectroscopy (REFINE). *Journal of Magnetic Resonance*, *170*(1), 88–96. https://doi.org/10.1016/j.jmr.2004.06.003.

Maly, T., Zwicker, K., Cernescu, A., Brandt, U., & Prisner, T. F. (2009). New pulsed EPR methods and their application to characterize mitochondrial complex I. *Biochimica et Biophysica Acta - Bioenergetics*, *1787*(6), 584–592. https://doi.org/10.1016/j.bbabio.2009.02.003.

Marko, A., Margraf, D., Yu, H., Mu, Y., Stock, G., & Prisner, T. (2009). Molecular orientation studies by pulsed electron-electron double resonance experiments. *Journal of Chemical Physics*, *130*(6). https://doi.org/10.1063/1.3073040.

Maryasov, A. G., & Bowman, M. K. (2013). Bloch equations for anisotropic paramagnetic centers with spin of 1/2. *Journal of Magnetic Resonance*, *233*, 80–86. https://doi.org/10.1016/j.jmr.2013.05.009.

Megarity, C. F., Siritanaratkul, B., Heath, R. S., Wan, L., Morello, G., Fitzpatrick, S. R., et al. (2019). Electrocatalytic volleyball: Rapid nanoconfined nicotinamide cycling for organic synthesis in electrode pores angewandte. *Angewandte Chemie, International Edition*, *58*, 4948–4952. https://doi.org/10.1002/anie.201814370.

Meiboom, S., & Gill, D. (1958). Modified spin-echo method for measuring nuclear relaxation times. *Review of Scientific Instruments*, *29*(8), 688–691. https://doi.org/10.1063/1.1716296.

Mersch, D., Lee, C. Y., Zhang, J. Z., Brinkert, K., Fontecilla-Camps, J. C., Rutherford, A. W., et al. (2015). Wiring of photosystem II to hydrogenase for photoelectrochemical water splitting. *Journal of the American Chemical Society*, *137*(26), 8541–8549. https://doi.org/10.1021/jacs.5b03737.

Meyer, A., Dechert, S., Dey, S., Höbartner, C., & Bennati, M. (2020). Measurement of angstrom to nanometer molecular distances with ^{19} F nuclear spins by EPR/ENDOR spectroscopy. *Angewandte Chemie International Edition*, *59*(1), 373–379. https://doi.org/10.1002/anie.201908584.

Milov, A. D., Ponomarev, A. B., & Tsvetkov, Y. D. (1984). Electron-electron double resonance in electron spin echo: Model biradical systems and the sensitized photolysis of decalin. *Chemical Physics Letters*, *110*(1), 67–72. https://doi.org/10.1016/0009-2614(84)80148-7.

Morrissey, S. R., Horton, T. E., & DeRose, V. J. (2000). Mn $^{2+}$ sites in the hammerhead ribozyme investigated by EPR and continuous-wave Q-band ENDOR spectroscopies. *Journal of the American Chemical Society*, *122*(14), 3473–3481. https://doi.org/10.1021/ja992989z.

Murray, P. R., Collison, D., Daff, S., Austin, N., Edge, R., Flynn, B. W., et al. (2011). An in situ electrochemical cell for Q- and W-band EPR spectroscopy. *Journal of Magnetic Resonance*, *213*(1), 206–209. https://doi.org/10.1016/j.jmr.2011.09.041.

Nakka, K. K., Tesiram, Y. A., Brereton, I. M., Mobli, M., & Harmer, J. R. (2014). Non-uniform sampling in EPR—Optimizing data acquisition for HYSCORE spectroscopy. *Physical Chemistry Chemical Physics*, *16*(31), 16378–16382. https://doi.org/10.1039/C4CP02172J.

Okada, Y., Uyama, M., Harada, M., & Okada, T. (2019). Quasichemical approach to pH shifts in frozen phosphate buffers. *Industrial & Engineering Chemistry Research*, *58*(39), 18422–18431. https://doi.org/10.1021/acs.iecr.9b03755.

Orii, Y., & Morjta, M. (1977). Measurement of the pH of frozen buffer solutions by using pH indicators. *The Journal of Biochemistry*, *81*(1), 163–168. https://doi.org/10.1093/oxfordjournals.jbchem.a131431.

Palmer, T., Santini, C.-L., Iobbi-Nivol, C., Eaves, D. J., Boxer, D. H., & Giordano, G. (1996). Involvement of the narJ and mob gene products in distinct steps in the biosynthesis of the molybdoenzyme nitrate reductase in Escherichia coli. *Molecular Microbiology*, *20*(4), 875–884. https://doi.org/10.1111/j.1365-2958.1996.tb02525.x.

Pannier, M., Veit, S., Godt, A., Jeschke, G., & Spiess, H. W. (2000). Dead-time free measurement of dipole-dipole interactions between electron spins. *Journal of Magnetic Resonance*, *142*(2), 331–340. https://doi.org/10.1006/jmre.1999.1944.

Pliotas, C. (2017). Chapter eight—Ion channel conformation and oligomerization assessment by site-directed spin labeling and pulsed-EPR. In C. Ziegler (Ed.), *Vol. 594. A structure-function toolbox for membrane transporter and channels* (pp. 203–242). Academic Press. https://doi.org/10.1016/bs.mie.2017.05.013.

Polyhach, Y., Bordignon, E., Tschaggelar, R., Gandra, S., Godt, A., & Jeschke, G. (2012). High sensitivity and versatility of the DEER experiment on nitroxide radical pairs at Q-band frequencies. *Physical Chemistry Chemical Physics*, *14*(30), 10762. https://doi.org/10.1039/c2cp41520h.

Ponti, A., & Schweiger, A. (1995). Nuclear coherence-transfer echoes in pulsed EPR. *The Journal of Chemical Physics*, *102*(13), 5207–5219. https://doi.org/10.1063/1.469246.

Reijerse, E., & Savitsky, A. (2017). Electron Paramagnetic Resonance Instrumentation. In *eMagRes* John Wiley & Sons, Ltd. https://doi.org/10.1002/9780470034590.emrstm1511.

Reuillard, B., Ly, K. H., Hildebrandt, P., Jeuken, L. J. C., Butt, J. N., & Reisner, E. (2017). High performance reduction of H_2O_2 with an electron transport decaheme cytochrome on a porous ITO electrode. *Journal of the American Chemical Society, 139*(9), 3324–3327. https://doi.org/10.1021/jacs.6b12437.

Richardson, K. H., Wright, J. J., Šimėnas, M., Thiemann, J., Esteves, A. M., McGuire, G., et al. (2021). Functional basis of electron transport within photosynthetic complex I. *Nature Communications, 12*(1). https://doi.org/10.1038/s41467-021-25527-1.

Riplinger, C., Kao, J. P. Y., Rosen, G. M., Kathirvelu, V., Eaton, G. R., Eaton, S. S., et al. (2009). Interaction of radical pairs through-bond and through-space: Scope and limitations of the point-dipole approximation in electron paramagnetic resonance spectroscopy. *Journal of the American Chemical Society, 131*(29), 10092–10106. https://doi.org/10.1021/ja901150j.

Roessler, M. M., Evans, R. M., Davies, R. A., Harmer, J., & Armstrong, F. A. (2012). EPR spectroscopic studies of the Fe–S clusters in the O2-tolerant [NiFe]-hydrogenase Hyd-1 from Escherichia coli and characterization of the unique [4Fe–3S] cluster by HYSCORE. *Journal of the American Chemical Society, 134*(37), 15581–15594. https://doi.org/10.1021/ja307117y.

Roessler, M. M., King, M. S., Robinson, A. J., Armstrong, F. A., Harmer, J., & Hirst, J. (2010). Direct assignment of EPR spectra to structurally defined iron-sulfur clusters in complex I by double electron-electron resonance. *Proceedings of the National Academy of Sciences of the United States of America, 107*(5), 1930–1935. https://doi.org/10.1073/pnas.0908050107.

Roessler, M. M., & Salvadori, E. (2018). Principles and applications of EPR spectroscopy in the chemical sciences. *Chemical Society Reviews, 47*(8), 2534–2553. https://doi.org/10.1039/C6CS00565A.

Rosano, G. L., & Ceccarelli, E. A. (2014). Recombinant protein expression in Escherichia coli: Advances and challenges. *Frontiers in Microbiology, 5*. https://doi.org/10.3389/fmicb.2014.00172.

Roscher, A., Emsley, L., & Roby, C. (1996). The effect of imperfect saturation in saturation-recovery T1 measurements. *Journal of Magnetic Resonance, Series A, 118*(1), 108–112. https://doi.org/10.1006/jmra.1996.0015.

Sabaty, M., Grosse, S., Adryanczyk, G., Boiry, S., Biaso, F., Arnoux, P., et al. (2013). *Detrimental effect of the 6 His C-terminal tag on YedY enzymatic activity and influence of the TAT signal sequence on YedY synthesis*. http://www.biomedcentral.com/1471-2091/14/28.

Salvadori, E., Fung, M. W., Hoffmann, M., Anderson, H. L., & Kay, C. W. M. (2015). Exploiting the symmetry of the resonator mode to enhance PELDOR sensitivity. *Applied Magnetic Resonance, 46*(4), 359–368. https://doi.org/10.1007/s00723-014-0621-8.

Santos, B. D., Morones-Ramirez, J. R., Balderas-Renteria, I., Casillas-Vega, N. G., Galbraith, D. W., & Zarate, X. (2019). Optimizing periplasmic expression in Escherichia coli for the production of recombinant proteins tagged with the small metal-binding protein SmbP. *Molecular Biotechnology, 61*(6), 451–460. https://doi.org/10.1007/s12033-019-00176-4.

Saveant, J. M. (2006). *Elements of molecular and biomolecular electrochemistry*. John Wiley & Sons, Inc.

Schiemann, O., Heubach, C. A., Abdullin, D., Ackermann, K., Azarkh, M., Bagryanskaya, E. G., et al. (2021). Benchmark test and guidelines for DEER/PELDOR experiments on nitroxide-labeled biomolecules. *Journal of the American Chemical Society, 143*(43), 17875–17890. https://doi.org/10.1021/jacs.1c07371.

Schweiger, A., & Jeschke, G. (2001). *Principles of pulse electron paramagnetic resonance*. Oxford, UK: Oxford University Press.

Seif-Eddine, M., Biaso, F., Arias-Cartin, R., Pilet, E., Rendon, J., Lyubenova, S., et al. (2017). Probing the menasemiquinone binding mode to nitrate reductase A by selective 2H and 15N labeling, HYSCORE spectroscopy, and DFT modeling. *ChemPhysChem*, *18*(19), 2704–2714. https://doi.org/10.1002/cphc.201700571.

Seif-Eddine, M., Biaso, F., Rendon, J., Pilet, E., Guigliarelli, B., Magalon, A., et al. (2020). 1,2H hyperfine spectroscopy and DFT modeling unveil the demethylmenasemiquinone binding mode to E. coli nitrate reductase A (NarGHI). *Biochimica et Biophysica Acta (BBA) - Bioenergetics*, *1861*(8), 148203. https://doi.org/10.1016/j.bbabio.2020.148203.

Serra, I., García Rubio, I., & van Doorslaer, S. (2021). Pitfalls in sample preparation of metalloproteins for low-temperature EPR: The example of alkaline myoglobin. *Applied Magnetic Resonance*. https://doi.org/10.1007/s00723-021-01428-w.

Shleev, S., Christenson, A., Serezhenkov, V., Burbaev, D., Yaropolov, A., Gorton, L., et al. (2005). Electrochemical redox transformations of T1 and T2 copper sites in native Trametes hirsuta laccase at gold electrode. *Biochemical Journal*, *385*(3), 745–754. https://doi.org/10.1042/BJ20041015.

Šimėnas, M., O'Sullivan, J., Zollitsch, C. W., Kennedy, O., Seif-Eddine, M., Ritsch, I., et al. (2021). A sensitivity leap for X-band EPR using a probe head with a cryogenic preamplifier. *Journal of Magnetic Resonance*, *322*. https://doi.org/10.1016/j.jmr.2020.106876.

Singh, V., Zhu, Z., Davidson, V. L., & McCracken, J. (2000). Characterization of the tryptophan tryptophyl-semiquinone catalytic intermediate of methylamine dehydrogenase by electron spin−echo envelope modulation spectroscopy. *Journal of the American Chemical Society*, *122*(5). https://doi.org/10.1021/ja9934246.

Sinnecker, S., Flores, M., & Lubitz, W. (2006). Protein–cofactor interactions in bacterial reaction centers from Rhodobacter sphaeroides R-26: Effect of hydrogen bonding on the electronic and geometric structure of the primary quinone. A density functional theory study. *Physical Chemistry Chemical Physics*, *8*(48), 5659–5670. https://doi.org/10.1039/B612568A.

Sinnecker, S., Reijerse, E., Neese, F., & Lubitz, W. (2004). Hydrogen bond geometries from electron paramagnetic resonance and electron−nuclear double resonance parameters: Density functional study of quinone radical anion−solvent interactions. *Journal of the American Chemical Society*, *126*(10), 3280–3290. https://doi.org/10.1021/ja0392014.

Siritanaratkul, B., Megarity, C. F., Roberts, T. G., Samuels, T. O. M., Winkler, M., Warner, J. H., et al. (2017). Transfer of photosynthetic NADP+/NADPH recycling activity to a porous metal oxide for highly specific, electrochemically-driven organic synthesis. *Chemical Science*, *8*(6), 4579–4586. https://doi.org/10.1039/c7sc00850c.

Smirnova, I., Kasho, V., Choe, J.-Y., Altenbach, C., Hubbell, W. L., & Ronald Kaback, H. (2007). Sugar binding induces an outward facing conformation of LacY. *Proceedings of the National Academy of Sciences of the United States of America*, *104*(42), 16504–16509.

Spindler, P. E., Schöps, P., Bowen, A. M., Endeward, B., & Prisner, T. F. (2016). Shaped Pulses in EPR. In *eMagRes* John Wiley & Sons, Ltd. https://doi.org/10.1002/9780470034590.emrstm1520.

Spindler, P. E., Waclawska, I., Endeward, B., Plackmeyer, J., Ziegler, C., & Prisner, T. F. (2015). Carr–Purcell pulsed electron double resonance with shaped inversion pulses. *The Journal of Physical Chemistry Letters*, *6*(21), 4331–4335. https://doi.org/10.1021/acs.jpclett.5b01933.

Stich, T. A., Whittaker, J. W., & Britt, R. D. (2010). Multifrequency EPR studies of manganese catalases provide a complete description of proteinaceous nitrogen coordination. *The Journal of Physical Chemistry B*, *114*(45), 14178–14188. https://doi.org/10.1021/jp908064y.

Stoll, S. (2017). Pulse EPR. In *eMagRes* John Wiley & Sons, Ltd. https://doi.org/10.1002/9780470034590.emrstm1510.

Stoll, S., & Britt, R. D. (2009). General and efficient simulation of pulse EPR spectra. *Physical Chemistry Chemical Physics*, *11*(31), 6614–6625. https://doi.org/10.1039/B907277B.
Stoll, S., & Kasumaj, B. (2008). Phase cycling in electron spin echo envelope modulation. *Applied Magnetic Resonance*, *35*(1), 15–32. https://doi.org/10.1007/s00723-008-0140-6.
Stoll, S., & Schweiger, A. (2006). EasySpin, a comprehensive software package for spectral simulation and analysis in EPR. *Journal of Magnetic Resonance*, *178*(1), 42–55. https://doi.org/10.1016/j.jmr.2005.08.013.
Sun, J., Velamakanni, B. V., Gerberich, W. W., & Francis, L. F. (2004). Aqueous latex/ceramic nanoparticle dispersions: colloidal stability and coating properties. *Journal of Colloid and Interface Science*, *280*(2), 387–399. https://doi.org/10.1016/j.jcis.2004.08.014.
Taguchi, A. T., O'Malley, P. J., Wraight, C. A., & Dikanov, S. A. (2014). Hyperfine and nuclear quadrupole tensors of nitrogen donors in the QA site of bacterial reaction centers: Correlation of the histidine Nδ tensors with hydrogen bond strength. *The Journal of Physical Chemistry B*, *118*(31), 9225–9237. https://doi.org/10.1021/jp5051029.
Tait, C. E., & Stoll, S. (2016). Coherent pump pulses in double electron electron resonance spectroscopy. *Physical Chemistry Chemical Physics*, *18*(27), 18470–18485. https://doi.org/10.1039/C6CP03555H.
Teucher, M., & Bordignon, E. (2018). Improved signal fidelity in 4-pulse DEER with Gaussian pulses. *Journal of Magnetic Resonance*, *296*, 103–111. https://doi.org/10.1016/j.jmr.2018.09.003.
Torrey, H. C. (1949). Transient nutations in nuclear magnetic resonance. *Physical Review*, *76*(8), 1059–1068. https://doi.org/10.1103/PhysRev.76.1059.
van Doorslaer, S. (2017). Hyperfine spectroscopy: ESEEM. In *eMagRes* John Wiley & Sons, Ltd. https://doi.org/10.1002/9780470034590.emrstm1517.
Varnado, C. L., & Goodwin, D. C. (2004). System for the expression of recombinant hemoproteins in Escherichia coli. *Protein Expression and Purification*, *35*(1), 76–83. https://doi.org/10.1016/j.pep.2003.12.001.
Vileno, B., Chamoun, J., Liang, H., Brewer, P., Haldeman, B. D., Facemyer, K. C., et al. (2011). Broad disorder and the allosteric mechanism of myosin II regulation by phosphorylation. *Proceedings of the National Academy of Sciences of the United States of America*, *108*(20). https://doi.org/10.1073/pnas.1014137108/-/DCSupplemental.
Ward, R., Bowman, A., El-Mkami, H., Owen-Hughes, T., & Norman, D. G. (2009). Long distance PELDOR measurements on the histone core particle. *Journal of the American Chemical Society*, *131*(4), 1348–1349. https://doi.org/10.1021/ja807918f.
Ward, R., Bowman, A., Sozudogru, E., El-Mkami, H., Owen-Hughes, T., & Norman, D. G. (2010). EPR distance measurements in deuterated proteins. *Journal of Magnetic Resonance*, *207*(1). https://doi.org/10.1016/j.jmr.2010.08.002.
Wright, J. J., Fedor, J. G., Hirst, J., & Roessler, M. M. (2020). Using a chimeric respiratory chain and EPR spectroscopy to determine the origin of semiquinone species previously assigned to mitochondrial complex I. *BMC Biology*, *18*(1). https://doi.org/10.1186/s12915-020-00768-6.
Wright, J. J., Salvadori, E., Bridges, H. R., Hirst, J., & Roessler, M. M. (2016). Small-volume potentiometric titrations: EPR investigations of Fe-S cluster N2 in mitochondrial complex I. *Journal of Inorganic Biochemistry*, *162*, 201–206. https://doi.org/10.1016/j.jinorgbio.2016.04.025.
Yang, Z., Kise, D., & Saxena, S. (2010). An approach towards the measurement of nanometer range distances based on Cu $^{2+}$ ions and ESR. *The Journal of Physical Chemistry B*, *114*(18). https://doi.org/10.1021/jp911637s.
Yang, Z., Wang, C., Zhou, Q., An, J., Hildebrandt, E., Aleksandrov, L. A., et al. (2014). Membrane protein stability can be compromised by detergent interactions with the extramembranous soluble domains. *Protein Science: A Publication of the Protein Society*, *23*(6), 769–789. https://doi.org/10.1002/pro.2460.

Yates, N. D. J., Fascione, M. A., & Parkin, A. (2018). Methodologies for "Wiring" redox proteins/enzymes to electrode surfaces. *Chemistry - A European Journal*, *24*(47), 12164–12182. https://doi.org/10.1002/chem.201800750.

Yeagle, G. J., Gilchrist, M. L., Jr., Walker, L. M., Debus, R. J., & Britt, R. D. (2008). Multifrequency electron spin-echo envelope modulation studies of nitrogen ligation to the manganese cluster of photosystem II. *Philosophical Transactions of the Royal Society of London. Series B, Biological Sciences*, *363*(1494), 1157–1166. https://doi.org/10.1098/rstb.2007.2211.

Yeh, P., & Kuwana, T. (1977). Reversible electrode reaction of cytochrome c. *Chemistry Letters*, *6*(10), 1145–1148. https://doi.org/10.1246/CL.1977.1145.

Young, C. L., Britton, Z. T., & Robinson, A. S. (2012). Recombinant protein expression and purification: A comprehensive review of affinity tags and microbial applications. *Biotechnology Journal*, 7(5), 620–634. https://doi.org/10.1002/biot.201100155.

Zhou, Y., Bowler, B. E., Eaton, G. R., & Eaton, S. S. (2000). Electron spin–lattice relaxation rates for high-spin Fe(III) complexes in glassy solvents at temperatures between 6 and 298 K. *Journal of Magnetic Resonance*, *144*(1), 115–122. https://doi.org/10.1006/jmre.2000.2047.

CHAPTER NINE

EPR of copper centers in the prion protein

Liliana Quintanar[a,]* and Glenn L. Millhauser[b,]*
[a]Department of Chemistry, Center for Research and Advanced Studies (Cinvestav), Mexico City, Mexico
[b]Department of Chemistry and Biochemistry, UC Santa Cruz, Santa Cruz, CA, United States
*Corresponding authors: e-mail address: lilianaq@cinvestav.mx; glennm@ucsc.edu

Contents

Abstract

Most proteins implicated in neurodegenerative diseases bind metal ions, notably copper and zinc. Metal ion binding may be part of the protein's function or, alternatively, may promote a deleterious gain of function. With regard to Cu^{2+} ions, electron paramagnetic resonance techniques have proven to be instrumental in determining the biophysical characteristics of the copper binding sites, as well as structural features of the coordinating protein and how they are impacted by metal binding. Here, the most useful methods are described as they apply to the prion protein, which serves as a model for the broader spectrum of neurodegenerative proteins.

1. Introduction

A wide range of neurodegenerative diseases associated with aging involve the aggregation of specific proteins of the central nervous system. Examples include the Aβ peptide and tau protein in Alzheimer's disease (Selkoe & Hardy, 2016), α-synuclein in Parkinson's disease (Proukakis et al., 2013; Spillantini et al., 1997) and the cellular prion protein (PrP^C) in the Transmissible Spongiform Encephalopathies (TSEs) (Prusiner,

Methods in Enzymology, Volume 666
ISSN 0076-6879
https://doi.org/10.1016/bs.mie.2022.02.003

1982), otherwise known as prion diseases. Definitive functions for most of these proteins are still being worked out. Curiously, however, many of them are metal binding proteins and coordinate copper ions in the form of Cu^{2+} (Aronoff-Spencer et al., 2000; Burns et al., 2002, 2003; Bush, 2003; Dudzik, Walter, Abrams, Jurica, & Millhauser, 2013; Jobling et al., 2001; Millhauser, 2004, 2007; Proukakis et al., 2013). Considerable effort has been expended on defining the precise details of the copper sites in these proteins with the expectation that such structural information will provide insight into both protein function and the processes involved in aggregation and subsequent neurodegeneration.

Because Cu^{2+} has a paramagnetic d^9 electron configuration, electron paramagnetic resonance (EPR) plays a key role in determining coordination, geometry and reactivity at copper centers. While each neurodegenerative protein exhibits its own unique coordination features and aggregation properties, PrP^C is unique insofar that it possesses both structured and unstructured protein domains (Riek, Hornemann, Wider, Glockshuber, & Wuthrich, 1997). This methods chapter will therefore focus on PrP^C as a model system for studying both prion and related neurodegerative peptides and proteins in their respective monomeric states. In addition, Zn^{2+} uptake by PrP^C also features in the protein's putative function. Although Zn^{2+} is diamagnetic, and therefore EPR silent, EPR methodologies have nevertheless proven useful in identifying its influence on global PrP^C structure (Spevacek et al., 2013).

We consider two complementary strategies for defining Cu^{2+} coordination modes, which, by analogy to modern proteomics (Han, Jin, Breuker, & McLafferty, 2006), we refer to as "bottom-up" and "top-down." With the bottom-up approach, short segments of a complicated protein are produced by either expression or Solid Phase Peptide Synthesis (SPPS) and examined by EPR in isolation from the full protein. Alternatively, the top-down strategy focuses on EPR of the full protein and select mutants. There are also intermediate strategies where one investigates intact protein domains. As we describe relevant EPR methodologies, the relative benefits of bottom-up and top-down will be elucidated.

The human prion protein is 209 residues in length, after removal of the N-terminal signal peptide, and is decorated by a C-terminal glycophosphatidylinositol (GPI)-anchor, two Asn-linked carbohydrate moieties (Hornemann, Schorn, & Wuthrich, 2004). There are small variations in length, depending on the specific mammalian species. As shown in Fig. 1, residues 23-125 are unstructured random coil (in the absence of metal ions), while the C-terminal domain, residues 126-231, forms a unique

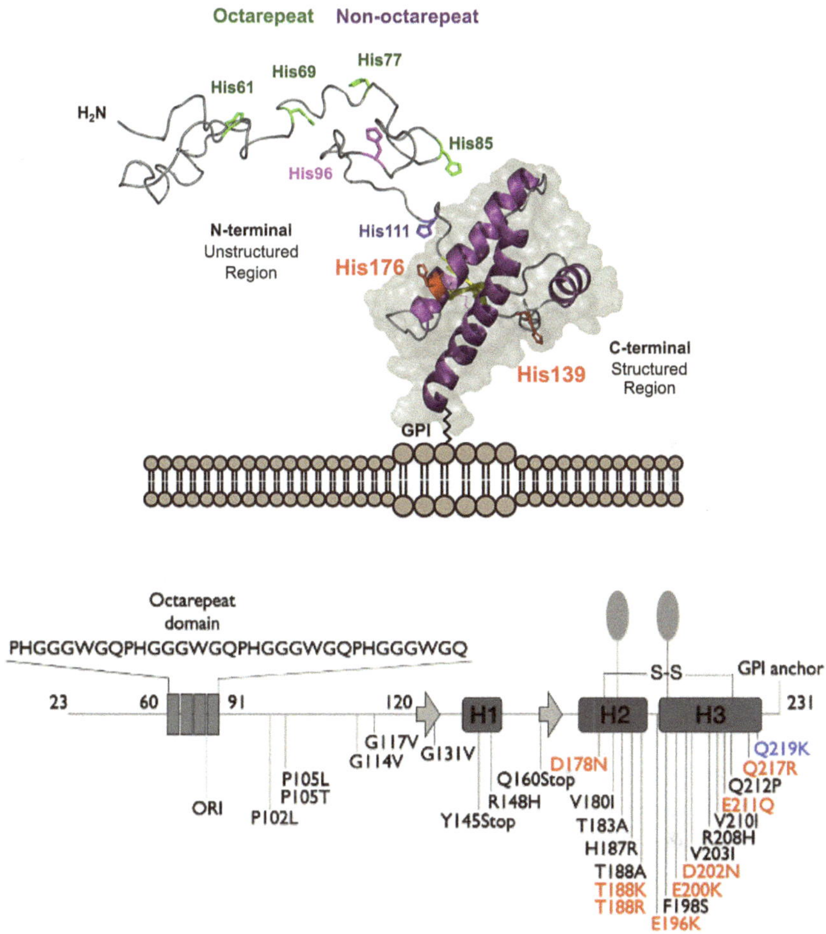

Fig. 1 Top, three-dimensional rendering of the cellular prion protein, PrP^C, noting the intrinsically disordered N-terminal domain and the helical C-terminal domain. Metal ion binding takes place primarily within the N-terminal domain involving octarepeat His residues (green) and non-octarepeat residues (purple). Bottom, PrP^C sequence features revealing locations of the octarepeat domain, secondary structure elements in the C-terminal domain, GPI-anchor, glycan sites, and mutations that confer familial prion disease. Mutations labeled red shift side chain charge to a more positive state. The mutation Q219K (blue) is protective against disease. *Figure adapted from Posadas, Y., Lopez-Guerrero, V., Segovia, J., Perez-Cruz, C., & Quintanar, L. (2022) Dissecting the copper bioinorganic chemistry of the functional and pathological roles of the prion protein: Relevance in Alzheimer's disease and cancer. Current Opinion in Chemical Biology 66, 102098.*

disulfide-stabilized structure composed of three α-helices and a short anti-parallel β-sheet (Riek et al., 1997). Also shown are familial mutations that give rise to inherited prion diseases (Prusiner, 1997; Spevacek et al., 2013).

PrP^C is localized to extracellular membrane surfaces through insertion of the GPI-anchor into the outer membrane leaflet (Stahl, Borchelt, Hsiao, & Prusiner, 1987). While PrP^C is localized predominantly in the central nervous system, it is also found distributed in peripheral tissues (Zomosa-Signoret, Arnaud, Fontes, Alvarez-Martinez, & Liautard, 2008). A notable feature of the unstructured segment of PrP^C is the octarepeat (OR) domain, so named because it is composed of tandem, eight-residue PHGGGWGQ repeats (Aronoff-Spencer et al., 2000; Hornshaw, McDermott, & Candy, 1995). Initial work demonstrating copper binding focused in the OR segment, given that it contains four His residues. However, more recent work finds that there are important copper-binding segments outside of this domain (Burns et al., 2003; Chaves et al., 2014; Giachin et al., 2015; Grande-Aztatzi, Rivillas-Acevedo, Quintanar, & Vela, 2013; Qin, Yang, Mastrangelo, & Westaway, 2002; Rivillas-Acevedo et al., 2011) and, moreover, that copper stabilizes global PrP^C structure through simultaneous coordination between OR and C-terminal His residues (Evans & Millhauser, 2017; Evans, Pushie, Markham, Lee, & Millhauser, 2016; McDonald et al., 2019; Schilling et al., 2020). Recent reviews describe the links among prion protein copper binding, physiological function and global structure (Evans & Millhauser, 2017; Posadas, Lopez-Guerrero, Segovia, Perez-Cruz, & Quintanar, 2022; Salzano, Giachin, & Legname, 2019).

This chapter will describe the EPR methods listed in Table 1, as they apply to characterizing Cu^{2+}, and to a lesser extent Zn^{2+}, binding to PrP^C. As we work through the approaches, we will note differential benefits to bottom-up vs top-down strategies.

Table 1 EPR methods and their applicability in Bottom up vs Top down investigations.

EPR method	Bottom up (peptides)	Top down (proteins)
Continuous Wave (CW)	✓✓	✓
ESEEM	✓✓	✓✓
Isotope edited ESEEM and HYSCORE	✓✓	*
DEER Nitroxide-Nitroxide	✓	✓✓
DEER Copper-Nitroxide	✓	✓✓
Pulsed ENDOR	✓	✓✓

✓✓: ideally suited; ✓: suitable but with limitations; *: requires specialized sample design.

2. Sample preparation

EPR samples typically contain Cu^{2+} concentrations from 10 μM to nearly a 1.0 mM. The sample concentration must be adapted to the specific technique. Continuous Wave (CW) EPR is the most sensitive and easily operates at the low end of the feasible concentration range. Buffers are selected to avoid those that directly bind copper and may be degassed before use to avoid line broadening from dissolved oxygen. We find that morpholine-based buffers, such as N-ethylmorpholine (NEM) or 2-(N-morpholino)ethanesulfonic acid (MES), work well for stabilizing solutions in the pH range below 7.0; while other noncomplexing tertiary amine buffers have been reported (Yu, Kandegedara, Xu, & Rorabacher, 1997). Most EPR experiments are performed in frozen solution and, therefore, require a cryoprotectant such as glycerol, which helps maintain a spatially uniform distribution of the proteins. We note that free copper at sample pH values above roughly 6.0 forms diamagnetically-coupled hydroxides and, thus, are EPR silent (Aronoff-Spencer et al., 2000). In other words, if one adds an excess of copper to a protein solution at pH above 6.0, any copper remaining in solution and not bound to the protein, will not give an EPR signal. On the other hand, if working below pH 6.0, signals from free Cu^{2+} may subtracted from acquired spectra. Finally, certain pulsed EPR methods benefit from sample preparation strategies that give long T_2 or T_1 relaxation times. In such cases, it is often beneficial to use deuterated solvents.

A typical EPR sample solution will have the following:

- 25 mM N-ethylmorpholine (NEM); pH adjusted with HCl or KOH to 6.0 or above (20 to 100 mM NaCl may be required for stability of certain proteins)
- 20–30% by volume glycerol
- protein or peptide solution (100 μM and above), degassed (if necessary) with nitrogen, helium, or by freeze–pump–thaw cycles
- paramagnetic metal ion salt in final concentration 100 μM and above, matched to appropriate protein stoichiometry

3. Continuous wave (CW) EPR

CW EPR, typically performed at X-band frequencies (~9.0 GHz) is a primary technique for determining EPR spectra. The great benefit of CW EPR for Cu^{2+} bound to neurodegenerative proteins is that one may quickly extract spectral parameters that report on the coordination environment. Correlations originally developed by Peisach and Blumberg (1974), applied to frozen aqueous solutions, distinguish the degree of nitrogen, oxygen, sulfur bound to the copper center. Moreover, if there are multiple coordination species, these are often represented as superimposed spectra. CW EPR is especially useful for bottom-up strategies since short peptide segments give sharp, well-resolved Cu^{2+} spectra.

An early experiment demonstrating the benefit of CW EPR applied to PrP^{C} is shown in Fig. 2 (Aronoff-Spencer et al., 2000). With the hypothesis that Cu^{2+} would coordinate to the OR domain (within residues 57–91), this experiment examined spectra obtained from a series of progressively shorter, OR-derived peptides (Aronoff-Spencer et al., 2000). By simple inspection, one can see that peptides down to the segment HGGGW yield spectra approximately equivalent to the full OR-domain. However, peptides shorter than this (HGGG and HGG) give spectra clearly distinguishable from the OR spectrum. In addition, analysis of the copper hyperfine splittings ($I=3/2$), as indicated by the grid markers, suggest that coordination arises from one oxygen and three nitrogen atoms (1O, 3N). These findings were eventually borne out by X-ray crystallography, thus demonstrating the power of this very straightforward technique.

Beyond the octarepeat region, PrP^{C} displays two additional Cu^{2+} binding sites associated to His96 and His111 that have been termed as non-octarepeat sites (Burns et al., 2003). These sites have also been modeled by peptide fragments, such as 92–96 (GGGTH) and 106–115 (KTNMKHMAGA). EPR studies of these peptides and several variants helped elucidate that metal binding to these sites is pH dependent and it involves deprotonated amides that precede the His residue in the sequence (Fig. 3) (Grande-Aztatzi et al., 2013; Rivillas-Acevedo et al., 2011). In both cases, drastic changes in the EPR spectra of the Cu^{2+} complexes reflect a change in the equatorial coordination mode from a 3N1O to a 4N shell. The fourth nitrogen-based ligand arises from the deprotonation of a backbone amide (pKa ~7.5–7.8) that enters the metal coordination sphere. Sometimes, a highly covalent nitrogen-rich environment may yield strong hyperfine nitrogen couplings that become evident in the EPR spectrum, as is the case of the Cu^{2+} complex with GGGTH at pH 8.5 (Fig. 3). Pulsed EPR techniques can be used to accurately measure nitrogen couplings (*vide infra*).

Using peptides to model Cu^{2+} binding sites in PrP^{C} has been a successful approach, given the fact that the N-terminal region is intrinsically disordered. When using peptide fragments to model metal binding sites, it is

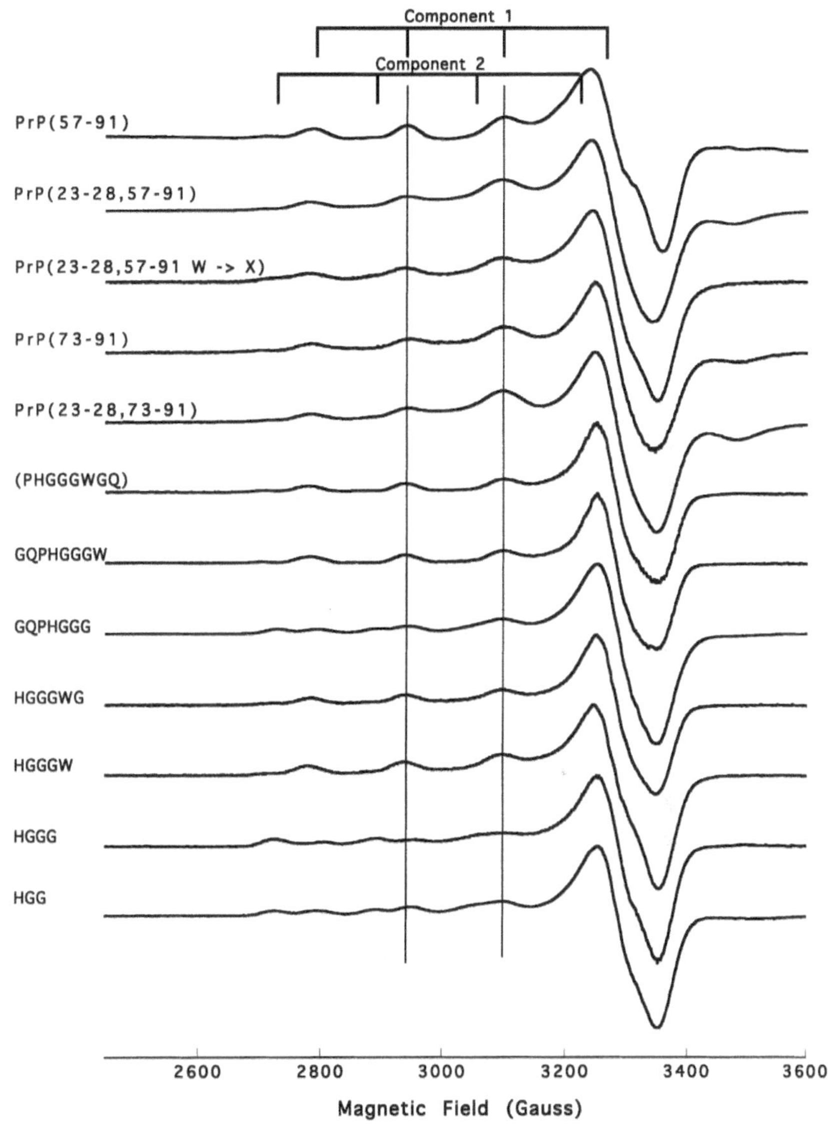

Fig. 2 X-band CW EPR spectra, displayed in the first-derivative mode, of OR-derived peptides obtained at pH 7.45 and 77 K. The grid markers identify the four hyperfine lines from the copper $I = 3/2$ nucleus. These spectra suggest the existence of two distinct species, initially referred to as Component 1 and Component 2.

important to acetylate the N-terminus and amidate the C-terminus to cap the free NH_2 and COO^- groups of the peptide and prevent them from coordinating to the metal ion. Indeed, having a free NH_2 group drastically impacts Cu^{2+} coordination to a peptide, as shown for the HMAGA

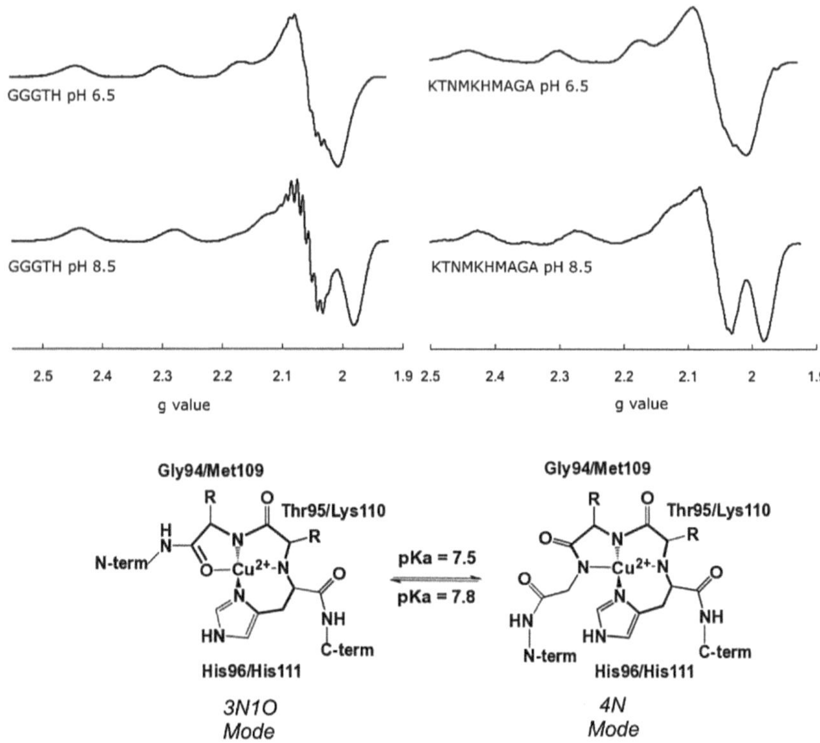

Fig. 3 X-band CW EPR spectra of Cu^{2+} complexes at the His96 (GGGTH) and His111 (KTNMKHMAGA) sites, at pH6.5 and 8.5. Spectral changes upon pH are associated to the conversion of a 3N1O equatorial coordination mode to a 4N, as drawn. The pKa associated to this conversion is 7.8 for the His96 site and 7.5 for the His111 site.

fragment, which models how the His111 site in PrP^C is impacted upon α-cleavage (Sanchez-Lopez, Fernandez, & Quintanar, 2018). The electron pair of a free-NH_2 group acts as a strong anchoring moiety for Cu^{2+}. Indeed, a titration of the NH_2-HMAGA peptide by Cu^{2+} reveals the presence of species where the free-NH_2 group dominates metal coordination (Fig. 4).

4. Electron spin echo envelope modulation (ESEEM)

Histidine residues, with their imidazole sidechains, invariably play critical roles in both copper and zinc coordination in proteins. ESEEM is a pulsed EPR technique uniquely diagnostic for imidazole coordination

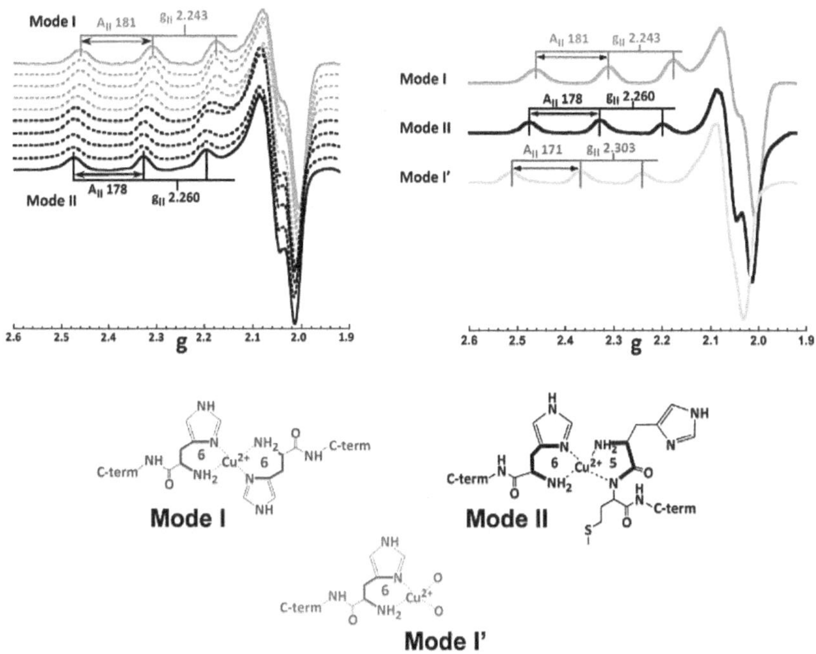

Fig. 4 X-band CW EPR spectra of Cu^{2+} complexes with the NH_2-HMAGA fragment (without acetylation), modeling metal binding to His111 in α-cleaved PrP^C. A titration of the NH_2-HMAGA peptide yields EPR spectra that reveals two species: Mode I favored up to 0.5 equiv. of Cu^{2+}, and Mode II that forms at a 1:1 metal: peptide ratio. At low equiv. of metal ion (0.5 equiv) and low pH (5.0) a third species is favored, termed Mode I′. Proposed coordination models for these species are shown. *Figure adapted from Sanchez-Lopez C, Fernandez CO, & Quintanar L (2018) Neuroprotective alpha-cleavage of the human prion protein significantly impacts Cu(II) coordination at its His111 site.* Dalton Transactions, *47(28), 9274–9282.*

to copper centers (Mims & Peisach, 1978). The measurement entails recording the intensity of a spin echo as a function of evolving delays between specific microwave pulses. There are several schemes, the benefits of which are thoroughly reviewed elsewhere (Kevan & Bowman, 1990; Sahu, McCarrick, & Lorigan, 2013). Regardless of specific pulse scheme, the Fourier transform of the time-dependent signal reveals frequencies arising from ^{14}N ($I = 1$) nuclear quadrupole transitions. With Cu^{2+} X-band EPR, the observed transitions are intense, arising from the remote nitrogen of the coordinated imidazole ring.

As shown in Fig. 5A, there are three characteristic transitions below 2.0 MHz arising from the individual quadrupole resonances among the three

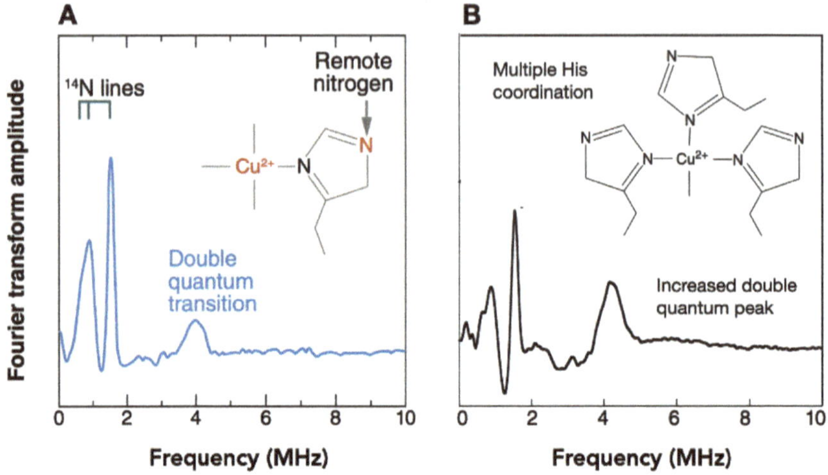

Fig. 5 Characteristic ESEEM spectra for (A) single His coordination and (B) multiple His coordination. The three lines below 2.0 MHz in each case arise from the remote ^{14}N ($I = 1$) nuclear quadrupole transitions. The intensity of the transition at ~4 MHz is indicative of the number of coordinated His residues.

^{14}N spin states. A fourth so-called "double quantum transition" occurs at approximately 4.0 MHz. Given the uniqueness of this spectrum, ESEEM is ideally diagnostic for His coordination.

ESEEM is readily applied in both bottom-up and top-down strategies. Copper proteins at typical EPR concentrations (100 μM and above), and temperatures 20 K or lower, give quality ESEEM spectra with fully resolvable lines (Chattopadhyay et al., 2005). Moreover, when developing bottom-up peptide models of protein copper centers, ESEEM spectra may be used to compare and validate preservation of the coordination environment. There are important benefits that extend beyond the simple identification of imidazole coordination. For example, Fig. 5B shows a resulting ESEEM spectrum for multiple His coordination with a diagnostic increase in the double quantum transition (Chattopadhyay et al., 2005). With PrP^{C}, this unique spectroscopic signature was combined with CW EPR measurements to demonstrate concentration-dependent changes in His coordination in the OR domain as a function of copper:domain ratio. Specifically, when saturated at high [Cu^{2+}], each His coordinates a single Cu^{2+} ion, which is referred to as component 1 binding. However, at low [Cu^{2+}], approximately 1:1 copper to protein, each copper ion is coordinated by multiple His residues (referred to as component 3 binding). These observations offer perspective on how to carry out characterization with other neurodegenerative proteins. If there are multiple, proximal His residues in a flexible, disordered polypeptide segment, concentration-dependent measurements combined with ESEEM are essential.

With the bottom-up approach, peptides prepared with selectively placed stable isotopes or unnatural amino acids may to used to directly test proposed coordination models. For example, ESEEM obtained from a peptide corresponding to the OR domain, fully Cu^{2+} saturated (Component 1), yields the ESEEM spectrum shown in Fig. 6. There are multiple lines beyond those expected from the transitions due to the single His imidazole. Molecular modeling, pH titrations and companion EPR measurements, suggested the structure shown in Fig. 6.

A necessary consequence of this structure is that the amide nitrogen of Gly(3) (blue) is rigidly held at a distance compatible with X-band ESEEM transitions. Therefore, a prediction from the structure in Fig. 6 is that disruption of Gly(2) coordination to Cu^{2+} would eliminate the fixed position of Gly(3). To test the feasibility of this structure, OR peptides prepared with *N*-methylglycine (sarcosine) at Gly(2) were examined by ESEEM (Chattopadhyay et al., 2005). The resulting spectra were equivalent to that of Fig. 5A, thus confirming assignment of the additional lines to ^{14}N from Gly(3) and, importantly, further validating the proposed component 1 structure in Fig. 6.

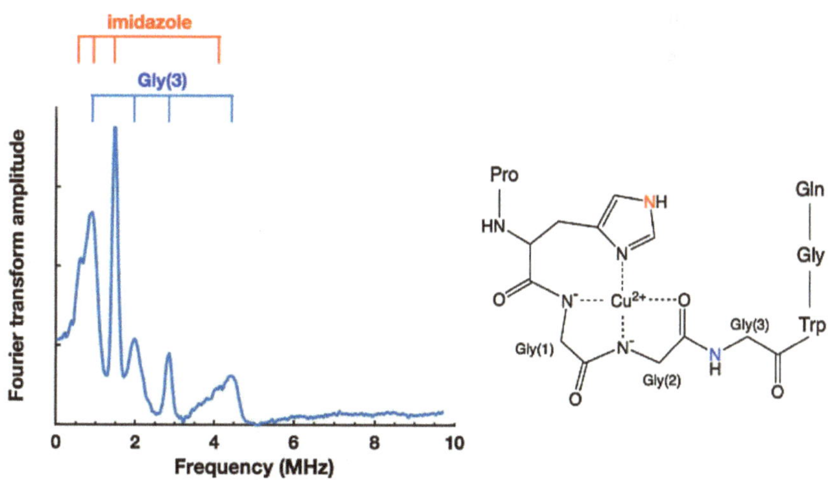

Fig. 6 ESEEM spectra obtained from Cu^{2+} coordinated by a single repeat segment. Transitions beyond those assigned to the remote nitrogen of the single His side chain (red) arise from the backbone nitrogen of Gly(3) (blue). The assignment of Gly(3) was verified by the introduction of a non-coordinating residue at Gly(2).

Hyperfine Sublevel Correlation (HYSCORE) spectroscopy is a two-dimensional variation of ESEEM that often reveals additional electron-nuclear couplings (Hofer, Grupp, & Mehring, 1986). For example, while ESEEM at X-band is mainly limited to weakly coupled $I=1$ nuclei, such as those demonstrated in the examples above, HYSCORE expands this capability to anisotropically broadened transitions observed for dipolar coupled $I=½$ nuclei (Burns et al., 2002). This capability was used in a bottom-up refinement of the component 1 structure shown in Fig. 6. Specifically, replacement of Gly(3) with its ^{15}N-labeled analog in the peptide HGGGW revealed pronounced cross peaks with frequencies commensurate of a non-liganded ^{15}N nucleus, and consistent with the distance between the copper center and the nitrogen of the third glycine. The significant advantage in this example is that, as opposed to editing out unassigned peaks by disruption of the coordination structure, as was done with the introduction of sarcosine, or by elimination of peaks with ^{15}N labels, HYSCORE provides positive spectroscopic confirmation with output that one can use to quantitatively assess dipolar-coupled distances.

5. Double electron-electron resonance (DEER) EPR

Intramolecular distance measurements are often crucial for identifying structural features and organization. The canonical methods for assessing monomeric peptide and protein structure are, of course, X-ray crystallography and NMR. However, in the case of proteins with large, disordered segments, or with bound paramagnetic species such as Cu^{2+}, application of these methods may not be feasible. In the case of PrP^{C}, both methods have yielded critical structural details for the ordered C-terminal domain, but neither has succeeded in resolving structure or detailed Cu^{2+}-binding properties of the intrinsically disordered N-terminal domain.

DEER EPR is a two frequency, pump-probe pulsed technique that provides detailed distances and distance distributions in the range from approximately 20 Å to 100 Å and beyond (Jeschke & Polyhach, 2007). Typically, DEER EPR is applied to distances between pairs of nitroxide spin labels, with each label engineered to a specific site within a protein or protein complex. Measurements are carried out at cryogenic temperatures of 80 K or below, depending on sample and instrument set up. The theory and detailed application of DEER EPR, along with benefits and shortcomings, has been described in numerous reviews (Schiemann et al., 2021). Modern instrumentation readily obtains quality distance data from samples of less than 80 μL and 30 μM doubly-labeled protein or peptide. The focus here will be on the application of DEER EPR to domain organization within PrP^C.

DEER EPR has thus far provided essential information for assessing how both Zn^{2+} (Spevacek et al., 2013) and Cu^{2+} (Evans et al., 2016) impact PrP^C structure, with measurements performed on the full protein, and hence, a top-down strategy. Regarding Zn^{2+} coordination to PrP^C, stoichiometry and chemical protection assays initially demonstrated that the ion coordinates to the four His imidazoles within the OR domain (Walter, Stevens, Visconte, & Millhauser, 2007). However, physiological experiments suggested that Zn^{2+} might introduce structural organization beyond OR coordination. To address this with DEER EPR, a family of PrP^C constructs were developed where each has one spin label in the C-terminal domain and a second in, or near, the OR domain. Nitroxide spin labeling vicinal to the OR domain used the standard approach of introducing a unique Cys residue, followed by reaction with the methanethiosulfonate spin label (MTSL) reagent. However, because the C-terminal domain possesses two conserved Cys residues that form a critical disulfide bond, spin labeling with MTSL is problematic. Consequently, labeling in this domain utilized an expanded genetic code approach to incorporate para-acetyl phenylalanine, followed by reaction with a hydroxylamine nitroxide reagent, to produce the ketoxime-linked K1 side chain (Fig. 7) (Fleissner et al., 2009).

Nitroxide-nitroxide X-band DEER EPR was performed on the family of doubly labeled proteins both in the absence and presence of one equivalent of Zn^{2+} (labeling scheme shown in Fig. 7A). Upon addition of zinc, a clear decrease in interdomain distances was observed, consistent with docking between the Zn^{2+}-occupied OR domain and a well-defined surface patch on the C-terminal domain, as shown in Fig. 7. Refined with energy minimization, DEER EPR distance restraints place the Zn^{2+}-occupied OR over a conserved cluster of negatively charged residues comprised of D177, E199, E206 and E210 (mouse sequence, Fig. 7C).

Interestingly, families who carry charge altering mutation at D177, E199, or E210 are predisposed to inherited prion disease. These findings led to the hypothesis that the Zn^{2+}-promoted interdomain interaction is protective against PrP^C-mediated neuronal toxicity. While the initial experiments described here were performed at X-band, it is now well recognized that Q-band (34 GHz) EPR is vastly more sensitive for nitroxide-nitroxide DEER EPR measurements.

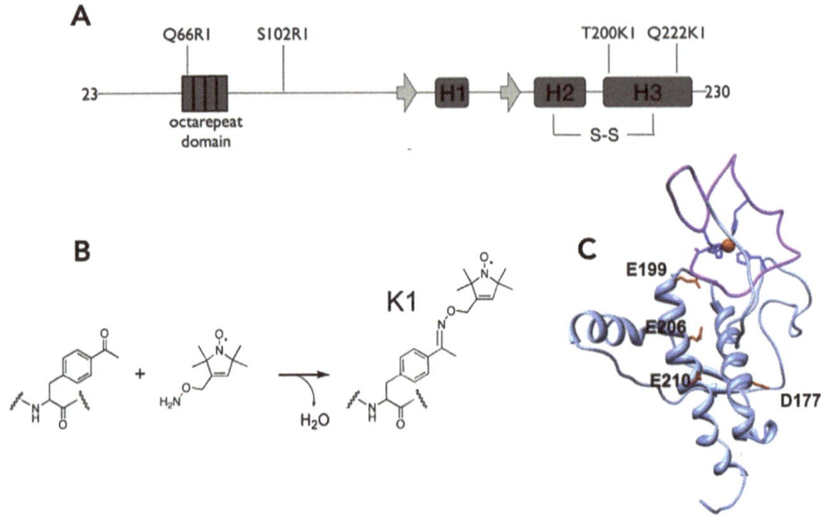

Fig. 7 Nitroxide labeling strategy and resulting PrP^C conformation resulting from Zn^{2+} coordination in the OR domain. (A) N-terminal labeling was achieved by introduction of a selectively placed Cys residue, followed by reaction with MTSL, resulting in the R1 side chain. (B) Labeling in the C-terminal domain involved incorporation of the unnatural amino acid para-acetyl phenylalanine, which was then reacted with hydroxylamine nitroxide reagent. (C) Energy minimization using distances derived from the nitroxide-nitroxide DEER EPR results identified a compact structure with the Zn^{2+}-OR forming an interdomain *cis* interaction with the regulatory C-terminal domain.

The physiological implications raised by the studies with Zn^{2+} motivated parallel investigations with copper, specifically probing whether Cu^{2+} would drive similar domain-domain contacts. The intrinsic paramagnetism of Cu^{2+} enables direct DEER EPR distance measurements between this ion and site-specifically attached spin labels. To keep both the DEER nitroxide pump and copper probe frequencies within the resonator bandwidth, this experiment is currently limited to X-band EPR. Detailed measurements from four PrP^C constructs, with K1 spin labels in the C-terminal domain, resulted in precise localization of the copper ion above the conserved, negatively charged patch described above (Evans et al., 2016).

Because DEER EPR does not resolve distances below ~15 Å, it may be important to consider the possibility that shorter Cu-nitroxide distances exist in the distributions derived from these spin-labeled PrP variants. To address this, a room temperature CW EPR on the same samples, along with careful measurements of nitroxide line intensity reduction that results from dipolar broadening, identifies conformations resulting from copper ion—nitroxide distances within 8–25 Å (Voss, Salwinski, Kaback, & Hubbell, 1995). When applied to the Cu^{2+}-PrP^C complex, the resulting distances supported the distributions derived from the copper-nitroxide DEER EPR measurements.

An exciting hypothesis from the studies above is that copper and zinc stabilize conformations of PrP^C that resist cellular toxicity, as implicated in neurodegenerative disease. Although beyond the scope of this EPR-centered chapter, it is noteworthy that subsequent electrophysiological experiments performed on cultured cells demonstrated that the metal ion-promoted a *cis* interaction that is indeed protective against PrP^C-caused membrane leakage and consequent cell death (Schilling et al., 2020; Wu et al., 2017).

6. Concluding remarks

The methods described in this chapter demonstrate how both basic and advanced EPR methods further the understanding and reveal critical features of partially ordered proteins, relevant to neurodegenerative diseases. Speciation and structural details of copper coordination sites in proteins can be obtained from EPR studies. Beyond those approaches described here, one may also apply pulsed Electron-Nuclear Double Resonance (ENDOR) to elucidate fine structural details of copper coordination sites. While specific metal binding sites in intrinsically disordered proteins may be successfully modeled by peptide fragments ("bottom-up" approach), correlation to studies of the full protein ("top-down" approach) provide important insights into the impact of metal binding in protein structure. What's truly exciting at this juncture is that EPR methods applied to PrP^C revealed unanticipated, metal ion-assisted conformations that regulate the protein's intrinsic toxicity. Given the remarkable spectrum of proteins involved in neurodegenerative processes, there is little doubt that EPR will uncover similar stories as they apply to other protein-driven dementias.

Acknowledgments

Authors would like to thank Yanahi Posadas and Carolina Sánchez-López for assistance with adapting Figs. 3 and 4. L.Q. thanks funding from Ministry of Education of Mexico (SEP-Cinvestav). G.L.M. gratefully acknowledges funding in support of this work from the National Institutes of Health (R35 GM131781).

References

Aronoff-Spencer, E., Burns, C. S., Avdievich, N. I., Gerfen, G. J., Peisach, J., Antholine, W. E., et al. (2000). Identification of the Cu^{2+} binding sites in the N-terminal domain of the prion protein by EPR and CD spectroscopy. *Biochemistry, 39*, 13760–13771.

Burns, C. S., Aronoff-Spencer, E., Dunham, C. M., Lario, P., Avdievich, N. I., Antholine, W. E., et al. (2002). Molecular features of the copper binding sites in the octarepeat domain of the prion protein. *Biochemistry, 41*, 3991–4001.

Burns, C. S., Aronoff-Spencer, E., Legname, G., Prusiner, S. B., Antholine, W. E., Gerfen, G. J., et al. (2003). Copper coordination in the full-length, recombinant prion protein. *Biochemistry, 42*(22), 6794–6803.

Bush, A. I. (2003). The metallobiology of Alzheimer's disease. *Trends in Neurosciences, 26*(4), 207–214.

Chattopadhyay, M., Walter, E. D., Newell, D. J., Jackson, P. J., Aronoff-Spencer, E., Peisach, J., et al. (2005). The Octarepeat domain of the prion protein binds cu(II) with three distinct coordination modes at pH 7.4. *Journal of the American Chemical Society, 127*(36), 12647–12656.

Chaves, J. A., Sanchez-Lopez, C., Gomes, M. P., Sisnande, T., Macedo, B., de Oliveira, V. E., et al. (2014). Biophysical and morphological studies on the dual interaction of non-octarepeat prion protein peptides with copper and nucleic acids. *Journal of Biological Inorganic Chemistry, 19*(6), 839–851.

Dudzik, C. G., Walter, E. D., Abrams, B. S., Jurica, M. S., & Millhauser, G. L. (2013). Coordination of copper to the membrane-bound form of alpha-synuclein. *Biochemistry, 52*(1), 53–60.

Evans, E. G. B., & Millhauser, G. L. (2017). Copper- and zinc-promoted Interdomain structure in the prion protein: A mechanism for autoinhibition of the neurotoxic N-terminus. *Progress in Molecular Biology and Translational Science, 150*, 35–56.

Evans, E. G., Pushie, M. J., Markham, K. A., Lee, H. W., & Millhauser, G. L. (2016). Interaction between prion Protein's copper-bound Octarepeat domain and a charged C-terminal pocket suggests a mechanism for N-terminal regulation. *Structure, 24*(7), 1057–1067.

Fleissner, M. R., Brustad, E. M., Kalai, T., Altenbach, C., Cascio, D., Peters, F. B., et al. (2009). Site-directed spin labeling of a genetically encoded unnatural amino acid. *Proceedings of the National Academy of Sciences of the United States of America, 106*(51), 21637–21642.

Giachin, G., Mai, P. T., Tran, T. H., Salzano, G., Benetti, F., Migliorati, V., et al. (2015). The non-octarepeat copper binding site of the prion protein is a key regulator of prion conversion. *Scientific Reports, 5*, 15253.

Grande-Aztatzi, R., Rivillas-Acevedo, L., Quintanar, L., & Vela, A. (2013). Structural models for cu(II) bound to the fragment 92–96 of the human prion protein. *The Journal of Physical Chemistry. B, 117*(3), 789–799.

Han, X., Jin, M., Breuker, K., & McLafferty, F. W. (2006). Extending top-down mass spectrometry to proteins with masses greater than 200 kilodaltons. *Science, 314*(5796), 109–112.

Hofer, P., Grupp, A., & Mehring, M. (1986). High-resolution time-domain electron-nuclear-sublevel spectroscopy by pulsed coherence transfer. *Phys Rev A Gen Phys, 33*(5), 3519–3522.

Hornemann, S., Schorn, C., & Wuthrich, K. (2004). NMR structure of the bovine prion protein isolated from healthy calf brains. *EMBO Reports, 5*(12), 1159–1164.

Hornshaw, M. P., McDermott, J. R., & Candy, J. M. (1995). Copper binding to the N-terminal tandem repeat regions of mammalian and avian prion protein. *Biochem. Biophys. Res. Comm., 207*, 621–629.

Jeschke, G., & Polyhach, Y. (2007). Distance measurements on spin-labelled biomacromolecules by pulsed electron paramagnetic resonance. *Physical Chemistry Chemical Physics*, *9*(16), 1895–1910.

Jobling, M. F., Huang, X., Stewart, L. R., Barnham, K. J., Curtain, C., Volitakis, I., et al. (2001). Copper and zinc binding modulates the aggregation and neurotoxic properties of the prion peptide PrP106-126. *Biochemistry*, *40*, 8073–8084.

Kevan, L., & Bowman, M. K. (1990). *Modern pulsed and continuous-wave electron spin resonance*. New York: Wiley.

McDonald, A. J., Leon, D. R., Markham, K. A., Wu, B., Heckendorf, C. F., Schilling, K., et al. (2019). Altered domain structure of the prion protein caused by cu(2+) binding and functionally relevant mutations: Analysis by cross-linking, MS/MS, and NMR. *Structure*, *27*(6), 907–922 e905.

Millhauser, G. L. (2004). Copper binding in the prion protein. *Accounts of Chemical Research*, *37*(2), 79–85.

Millhauser, G. L. (2007). Copper and the prion protein: Methods, structures, function, and disease. *Annual Review of Physical Chemistry*, *58*, 299–320.

Mims, W. B., & Peisach, J. (1978). The nuclear modulation effect in electron spin echoes for complexes of Cu^{2+} and imidazole with ^{14}N and ^{15}N. *The Journal of Chemical Physics*, *69*, 4921–4930.

Peisach, J., & Blumberg, W. E. (1974). Structural implications derived from the analysis of electron paramagnetic resonance spectra of natural and artificial copper proteins. *Archives of Biochemistry and Biophysics*, *165*, 691–708.

Posadas, Y., Lopez-Guerrero, V., Segovia, J., Perez-Cruz, C., & Quintanar, L. (2022). Dissecting the copper bioinorganic chemistry of the functional and pathological roles of the prion protein: Relevance in Alzheimer's disease and cancer. *Current Opinion in Chemical Biology*, *66*, 102098.

Proukakis, C., Dudzik, C. G., Brier, T., MacKay, D. S., Cooper, J. M., Millhauser, G. L., et al. (2013). A novel alpha-synuclein missense mutation in Parkinson disease. *Neurology*, *80*(11), 1062–1064.

Prusiner, S. B. (1982). Novel proteinaceous infectious particles cause scrapie. *Science*, *216*, 136–144.

Prusiner, S. B. (1997). Prion diseases and the BSE crisis. *Science*, *278*, 245–251.

Qin, K., Yang, Y., Mastrangelo, P., & Westaway, D. (2002). Mapping cu(II) bindings sites in prion proteins by diethylpyrocarbonate modification and MALDI-TOF mass spectrometric "footprinting". *The Journal of Biological Chemistry*, *277*, 1981–1990.

Riek, R., Hornemann, S., Wider, G., Glockshuber, R., & Wuthrich, K. (1997). NMR characterization of the full-length recombinant murine prion protein, mPrP(23 − 231). *FEBS Letters*, *413*, 282–288.

Rivillas-Acevedo, L., Grande-Aztatzi, R., Lomeli, I., Garcia, J. E., Barrios, E., Teloxa, S., et al. (2011). Spectroscopic and electronic structure studies of copper(II) binding to His111 in the human prion protein fragment 106-115: Evaluating the role of protons and methionine residues. *Inorganic Chemistry*, *50*(5), 1956–1972.

Sahu, I. D., McCarrick, R. M., & Lorigan, G. A. (2013). Use of electron paramagnetic resonance to solve biochemical problems. *Biochemistry*, *52*(35), 5967–5984.

Salzano, G., Giachin, G., & Legname, G. (2019). Structural consequences of copper binding to the prion protein. *Cell*, *8*(8).

Sanchez-Lopez, C., Fernandez, C. O., & Quintanar, L. (2018). Neuroprotective alpha-cleavage of the human prion protein significantly impacts Cu(II) coordination at its His111 site. *Dalton Transactions*, *47*(28), 9274–9282.

Schiemann, O., Heubach, C. A., Abdullin, D., Ackermann, K., Azarkh, M., Bagryanskaya, E. G., et al. (2021). Benchmark test and guidelines for DEER/PELDOR experiments on nitroxide-labeled biomolecules. *Journal of the American Chemical Society*, *143*(43), 17875–17890.

Schilling, K. M., Tao, L., Wu, B., Kiblen, J. T. M., Ubilla-Rodriguez, N. C., Pushie, M. J., et al. (2020). Both N-terminal and C-terminal histidine residues of the prion protein are essential for copper coordination and neuroprotective self-regulation. *Journal of Molecular Biology*.

Selkoe, D. J., & Hardy, J. (2016). The amyloid hypothesis of Alzheimer's disease at 25 years. *EMBO Molecular Medicine*, *8*(6), 595–608.

Spevacek, A. R., Evans, E. G., Miller, J. L., Meyer, H. C., Pelton, J. G., & Millhauser, G. L. (2013). Zinc drives a tertiary fold in the prion protein with familial disease mutation sites at the interface. *Structure*, *21*(2), 236–246.

Spillantini, M. G., Schmidt, M. L., Lee, V. M., Trojanowski, J. Q., Jakes, R., & Goedert, M. (1997). Alpha-synuclein in Lewy bodies. *Nature*, *388*(6645), 839–840.

Stahl, N., Borchelt, D. R., Hsiao, K., & Prusiner, S. B. (1987). Scrapie prion protein contains a phosphatidylinositol glycolipid. *Cell*, *51*(2), 229–240.

Voss, J., Salwinski, L., Kaback, H. R., & Hubbell, W. L. (1995). A method for distance determination in proteins using a designed metal ion binding site and site-directed spin labeling: Evaluation with T4 lysozyme. *Proceedings of the National Academy of Sciences of the United States of America*, *92*(26), 12295–12299.

Walter, E. D., Stevens, D. J., Visconte, M. P., & Millhauser, G. L. (2007). The prion protein is a combined zinc and copper binding protein: Zn2+ alters the distribution of Cu2+ coordination modes. *Journal of the American Chemical Society*, *129*(50), 15440–15441.

Wu, B., McDonald, A. J., Markham, K., Rich, C. B., McHugh, K. P., Tatzelt, J., et al. (2017). The N-terminus of the prion protein is a toxic effector regulated by the C-terminus. *eLife*, *6*.

Yu, Q., Kandegedara, A., Xu, Y., & Rorabacher, D. B. (1997). Avoiding interferences from Good's buffers: A contiguous series of noncomplexing tertiary amine buffers covering the entire range of pH 3-11. *Analytical Biochemistry*, *253*(1), 50–56.

Zomosa-Signoret, V., Arnaud, J. D., Fontes, P., Alvarez-Martinez, M. T., & Liautard, J. P. (2008). Physiological role of the cellular prion protein. *Veterinary Research*, *39*(4), 9.

CHAPTER TEN

Paramagnetic resonance investigation of mono- and di-manganese-containing systems in biochemistry

Joshua Telser*
Department of Biological, Physical and Health Sciences, College of Science, Health and Pharmacy, Roosevelt University, Chicago, IL, United States
*Corresponding author: e-mail address: jtelser@roosevelt.edu

Contents

Abstract

Although not as important as iron, manganese is one of the key transition metal (d block) elements in biochemistry. The role of Mn in photosynthesis is perhaps its "starring role," but has been extensively covered elsewhere and will not be discussed in this chapter. In contrast to iron, for which ^{57}Fe Mössbauer effect (nuclear gamma resonance) spectroscopy is available, ^{55}Mn (100% abundant) lacks this ability. Instead, the defining technique for Mn is electron paramagnetic resonance (EPR) spectroscopy, as Mn(II), Mn(IV), and even Mn(III), as well as certain di-Mn systems are generally EPR active. This chapter describes how EPR and its advanced derivatives, electron spin echo envelope modulation (ESEEM) and electron–nuclear/electron–electron double resonance (ENDOR/ELDOR) spectroscopies, provide valuable biochemical information on many Mn-containing systems. These include endogenous Mn in mono- and di-Mn (protein) enzymes as well as brief mention of applications to Mn in non-protein biological systems.

Methods in Enzymology, Volume 666
ISSN 0076-6879
https://doi.org/10.1016/bs.mie.2022.02.005

1. Introduction

Freisinger and Sigel, organizers of the 19th International Conference on Biological Inorganic Chemistry (ICBIC-19), Interlaken, Switzerland, (19th International Conference on Biological Inorganic Chemistry (ICBIC-19), 2019) toward the end of the "Before Time" (11–16 August 2019), describe "The Bioinorganic Periodic Table" (Freisinger & Sigel, 2019). This shows that most of the first row transition metal elements, i.e., those with $3d^1$–$3d^{10}$ ground state elemental electronic configurations have a biochemical role, if not necessarily in humans (mammals), then in other organisms (e.g., vanadium (Group 5) in marine organisms (Crans, Smee, Gaidamauskas, & Yang, 2004) and in a nitrogenase from *Azotobacter* species (Hu, Lee, & Ribbe, 2012)). Exceptions are scandium (Group 3, $3d^1$) and titanium (Group 4, $3d^2$), but Ti is an exception only in the sense that there are no Ti-containing enzymes; this elements has biological roles as reviewed in detail by Buettner and Valentine (Buettner & Valentine, 2012). Even though Sc has not been identified as having a biological role (and is a rather scarce element), its close relatives in the f block (i.e., "the inner transition," or lanthanoid elements, which are "honorary" Group 3 elements) are found at the active site of metalloenzymes (Cotruvo, 2019). The situation with chromium (Group 6, "$3d^4$" since its ground state configuration is actually $4s^1 3d^5$) is ambiguous (Vincent, 2021). This bring us finally to manganese (Group 7, $3d^5$), the subject of this chapter. The d block elements following Mn (Fe, Co, Ni, Cu, and Zn) all have major biological roles (Fiedler & Fischer, 2017), with iron being the "800-pound gorilla" of d block biochemistry, in the many enzymes with Fe_2 active sites (Fox, 2021; Rajakovich et al., 2020). Thus, manganese is the "earliest" transition metal to have an unequivocally important role in biochemistry (Smith, Hadler, Schenk, Hanson, & Mitić, 2010).

What then are these important biochemical roles of manganese? The "starring role" of Mn is not as a solo actor, but in a cubane-like cluster, (Koua, Umena, Kawakami, & Shen, 2013; Umena, Kawakami, Shen, & Kamiya, 2011) Mn_3CaO_4MnO, wherein the fourth Mn (Mn4) is separately indicated as it is a "dangler" ion (Mn1, 2, 3; O1, 2, 3, 5 and Ca comprise the cubane, and the separate oxygen is O4) as shown in Fig. 1.

This unusual cluster is the active site of the oxygen evolving complex (OEC) of Photosystem II (PS II) (Debus, 2021) and is thus a candidate for "most important enzyme for life on Earth," although nitrogenase has

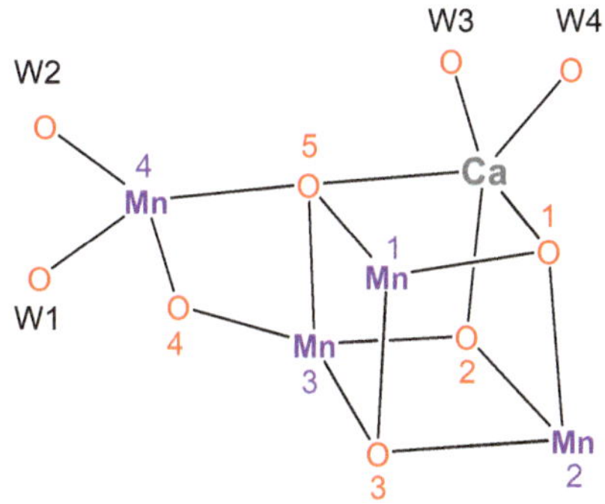

Fig. 1 Cartoon of the Mn-containing active site, Mn_4CaO_5 cluster, of the oxygen evolving complex (OEC) of Photosystem II (PS II), based on the actual crystallographic structures (PDB: 3WU2) reported by Umena et al. (Umena et al., 2011) with their atomic numbering scheme. The hydronation state of the water molecules W1–4) is not indicated. The strontium-substituted cluster (i.e., Sr for Ca; PDB: 4IL6) is structurally similar (Koua et al., 2013).

its proponents for that status (Hoffman, Dean, & Seefeldt, 2009). Perhaps, the OEC is the most important for aerobic life and nitrogenase for anaerobic. As befitting its importance, the OEC and PS II in general have been extensively reviewed (Barber, 2016; Brynda & Britt, 2010; Cox, Pantazis, Neese, & Lubitz, 2013) and thus will not be discussed further in this chapter, except to note that the spectroscopic technique of focus here, electron paramagnetic resonance (EPR), continues to be applied to this system, in both conventional (Debus, 2021) and advanced (Lohmiller et al., 2017; Marchiori, Debus, & Britt, 2020) modalities along with X-ray methods (Suga et al., 2019).

There are, however, many other important biological roles for manganese. These include metalloenzymes with a single (i.e., mononuclear) Mn active site (Sheng et al., 2014), which have inspired much model chemistry, (Guo, Corona, Ray, & Nam, 2019; Kovacs, 2015; Rice, Massie, & Jackson, 2017; Sahu & Goldberg, 2016). Representative structures are shown in Fig. 2: the Mn superoxide dismutase (MnSOD) from *Homo sapiens* (Hearn et al., 2003) and the Mn homoprotocatechuate 2,3-dioxygenase (MnMndD) from *Arthrobacter globiformis* (Vetting, Wackett, Que Jr, Lipscomb, & Ohlendorf, 2004).

The biological requirement for Mn by many organisms means that there needs to be a mechanism for Mn transport, including MntR (Golynskiy, Gunderson, Hendrich, & Cohen, 2006), MntC (Gagnon, Hadley, Ozarowski, Nolan, & Britt, 2019), and MntE (Martin & Giedroc, 2016).

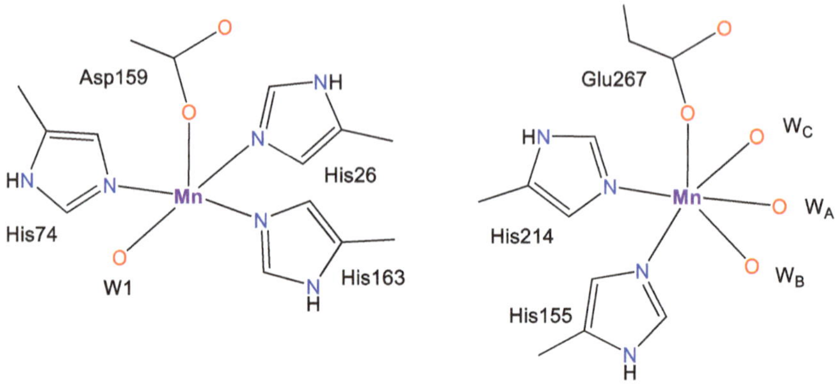

Fig. 2 Left: cartoon of the mono-Mn containing active site of manganese superoxide dismutase (MnSOD) from *Homo sapiens*, based on the actual crystallographic structures (PDB: 1LUV) reported by Hearn et al. (Hearn et al., 2003). The crystal structure of oxalate oxidase (OxOx) from *Hordeum vulgare* is very similar (PDB: 1FI2): Mn coordination by His88, His90, His137, Glu95, and a water (Woo, Dunwell, Goodenough, Marvier, & Pickersgill, 2000), as is the structure of oxalate decarboxylase (OxDC) from *Bacillus subtilis* (PDB: 1UW8): Mn coordination by His273, His275, His319, Glu280, and a water (Just et al., 2004). Right: cartoon of the mono-Mn containing active site of manganese homoprotocatechuate 2,3-dioxygenase (MnMndD) from *Arthrobacter globiformis*, based on the actual crystallographic structures (PDB: 1F1R and 1F1U; 1F1V has bound (3,4-dihydroxyphenyl)acetate) reported by Vetting et al. (Vetting et al., 2004). The hydronation states of the terminal water molecules (left, W1; right, W_A, W_B, W_C) are not indicated. The many non-heme Fe enzymes have analogous structures to these Mn enzymes.

Manganese transport and its role in virulence by microbial pathogens have been reviewed (Brunjes Brophy & Nolan, 2015; Juttukonda & Skaar, 2015; Papp-Wallace & Maguire, 2006), as has been the intimately related topic of bacterial manganese sensing and homeostasis (Waters, 2020). In the endless "sword vs shield" struggle of pathogens and hosts, there is a mammalian protein, calprotectin, that binds Mn(II) (via hexahistidine coordination) to deny pathogens their needed Mn (Brunjes Brophy, Nakashige, Gaillard, & Nolan, 2013; Hadley et al., 2018; Hadley, Gagnon, Ozarowski, Britt, & Nolan, 2019).

There are also enzymes with Mn_2 (i.e., dinuclear) sites (Christianson, 2005; Dismukes, 1996; Wu, Penner-Hahn, & Pecoraro, 2004), which have also inspired extensive model chemistry (Signorella, Palopoli, & Ledesma, 2018). Two representative structures, based on actual crystal structures (Barynin et al., 2001; Kanyo, Scolnick, Ash, & Christianson, 1996), are shown in Fig. 3. This motif is analogous to that found in di-Fe enzymes

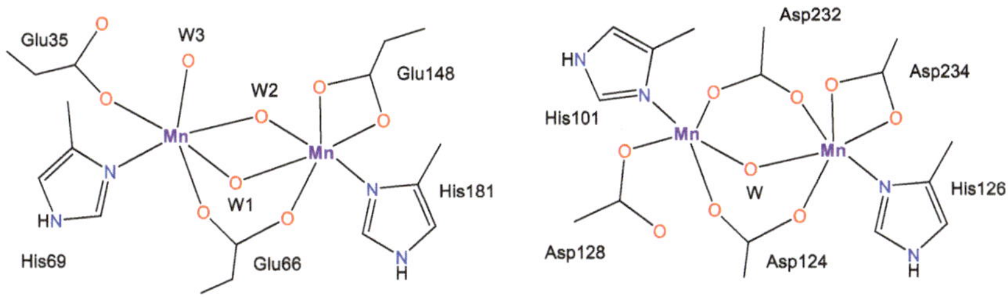

Fig. 3 Left: cartoon of the di-Mn containing active site of manganese catalase (MnCat) from *Lactobacillus plantarum*, based on the actual crystallographic structure (PDB: 1JKU) reported by Barynin et al. (Barynin et al., 2001). The many di-Fe and comparatively fewer MnFe enzymes have analogous structures, but with some differences as shown, for example, below in Fig. 29. Right: cartoon of the di-Mn containing active site of arginase I from *Rattus norvegicus*, based on the actual crystallographic structure (PDB: 1RLA) reported by Kanyo et al. (Kanyo et al., 1996). The hydronation states of the bridging (W1, W2; W) and terminal (W3) water molecules are not indicated in either structure.

(Fox, 2021; Rajakovich et al., 2020). There are also mixed-metal centers that contain MnFe active sites (Högbom, 2010; Kisgeropoulos et al., 2020; Kositzki et al., 2016; Roth & Plass, 2008).

This chapter is not intended to be a comprehensive listing of every naturally Mn-containing enzyme (protein), but will identify key examples in which EPR spectroscopy and related, advanced paramagnetic resonance techniques such as electron spin echo envelope modulation (ESEEM) and electron nuclear double resonance (ENDOR) spectroscopies are used to enhance understanding of the structure and mechanism of these enzymes. In addition, the role of EPR in elucidating the function of small molecule complexes of Mn in bacterial resistance to ionizing radiation will also be discussed (Gaidamakova et al., 2022; McNaughton et al., 2010; Sharma et al., 2017, 2013). These Mn systems are not enzymes, but are relevant here because they supplement or even replace the Mn superoxide dismutase (MnSOD) enzyme that has this presumed physiological function. Because these small molecules are predominantly Mn-nucleotide complexes (mono-, di-, and triphosphates), which can of course also be studied ex vivo (Potapov & Goldfarb, 2006; Un & Bruch, 2015), these whole cell studies are also relevant to investigation of metal-nucleic acid interactions (DeRose & Yglesias, 2021), whether in isolated DNA-binding proteins, ribozymes (Frederiksen & Piccirilli, 2009), or other systems that involve DNA or RNA and metal ions. Such systems do not normally employ Mn, but Mn(II) ion can be readily substituted for native Mg(II) ion and

EPR/ESEEM/ENDOR can then be fruitfully employed as has been shown in ribozymes and other systems (Hoogstraten, Grant, Horton, DeRose, & Britt, 2002; Horton, Clardy, & DeRose, 1998; Hunsicker-Wang, Vogt, & DeRose, 2009; Krstić, Endeward, Margraf, Marko, & Prisner, 2012; Morrissey et al., 1999; Morrissey, Horton, & DeRose, 2000; Stich et al., 2007; Vogt, Lahiri, Hoogstraten, Britt, & DeRose, 2006). The use of EPR and related techniques to study metal-ion binding in RNA has already been expertly reviewed in this series (Hunsicker-Wang et al., 2009), so this topic will not be covered here. Nor will the use of EPR/ESEEM/ENDOR to study analogous metal-DNA/(oligo)nucleotide interactions (Tsednee et al., 2019), or nucleotide-binding enzymes/proteins, such nitrogenase (Petersen et al., 2005), pyruvate kinase (Tan, Poyner, Reed, & Scholes, 1993; Tipton, McCracken, Cornelius, & Peisach, 1989) and *ras* p21, (Bennati et al., 2006; Halkides, Farrar, Larsen, Redfield, & Singel, 1994) be covered in this chapter. Related enzymes (i.e., by virtue of phosphate-containing substrates) such as enolase that employ Mg(II) can be studied in Mn(II)-substituted forms by EPR, but will not be covered here (Carmieli et al., 2007). EPR and related techniques have also been used to study the role of Mn in such diverse areas as biogeology (Lingappa et al., 2021; Tao et al., 2018) and amphibian fertilization (Seeler et al., 2021) will not be further mentioned. Another more traditionally biochemical area that involves Mn, in this case saccharides rather than nucleotides, is glycosyltransferases (Williams & Thorson, 2008). These will not be covered in this chapter and it appears that EPR has not been widely applied to their study, only theory (Sladek & Tvaroška, 2017).

2. Overview of EPR and related techniques

2.1 Overview of EPR in the context of Mn ions in biological environments

EPR spectroscopy has been widely applied to paramagnetic transition metal (d block) ions since the 1950s, and to metal ions in biological systems since the 1960s. A large number of classic references is available (Carrington & McLachlan, 1967; Pake & Estle, 1973), with some of them being reissued in recent years (Abragam & Bleaney, 2012), or even with updated editions (Atherton, 1993; Weil & Bolton, 2007). For pulse EPR, which was developed after above classics were originally written, the text by Schweiger and Jeschke is the definitive one (Schweiger & Jeschke, 2001). More recently, a thorough EPR text is available both as a book (Goldfarb & Stoll, 2018), and electronically, by chapter, through the journal *eMagRes* (eMagRes, 2007). There is also a text by Hagen specifically on "Biomolecular EPR Spectroscopy," (Hagen, 2008) which, in addition to theory, discusses sample preparation and many other practical aspects of EPR as related to biomolecules that are not covered in more physics-oriented texts. The present series has a recent chapter on "Pulse Double-Resonance EPR Techniques for the Study of Metallobiomolecules," (Cox, Nalepa, Pandelia, Lubitz, & Savitsky, 2015) which is the ideal one for "one-stop shopping".

Thanks to this background, only a few points will be made here, which are specifically relevant to the metalloenzyme systems of interest. EPR samples can be in many forms, such as whole cells or membrane protein/detergent mixtures, but the focus here is on isolated, purified metalloenzymes in aqueous buffer solutions. An optical quality glass is not required, but it is generally preferable to have a glassing agent present, typically ethylene glycol or glycerol. Optical quality glasses generally employ 1:1 *v*/v (i.e., 50%) glassing agent (ethylene glycol or glycerol), but much lower volume ratios (20%, or even 10% by volume) are sufficient so that a non-crystalline frozen solution obtains. The glassy frozen solution provides a truly random distribution of solute molecules (single-crystal EPR, even of metalloenzymes (Gurbiel, Bolin, Ronco, Mortenson, & Hoffman, 1991), is possible, but will not be discussed here). Equally important, disruption of the crystalline ice matrix by the glassing agent prevents "spin diffusion," i.e., fast spin-spin relaxation through the ordered matrix. Absent a glassing agent, relaxation is generally too fast to allow successful application of the pulse EPR techniques to be discussed below, although a continuous wave (CW) spectrum can generally be obtained. A caveat to this requirement of a glassing agent is that high-field and -frequency EPR (HFEPR), at least as practiced at the National High Magnetic Field Laboratory (NHMFL) in Tallahassee, Florida (USA), is relatively insensitive to sample configuration. Indeed, most of the studies performed have been on powdered samples of paramagnetic small molecules in undiluted form (Krzystek, Ozarowski, & Telser, 2006; Krzystek & Telser, 2016; Telser, Krzystek, & Ozarowski, 2014). As will be discussed later, there are several studies by HFEPR of Mn-containing enzymes and a goal of this chapter is to promote further applications of HFEPR to Mn-enzymes in particular and to metalloenzymes in general.

The concentration requirements of Mn-containing EPR samples is highly variable. Adventitious Mn(II) is often present in buffer solutions originating from metal salts or mineral acids used in their preparation.

EPR, particularly at Q-band (35 GHz) and W-band (95 GHz) frequencies is very sensitive to Mn(II), more so than X-band (~9.5 GHz) is. This is because at higher microwave frequencies (and higher corresponding resonant magnetic fields) forbidden transitions (i.e., those that are not the allowed EPR transitions: $\Delta m_S = \pm 1$, $\Delta m_I = 0$; see Fig. 4) are even more "forbidden". This in turn is because at higher fields/frequencies, the m_S and m_I levels are well separated and are not mixed. It is the mixing of m_S and of m_I levels that makes these "not good quantum numbers" so that the EPR (and NMR: $\Delta m_S = 0$, $\Delta m_I = \pm 1$) selection rules do not apply as strongly. The spin levels for Mn(II) are shown in Fig. 4, which is based on one given in this series (Hunsicker-Wang et al., 2009). Mixing in of electronic excited states (i.e., those with $S = 3/2$ and $1/2$) into the $S = 5/2$ ground state lead to zero-field splitting (ZFS) (Telser, 2017). In the case of Mn(II), ZFS is generally small ($|D| < 1\ \text{cm}^{-1} = 30$ GHz), and has been thoroughly analyzed by experiment and theory (Duboc, Collomb, & Neese, 2010; Duboc, Collomb, Pécaut, Deronzier, & Neese, 2008; Zein, Duboc, Lubitz, & Neese, 2008). Nevertheless, the presence of ZFS in Mn(II) has profound influence on its EPR spectra as will be discussed below.

An example of such adventitious Mn(II) is shown in Fig. 5, which presents the 35 GHz EPR spectrum of CooA (Clark et al., 2004), which is the first CO-sensing heme protein shown to be physiologically important and was originally found in the purple photosynthetic bacterium *Rhodospirillum rubrum*. This spectrum is relevant here because it is representative of metalloprotein EPR wherein the system of interest is dominant and has broad g value dispersion, as is typical for such low spin d^5 systems (McGarvey, 1998), but the sextet from trace $^{55}Mn^{2+}$ ($I = 5/2$, 100% abundance) is apparent, particularly in first derivative presentation, which is the standard form for EPR (Fig. 5, top). The experimental spectrum is in absorption mode (Fig. 5, bottom), which is the consequence of passage (i.e., relaxation related) effects (Mailer & Taylor, 1973). As will be shown below, absorption mode is extremely helpful for analyzing EPR spectra from Mn(II) species of direct interest. The oxidized CooA (i.e., low-spin Fe(III) heme) concentration is ~180 μM. Spin quantitation (by integration and comparison with a Mn(II) standard) is pointless in such a case, but the Mn(II) concentration is likely in the tens of nM range. Systems of interest that contain Mn(II) should be in the tens of mM range for successful EPR investigation, although higher concentrations (hundreds of mM) is helpful.

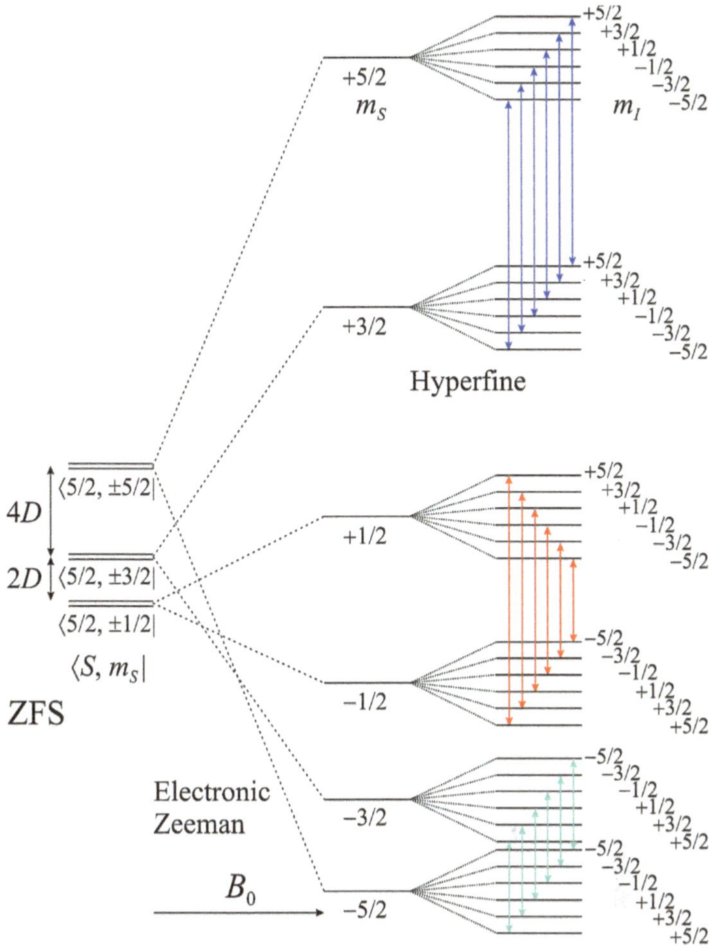

Fig. 4 Energy level diagram for $^{55}Mn^{2+}$ ion ($S=5/2$, $I=5/2$) showing (left to right) the effects of zero-field splitting (ZFS, with $D>0$; the $\langle S,m_S|$ ordering would be reversed with $D<0$; the complication of rhombic ZFS ($E\neq0$) is not shown), the electronic Zeeman interaction (B_0 is the external magnetic field), and hyperfine coupling between the electronic and nuclear spins (with $A>0$; the signs of the m_I values would be reserved with $A<0$). The nuclear Zeeman interaction is relatively weak and is not shown, nor is the nuclear quadrupole coupling interaction. The six red lines indicate EPR-allowed ($\Delta m_S=\pm1$, $\Delta m_I=0$) transitions within the $\langle S,\ m_S|=\langle 5/2,\ \pm1/2|$ spin manifolds, here $\langle S,m_S|=|5/2,-1/2\rangle\leftrightarrow|5/2,+1/2\rangle$. These correspond to the central (inner lines) sextet EPR pattern seen for typical Mn(II) species as evidenced in subsequent figures in this chapter. The six blue and six cyan lines indicate EPR-allowed transitions for two of four sets of "outer lines" that are often broadened out in typical Mn(II) EPR spectra. Those shown in blue correspond to $\langle S,m_S|=|5/2,+3/2\rangle\rightarrow|5/2,+5/2\rangle$, which would be relatively unpopulated except at higher temperatures; those shown in cyan correspond to $\langle S,m_S|=|5/2,-5/2\rangle\rightarrow|5/2,-3/2\rangle$, which would be the most populated at lower temperatures. Not shown are the transitions corresponding to $\langle S,m_S|=|5/2,-3/2\rangle\rightarrow|5/2,-1/2\rangle$ and $\langle S,m_S|=|5/2,+1/2\rangle\rightarrow|5/2,+3/2\rangle$. *This figure is based on Fig. 16.1 in Hunsicker-Wang, L., Vogt, M., & DeRose, V. J. (2009). Chapter 16 - EPR methods to study specific metal-ion binding sites in RNA. In Methods in Enzymology (Vol. 468, pp. 335–367): Academic Press, and is adapted with permission from it. Copyright 2009 Elsevier.*

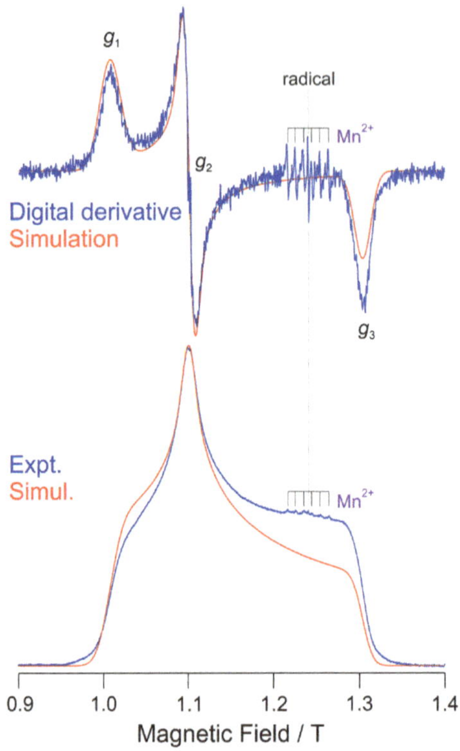

Fig. 5 Q-band EPR spectrum (blue trace, bottom) of oxidized (low-spin Fe(III) heme) CooA showing trace presence of adventitious Mn(II) (easily identified by the characteristic sextet pattern, along with an even smaller amount of organic radial of unknown provenance). Sample conditions: heme protein concentration, ~180 μM in 50 mM phosphate buffer, pH 7.4, with 500 mM NaCl (a possible source of trace Mn(II) ions), with 20% (*v*/v) glycerol. Experimental conditions: temperature, 2 K; microwave frequency, 34.908 GHz; microwave power, 2 mW; 100 kHz field modulation amplitude, 0.2 mT; the sample consisted of a fused silica tube (2 mm i.d.) with sample volume ~60 μL. Simulation (red traces) parameters: $S = 1/2$, $\mathbf{g} = [2.475, 2.265, 1.912]$, which are indicated on the figure; single-crystal Gaussian linewidths = 500, 220, 300 MHz (half-width at half-maximum). The intensity match between simulation and experiment at higher field is imperfect due to not having an ideal "powder pattern." This in turn is because the passage conditions are the same across the EPR envelope for such an anisotropic (in terms of both *g* values and relaxation properties) spin system. The upper blue trace is a digital derivative of the experimental spectrum to provide the standard EPR presentation mode. In this presentation, the amount of Mn(II) (and radical) is overemphasized because the nature of a first derivative emphasizes local maxima. Oxidized CooA sample provided by Mr. Brian R. Weaver and Prof. Judith N. Burstyn, University of Wisconsin-Madison.

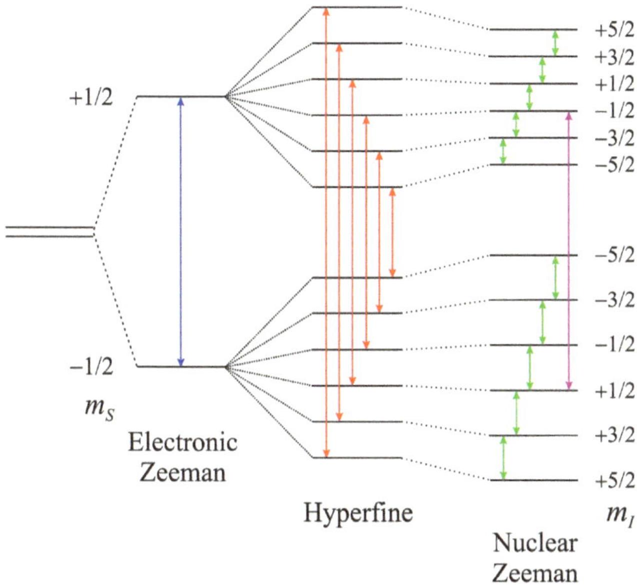

Fig. 6 Energy level diagram for $^{55}Mn^{2+}$ ion ($S=5/2$, $I=5/2$) but showing only the $m_S=\pm 1/2$ spin manifold with (left to right) the electronic Zeeman interaction, hyperfine coupling (with $A>0$), and the nuclear Zeeman interaction (for ^{55}Mn, which has $g_N>0$). Nuclear quadrupole coupling is not included. The blue line corresponds to a simple EPR transition (i.e., a single line at $g=2$), the red lines correspond to hyperfine split allowed EPR transitions (as in Fig. 4, giving the sextet seen for Mn(II); $\Delta m_S=\pm 1$, $\Delta m_I=0$), and the green lines correspond to ENDOR transitions (i.e., NMR allowed: $\Delta m_S=0$, $\Delta m_I=\pm 1$), those within the $m_S=-1/2$ (lower) manifold are at higher RF and are denoted ν_+; those within the $m_S=+1/2$ (upper) manifold are at higher RF and are denoted ν_-. The magenta line shows a single (among many possible), nominally forbidden transition (i.e., $\Delta m_S=\pm 1$, $\Delta m_I=\pm 1$) that can be observed by EDNMR (ELDOR-detected NMR (Cox et al., 2015; Cox, Nalepa, Lubitz, & Savitsky, 2017)).

Given that the central sextet (Fig. 4) is the most apparent feature in a typical EPR spectrum of a Mn(II) species, Fig. 6 presents the relevant energy levels in an expanded view. The $m_S=\pm 5/2$ and $\pm 3/2$ spin manifolds are off-scale, but the effect of the nuclear Zeeman interaction is included, which has a major role in ENDOR spectral positions. Nuclear quadrupole splitting is still ignored.

Mn(II) is by far the most important oxidation state of this element, both in coordination chemistry and in biological applications. But this oxidation state is not found at all in the OEC, which comprises Mn(III) and Mn(IV) (Petrie, Stranger, & Pace, 2017), nor is it the only one in other systems such as manganese superoxide dismutase (MnSOD), which converts between

Mn(II) and Mn(III) (Jackson, Gutman, Maliekal, Miller, & Brunold, 2013). Mn(III) ($3d^4$, typically, $S=2$, but $S=1$ is also possible) is traditionally thought of as "EPR-silent," due to the effects of ZFS. This "silence" can indeed be the case particularly for higher symmetry (cubic: $x=y=z$, or axial: $x=y\neq z$) systems, when conventional (i.e., X- or Q-band) microwave frequencies and corresponding magnetic fields are used. However, when high frequencies and fields are used (W-band (95 GHz) and above (up to ~1 THz), then a wealth of EPR transitions are possible (Krzystek et al., 2006; Krzystek & Telser, 2016; Telser et al., 2014). Unfortunately, thus far HFEPR has seen little application in bona fide metalloprotein systems. A notable exception is the study by Horitani et al. on *Physeter macrocephalus* myoglobin (Mb) reconstituted with manganese(*III*)protoporphyrin IX (Horitani, Yashiro, Hagiwara, & Hori, 2008). Frequencies from 33 GHz up to 122 GHz and swept fields of zero to 14 T allowed accurate determination of the ZFS parameters of this $S=2$ system: $D=-3.79\pm0.03\,\text{cm}^{-1}$ and $E=0.08\pm0.01\,\text{cm}^{-1}$ (with $g=2.00$). At 122 GHz (i.e., "true" HFEPR), the sample volume in a resonant cavity structure was only 0.3 μL, but the Mb concentration was ~10 mM. Table 1 in Horitani et al. provides a very helpful summary of the sample physical parameters (size, volume) used in their study (Horitani et al., 2008). The HFEPR spectrometer at NHMFL does not use a resonant structure, so frozen solution sample volumes are larger (~100 mL), but the sensitivity is much lower than for a resonant cavity structure. Regardless of sample volume, metalloenzyme concentrations of 10 mM are unlikely to be achievable, thus limiting the application of HFEPR to such integer-spin (non-Kramers) systems. Mn(III), however, in rhombic symmetry ($x\neq y\neq z$), the situation that typically obtains for the low symmetry active sites in metalloenzymes, can readily exhibit EPR spectra at X- (or Q-) band frequencies. This is explained in Fig. 7 using a diagram similar to that in

Table 1 Overview of ENDOR-relevant nuclei.

	Non-quadrupolar ($I=1/2$)	**Quadrupolar ($I\geq1$)**
Small hyperfine coupling relative to Larmor frequency (Larmor centered)	^{1}H, ^{13}C, ^{15}N, ^{31}P	^{2}H
Comparable hyperfine coupling and Larmor frequency (hyperfine centered)	^{13}C (in some biorganometallic cases), ^{57}Fe	^{14}N, ^{17}O, ^{55}Mn

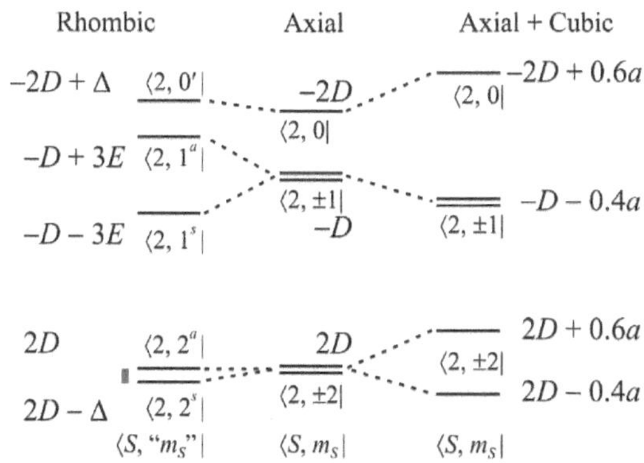

Fig. 7 Energy level diagram for $S=2$ (spin quintet) system with zero applied magnetic field and no hyperfine splitting (i.e., m_I levels not indicated, unlike in Fig. 4). Also in contrast to Fig. 4 (which is for $S=5/2$), the state ordering is for $D<0$, as is often the case for Mn(III) systems. The central column gives the energy levels for an axial (only) system: $D<0$, $E=0$, $a=0$. The energies are at $2D$, $-D$, and $-2D$ and the spin quantum number description of each is also given. The right column gives the energy levels for a system with both axial and cubic ZFS: $D<0$, $E=0$, $a>0$. Typically, the magnitude of the cubic ZFS (a) is very small, so it is exaggerated in this diagram for clarity, but it does lead to splitting within the ground state doublet. The left column is the most relevant to biological systems and gives the energy levels for a system with both axial and rhombic ZFS: $D<0$, $E>0$, $a=0$. The spin quantum number description of these states is complicated in that the $m_S=0$ state now has some $m_S=\pm 2$ mixed in (hence the nomenclature $\langle 0'|$ and the use of quotation marks on m_S) and correspondingly, one $m_S=\pm 2$ state now has some $m_S=0$ mixed in ($\langle 2^s|$), while one remains pure $m_S=\pm 2$ ($\langle 2^a|$). These labels arise from the wavefunctions' symmetry (ignoring the contribution from $\langle 0|$ to $\langle 2^s|$: $\langle 2^s| \approx \frac{1}{\sqrt{2}}[\langle +2|+\langle -2|]$, $\langle 2^a| = \frac{1}{\sqrt{2}}[\langle +2|-\langle -2|]$. The splitting between these two lowest energy levels can be in the X-band microwave region so that a low field signal is observed, as indicated by the small, vertical red line. Cubic ZFS may contribute to the total splitting within these $m_S=$"± 2" levels. The splitting between the two $m_S=\pm 1$ levels (similarly named: $\langle 1^s| = \frac{1}{\sqrt{2}}[\langle +1|+\langle -1|]$, $\langle 1^a| = \frac{1}{\sqrt{2}}[\langle +1|-\langle -1|]$) is relatively larger, but can be observed by HFEPR, as well as, potentially and more informative, transitions from $m_S=\pm 2$ to $m_S=\pm 1$ and $m_S=0$.

Fig. 4, but for $S=2$ Mn(III), with only zero magnetic field and omitting m_I levels for simplicity, but including ZFS effects not shown in Fig. 4.

The key point is that rhombic ZFS mixes $m_S=0$ and $m_S=\pm 2$ levels so that an EPR transition can be observed within the two lowest energy levels that are degenerate with only axial ZFS. The EPR selection rules are such that this transition is enhanced when the microwave field (B_1) is parallel to the external field (B_0), which is not the standard configuration, that being B_1 perpendicular to B_0. EPR spectrometers are available that have a "dual-mode" cavity so that the user can simply switch from the standard perpendicular mode to parallel mode. A "home-built" approach by Hoffman and coworkers was to magnetic field, rather than microwave structure. Rather than using the massive spectrometer electromagnet perpendicular to the microwave field, they slid the lHe immersion dewar out from the magnet and placed a Helmholtz coil with a DC power supply around the dewar tail, thus parallel to the microwave field (Song, Doan, Gurbiel, Sturgeon, & Hoffman, 1999). Doan and Hoffman have also shown that for ESEEM, the advantages of parallel-mode over perpendicular-mode ESEEM are minimal at best (Doan & Hoffman, 2000). The metalloproteins studied by Hoffman and coworkers using this approach include di-Fe systems such as hemerythrin (Hr) and soluble methane monoxygenase (sMMO) (Doan & Hoffman, 2000; Sturgeon et al., 1997), but no Mn-containing systems. Britt and coworkers have studied Mn in PS II and in MnSOD by parallel mode EPR (Britt, Peloquin, & Campbell, 2000; Campbell et al., 1999; Campbell et al., 2000). $Mn(III)_2$-substituted (the two Mn(III) ions are well separated) transferrin has very recently been shown to exhibit such a signal (Lockart, Edwards, Vincent, & Pierce, 2021). The bottom line is that for system that might contain $S=2$ Mn(III), the low field region should be examined even if only with a conventional X-band EPR spectrometer, and especially so if a pulse system is available.

Manganese(IV) ($3d^3$, $S=3/2$) is also EPR active. Its relevance in the present context stems only from bioinspired model chemistry, both in mononuclear systems (Fiedler & Fischer, 2017; Guo et al., 2019; Rice et al., 2017), and in multinuclear models for the OEC (Lee et al., 2020; Nguyen et al., 2017). Concerning Mn-containing enzymes, it is possible to prepare a "superoxidized" manganese catalase (MnCat) that contains Mn(III)Mn(IV) as opposed to the physiologically relevant $Mn_2(II)$ and $Mn_2(III)$ oxidation state forms. This species has an $S=1/2$ ground state resulting from antiferromagnetic coupling of the two Mn ions: $S_{tot}=S_{Mn(III)} - S_{Mn(IV)}=4/2 - 3/2=1/2$ and is of interest not only as a model for (part of) the OEC of PS II (Beal & O'Malley, 2016; Milikisiyants, Chatterjee, & Lakshmi, 2011), but also serves as an EPR active probe of MnCat itself (Stich, Whittaker, & Britt, 2010).

2.2 Overview of EPR techniques in the context of Mn ions

CW EPR requires no further discussion, but the advanced techniques, ENDOR (whether CW or pulse) and ESEEM merit a brief overview noting that expert coverage is given by Schweiger and Jeschke (Schweiger & Jeschke, 2001) and by the appropriate chapters in Goldfarb and Stoll's recent text (Goldfarb & Stoll, 2018), specifically those by Bowman et al. on FT-EPR (Bowman, Chen, & Maryasov, 2017), Harmer on ENDOR (Harmer, 2017), and van Doorslaer on ESEEM (Van Doorslaer). The chapter in this series by DeRose and coworkers is also recommended for a brief, but lucid overview (Hunsicker-Wang et al., 2009), as is the more extensive one by Cox et al (Cox et al., 2015).

ENDOR is EPR-detected NMR, which is achieved by monitoring an EPR signal, whether CW or a spin echo generated by the appropriate pulse sequence with concurrent application of radiofrequency (RF) radiation. Magnetically active nuclei such as $^{1,2}H$, ^{13}C, $^{14,15}N$, ^{17}O, ^{31}P, and ^{55}Mn undergo NMR transitions that affect the EPR signal intensity—*if* they are interacting (via hyperfine coupling) with the paramagnetic site (i.e., the unpaired electron(s)) that make up the active site in a metalloenzyme. Naturally abundant nuclei such as ^{1}H, ^{14}N, ^{31}P, and ^{55}Mn can in principle be observed in any sample in which these elements are present (i.e., all samples for ^{1}H), while use of D_2O buffer can introduce ^{2}H into exchangeable hydrogens and for those that can afford it, $H_2^{17}O$ can introduce this isotope into enzyme bound water/hydroxide sites, or $^{17}O_2$ into dioxygen-activating enzymes. Suitably isotopically labeled substrates/inhibitors can introduce ^{13}C or ^{15}N into the active site and the tools of molecular biology can in certain cases introduce protein residues that are isotopically labeled. ENDOR of ^{55}Mn is not typically attempted or especially needed in most Mn-containing systems, as hyperfine coupling to a single Mn(II) is often large enough to be resolved by EPR and relatively uninformative, but Britt and coworkers achieved remarkable success early on with this technique applied to the much more challenging situation found in PS II (Peloquin et al., 2000). Lubitz and coworkers have recently made significant advances using ^{55}Mn Q-band ENDOR on the OEC of PS II (Lohmiller et al., 2017).

Magnetically active nuclei can be loosely classified with respect to their ENDOR behavior as shown in Table 1, which is most appropriate for Q-band EPR conditions with a signal at $g \approx 2$ (35 GHz, 1.25 T). Fig. 8 depicts this behavior graphically. HFEPR can in principle be adapted for ENDOR, so that most nuclei become Larmor centered. Examples of nuclei that would normally be hyperfine dominated are ^{57}Fe (non-quadrupolar) ^{17}O (quadrupolar) so that high frequencies can be desirable to simplify the spectra. W-band ENDOR has been performed on a ^{57}Fe-enriched iron-sulfur enzyme (heterodisulfide reductase, Hdr) (Bennati, Weiden, Dinse, & Hedderich, 2004) and ^{17}O W-band ELDOR-detected NMR (EDNMR) (Cox et al., 2017, 2015) has been applied to the OEC of PS II,

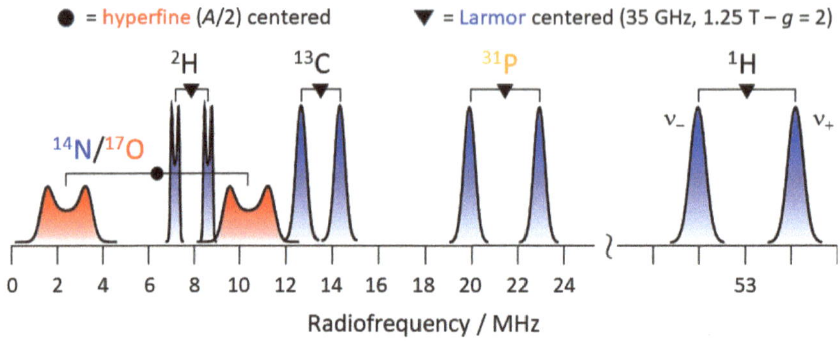

Fig. 8 Cartoon of ENDOR signals from biologically relevant nuclei as they would ideally appear (i.e., a "single-crystal-like" orientation (Hoffman, 2003)) for a Q-band EPR signal at $g=2$ (35 GHz, 1.25 T). Signals in blue correspond to nuclei with ENDOR signals centered at the respective Larmor (NMR) frequency (e.g., $\nu(^1\text{H}) \approx 53$ MHz) wherein the splitting between the two signals (ν_+ and ν_-) equals $|A|$. Signals in red correspond to ENDOR signals centered at $|A|/2$ (hyperfine centered) wherein the splitting between ν_+ and ν_- equals $2\nu_N$. Quadrupole splitting is also shown (but not labeled) for those nuclei with $I \geq 1$ (^{2}H, ^{14}N, ^{17}O). In the case where the nuclear quadrupole coupling is less than the hyperfine coupling (i.e., a first order situation), then each $\nu_\pm$ branch splits into $2I$ peaks, thus a doublet for ^{2}H and ^{14}N, as shown. If the hyperfine and quadrupole couplings are comparable, then the pattern is more complicated. The pattern is also more complicated for a nucleus of any I if there is hyperfine anisotropy.

(Lohmiller et al., 2017) and to di-Mn and MnFe systems (Rapatskiy et al., 2015) that will be described in detail later. EDNMR—wherein the "E" is an abbreviation of an acronym, namely, ELDOR, derived from electron–electron double resonance. ELDOR is similar to ENDOR, but with the second resonance coming from another microwave source so that EPR and NMR transitions are simultaneously excited (see Fig. 5). This technique is also known as DEER (acronym from Double Electron–Electron Resonance), which name is popular perhaps because of bringing up fond childhood memories of Bambi or Rudolph.

The question then is, how best to observe these signals? CW ENDOR has the advantage of giving higher signal-to-noise (S/N), but the signals are often distorted (Doan, Gurbiel, & Hoffman, 2007). Pulse ENDOR requires more time (greater signal averaging and even so, the S/N of CW ENDOR is hard to achieve) and is less forgiving to a novice spectroscopist, but generally provided better resolution and more accurate information. Within pulse ENDOR, there are two widely used pulse sequences, eponymously called Mims and Davies ENDOR. A derivative of Mims ENDOR, refocused

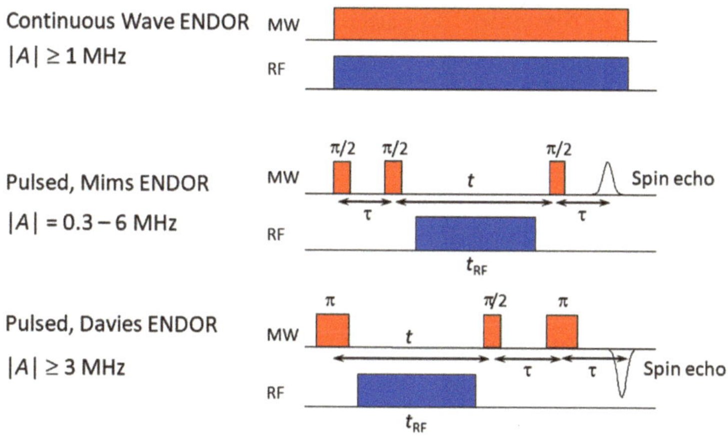

Fig. 9 Schematic of the pulse sequences used for Mims (center) and Davies (bottom) ENDOR with CW (top) for comparison. The approximate optimal hyperfine coupling values for each technique are given. Red indicates irradiation by microwaves and blue by radiofrequency. The time scale (the abscissa—the time domain—is not meant to be quantitative). Typical times for these sequences will be given in subsequent figures taken from the original literature.

Mims (ReMims) could be eponymously called Doan ENDOR (Doan & Hoffman, 1997), but this has not caught on except with the author. Fig. 9 provides a simplified pulse sequence picture as to these two techniques, with CW ENDOR for comparison. The basis for these techniques is the (electron) spin echo (Hahn, 1950), discovered by Erwin Hahn and often called a "Hahn spin-echo." A beautiful animation of the Hahn spin-echo created by Gavin Morley is available on Wikipedia (Morley, 2011). The original Hahn spin-echo is created by a two-pulse sequence. The ENDOR (and ESEEM) techniques employ three-(microwave) pulse sequences, but the principle is the same.

The reader is directed elsewhere, conveniently within this series for a "blow-by-blow" description of spin phenomena that give rise to ENDOR spectra (Cox et al., 2015; Hunsicker-Wang et al., 2009), but the spin echo intensity changes as a result of NMR transitions when resonance with the applied RF occurs. A key point that is not evident in the cartoon of ENDOR spectra shown in Fig. 8 is that regardless of ENDOR method the spectrum is obtained at a particular magnetic field position (g_{obs} value) on the EPR envelope. Each such g_{obs} value corresponds to a particular set of orientations of the spin system with respect to the external magnetic field and simulation of the entire 2D pattern (ENDOR signal intensity vs g_{obs} value) provides a complete picture of the hyperfine (and quadrupole for $I \geq 1$) tensors with respect to the **g** tensor (Broderick, Hoffman, & Broderick, 2018; Hoffman, 2003). As will be shown below for Mn-containing systems, this information can provide detailed metrics on substrate-active site interactions.

ENDOR, by definition a double resonance technique, requires RF equipment (synthesizer, amplifier, RF coils in the microwave cavity) whether in CW or pulse mode. However, nuclear information can be obtained absent actual application of RF excitation. Hyperfine couplings create modulations on the amplitude of the electron spin echo that depend on both hyperfine coupling and the Larmor frequency. This effect is electron spin echo envelope modulation (ESEEM) (Van Doorslaer, 2017). The relevant pulse sequence is not explicitly shown in Fig. 9, but simply corresponds to the Mims sequence absent the RF excitation. The mixing time (t) is varied and a plot of spin-echo intensity vs t shows a modulating amplitude superimposed on the overall decay of the echo. This time domain phenomenon can be Fourier transformed to give a frequency domain spectrum, comparable to that obtained directly from ENDOR. Generally, only small hyperfine couplings can be observed, and the most common application is to investigate weakly coupled ^{14}N, whether directly coordinated to Mn, or nearby (as in the various nitrogen atoms of nucleobases) (Morrissey et al., 1999; Vogt et al., 2006).

3. Overview of EPR and related techniques applied to biological Mn-containing systems

3.1 Mononuclear Mn-containing enzymes

Extradiol Dioxygenases. Extradiol catecholic dioxygenases catalyze the cleavage of the aromatic ring of the substrate with incorporation of both oxygen atoms from O_2 (Fiedler & Fischer, 2017), as shown in Fig. 10.

The active sites of these enzymes contain either Mn or Fe, coordinated by a two histidine, one carboxylate facial triad (see Fig. 2, right) with three labile (water) ligands that can be displaced by O_2 and catechol from substrate. A well-studied example is homoprotocatechuate 2,3-dioxygenase from *Arthrobacter globiformis* (MnMndD), which catalyzes the ring opening of homoprotocatechuate (HPCA, 2-(3,4-dihydroxyphenyl)acetate) and contains Mn(II) as the native metal (Emerson, Kovaleva, Farquhar, Lipscomb, & Que, 2008; Gunderson et al., 2008; Vetting et al., 2004).

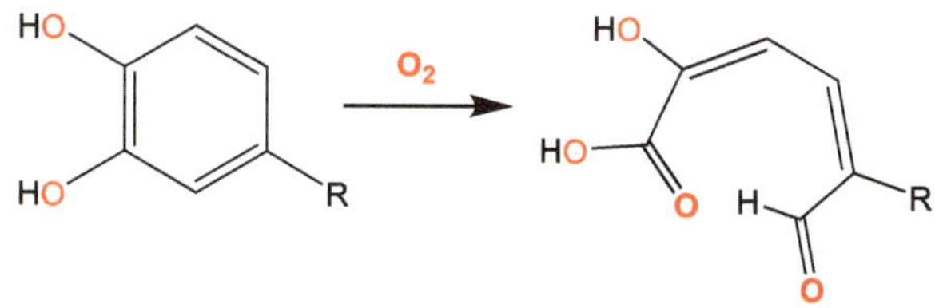

Fig. 10 Reaction catalyzed by Mn (and Fe) extradiol catechol dioxygenases such as MndD. The product is drawn so as to retain structural similarity with the original benzene ring and it is not meant to be the actual conformation. Hydronation state is analogously not specified. It is also clear that this reaction is an *extra*diol dioxygenase in that the C—C bond in the catechol (1,2-dihydroxybenzene moiety) is retained, while an adjacent C—C bond is broken. Homoprotocatechuic acid is 2-(3,4-dihydroxyphenyl) acetic acid) so R=-CH_2CO_2H in this archetypal case.

Despite the redox reactions involving O_2 reduction and hydrocarbon oxidation, the proposed mechanism retains formally Mn(II) throughout: a key intermediate is a $[Mn^{II}(O_2^{\cdot-})(SQ^{\cdot-})]$ where $SQ^{\cdot-}$ is a semiquinone radical. EPR was key in the analysis of MndD as the spectrum changed dramatically upon addition of substrate (Emerson et al., 2008; Gunderson et al., 2008). Rapid freeze-quench (RFQ) experiments allowed EPR observation of enzymatic intermediates in MndD catalysis.

Extensive discussion of the RFQ technique is beyond the scope of this chapter, but it merits emphasis that EPR can be applied to transient and not only steady state species. The rapidly mixed species (e.g., enzyme and substrate solutions) are quenched by a cryo-solvent (or simply lN_2) and the aqueous reaction mixture (ice) is packed into an EPR tube (the most challenging aspect of the experiment, particularly for Q-band EPR/ENDOR). The rapid freezing can be achieved by spinning metal wheels as seen in Fig. 11. RFQ has been most notably applied in recent years to the study of enzymes in the radical SAM superfamily that contain an F4S4 active site cluster that binds SAM via an organometallic intermediate (Byer et al., 2018; Horitani et al., 2016).

Lipoxygenase. Lipoxygenase (LO; alternatively, LOX), best characterized from *Glycine max* (soy(a) bean; hence SLO) catalyzes the specific oxidation of long chain unsaturated fatty acids (linoleic and arachidonic acids) to hydroperoxides. The active metal in SLO is iron and EPR studied have been performed on native SLO with nitroxide spin-labeled substrates (Gaffney, 2014). The high-spin ($S=2$) Fe(II) active site in SLO is not well-suited to EPR/ENDOR investigation (SLO in the oxidized, Fe(III) form can be studied by EPR (Gaffney, 2014)); however, Mn(II) can be substituted for Fe(II). Although Mn-SLO is inactive, the active site structures of

Fig. 11 Rapid freeze-quench (RFQ) apparatus in the laboratory of Prof. Brian M. Hoffman, Northwestern University, Evanston, IL (photo is of system not in operation). The two syringes, typical for rapid mixing kinetics studies are seen at the top and the two wheels that rapidly cool the solution are seen near the center. An insulated container for lN_2 is seen at the bottom. (Photo by the author).

Mn-substituted and native SLO are nearly superimposable (Horitani et al., 2017). (A structure of Mn-SLO is available on the PDB, code 4WFO, but has yet to be otherwise published). Extensive 35 GHz EPR/ENDOR studied were performed by Klinman, Hoffman and coworkers on Mn-SLO taking advantage of the availability of specifically ^{13}C-labeled linoleic acid (i.e., at C10 and C11—the latter being the site of C-H → C-OOH oxidation). Mn(II) as a spin system for ENDOR differs from typical $S = 1/2$ systems such as low-spin Fe(III) or Cu(II). As shown by the energy levels in Fig. 4, there is the "inner sextet": $\langle S,\ m_S,\ I,\ m_I| = \langle 5/2,\ -1/2,\ 5/2,\ m_I| \rightarrow \langle 5/2,\ +1/2,\ 5/2,\ m_I|$, which is analogous to the transitions in, say,

Cu(II) or VO^{2+} (V(IV), $S=1/2$), but then there are the four "outer sextets": $\langle 5/2, -5/2, 5/2, m_I| \rightarrow \langle 5/2, -3/2, 5/2, m_I|$, $\langle 5/2, -3/2, 5/2, m_I| \rightarrow \langle 5/2, -1/2, 5/2, m_I|$, $\langle 5/2, +1/2, 5/2, m_I| \rightarrow \langle 5/2, +3/2, 5/2, m_I|$, and $\langle 5/2, +3/2, 5/2, m_I| \rightarrow \langle 5/2, +5/2, 5/2, m_I|$, which often do not show resolved hyperfine splitting. The presence of these sets of EPR transitions would complicate the corresponding ENDOR signals, whether from ^{13}C or any other probe nucleus, but at low temperature (2 K), even with the small magnitude of ZFS in these systems, the $-5/2 \rightarrow -3/2$ manifold is dominant, with a distinguishable contribution from $-3/2 \rightarrow -1/2$; the other two, $+1/2 \rightarrow +3/2$ and $+3/2 \rightarrow +5/2$, respectively, being negligible or zero. At 35 GHz and $g=2.00$, ^{13}C ENDOR transitions for the relatively distant labels would be centered at the Larmor frequency ($\nu(^{13}C)=13.4$ MHz) and split by $|A/2|$. These A values, however, are scaled by m_S, so that ENDOR transitions within the $m_S=\pm 3/2$ manifold would be at $\nu_{\pm 3/2}=[\nu(^{13}C) \pm 3|A/2|]$ and within the $m_S=\pm 5/2$ manifold at $\nu_{\pm 5/2}=[\nu(^{13}C) \pm 5|A/2|]$. There are two consequences to this, both providing "more bang for the buck" in that: multiple signals for a given label allows more precise determination of ENDOR parameters, and the m_S scaling means that a nuclear spin is effectively greater in a high-spin versus a "normal" $S=1/2$ paramagnet. Since the distance sensitivity of ENDOR, as with ELDOR and NMR itself, is proportional to magnetic moment magnitude, a larger effective spin means a longer distance can be measured. Fig. 12 shows 35 GHz Mims ENDOR spectra for wild type (mutants were also studied) Mn-SLO with bound ^{13}C10,11-labeled linoleic acid (LA) (Horitani et al., 2017). Additionally, Davies ENDOR of ^{14}N of Mn-bound histidine and ^{1}H of Mn-bound water/hydroxide (by comparison of spectra in H_2O and D_2O buffer) were recorded.

The metrical information derived from the two ^{13}C labels (along with resting state crystal structures and ^{1}H ENDOR) allowed a quantitative picture of substrate binding to Mn-SLO to be made. This is shown in Fig. 13. The results of this study and a subsequent one (Offenbacher, Sharma, Doan, Klinman, & Hoffman, 2020) allowed fundamental conclusions to be drawn not only about SLO, but of relevance hydrogen tunneling and to isotope effects in enzymes in general. Of relevance here is the efficacy of Mn(II) as a spin-probe of metalloenzyme mechanism. A skeptic would point out that Mn-SLO is not actually active, but there have recently been reported native Mn-LOs, not from a useful plant, but from harmful (from a human point of view) fungi: *Fusarium oxysporum* (Wennman, Magnuson, Hamberg, & Oliw, 2015), which is a devastating pathogen to a wide range of fruits

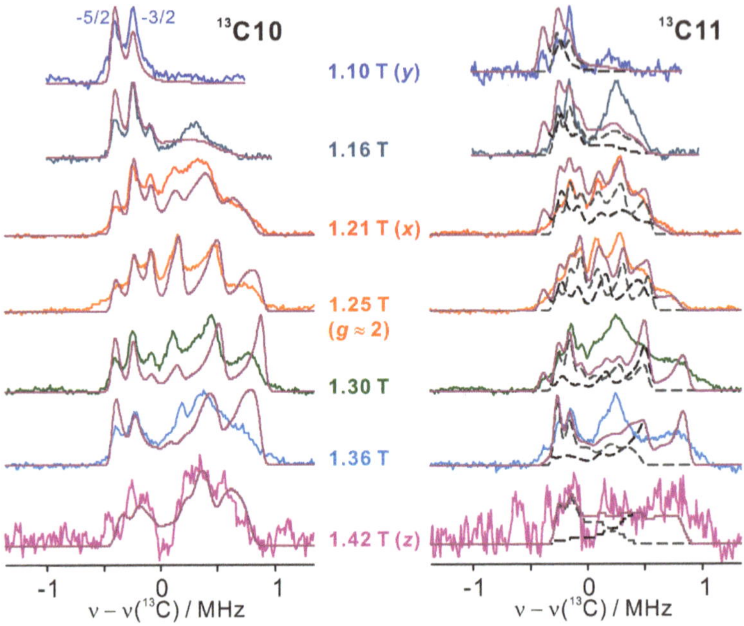

Fig. 12 35 GHz 2D field-frequency pattern ^{13}C Mims ENDOR recorded at 2 K for wild type (WT) Mn-SLO with ^{13}C10-LA (left) and ^{13}C11-LA (right) each with simulations (purple traces). Conditions: microwave frequency ~ 34.8 GHz, MW pulse length ($\pi/2$) = 30 ns, τ = 500 ns, repetition rate = 100 Hz. The magnetic field positions (T) are indicated along with the corresponding molecular orientation, where applicable. At the extrema, ENDOR transitions from only the outer manifold are seen (ground state m_S values given for ^{13}C10). Simulations are the sum of two sites, as shown for ^{13}C11 (high occupancy, 80%, black dashed lines; and low occupancy, 20%, gray dashed lines) using a point-dipole model with a fixed Mn—^{13}C distance and orientation (e.g., $d_{Mn\text{-}C}$ = 4.89 Å; ϕ = 0, θ = 40°; values for ^{13}C11, high occupancy site). *Adapted with permission from Fig. 5 of Horitani, M., Offenbacher, A. R., Carr, C. A. M., Yu, T., Hoeke, V., Cutsail III, G. E., ... Hoffman, B. M. (2017).* ^{13}*C ENDOR spectroscopy of lipoxygenase–substrate complexes reveals the structural basis for C–H activation by tunneling.* Journal of the American Chemical Society, *139(5), 1984–1997. doi:10.1021/jacs.6b11856. Copyright 2017 American Chemical Society.*

and vegetables and *Magnaporthe oryzae* (Kostenko, Ray, Iavarone, & Offenbacher, 2019), which is a serious pathogen to rice plants. In these cases, one hopes that similar results will obtain for a lipoxygenase with its native metal ion, which could aid in the development of specific fungicides.

Fosfomycin Resistance Proteins (FosA, FosB, FosX). These are metalloglutathione transferases that confers resistance to the broad-spectrum antibiotic fosfomycin, (1*R*,2*S*)-epoxypropylphosphonic acid. FosA and FosB catalyze the addition of thiols, glutathione (GSH) in the case of FosA and

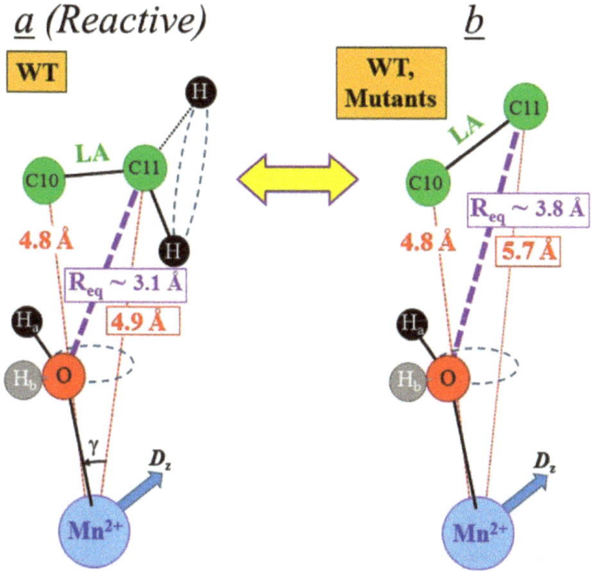

Fig. 13 Model structures for active site ground state (A) and (B) conformers of Mn-SLO with bound linoleic acid (LA). The distances represented by the red dotted lines are ENDOR-derived distances; the Mn – O distance, obtained from the X-ray crystal structure, is 2.23 Å. For the (A) conformer in WT enzyme, the angle, $\gamma = 26°$, is calculated by MD simulations using ENDOR-derived distances as restraints. The Euler angles (ϕ, θ) relative to ZFS axes for ^{13}C10, ^{13}C11, and $^{1}H_a$ in the WT (A) conformer are (0°, 50°), (0°, 40°), and (0°, 70°), respectively; the shared values of $\phi \sim 0$ imply that all three lie, within error, in the $x-z$ plane of the ZFS tensor. The purple dashed line represents the ground state equilibrium reactive hydrogen donor – acceptor distance (R_{eq}) calculated trigonometrically. Dashed circles suggest the nuclei (oxygen of water and C11 hydrogens in (a) conformer) lie out of the $x-z$ plane of the ZFS axes. *Reprinted with permission from Fig. 7 of Horitani, M., Offenbacher, A. R., Carr, C. A. M., Yu, T., Hoeke, V., Cutsail III, G. E., ... Hoffman, B. M. (2017). ^{13}C ENDOR spectroscopy of lipoxygenase–substrate complexes reveals the structural basis for C–H activation by tunneling.* Journal of the American Chemical Society, *139(5), 1984–1997. doi:10.1021/jacs.6b11856. Copyright 2017 American Chemical Society.*

L-cysteine in the case of FosB (Fillgrove, Pakhomova, Newcomer, & Armstrong, 2003), to carbon-1 of fosfomycin to yield 1-(*S*-glutathionyl)-2-hydroxypropylphosphonic acid (GSfos), which is inactive as an antibiotic (Thompson et al., 2014). This reaction is illustrated in Fig. 14. FosX catalyzes the addition of water and thus is a simple hydrolytic ring opening reaction.

Pseudomonas aeruginosa FosA contains a single Mn(II) ion at its active site that binds fosfomycin and activates it for the ring opening reaction that renders the antibiotic impotent (Bernat, Laughlin, & Armstrong, 1997;

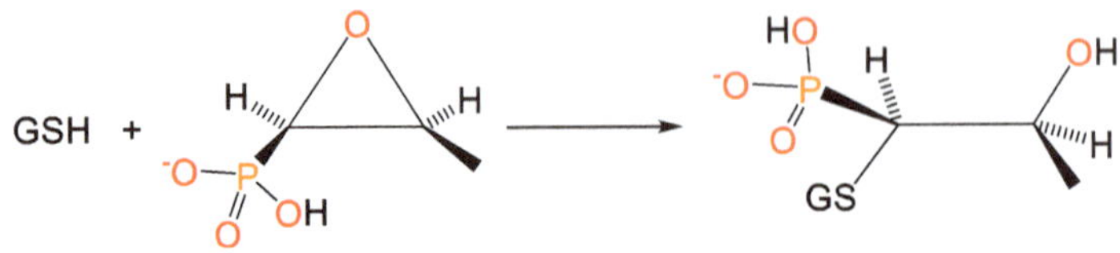

Fig. 14 Reaction catalyzed by the mononuclear Mn(II) enzyme FosA. Glutathione (GSH) opens the oxirane ring of fosfomycin ((1*R*,2*S*)-epoxypropylphosphonic acid) to give the addition product 1-(*S*-glutathionyl)-2-hydroxypropylphosphonic acid (GSfos). FosB catalyzes the analogous addition of the thiol group of cysteine to give 1-(*S*-L-cysteinyl)-2-hydroxypropylphosphonic acid and FosX uses simply water to produce 1,2-dihydroxypropylphosphonic acid.

Smoukov et al., 2002). The Mn(II) ion does not undergo any redox changes, but its EPR spectrum changes dramatically upon substrate binding. Without substrate, FosA exhibits a 35 GHz EPR spectrum not unlike that for simply aqueous Mn(II), albeit with a slightly larger ZFS ($|D| \approx 0.03\,\text{cm}^{-1}$ for $Mn^{2+}_{(aq)}$, in the presence or absence of fosfomycin; $|D| \approx 0.06\,\text{cm}^{-1}$ for FosA). Upon addition of substrate (or of product—GSfos), the EPR spectrum changes dramatically, as shown in Fig. 15. This change is the result of oxidation of a large increase in ZFS: $|D| \approx 0.23\,\text{cm}^{-1}$ for FosA with fosfomycin; $|D| \approx 0.27\,\text{cm}^{-1}$ for FosA with GSfos. When only higher temperature EPR was used earlier (273 K at 35 GHz and 293 K at X-band) (Bernat et al., 1997), addition of fosfomycin (or GSfos) led to disappearance of the typical sextet EPR signal, leading to the possibility that Mn(II) had been oxidized to "EPR-silent" Mn(III). Low temperature (2 K), rapid-passage (Mailer & Taylor, 1973) EPR revealed that all the EPR signal was present, but changed due to direct substrate/product binding to the Mn(II) active site, which allowed a new proposal as to enzyme mechanism (Smoukov et al., 2002).

A subsequent study on FosA employed pulse EPR/ENDOR methods as well as an inhibitor, phosphonoformate (Pf, $(HO)_2P(=O)CO_2^-$) (Walsby, Telser, Rigsby, Armstrong, & Hoffman, 2005). The key aspect of this study was the use of ^{31}P Davies ENDOR on the suite of samples that allowed metrical information to be obtained from the "non-local" dipolar part of the hyperfine coupling ($T_{\text{non-loc}}$—so-called because it is through space, using a simple electron point dipole-nuclear point dipole model), also taking into account the anisotropic contribution from ^{31}P 3p spin density (T_{loc}; for sp^3 hybridization in this case). The resulting structural description, for $E{\bullet}Mn^{2+}{\bullet}S$, taking also into account the results of X-ray crystallography (Rife, Pharris, Newcomer, & Armstrong, 2002), which were reported after

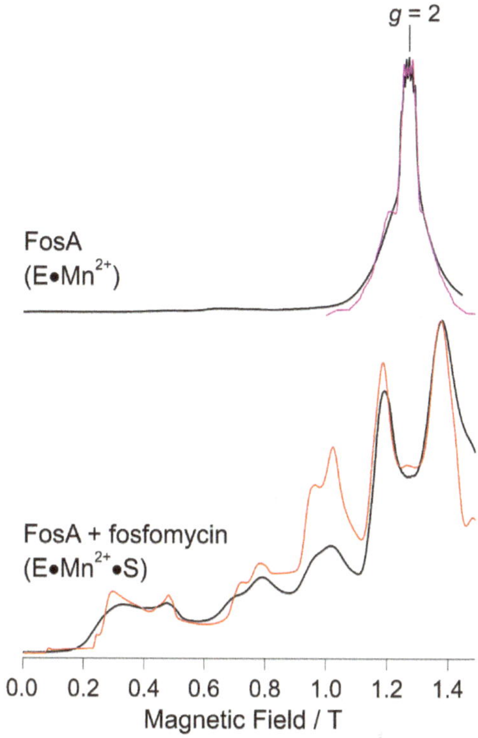

Fig. 15 CW EPR of FosA (E•Mn^{2+}; top) and FosA plus fosfomycin (substrate; E•Mn^{2+}•S, bottom). Conditions: microwave frequency, 35.240 GHz (35.130 GHz for E•Mn^{2+}•S); MW power, 20 dB (2 mW); 100 kHz modulation amplitude, 0.17 mT (0.068 mT for E•Mn^{2+}•S), temperature, 2 K; time constant, 64 ms; scan speed, 0.2 T/min; simulation parameters: magenta trace (E•Mn^{2+}): $S=5/2$, $D=0.060\,\text{cm}^{-1}$, $E=0.06\,\text{cm}^{-1}$; red trace (E•Mn^{2+}•S): $D=0.23\,\text{cm}^{-1}$, $E=0.020\,\text{cm}^{-1}$. *Adapted with permission from Fig. 2 of Smoukov, S. K., Telser, J., Bernat, B. A., Rife, C. L., Armstrong, R. N., & Hoffman, B. M. (2002). EPR study of substrate binding to the Mn(II) active site of the bacterial antibiotic resistance enzyme FosA: A better way to examine Mn(II).* Journal of the American Chemical Society, *124(10), 2318–2326. doi:10.1021/ja012480f. Copyright 2002 American Chemical Society.*

the original EPR study (Smoukov et al., 2002), is shown in Fig. 16. EPR/ENDOR studies on FosB and FosX were also undertaken, but the collaborative project came to an end with the tragic death of Richard N. Armstrong (Vanderbilt U.) in 2015 (Armstrong, 2015).

Oxalate Oxidase (OxOx) and Oxalate Decarboxylase (OxDC). These two closely related enzymes, Oxalate Oxidase (OxOx) and Oxalate Decarboxylase (OxDC) each contain a mono-Mn(II) active site (Campomanes et al., 2014; Chang et al., 2004; Moomaw et al., 2009). The function of these enzymes is to destroy oxalate ion, a potentially harmful byproduct of metabolism in

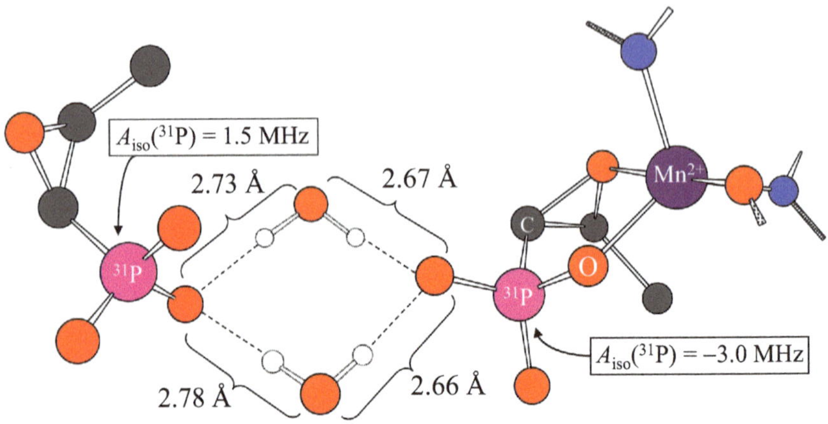

Fig. 16 Proposed hydrogen bonding pathway for electron spin density to ^{31}P of second fosfomycin substrate molecule in FosA active site. Atomic coordinates as reported by Rife et al. (Rife et al., 2002). Hydrogens atoms are positioned so as to minimize the P-O-H-O path length. *Adapted with permission from Fig. 5 of Walsby, C.J., Telser, J., Rigsby, R.E., Armstrong, R.N., & Hoffman, B.M. (2005). Enzyme control of small-molecule coordination in FosA as revealed by 31P pulsed ENDOR and ESE-EPR. Journal of the American Chemical Society, 127 (23), 8310–8319. doi:10.1021/ja044094e. Copyright 2005 American Chemical Society.*

many organisms (microbes, fungi, and plants). Humans lack this ability and oxalate precipitates can cause kidney and urinary tract “stones”. The reactions catalyzed by OxOx and OxDC are shown in Fig. 17.

OxOx obliterates oxalate into two CO_2 molecules, and does so aerobically producing hydrogen peroxide, which itself may need to be dealt with. OxDC catalyzes a disproportionation reaction ($2C^{3+} \rightarrow C^{4+} + C^{2+}$), but nevertheless requires O_2 for activity. As a result, a similar catalytic mechanism is proposed for both enzymes, involving Mn-bound oxalate and dioxygen (as $[Mn^{III}O_2^{\bullet -}]$—both monodentate (Zhu et al., 2016; Zhu, Wilcoxen, Britt, & Richards, 2016). The active site structure is similar to that shown above for MnSOD (see Fig. 2) (Woo et al., 2000). The presence of mononuclear Mn(II) as the active site of these two enzymes makes them a rich vein for EPR “mining” (Angerhofer et al., 2007; Campomanes et al., 2014; Chang et al., 2004; Moomaw et al., 2009; Zhu, Easthon, et al., 2016; Zhu, Wilcoxen, et al., 2016). Presenting many such Mn(II) EPR spectra, however, informative as to these enzymes, is superfluous at this point. However, two aspects of these extensive studies will be highlighted here. The first is the powerful use of parallel mode EPR detection (Twahir, Ozarowski, & Angerhofer, 2016; Zhu, Wilcoxen, et al., 2016). This is

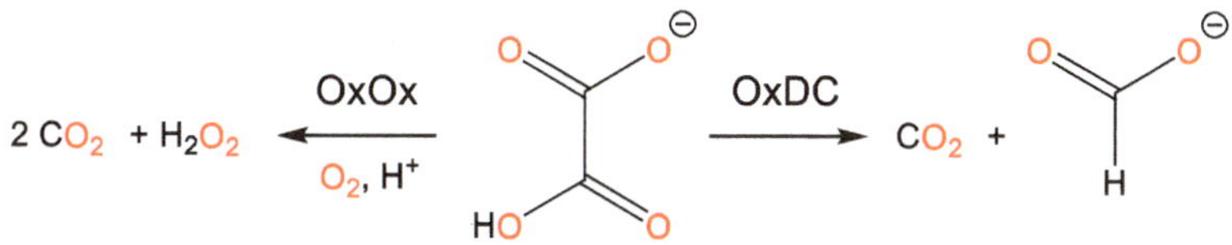

Fig. 17 Reactions catalyzed by oxalate oxidase (OxOx; left) and oxalate decarboxylase (OxDC; right). The hydronation state is appropriate for neutral pH conditions.

demonstrated in Fig. 18, which compares the perpendicular (top row) and parallel (bottom row) mode X-band EPR spectra of wild type OxDC in the absence (left column) and presence (right column) of oxalate ion (different pH values were investigated in each case as well, as the reaction has a strong pH dependence). Forbidden ($\Delta m_S = \pm 2$) transitions of Mn(II) are seen absent oxalate, but no Mn(III) signal. In contrast, the presence of oxalate at acidic pH, where OxDC exhibits catalytic activity, caused substantial changes in both the perpendicular and parallel mode EPR spectra (Zhu, Wilcoxen, et al., 2016). Most importantly, a new sextet signal centered near 75 mT appeared in the parallel mode spectrum at pH 4.2 and 5.7 (Fig. 18D), which arises from enzyme-bound Mn(III). The gradual disappearance of this signal with increasing pH suggests that this species is correlated with enzyme catalysis, and was in a parallel (no pun intended) study associated with bound formate (product) (Twahir et al., 2016).

The second exciting application of EPR in the study of OxOx and OxDc is the use of HFEPR (Angerhofer et al., 2007; Campomanes et al., 2014; Moomaw et al., 2013; Tabares et al., 2009; Twahir et al., 2016). This technique (at 406.4 GHz) yielded **g** tensor characterization of a radical that would be a mere "spike" at lower frequencies and allowed its assignment as a tyrosyl radical that is a side product of OxDC catalysis at $5 < \text{pH} < 6$. The Mn(II) signal could also be readily observed at such high frequencies, allowing precise determination of ZFS parameters of two Mn(II) sites in OxDC (Angerhofer et al., 2007). Fig. 19 presents the HFEPR spectrum of wild type OxDC at two temperatures recorded using the spectrometer at NHMFL (Campomanes et al., 2014). The difference between the two Mn(II) sites is apparent, but would not be using conventional (i.e., W-band or lower frequency) EPR. The enzyme concentration is provided (27 mg/mL; so ~0.62 mM using ~44 kDa (Crystal structure of oxalate decarboxyl ase, 2022). Such a concentration seems feasible for other metalloenzymes and thus HFEPR should be more widely attempted. HFEPR of OxOx from *Ceriporiopsis subvermispora* (a white rot fungus) gave similar results (Moomaw et al., 2013).

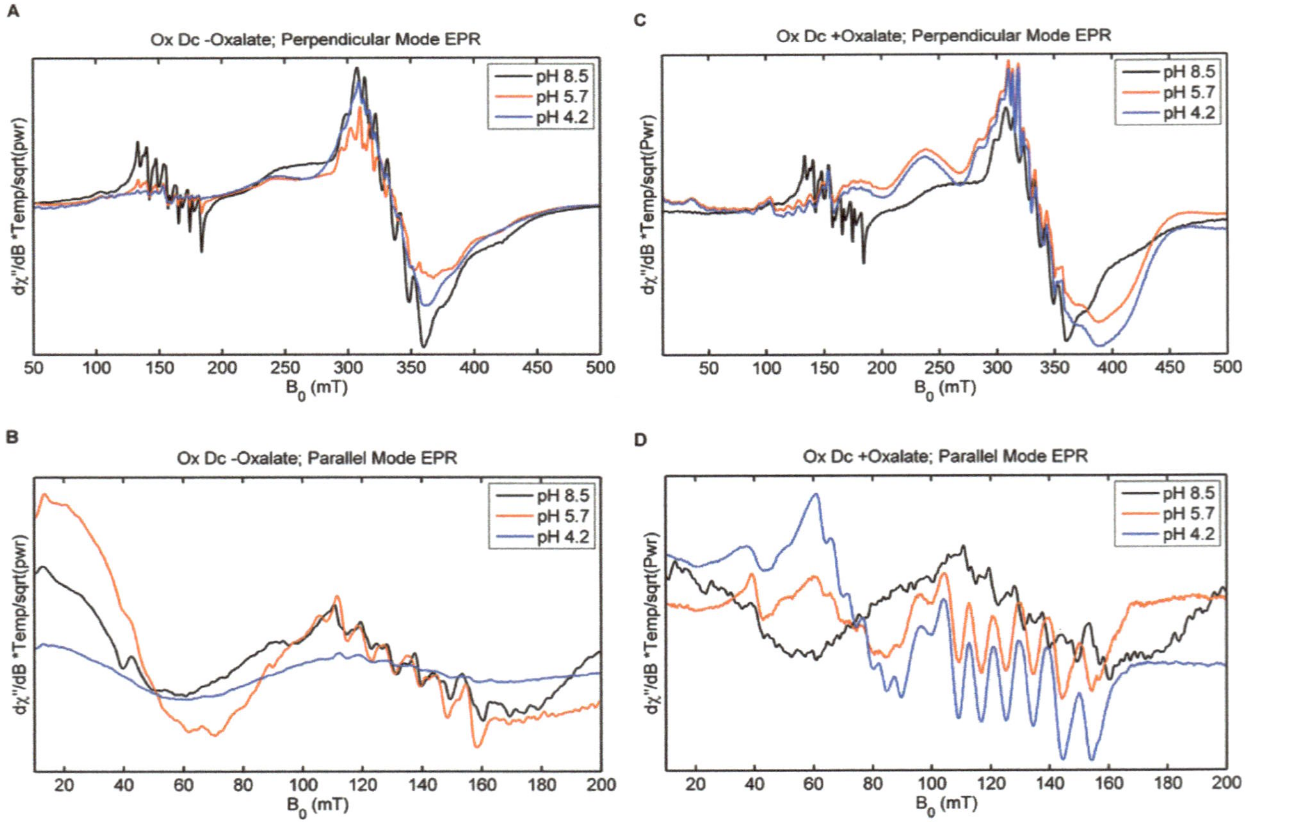

Fig. 18 (A) Perpendicular mode and (B) parallel mode EPR spectra of wt OxDC in the *absence* of oxalate at pH 8.5 (black), pH 5.7 (red), and pH 4.2 (blue). (C) Perpendicular mode and (D) parallel mode EPR spectra of wt OxDC in the *presence* of 100 mM oxalate at pH 8.5 (black), pH 5.7 (red), and pH 4.2 (blue). Perpendicular mode EPR spectra were recorded at a 9.37 GHz microwave frequency, a 0.5 mW power, and a 5 G modulation amplitude, at 10 K. Parallel mode EPR spectra were recorded at a 9.40 GHz microwave frequency, a 10 mW power, and an 8 G modulation amplitude, at 5.5 K. *Reprinted with permission from Fig. 1 in Zhu, W., Wilcoxen, J., Britt, R. D., & Richards, N. G. J. (2016). Formation of hexacoordinate Mn(III) in* Bacillus subtilis *oxalate decarboxylase requires catalytic turnover.* Biochemistry, *55*(3), 429–434. doi:10.1021/acs.biochem.5b01340. Copyright 2016 American Chemical Society.

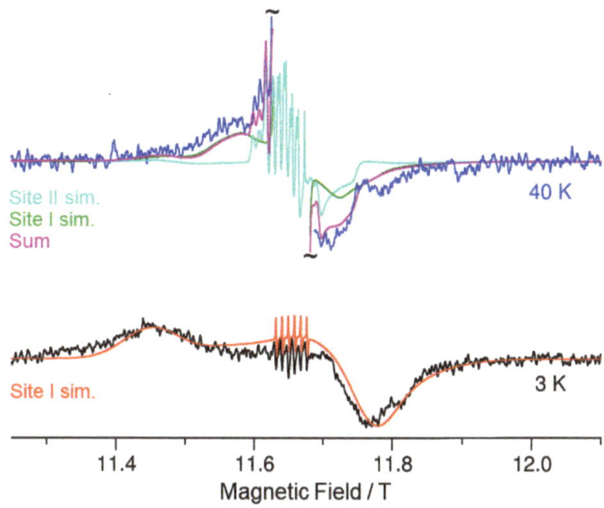

Fig. 19 HFEPR spectra of WT OxDC (in 50 mM Tris buffer, 500 mM NaCl, pH 8.5) taken at 326.4 GHz and 3 K (lower, black trace) and 40 K (upper, blue trace) together with spectral simulations demonstrating the two Mn sites present (I and II). The central part of the experimental spectrum at 40 K together with the corresponding spectral simulations involving site I (green and magenta traces) are removed for clarity. Spectral simulations correspond to a single Mn(II) species with small magnitude ZFS (site I: $D = -1.35$ GHz, $|E| = 0.20$ GHz; $|E/D| = 0.15$) at 3 K (red trace) and 40 K (green trace) and a single Mn(II) species with large ZFS (site II: $|D| = 10.43$ GHz, $|E| = 2.1$ GHz; $|E/D| = 0.20$) at 40 K (cyan trace). The sum of site I and site II simulations at 40 K is also shown (magenta trace). *Adapted from an original figure kindly provided by Prof. Alexander Angerhofer, University of Florida, Gainesville, FL, USA.*

Manganese Superoxide Dismutase (MnSOD). MnSOD has been mentioned above, but merits a brief section of its own. SODs catalyze the disproportionation of superoxide radical ion into dioxygen and hydrogen peroxide. The first, CuZnSOD, was discovered by Irwin Fridovich (Duke Today Staff, 2019) but Fe, Mn, and Ni SODs are known. The reaction for MnSOD is: (a) $Mn^{III} + O_2^{\bullet -} \rightarrow [Mn^{III}O_2^{\bullet -}] \rightarrow Mn^{II} + O_2$; (b) $Mn^{III} + O_2^{\bullet -} \rightarrow [Mn^{II}O_2^{\bullet -}] + 2H^+ \rightarrow Mn^{III} + H_2O_2$ (Hearn et al., 2003). Of relevance here are the HFEPR studies by Sun Un and coworkers using a spectrometer at CEA Saclay (France) originally designed for photosynthesis research that operates at 190 and 285 GHz. These workers were able to make accurate determinations of ZFS in a wide range of reduced (i.e., Mn(II)) SODs: Mn, Mn-substituted Fe, and cambialistic (active with either Mn or Fe) (Un, Tabares, Cortez, Hiraoka, & Yamakura, 2004). HFEPR on oxidized (i.e., Mn(III)) in SOD appears yet to be reported.

Parallel mode EPR, however, has been successfully applied to this form of MnSOD and ^{55}Mn hyperfine coupling was determined along with the ZFS (Campbell et al., 1999). Note that determination of ZFS from only the low field signal seen in low-symmetry $S=2$ systems (see Fig. 7) is less preferable than a full multifield/multifrequency "2D" HFEPR methodology that thus far seems possible only for model compounds (Krzystek et al., 2006).

Hydroxynitrile Lyase (HNL). Hydroxynitrile lyase (HNL) is relatively recently identified metalloenzyme isolated from *Granulicella tundricola* (*Gt*), a soil bacterium found in the tundra of northern Finland (Hajnal et al., 2013). *Gt*HNL is unusual not only in that mononuclear Mn enzymes are relatively rare, but also it exhibits a novel type of biological Mn coordination. In contrast to the active site structures of MnSOD and OxOx/OxDC (Fig. 2), as shown in Fig. 20, *Gt*HNL has an additional amino acid ligand, histidine 96, which is required for enzyme activity.

Moreover, the reaction catalyzed by *Gt*HNL is also unusual and potentially of commercial interest in pharmaceutical and other types of high value added organic synthesis (Wiedner et al., 2015). *Gt*HNL catalyzes the reversible cleavage of cyanohydrins to carbonyl compounds and HCN, but it is the formation of cyanohydrins that is of interest, a specific example being the synthesis of (*R*)-2-chloromandelonitrile ((*R*)-2-hydroxy-(2-chlorophenyl) acetonitrile) from 2-chlorobenzaldehyde, as shown in Fig. 21.

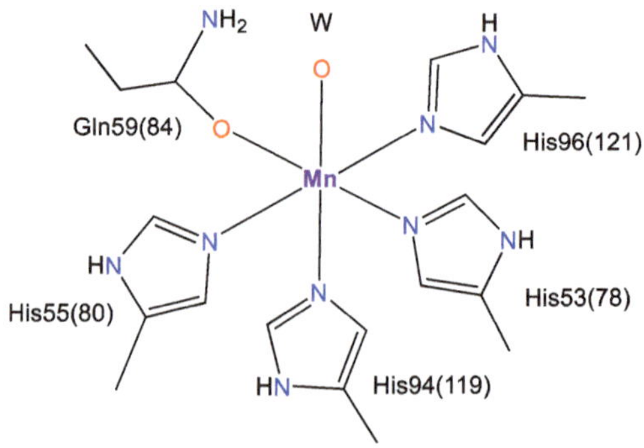

Fig. 20 Cartoon of the mono-Mn containing active site of hydroxynitrile lyase (HNL) from *Granulicella tundricola*, based on the actual crystallographic structures (PDB: 4BIF) reported by Hajnal et al. (Hajnal et al., 2013). The originally reported residue numbers are given along with those reported in the PDB in parentheses. The hydronation state of the terminal water molecule (W) is not indicated.

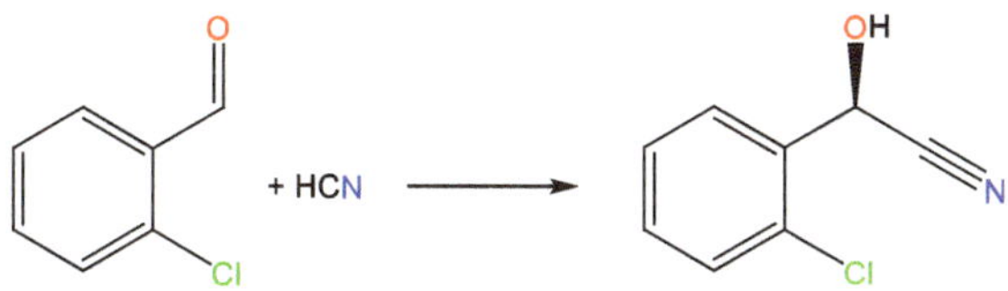

Fig. 21 Synthesis of (*R*)-2-chloromandelonitrile ((*R*)-2-hydroxy-(2-chlorophenyl)acetonitrile) from 2-chlorobenzaldehyde and HCN, which can be catalyzed by the mononuclear Mn-enzyme hydroxynitrile lyase (HNL) from *Granulicella tundricola.*

X-band EPR studies of *Gt*HNL suggested a mechanism in which the aldehyde is activate by O-coordination to Mn(II) (displacing bound water) and is thus activated for nucleophilic attack by cyanide ion (Vertregt et al., 2016).

3.2 Mononuclear Mn-containing biological non-enzymatic systems

Manganese Uptake/Transport. Gram-positive human pathogens such as *Staphylococcus aureus* and *Streptococcus pneumonia* require Mn(II) and acquire this ion from the host using the Mn-binding proteins MntC and PsaA, respectively (Brunjes Brophy & Nolan, 2015; Hadley et al., 2018). Naturally enough, EPR is an important tool in studying these proteins. The Mn(II) in MntC is coordinated similarly to what is found in other mononuclear-Mn proteins, yet distinctly differently, in that there are only four protein-derived ligands and no bound water, as shown in Fig. 22. This X-ray crystallographic study (Gribenko et al., 2013) is also notable to the author, a one-time researcher in industry, in that the work was done at a pharmaceutical company (i.e., privately funded, for-profit), rather than at a university or national laboratory (i.e., often publicly funded, always non-profit). The structure of PsaA is similar (McDevitt et al., 2011). The exclusion of bound water may be a function of a binding/transport/uptake type of protein as opposed to an enzyme, where (a) labile (i.e., water) site(s) are needed.

Standard X-band EPR studies were nevertheless extremely rewarding in providing information on Mn(II) binding (Hayden, Brophy, Cunden, & Nolan, 2013), particularly in competition with calprotectin (see next section). What is most notable here, is that, analogously to the studies on OxDC, HFEPR was successfully applied to both MntC and PsaA (Gagnon et al., 2019). Not shown here is the 388 GHz spectrum recorded at 30 K, which has the typical Mn(II) sextet EPR signature (i.e., to a casual observer, the

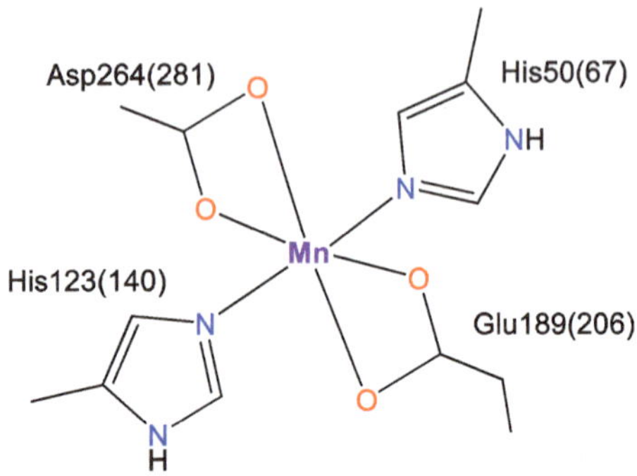

Fig. 22 Cartoon of the mono-Mn containing active site of the Mn transporter MntC from *Staphylococcus aureus*, based on the actual crystallographic structures (PDB: 4K3V) reported by Gribenko et al. (Gribenko et al., 2013). The originally reported residue numbers are given along with those reported in the PDB in parentheses. The carboxylate groups of Asp264 and Glu89 are both bidentate, but each is involved in H-bonding with a water and separately with other residues (none of these interactions is shown). The Mn transporter PsaA from *Streptococcus pneumonia* (PDB: 3ZTT) is essentially the same with coordination to Mn(II) by His67(61), His139(133), Glu205(199), and Asp280(274) (McDevitt et al., 2011), where the residue numbers in the PDB are in parentheses.

spectrum looks the same as at 35 GHz—more than an order of magnitude lower frequency). Simulation of this higher temperature spectrum allowed determination of $g_{iso} = 2.0011$ for MntC (2.0007 for PsaA) and $a_{iso}(^{55}Mn) =$ 241 MHz (236 MHz for PsaA). For comparison, $Mn^{2+}_{(aq)}$ has $a_{iso}(^{55}Mn) \approx$ 265 MHz (the value is slightly sensitive to solution conditions) (Stich et al., 2007). This difference (10(1)%) may seem trivial, but it is not in that $a_{iso}(^{55}Mn(II))$ is relatively invariant (as is $g_{iso}(Mn(II))$) due to the spherical electron distribution of this high-spin $3d^5$ ion (^6S free-ion). The authors attribute the lower $a_{iso}(^{55}Mn)$ value in these proteins to covalent interactions with the two carboxylate ligands (Fig. 22). For comparison, OxDC with only one carboxylate ligand (Fig. 2), has $a_{iso}(^{55}Mn) = 250$ MHz (for site I; 253 MHz for site II) (Angerhofer et al., 2007), so only ~5% lower than the hexaaqua ion. However, OxOx has $a_{iso}(^{55}Mn) = 243$ MHz (with $g_{iso} = 2.00108$) (Moomaw et al., 2013).

HFEPR at low temperature (3 K, Fig. 23), however, made the lowest lying transitions (see Fig. 4, namely, the cyan transitions) relatively dominant which allowed accurate determination of the ZFS of Mn(II) in these proteins. These values were compared with those in other Mn(II) sites and given in Table 1 in Gagnon et al. (Gagnon et al., 2019), which is a fantastic reference of ZFS (and $a_{iso}(^{55}Mn)$) for biological Mn(II) centers, although

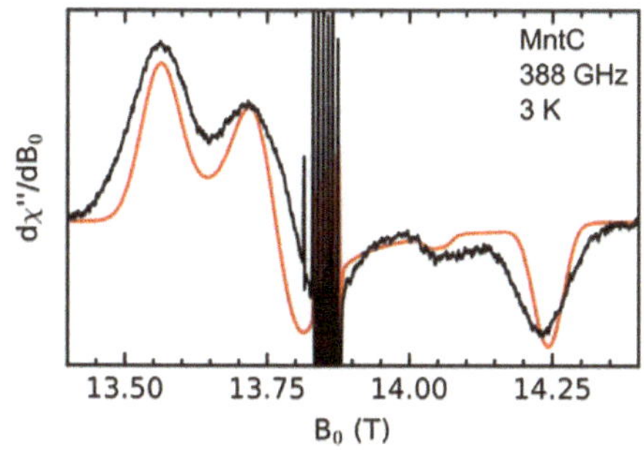

Fig. 23 HFEPR spectrum (black trace) of Mn(II)-MntC at $T=3$ K, which are conditions where the spectrum is dominated by the $m_S=-5/2 \leftrightarrow -3/2$ and $m_S=-3/2 \leftrightarrow -1/2$ transitions. The narrow and thus intense in first derivative presentation central sextet from the $m_S=\pm 1/2 \leftrightarrow \mp 1/2$ transition with resolved ^{55}Mn hyperfine splitting is seen at ~13.85 T ($g_{iso}=2.0011$ from simulation at 30 K). Buffer: 75 mM HEPES, 100 mM NaCl, pH 7.5 (no glassing agent is needed). Experimental settings: spectrometer frequency, 388 GHz; 50 kHz field modulation amplitude, 2.5 mT. Simulation (red trace) parameters: $S=5/2$, $D=+2.72$ GHz (0.0907 cm^{-1}), $E=+0.48$ GHz (0.016 cm^{-1}; positive sign assumed), $a_{iso}(^{55}\text{Mn})=241$ MHz. The simulation reproduces the resolved hyperfine, albeit difficult to see on this broader field range spectrum (but easily seen in narrower field range spectra shown in their Figs. 3 and 4, not reproduced here). *Reprinted with permission from Fig. 5 in Gagnon et al.* J. Phys. Chem. B ***2019***, 123, *4929–4934. doi:10.1021/acs.jpcb.9b03633. Copyright 2019 American Chemical Society.*

strangely and disappointingly, it omits any forms of FosA (see (Smoukov et al., 2002)) as well as OxOx (Moomaw et al., 2013). The low symmetry of these Mn transport protein sites (see Fig. 22) may be the reason for their Mn(II) centers having among the largest magnitude D values (Gagnon et al., 2019) and are relatively rhombic as well ($|E/D|=0.122$ in PsaA, 0.177 in MntC). However, the ZFS for *Ceriporiopsis subvermispora* OxOx is reasonably close to that for MntC and PsaA: $|D|=2.850$ GHz, $|E|=280$ MHz for OxOx (Moomaw et al., 2013) vs $|D|=$ 2.80(7) GHz, $|E|=$ 0.42(7) as the average values for the two transport proteins (Gagnon et al., 2019).

Calprotectin (CP). As thoroughly reviewed by Brophy and Nolan (Brunjes Brophy & Nolan, 2015) and by Juttukonda and Skaar (Juttukonda & Skaar, 2015), Mn uptake is vital for the nefarious (from a anthrocentric view) existence of numerous pathogenic microorganisms. Calprotectin (CP) appears to be the only "shield" available to humans against these "swords" (Zygiel & Nolan, 2018). Calprotectin, as its name implies, binds calcium(II), but also other metal dications: Mn(II), Fe(II), Ni(II), and Zn(II). The active site of CP is very unusual in bioinorganic chemistry that it is homoleptic, having hexahistidine coordination, as shown in Fig. 24.

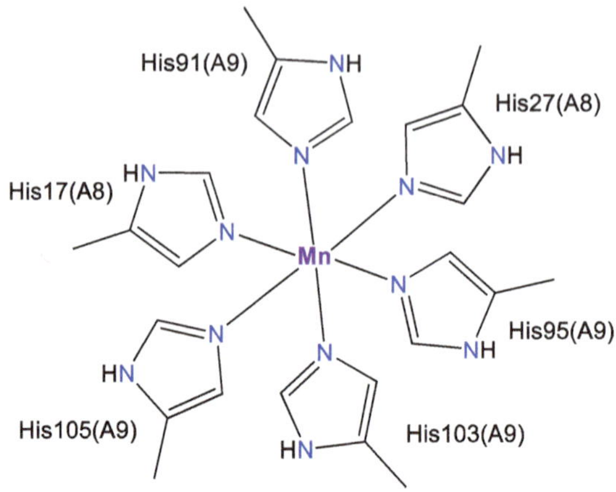

Fig. 24 Cartoon of the mono-Mn containing active site of calprotectin (CP) from *Homo sapiens*, based on the actual crystallographic structure (PDB: 4XJK) reported by Gagnon et al. (Gagnon et al., 2015). The S100 protein subunits are indicated along with the residue number. Two histidine ligands are from one subunit (A8) and four from the other (A9), so Mn coordination helps bind these subunits.

Conventional X-band EPR, which among other findings (Brunjes Brophy et al., 2013; Gagnon et al., 2015, 2019; Hadley et al., 2018, 2019) demonstrated the very small ZFS as expected for this homoleptic Mn(II) site ($|D|=0.525$ GHz for Mn-CP from *Mus musculus*; (Hadley et al., 2019) $|D|=0.485$ GHz for Mn-CP from *H. sapiens* (Gagnon et al., 2015). Is one man or mouse? The difference is only 40 MHz, for calprotectin at least. The rhombicity in both is very high ($|E/D|=0.30$; recall that 0.333 is the maximum rhombicity), which interestingly, is the case for homoleptic Fe(III) as in its hexaaqua cation. Of particular relevance here are the beautiful studies by Nolan, with Britt and coworkers on Mn-CP that included HFEPR (Hadley et al., 2019), but more importantly, ENDOR (Gagnon et al., 2015, 2019; Hadley et al., 2018). Previous sections have noted ENDOR from ^{1,2}H, ^{13}C, and ^{31}P, but the study on Mn-CP allows highlighting another important NMR-active nucleus, namely, ^{15}N. ENDOR can be readily observed from the nearly 100% abundant (99.632% to be precise) ^{14}N, but this nucleus has two disadvantages compared to ^{15}N. First, quadrupole coupling in ^{14}N, while potentially informative, leads to broader and more complicated ENDOR spectra than for non-quadrupolar ^{15}N. Second, ^{15}N has a larger g_N value (by a factor of 1.40) so the hyperfine coupling and Larmor frequency are proportionately larger, which facilitates observation of ENDOR signals. Nolan and coworkers prepared globally ^{15}N-labeled calprotectin murine S100A8 and S100A9 subunits (human (Gagnon et al., 2015) or murine (Hadley et al., 2019)) by overexpression in a M9 minimal medium with $^{15}NH_4Cl$ as the (sole) nitrogen source. Ammonium chloride 99% enriched in ^{15}N is relatively inexpensive (USD 86.00 for 1 g (Ammonium Chloride (^{15}N, 99%), 2022)) so globally ^{15}N-enriched metalloprotein samples are not unattainable (globally ^{17}O-enriched samples are fiscally unattainable, unfortunately). The paramagnetic site specificity of ENDOR means that ^{15}N distant from the $[Mn^{II}(His)_6]^{2+}$ center in Mn-CP are irrelevant, which would not be the case using ^{15}N NMR. Moreover, once the $A(^{15}N)$ values are determined, then the corresponding $A(^{14}N)$ values can easily be calculated from the g_N ratio and the quadrupole coupling extracted. These parameters can then be used for theoretical studies of the electronic structure of the active site. Britt, Nolan and coworkers were able to extract the $A(^{14,15}N)$ values of all histidine imidazole nitrogen atoms in the Mn(II) coordination sphere of CP (Gagnon et al., 2015; Hadley et al., 2019). A representative ^{15}N ENDOR spectrum is shown in Fig. 25.

ENDOR and ESEEM of Mn-CP in H_2O vs D_2O buffer also provided information on H-bonding interactions and solvent accessibility in the active site (Gagnon et al., 2015). A "bottom line" from these studies is that murine and human Mn-CP are remarkably similar (Hadley et al., 2019).

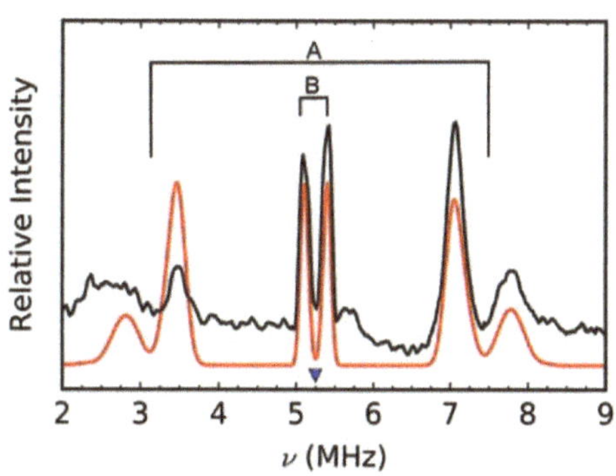

Fig. 25 ^{15}N Mims ENDOR spectrum (black) of Mn(II) and Ca(II) bound to globally ^{15}N-labeled murine CP. The red trace is a simulation. The τ value is 700 ns. A and B indicate the two sets of resonances centered at the ^{15}N Larmor frequency (5.25 MHz), which is indicated by the inverted triangle. The buffer was 75 mM HEPES, 100 mM NaCl, pH 7.5. Experimental settings: microwave frequency = 34.0 GHz; magnetic field = 1216.6 mT; T = 10 K. *Reprinted with permission from Fig. 8 in Hadley et al.* Inorg. Chem. ***2019***, 58, *13578–13590. doi:10.1021/acs.inorgchem.9b00763. Copyright 2019 American Chemical Society.*

Biological Mn(II) as a Radiation Protectant. Bacteria and yeast are far more resistant to ionizing radiation than mammals are. A source of this radiation resistance is Mn(II) which can reduce reactive oxygen species (ROS) and other cytotoxic products of ionizing radiation. Among bacteria, *Deinococcus radiodurans*, a nonsporulating bacterium, is notable in that it is capable of surviving doses (12–16 kGy) of X-rays and γ-rays that are 10-fold greater than the yeast *Saccharomyces cerevisiae*, 20-fold greater than the bacterium *Escherichia coli*, and 3000-fold greater than human cells (Sharma et al., 2017). Hoffman, in collaboration with Culotta and Valentine (McNaughton et al., 2010) and with Daly (Gaidamakova et al., 2022; Sharma et al., 2017, 2013) have shown that EPR, especially at low temperature and 35 GHz, is an ideal in vivo (i.e., of intact cells) probe of this important form of Mn(II). The effectiveness of 35 GHz EPR at 2 K is shown in Fig. 26, in this case for a strain of yeast (*S. cerevisiae* EXF-6218). There are broadly two types of Mn(II), denoted ***H*** for high-symmetry (i.e., narrower EPR envelope due to smaller $|D|$), and ***L*** for low-symmetry (i.e., broader EPR envelope due to larger $|D|$) and the relative amounts of the two can be quantified as shown. The fraction of ***H*** (not of ***L***!) correlates with radiation resistance.

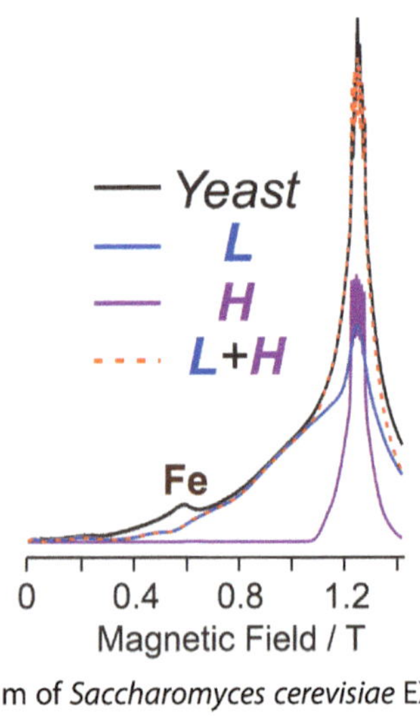

Fig. 26 35 GHz EPR spectrum of *Saccharomyces cerevisiae* EXF-6218 (black trace), along with a simulation (dashed red trace) obtained by appropriately summing the contributions from simulations of exemplar spectra representing the ***H***-Mn^{2+} (violet trace) and ***L***-Mn^{2+} (blue trace) pools (fraction of ***H*** $=0.20$). Conditions: microwave frequency of 34.8–34.9 GHz, $T=2$ K, scan time $=8$ min, 100 kHz field modulation amplitude $=2$ G. *Adapted from an original figure kindly provided by Dr. Ajay Sharma, Northwestern University, Evanston, IL, USA.*

Manganese(II) in this context, is not protein-bound, in contrast to the cases described above, but is bound by small molecules, primarily nucleotides, for which the Mn-O_nP interaction can be seen by ^{31}P ENDOR (McNaughton et al., 2010; Sharma et al., 2013). ^{14}N ESEEM is also informative in demonstrating that radiation leads to Mn—N coordination in *E. coli*, but not in *D. radiodurans* (Sharma et al., 2013). That small molecule Mn(II) (***H***), rather than MnSOD is the major defense against ionizing radiation not only in the microorganism *D. radiodurans*, but also in the radiation-resistant nematode *Caenorhabditis elegans* (Gaidamakova et al., 2022) is one of many significant findings in these studies. Mn(*II*)SOD is indeed detectable in these samples, in particular by HFEPR, as shown in Fig. 27.

3.3 Dinuclear Mn-containing enzymes

In contrast to the situation for mononuclear Mn sites, whether in enzymes, transport proteins, DNA/RNA, et al., the application of EPR to dinuclear systems is less fruitful (or "fruitworthy," a term coined, in the Irish literary tradition, by the late Mayor of Chicago, Jane M. Byrne, neé Burke). The problem is exchange coupling between the two Mn ions (Bencini & Gatteschi, 1990), which are relatively close in space and directly connected by bridging ligand(s), as shown in Fig. 3. Integer spin (non-Kramers) Mn ions, essentially Mn(III) ($S = 2$) are already problematic for conventional

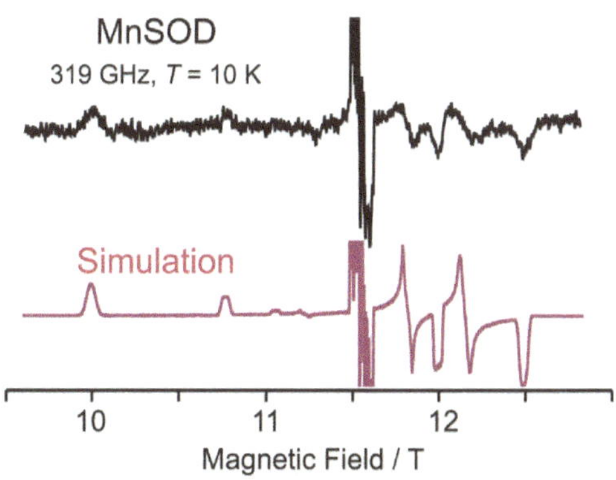

Fig. 27 HFEPR derivative-display spectrum of Mn(*II*)SOD (black trace), plus simulation (violet trace) using the program SPIN by A. Ozarowski. Conditions: $T = 10$ K, microwave frequency = 319.2 GHz, scan time = 2 mT/s, 50 kHz field modulation amplitude = 2.2 mT. Simulation parameters: $S = 5/2$, $D = -10.493$ GHz, $E = -0.779$ GHz (given the same sign by convention); the large central signal ($mS = \pm 1/2 \leftrightarrow \mp 1/2$) is truncated in both experiment and simulation. *Adapted from an original figure kindly provided by Dr. Ajay Sharma, Northwestern University, Evanston, IL, USA.*

EPR, but strong antiferromagnetic coupling between then can lead to a total (giant) spin ground state that is diamagnetic ($S_{tot} = S_1 - S_2 = 2 - 2 = 0$). But this can also be the case for half-integer spin (Kramers) Mn ions (Mn(II), $S = 5/2$ or Mn(IV), $S = 3/2$: $S_{tot} = S_1 - S_2 = n/2 - n/2 = 0$; $n = 3, 5$). The "ladder" of excited states ($S_{tot} = j, j = 1, 2, ..., n$) can give EPR spectra at higher temperature, but these can be difficult to obtain and interpret, requiring HFEPR and true experts, as in these di-Mn(III) model systems (Liu, Krzystek, Hill, Barrios, & Aromi, 2013; Retegan, Collomb, Neese, & Duboc, 2013). For the purposes of enzymology, what is preferable is a mixed oxidation state, one integer-spin Mn ion and one half-integer. It is also possible for there to be an additional paramagnet, such as an organic (protein-derived) radical, as can occur in ribonucleotide reductase (RNR) (Cotruvo & Stubbe, 2010). Nevertheless, if the exchange coupling is weak, then EPR can be useful in di-Mn (or mixed FeMn) enzymes as shown below.

Mn Catalase (MnCat). A key example of an "EPR/ENDOR-friendly" mixed oxidation state are the studies on Mn catalase (MnCat) in its "super-oxidized" (Mn(II*I*)Mn(IV)) state (McConnell et al., 2012; Stich et al., 2010). ESEEM and HYSCORE (i.e., two-dimensional ESEEM) provided a complete characterization of the N-donor ligands to both Mn ions (Stich et al., 2010). A beautifully complementary study on this system used $H_2^{17}O$ (as well as D_2O) to introduce this expensive, but rewarding isotope into the terminal and bridging O sites (see Fig. 3) in MnCat (McConnell et al., 2012). It is, however, possible to record EPR spectra for the red, Mn(II)Mn(II) form of MnCat as shown many years ago (Khangulov, Barynin, Voevodskaya, & Grebenko, 1990).

Sulfate Thiohydrolase (SoxB). Various microorganisms can use elemental sulfur or reduced sulfur compounds as energy (electron) sources (Friedrich, Rother, Bardischewsky, Quentmeier, & Fischer, 2001). One such organism is the Gram-negative bacterium *Paracoccus pantotrophus*, which can be used to remove H_2S from water. The sulfur oxidizing machinery of *P. pantotrophus* includes an enzyme, SoxB, that acts as a sulfate thiohydrolase (EC 3.1.6.20), with the following in vitro redox reaction: $O_3SS^{2-} + 5\ H_2O \rightarrow 2\ SO_4^{2-} + 10H^+ + 8\ e^-$. The specific in vivo reaction catalyzed by SoxB is:

$$\mathrm{SoxY-(S)-SSO_3^- + H_2O \rightarrow SoxY-(S)-SH + SO_4^{2-} + H^+},$$

where SoxY can have either a thiol(ate) or (hydro)disulfide (given by (S) in the equation) bound. The crystal structure of SoxB from *Thermus thermophilus* HB27 has been reported (PDB: 2WDF; 2WDE with thiosulfate; 2WDD with sulfate) (Sauvé et al., 2009). The active site of SoxB contains two Mn(II) ions and a beautiful X- and Q-band EPR study was performed on this electronically challenging system by Epel et al. (Epel, Schäfer, Quentmeier, Friedrich, & Lubitz, 2005). They were able to extract the exchange coupling (*J*) parameter between the two Mn(II) ions which gave a Mn—Mn distance.

Formiminoglutamase (FIGase). Formiminoglutamase (FIGase, also known as *N*-formimino-L-glutamate formiminohydrolase) catalyzes the final step of a histidine utilization pathway (Hai, Dugery, Healy, & Christianson, 2013).

FIGase is part of the same enzyme superfamily as arginase (Christianson, 2005). The active site structure of FIGase (PDB: 4MXR) is very similar to that of arginase I (PDB: 1RLA; see Fig. 3), except one Mn(II) ion has no histidine ligand; instead, asparagine (Asn114) is bound via Oδ. Indeed, the active site structure of the H101N mutant of arginase I (PDB: 1P8P) (Cama, Emig, Ash, & Christianson, 2003) is very similar to that of native FIGase (Hai et al., 2013). The reaction catalyzed has some similarity as well, as shown by comparison of the arginase (Fig. 28, top) and FIGase (Fig. 28, bottom) reactions. FIGase has been isolated from *Trypanosoma cruzi*, which

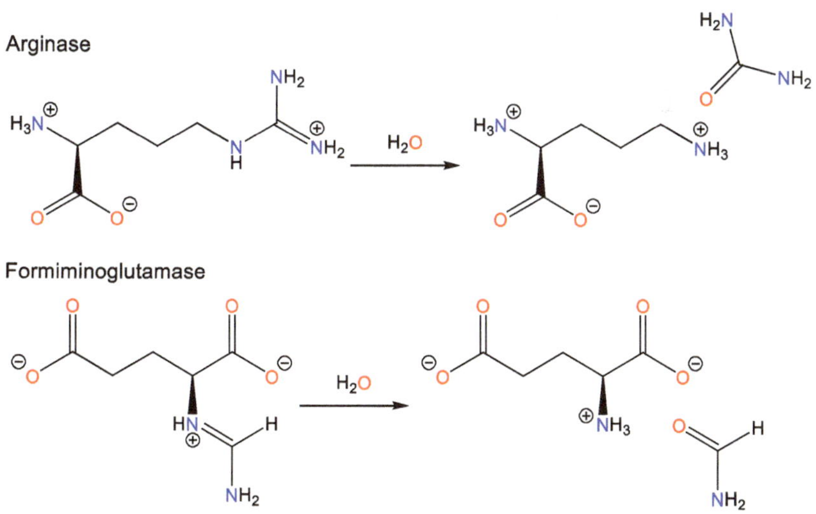

Fig. 28 Reactions catalyzed by arginase I (top) and formiminoglutamase (FIGase; bottom). In arginase, the reactant is L-arginine and the products are L-ornithine and urea. In FIGase, the reactant is *N*-formimino-L-glutamate and the products are L-glutamate and formamide. This is the final step in histidine utilization pathway I; for the complete picture see Hai et al. (Hai et al., 2013).

is the parasite that causes Chagas disease, a scourge of the tropical Americas, such as much of Brazil (the eponym is Carlos Chagas, a Brazilian physician who discovered the disease in 1909). As an aside, the structures of arginase I are of the enzyme from *Rattus norvegicus* (Cama et al., 2003; Kanyo et al., 1996) a scourge of much of the world, especially cities such as Philadelphia, where the work was performed (and Chicago, where the author is located). In addition to structural studies on *Trypanosoma cruzi* FIGase, Christianson and coworkers definitively demonstrated by enzymological work that FIGase is a di-Mn(II) enzyme (Hai et al., 2013). This important enzyme would seem an ideal candidate for EPR investigation (e.g., as described above for SoxB), but no such studies have been performed to date.

Ribonucleotide Reductase (RNR). Class I ribonucleotide reductases (RNRs) are found in di-Fe (I-a), di-Mn (I-b), and FeMn (I-c) forms (Cotruvo & Stubbe, 2012). Because of its fundamental importance, namely, catalyzing the formation of deoxyribonucleotides from ribonucleotides, RNR has been intensely studied (Ruskoski & Boal, 2021). Class I RNR consists of two homodimeric proteins, a larger subunit R1 that contains the site of ribonucleotide reduction and a smaller subunit R2 with the metalloradical cofactor, an M_2 (M = Fe or Mn) cluster that stabilizes a free tyrosyl radical, which initiates enzymatic function (Zhang & Stubbe, 2011). The active site structure of di-Mn RNR R2 from *Corynebacterium ammoniagenes* (a Gram-positive soil bacterium used industrially to produce nucleotides) has been determined by Cox et al. (2010) and is shown in Fig. 28.

Only a few EPR-related aspects of studies on Class I-b RNRs will be highlighted here. A study by Cotruvo and Stubbe (no "et al." in this case) reported the EPR spectra of the di-Mn(II) form of *Escherichia coli* Class I-b RNR (Cotruvo & Stubbe, 2010). Weak exchange coupling, analogous to that described above for SoxB (Epel et al., 2005) and seen earlier for MnCat (Khangulov et al., 1990), led to a multiline pattern demonstrating the di-Mn(II) state. The di-Mn(III) alone form is EPR-silent, but the Mn(III)Mn(IV) form is EPR active ($S_{tot} = S_{Mn(III)} - S_{Mn(IV)} = 4/2 - 3/2 = 1/2$), and the resulting multiline EPR pattern has been thoroughly analyzed (Cotruvo, Stich, Britt, & Stubbe, 2013). More importantly, Cotruvo and Stubbe showed the enzymatically active form was di-Mn(III)-$Y^{\bullet}$, where $Y^{\bullet}$ is a tyrosyl radical (Cotruvo & Stubbe, 2010, 2011). A detailed analysis of the multifrequency (9, 34, and 244 GHz)

EPR spectra of *Corynebacterium ammoniagenes* Mn_2-RNR was performed by Cox et al. (Cox et al., 2010). That an $S=1/2$ radical type of EPR signal obtains indicates that the di-Mn center must have integer spin, which is possible only if the two Mn ions have the same oxidation state, otherwise a half-integer total spin would result, which when coupled to the radical would give an integer spin (or at least. Cox et al. therefore used two models: Mn(III)Mn(III) and Mn(IV)Mn(IV); Mn(I) being too reduced and Mn(V) being too oxidized to be viable. A set of spin Hamiltonian parameters was extracted for each of these two models. These includes single-ion ZFS (*D* and *E*) and g_{iso} values for each of Mn_A and Mn_B (see Fig. 29), the (relatively large magnitude) exchange coupling between the two and the (relatively small magnitude) exchange coupling between the $[Mn_2]^{6+,8+}$ unit and the radical. The intrinsic *g* values of the tyrosyl (Tyr115, see Fig. 29) were known from many studies on isolated $Y^{\bullet}$ and other phenolic radicals. A further consideration was the orientation of $Y_{115}{}^{\bullet}$ and the di-Mn center in a brilliant analysis that made use of the enzyme crystal structure, the known **g** tensor orientation of tyrosyl/phenol radical species,

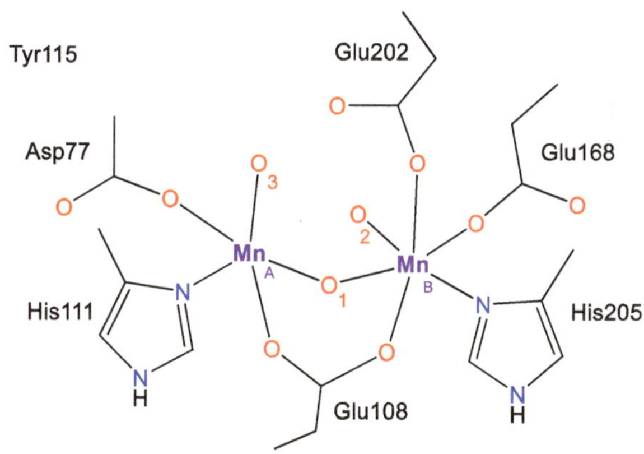

Fig. 29 Cartoon of the di-Mn containing active site of RNR subunit R2F from *Corynebacterium ammoniagenes*, based on the actual crystallographic structure (PDB: 3MJO) reported by Cox et al. (Cox et al., 2010). The hydronation states of the bridging (O1) and terminal (O2, O3) water/hydroxide/oxide moieties are not indicated. The approximate location of Ty115, the site of the catalytically relevant radical is roughly indicated by label only. The PDB residues are decremented by one from those given here, which are the authors' original. This structure can be contrasted with that of MnCat shown above in Fig. 3.

and obviously the multifrequency EPR spectra of the spin coupled system. The "bottom line" of this tour de force was that the active form is indeed di-Mn(III)-Y$^{\bullet}$ (Cox et al., 2010).

Another important type of RNR is Class I-c RNR, which employs a MnFe active site. The Class I-c RNR from *Chlamydia trachomatis* (*Ct*RNR) has been extensively studied by Mössbauer spectroscopy, thanks to its Fe site, in its active, Mn(IV)Fe(III) state (Dassama, Boal, Krebs, Rosenzweig, & Bollinger, 2012), which is not amenable to EPR due to its $S=1$ ground state (from $S_{tot}=S_{Fe(III)}-S_{Mn(IV)}=5/2-3/2=1$). There is also the Mn(III)Fe(III) state wherein $S_{tot}=S_{Fe(III)}-S_{Mn(III)}=5/2-4/2=1/2$, making it well suited for EPR, albeit less important for understanding enzyme mechanism. With ^{57}Fe-enrichment (needed for Mössbauer), and interesting multiline hyperfine structure results from coupling to both ^{57}Fe and ^{55}Mn (Dassama et al., 2012). Another EPR-active form of *Ct*RNR is Mn(IV)Fe(IV) wherein $S_{tot}=S_{Fe(IV)}-S_{Mn(IV)}=4/2-3/2=1/2$. Thus far, ^{17}O as either $^{17}O_2$ or $H_2^{17}O$ has not been used on this system, but Martinie et al. used ^{1,2}H ENDOR to provide information on the proton/deuteron-ation state of the active site (Martinie et al., 2017). Previous figures in this chapter have reproduced ENDOR spectra from ^{13}C (Fig. 12) and ^{15}N (Fig. 25) labeled systems (^{1}H ENDOR, while important, is not deemed sufficiently exotic to be shown here, and ^{31}P ENDOR is amply documented in studies on nucleic acid systems that are not covered here (Hunsicker-Wang et al., 2009)). Fig. 30 presents ^{2}H ENDOR spectra reported by Martinie et al. (Martinie et al., 2017), which is the first quadrupolar nucleus ($I=1$ for ^{2}H) to be depicted in this chapter, and the spectra exhibit resolved quadrupole splitting (see also Fig. 8).

The key result of this study is that this intermediate comprises a di-μ-oxo diamond core structure with a terminal hydroxide ligand to the Mn(IV) ion (Martinie et al., 2017). This and other catalytically relevant states of MnFe-RNR are shown in Fig. 31.

Along the lines of the studies on Class I-b RNR involving radical sites (Zhang & Stubbe, 2011), Livada et al. used ELDOR (DEER) to probe *Ct*RNR (Livada et al., 2015), wherein the initiating event converts the $S=1$ Mn(IV)/Fe(III) cofactor to the $S=1/2$ Mn(III)/Fe(III) form in the β subunit (analogous to R2 in di-Fe RNR) and generates a cysteinyl radical in the α active site (analogous to R1 in di-Fe RNR). The radical can be trapped by use of 2′-azido-2′-deoxyuridine-5′-diphosphate, which generates a long-lived, nitrogen-centered radical (N$^{\bullet}$) in α. The distance

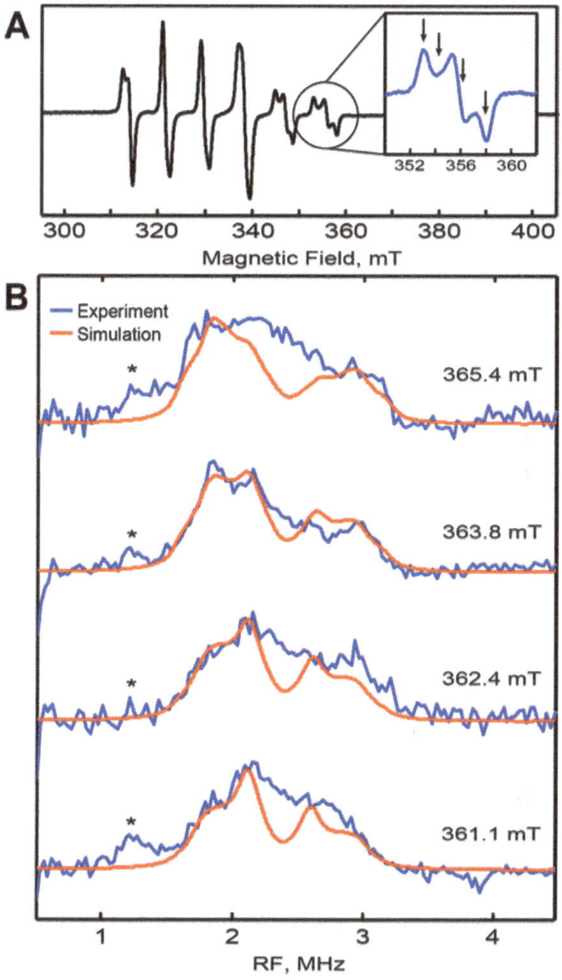

Fig. 30 Interrogation of the Mn(*IV*)Fe(IV) form of *Chlamydia trachomatis* Class I-c RNR (*Ct*RNR) by EPR/ENDOR. (A) Continuous-wave, X-band spectrum (the sextet from ^{55}Mn hyperfine coupling is apparent) with expanded view of the packet at the highest magnetic field (inset). (B) X-band, field dependent, ^{2}H-ENDOR spectra (blue) and simulations (red) enzyme prepared in D_2O. The field positions for the ENDOR spectra are indicated by arrows in the inset of panel (A). The peaks marked with an asterisk are artifacts. The deviations between the experimental and simulated spectra near the ^{2}H Larmor frequency ($\nu(^{2}H) = 2.36$, 2.37, 2.38, and 2.39 MHz at 361.1, 362.4, 363.8, and 365.4 mT, respectively) are attributable to matrix deuterons (i.e., bulk solvent D_2O). All measurements were performed at 15 K. CW and ENDOR measurements were performed at 9.628 and 9.702 GHz, respectively. Simulations used: $\mathbf{A}(^{2}H) = [-0.90, -0.43, +1.41]$ MHz with Euler angles [0, 25, 0]° and $\mathbf{P}(^{2}H) = [-0.07, -0.07, 0.14]$ MHz with Euler angles [0,110,0]°. *Reprinted with permission from Fig. 2 in Martinie et al.* J. Am. Chem. Soc. ***2017***, 139, *1950–1957. doi:10.1021/jacs.6b11563. Copyright 2017 American Chemical Society.*

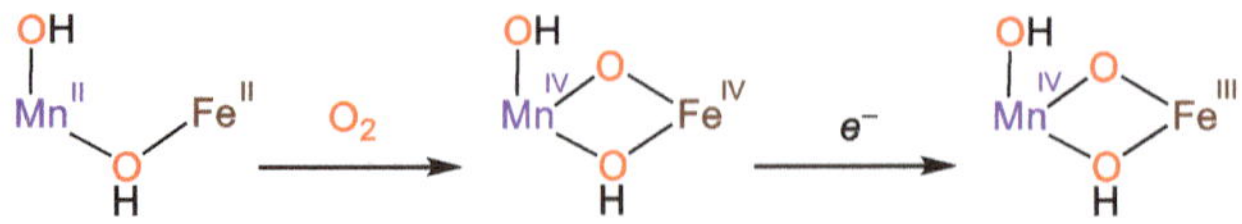

Fig. 31 Activation pathway of the Class I-c ribonucleotide reductase (RNR). The hydronation state of all OH species is uncertain, except for that of the bridging hydroxide in the center structure ($S = 1/2$ ground state), which was determined using $^{1,2}H$ ENDOR (Martinie et al., 2017). The rightmost structure has a diamagnetic ($S = 0$) ground state.

between the Mn(III)/Fe(III) cofactor in β and $N^{\bullet}$ in α was found to be 43 ± 1 Å via ELDOR. Livada et al. thus provided the first structural data on the *Ct*RNR holoenzyme complex and the first direct experimental measurement of the inter-subunit radical translocation distance in any Class I RNR (Livada et al., 2015).

R2lox. A new class of oxidases (R2lox), identified in the pathogenic bacterium *Mycobacterium tuberculosis* and a number of other bacteria (e.g., *Geobacillus kaustophilus*) and archaea (Högbom, 2010) is homologous to the metal-containing subunit of Class I-c RNR, but has a different in vivo function (Shafaat et al., 2014). The EPR-active Mn(III)Fe(III) ($S_{tot} = S_{Fe(III)} - S_{Mn(III)} = 5/2 - 4/2 = 1/2$) form of R2lox (including ^{57}Fe-enriched) has been investigated by a wealth of advanced EPR techniques, including Q-band ^{55}Mn and ^{1}H ENDOR, ^{2}H-HYSCORE, and W-band EDNMR (Shafaat et al., 2014). Most notable is that ^{17}O EDNMR was successfully performed not only for several Mn(*III*)Mn(IV) model compounds, but also for MnCat in both its Mn(II)Mn(III) and Mn(III)Mn(IV) states and for R2lox in its Mn(III)Fe(III) form (Rapatskiy et al., 2015). Fig. 32 presents ^{17}O EDNMR results for R2lox. In the case of ^{17}O with its high nuclear spin ($I = 5/2$), quadrupole splitting is not resolved unlike for ^{2}H (Fig. 30), but important biochemical information is obtained.

These studies showed that there is a μ-hydroxo/bis-μ-carboxylato bridge between the two metal ions, with a terminal water on the Mn(III) ion. EPR has played an important role in subsequent studies by Shafaat and coworkers on R2lox (Kisgeropoulos et al., 2020), which addressed fundamental questions such as cluster assembly (Miller, Trivelas, Maugeri, Blaesi, & Shafaat, 2017), and photoactivation (Maugeri et al., 2018). Current work involves the use of multifrequency EPR to address the reactivity of R2lox (Kisgeropoulos, Gan, Greer, Hazel, & Shafaat, 2022).

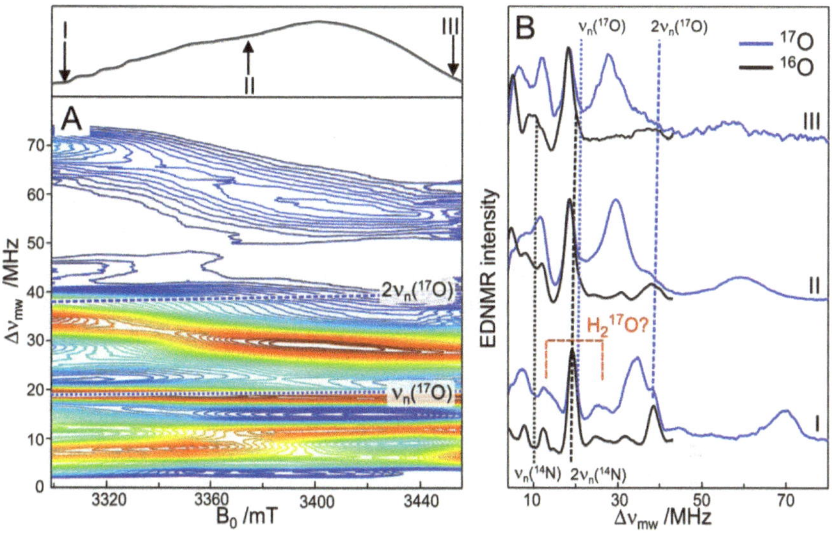

Fig. 32 W-band (~95 GHz) 2D-EDNMR spectra of the ^{17}O labeled (introduced via exchange with $H_2{}^{17}O$) R2lox cofactor (Mn(III)Fe(III), $S=1/2$ form). (A) The upper part of (A) is the W-band EPR envelope of R2lox that indicates three resonant field (B_0) positions, I, II, and III, which correspond to the fields at which the three EDNMR spectra shown in (B) were recorded. The lower part of (A) shows the EDNMR signal, namely, the difference between the two microwave frequencies ($\Delta\nu_{mw}$; i.e., the resonant NMR frequency) in this double EPR technique plotted vs B_0 in a color contour "3D" intensity scale. In (B), standard plots of EDNMR signal intensity vs $\Delta\nu_{mw}$ are shown, comparing spectra for the ^{17}O labeled cofactor (blue lines) to unlabeled cofactor (black lines). Signals from ^{14}N are present in both (including an intense "double quantum" transition at $2\nu(^{14}N) \approx 20$ MHz), but signals due to ^{17}O are present only in the labeled sample. *Reprinted with permission from Fig. 7 in Rapatskiy et al.* J. Phys. Chem. B ***2015***, 119, *13904–13921. doi:10.1021/acs.jpcb.5b04614. Copyright 2015 American Chemical Society.*

4. Summary

As mentioned at the outset, no transition metal can compete with iron for biological relevance. Yet manganese is strong contender for the "silver medal," with apologies for the mixed metal metaphor. From a spectroscopic point of view, mononuclear Mn(II) is ideal for EPR and its advanced derivatives, ENDOR and ESEEM/HYSCORE, particularly at Q-band. This point is well known, but merits repeating to the metallobiochemistry community. One aspect that is less widely known is the benefits of Mn(II) substitution not only for ions such as Mg(II), as done in the nucleotide-binding, kinase, etc., area but also for Fe(II), as shown in the work on SLO, which ends up being "legitimized" by the discovery of Mn-lipoxygenases. In another direction, the wide range of microwave frequencies now available in EPR needs to be better publicized and promoted to this community. HFEPR has been shown to be useful in several Mn-containing systems already and has the potential for wider application. One possibility is in the area of di-Mn systems with total spin states that are necessarily integer, such as Mn(III)Mn(III). From a biochemistry point of view, the number of Mn-containing systems continues to increase as evidenced by the relatively recent discovery of, e.g., Classes I-b,c RNR and R2lox, and of SoxB, and formiminoglutamase in the dinuclear and hydroxynitrile lyase

Acknowledgments

The author wishes to thank Prof. Brian M. Hoffman and Dr. Ajay Sharma, Northwestern University, Evanston, IL, USA; Prof. Masaki Horitani, Saga University, Japan; Prof. Alexander Angerhofer, University of Florida, Gainesville, FL, USA; and Dr. Andrew Ozarowski (né Andrzej Ożarowski), NHMFL/Florida State University, Tallahassee, FL, USA for their massive help with figures used in this chapter. The author is also very grateful for helpful discussions with them and also with Prof. Victoria J. DeRose, University of Oregon, Eugene, OR, USA; Prof. Wilfred R. Hagen, Technische Universiteit Delft, The Netherlands; Prof. Timothy A. Jackson, University of Kansas, Lawrence, KS, USA; Prof. Adam R. Offenbacher, East Carolina University, Greenville, NC, USA; Prof. Brad S. Pierce, University of Alabama, Tuscaloosa, AL, USA; Prof. Hannah S. Shafaat, The Ohio State University, Columbus, OH, USA; Prof. Alexey Silakov, Pennsylvania State University, University Park, PA, USA; Prof. Jon S. Thorson, University of Kentucky, Lexington, KY, USA; and the esteemed editor of this volume, Prof. R. David Britt, University of California, Davis, CA, USA. The author also thanks Prof. Judith N. Burstyn and Mr. Brian R. Weaver, University of Wisconsin, Madison, WI, USA for kindly providing the CooA sample used for illustrative purposes only in Fig. 5, which sample is part of a collaborative project on CooA unrelated to the topic of this chapter. Lastly, the author dedicates this chapter to the memory of Prof. Richard N. Armstrong, Vanderbilt University, with whom the author had the privilege of collaborating on Mn enzymes.

References

Abragam, A., & Bleaney, B. (2012). *Electron paramagnetic resonance of transition ions (Oxford classic texts in the physical sciences)*. Oxford, UK: Oxford University Press.

AMMONIUM. (2022). *CHLORIDE (^{15}N, 99%)*. Retrieved from https://shop.isotope.com/productdetails.aspx?itemno=NLM-467-PK.

Angerhofer, A., Moomaw, E. W., García-Rubio, I., Ozarowski, A., Krzystek, J., Weber, R. T., et al. (2007). Multifrequency EPR studies on the Mn(II) centers of oxalate decarboxylase. *The Journal of Physical Chemistry. B, 111*(19), 5043–5046. https://doi.org/10.1021/jp0715326.

Armstrong, D. R. (2015). *RICHARD NEIL*. Retrieved from https://medschool.vanderbilt.edu/wp-content/uploads/sites/7/public_files/Obits/Armstrong/ARMSTRONG-Obituary.pdf.

Atherton, N. M. (1993). *Electron spin resonance. Theory and applications*. Chichester, UK: Ellis Horwood.

Barber, J. (2016). Mn_4Ca cluster of photosynthetic oxygen-evolving center: Structure, function and evolution. *Biochemistry*, *55*(42), 5901–5906. https://doi.org/10.1021/acs.biochem.6b00794.

Barynin, V. V., Whittaker, M. M., Antonyuk, S. V., Lamzin, V. S., Harrison, P. M., Artymiuk, P. J., et al. (2001). Crystal structure of manganese catalase from *lactobacillus plantarum*. *Structure*, *9*(8), 725–738. https://doi.org/10.1016/S0969-2126(01)00628-1.

Beal, N. J., & O'Malley, P. J. (2016). Manganese oxidation state assignment for manganese catalase. *Journal of the American Chemical Society*, *138*(13), 4358–4361. https://doi.org/10.1021/jacs.6b02600.

Bencini, A., & Gatteschi, D. (1990). *Electron paramagnetic resonance of exchange coupled systems*. Berlin, Germany: Springer Verlag.

Bennati, M., Hertel, M. M., Fritscher, J., Prisner, T. F., Weiden, N., Hofweber, R., et al. (2006). High-frequency 94 GHz ENDOR characterization of the metal binding site in wild-type Ras·GDP and its oncogenic mutant G12V in frozen solution. *Biochemistry*, *45*(1), 42–50. https://doi.org/10.1021/bi051156k.

Bennati, M., Weiden, N., Dinse, K.-P., & Hedderich, R. (2004). ^{57}Fe ENDOR spectroscopy on the iron−sulfur cluster involved in substrate reduction of Heterodisulfide reductase. *Journal of the American Chemical Society*, *126*(27), 8378–8379. https://doi.org/10.1021/ja0498179.

Bernat, B. A., Laughlin, L. T., & Armstrong, R. N. (1997). Fosfomycin resistance protein (FosA) is a manganese Metalloglutathione transferase related to glyoxalase I and the Extradiol dioxygenases. *Biochemistry*, *36*(11), 3050–3055. https://doi.org/10.1021/bi963172a.

Bowman, M. K., Chen, H., & Maryasov, A. G. (2017). Fourier-transform EPR. In D. Goldfarb, & S. Stoll (Eds.), *eMagRes* (pp. 387–406). John Wiley & Sons, Ltd.

Britt, R. D., Peloquin, J. M., & Campbell, K. A. (2000). Pulsed and parallel-polarization EPR characterization of the photosystem II oxygen-evolving complex. *Annual Review of Biophysics and Biomolecular Structure*, *29*, 463–495. https://doi.org/10.1146/annurev.biophys.29.1.463.

Broderick, W. E., Hoffman, B. M., & Broderick, J. B. (2018). Mechanism of radical initiation in the radical S-adenosyl-L-methionine superfamily. *Accounts of Chemical Research*, *51*(11), 2611–2619. https://doi.org/10.1021/acs.accounts.8b00356.

Brunjes Brophy, M., Nakashige, T. G., Gaillard, A., & Nolan, E. M. (2013). Contributions of the S100A9 C-terminal tail to high-affinity Mn(II) chelation by the host-defense protein human calprotectin. *Journal of the American Chemical Society*, *135*(47), 17804–17817. https://doi.org/10.1021/ja407147d.

Brunjes Brophy, M., & Nolan, E. M. (2015). Manganese and microbial pathogenesis: Sequestration by the mammalian immune system and utilization by microorganisms. *ACS Chemical Biology*, *10*(3), 641–651. https://doi.org/10.1021/cb500792b.

Brynda, M., & Britt, R. D. (2010). The manganese-calcium cluster of the oxygen-evolving system: Synthetic models, EPR studies, and electronic structure calculations. In G. Hanson, & L. Berliner (Eds.), *Metals in biology: Applications of high-resolution EPR to metalloenzymes* (pp. 203–271). New York, NY: Springer New York.

Buettner, K. M., & Valentine, A. M. (2012). Bioinorganic chemistry of titanium. *Chemical Reviews*, *112*(3), 1863–1881. https://doi.org/10.1021/cr1002886.

Byer, A. S., Yang, H., McDaniel, E. C., Kathiresan, V., Impano, S., Pagnier, A., et al. (2018). Paradigm shift for radical *S*-adenosyl-L-methionine reactions: The organometallic intermediate **Ω** is central to catalysis. *Journal of the American Chemical Society*, *140*(28), 8634–8638. https://doi.org/10.1021/jacs.8b04061.

Cama, E., Emig, F. A., Ash, D. E., & Christianson, D. W. (2003). Structural and functional importance of first-Shell metal ligands in the binuclear manganese cluster of arginase I. *Biochemistry*, *42*(25), 7748–7758. https://doi.org/10.1021/bi030074y.

Campbell, K. A., Force, D. A., Nixon, P. J., Dole, F., Diner, B. A., & Britt, R. D. (2000). Dual-mode EPR detects the initial intermediate in Photoassembly of the photosystem II Mn cluster: The influence of amino acid residue 170 of the D1 polypeptide on Mn coordination. *Journal of the American Chemical Society*, *122*(15), 3754–3761. https://doi.org/10.1021/ja000142t.

Campbell, K. A., Yikilmaz, E., Grant, C. V., Gregor, W., Miller, A.-F., & Britt, R. D. (1999). Parallel polarization EPR characterization of the Mn(III) Center of Oxidized Manganese Superoxide Dismutase. *Journal of the American Chemical Society*, *121*(19), 4714–4715. https://doi.org/10.1021/ja9902219.

Campomanes, P., Kellett, W. F., Easthon, L. M., Ozarowski, A., Allen, K. N., Angerhofer, A., et al. (2014). Assigning the EPR fine structure parameters of the Mn(II) centers in *Bacillus subtilis* oxalate decarboxylase by site-directed mutagenesis and DFT/MM calculations. *Journal of the American Chemical Society*, *136*(6), 2313–2323. https://doi.org/10.1021/ja408138f.

Carmieli, R., Larsen, T. M., Reed, G. H., Zein, S., Neese, F., & Goldfarb, D. (2007). The catalytic Mn^{2+} sites in the enolase – inhibitor complex: Crystallography, single-crystal EPR, and DFT calculations. *Journal of the American Chemical Society*, *129*(14), 4240–4252. https://doi.org/10.1021/ja066124e.

Carrington, A., & McLachlan, A. D. (1967). *Introduction to magnetic resonance with applications to chemistry and chemical physics*. New York: Harper & Row.

Chang, C. H., Svedruzic, D., Ozarowski, A., Walker, L., Yeagle, G., Britt, R. D., et al. (2004). EPR spectroscopic characterization of the manganese center and a free radical in the oxalate decarboxylase reaction: Identification of a tyrosyl radical during turnover. *The Journal of Biological Chemistry*, *279*(51), 52840–52849. https://doi.org/10.1074/jbc.M402345200.

Christianson, D. W. (2005). Arginase: Structure, mechanism, and physiological role in male and female sexual arousal. *Accounts of Chemical Research*, *38*(3), 191–201. https://doi.org/10.1021/ar040183k.

Clark, R. W., Youn, H., Parks, R. B., Cherney, M. M., Roberts, G. P., & Burstyn, J. N. (2004). Investigation of the role of the N-terminal proline, the distal heme ligand in the CO sensor CooA. *Biochemistry*, *43*(44), 14149–14160. https://doi.org/10.1021/bi0487948.

Cotruvo, J. A., Jr. (2019). The chemistry of lanthanides in biology: Recent discoveries, emerging principles, and technological applications. *ACS Central Science*, *5*(9), 1496–1506. https://doi.org/10.1021/acscentsci.9b00642.

Cotruvo, J. A., Jr., Stich, T. A., Britt, R. D., & Stubbe, J. (2013). Mechanism of assembly of the dimanganese-tyrosyl radical cofactor of class Ib ribonucleotide reductase: Enzymatic generation of superoxide is required for tyrosine oxidation via a Mn(III)Mn(IV) intermediate. *Journal of the American Chemical Society*, *135*(10), 4027–4039. https://doi.org/10.1021/ja312457t.

Cotruvo, J. A., Jr., & Stubbe, J. (2010). An active Dimanganese(III)−tyrosyl radical cofactor in *Escherichia coli* class Ib ribonucleotide reductase. *Biochemistry*, *49*(6), 1297–1309. https://doi.org/10.1021/bi902106n.

Cotruvo, J. A., Jr., & Stubbe, J. (2011). *Escherichia coli* class Ib ribonucleotide reductase contains a Dimanganese(III)-tyrosyl radical cofactor in vivo. *Biochemistry*, *50*(10), 1672–1681. https://doi.org/10.1021/bi101881d.

Cotruvo, J. A., Jr., & Stubbe, J. (2012). Metallation and mismetallation of iron and manganese proteins in vitro and in vivo: The class I ribonucleotide reductases as a case study. *Metallomics*, *4*(10), 1020–1036. https://doi.org/10.1039/c2mt20142a.

Cox, N., Nalepa, A., Lubitz, W., & Savitsky, A. (2017). ELDOR-detected NMR: A general and robust method for electron-nuclear hyperfine spectroscopy? *Journal of Magnetic Resonance*, *280*, 63–78. https://doi.org/10.1016/j.jmr.2017.04.006.

Cox, N., Nalepa, A., Pandelia, M.-E., Lubitz, W., & Savitsky, A. (2015). Chapter nine - pulse double-resonance EPR techniques for the study of Metallobiomolecules. In P. Z. Qin, & K. Warncke (Eds.), *Vol. 563. Methods in enzymology* (pp. 211–249). Academic Press.

Cox, N., Ogata, H., Stolle, P., Reijerse, E., Auling, G., & Lubitz, W. (2010). A tyrosyl – Dimanganese coupled spin system is the native Metalloradical cofactor of the R2F subunit of the ribonucleotide reductase of *Corynebacterium ammoniagenes*. *Journal of the American Chemical Society*, *132*(32), 11197–11213. https://doi.org/10.1021/ja1036995.

Cox, N., Pantazis, D. A., Neese, F., & Lubitz, W. (2013). Biological water oxidation. *Accounts of Chemical Research*, *46*(7), 1588–1596. https://doi.org/10.1021/ar3003249.

CR. (2022). *Crystal structure of oxalate decarboxylase*. Retrieved from https://pdbj.org/mine/structural_details/1uw8.

Crans, D. C., Smee, J. J., Gaidamauskas, E., & Yang, L. (2004). The chemistry and biochemistry of vanadium and the biological activities exerted by vanadium compounds. *Chemical Reviews*, *104*(2), 849–902. https://doi.org/10.1021/cr020607t.

Dassama, L. M. K., Boal, A. K., Krebs, C., Rosenzweig, A. C., & Bollinger, J. M. (2012). Evidence that the β subunit of *chlamydia trachomatis* ribonucleotide reductase is active with the manganese ion of its manganese(IV)/iron(III) cofactor in site 1. *Journal of the American Chemical Society*, *134*(5), 2520–2523. https://doi.org/10.1021/ja211314p.

Debus, R. J. (2021). Alteration of the O_2-producing Mn_4Ca cluster in photosystem II by the mutation of a metal ligand. *Biochemistry*, *60*(51), 3841–3855. https://doi.org/10.1021/acs.biochem.1c00504.

DeRose, V. J., & Yglesias, M. V. (2021). 8.39 - metal ion interactions with DNA, RNA, and nucleic acid enzymes. In E. C. Constable, G. Parkin, & L. Que Jr., (Eds.), *Comprehensive coordination chemistry III* (pp. 968–993). Oxford: Elsevier.

Dismukes, G. C. (1996). Manganese enzymes with binuclear active sites. *Chemical Reviews*, *96*(7), 2909–2926. https://doi.org/10.1021/cr950053c.

Doan, P. E., Gurbiel, R. J., & Hoffman, B. M. (2007). The ups and downs of Feher-style ENDOR. *Applied Magnetic Resonance*, *31*(3), 649. https://doi.org/10.1007/BF03166608.

Doan, P. E., & Hoffman, B. M. (1997). Making hyperfine selection in Mims ENDOR independent of deadtime. *Chemical Physics Letters*, *269*(3–4), 208–214. https://doi.org/10.1016/S0009-2614(97)00293-5.

Doan, P. E., & Hoffman, B. M. (2000). How important is parallel-mode EPR in electron spin echo envelope modulation studies of non-Kramers systems? *Inorganica Chimica Acta*, *297*(1), 400–403. https://doi.org/10.1016/S0020-1693(99)00386-2.

Duboc, C., Collomb, M.-N., & Neese, F. (2010). Understanding the zero-field splitting of mononuclear manganese(II) complexes from combined EPR spectroscopy and quantum chemistry. *Applied Magnetic Resonance*, *37*(1–4), 229–245. https://doi.org/10.1007/s00723-009-0085-4.

Duboc, C., Collomb, M.-N., Pécaut, J., Deronzier, A., & Neese, F. (2008). Definition of magneto-structural correlations for the Mn^{II} ion. *Chemistry - A European Journal*, *14*(21), 6498–6509. https://doi.org/10.1002/chem.200800426.

Duke Today Staff. (2019). *Duke flags lowered: Biochemist Irwin Fridovich dies at age 90*. Retrieved from https://today.duke.edu/2019/11/duke-flags-lowered-biochemist-irwin-fridovich-dies-age-90.

Emerson, J. P., Kovaleva, E. G., Farquhar, E. R., Lipscomb, J. D., & Que, L., Jr. (2008). Swapping metals in Fe- and Mn-dependent dioxygenases: Evidence for oxygen activation without a change in metal redox state. *Proceedings of the National Academy of Sciences of the United States of America*, *105*(21), 7347–7352. https://doi.org/10.1073/pnas.0711179105.

Epel, B., Schäfer, K.-O., Quentmeier, A., Friedrich, C., & Lubitz, W. (2005). Multifrequency EPR analysis of the dimanganese cluster of the putative sulfate thiohydrolase SoxB of *Paracoccus pantotrophus*. *Journal of Biological Inorganic Chemistry*, *10*(6), 636–642. https://doi.org/10.1007/s00775-005-0015-3.

Fiedler, A. T., & Fischer, A. A. (2017). Oxygen activation by mononuclear Mn, co, and Ni centers in biology and synthetic complexes. *Journal of Biological Inorganic Chemistry*, *22*(2), 407–424. https://doi.org/10.1007/s00775-016-1402-7.

Fillgrove, K. L., Pakhomova, S., Newcomer, M. E., & Armstrong, R. N. (2003). Mechanistic diversity of Fosfomycin resistance in pathogenic microorganisms. *Journal of the American Chemical Society*, *125*(51), 15730–15731. https://doi.org/10.1021/ja039307z.

Fox, B. G. (2021). 8.18 - diiron enzyme structure and catalysis. In E. C. Constable, G. Parkin, & L. Que Jr., (Eds.), *Comprehensive coordination chemistry III* (pp. 455–499). Oxford: Elsevier.

Frederiksen, J. K., & Piccirilli, J. A. (2009). Identification of catalytic metal ion ligands in ribozymes. *Methods*, *49*(2), 148–166. https://doi.org/10.1016/j.ymeth.2009.07.005.

Freisinger, E., & Sigel, R. K. O. (2019). The bioinorganic periodic table. *Chimia International Journal for Chemistry*, *73*(3), 185–193. https://doi.org/10.2533/chimia.2019.185.

Friedrich, C. G., Rother, D., Bardischewsky, F., Quentmeier, A., & Fischer, J. (2001). Oxidation of reduced inorganic sulfur compounds by bacteria: Emergence of a common mechanism? *Applied and Environmental Microbiology*, *67*(7), 2873–2882. https://doi.org/10.1128/AEM.67.7.2873-2882.2001.

Gaffney, B. J. (2014). Connecting lipoxygenase function to structure by electron paramagnetic resonance. *Accounts of Chemical Research*, *47*(12), 3588–3595. https://doi.org/10.1021/ar500290r.

Gagnon, D. M., Brophy, M. B., Bowman, S. E. J., Stich, T. A., Drennan, C. L., Britt, R. D., et al. (2015). Manganese binding properties of human calprotectin under conditions of high and low calcium: X-ray crystallographic and advanced electron paramagnetic resonance spectroscopic analysis. *Journal of the American Chemical Society*, *137*(8), 3004–3016. https://doi.org/10.1021/ja512204s.

Gagnon, D. M., Hadley, R. C., Ozarowski, A., Nolan, E. M., & Britt, R. D. (2019). High-field EPR spectroscopic characterization of Mn(II) bound to the bacterial solute-binding proteins MntC and PsaA. *The Journal of Physical Chemistry. B*, *123*(23), 4929–4934. https://doi.org/10.1021/acs.jpcb.9b03633.

Gaidamakova, E. K., Sharma, A., Matrosova, V. Y., Grichenko, O., Volpe, R. P., Tkavc, R., et al. (2022). Small-molecule Mn antioxidants in *Caenorhabditis elegans* and *Deinococcus radiodurans* supplant MnSOD enzymes during aging and irradiation. *MBio*. https://doi.org/10.1128/mbio.03394-21.

Goldfarb, D., & Stoll, S. (2018). *EPR spectroscopy: Fundamentals and methods*. John Wiley & Sons, Ltd.

Golynskiy, M. V., Gunderson, W. A., Hendrich, M. P., & Cohen, S. M. (2006). Metal binding studies and EPR spectroscopy of the manganese transport regulator MntR. *Biochemistry*, *45*(51), 15359–15372. https://doi.org/10.1021/bi0607406.

Gribenko, A., Mosyak, L., Ghosh, S., Parris, K., Svenson, K., Moran, J., et al. (2013). Three-Dimensional Structure and Biophysical Characterization of *Staphylococcus aureus* Cell Surface Antigen–Manganese Transporter MntC. *Journal of Molecular Biology*, *425*(18), 3429–3445. https://doi.org/10.1016/j.jmb.2013.06.033.

Gunderson, W. A., Zatsman, A. I., Emerson, J. P., Farquhar, E. R., Que, L., Jr., Lipscomb, J. D., et al. (2008). Electron paramagnetic resonance detection of intermediates in the enzymatic cycle of an Extradiol dioxygenase. *Journal of the American Chemical Society*, *130*(44), 14465–14467. https://doi.org/10.1021/ja8052255.

Guo, M., Corona, T., Ray, K., & Nam, W. (2019). Heme and nonheme high-valent iron and manganese oxo cores in biological and Abiological oxidation reactions. *ACS Central Science*, *5*(1), 13–28. https://doi.org/10.1021/acscentsci.8b00698.

Gurbiel, R. J., Bolin, J. T., Ronco, A. E., Mortenson, L., & Hoffman, B. M. (1991). Single-crystal EPR and ENDOR study of nitrogenase from *clostridium pasteurianum*. *Journal of Magnetic Resonance*, *91*(2), 227–240. https://doi.org/10.1016/0022-2364(91)90188-Y.

Hadley, R. C., Gagnon, D. M., Brophy, M. B., Gu, Y., Nakashige, T. G., Britt, R. D., et al. (2018). Biochemical and spectroscopic observation of Mn(II) sequestration from bacterial Mn(II) transport machinery by calprotectin. *Journal of the American Chemical Society*, *140*(1), 110–113. https://doi.org/10.1021/jacs.7b11207.

Hadley, R. C., Gagnon, D. M., Ozarowski, A., Britt, R. D., & Nolan, E. M. (2019). Murine calprotectin coordinates Mn(II) at a Hexahistidine site with ca(II)-dependent affinity. *Inorganic Chemistry*, *58*(20), 13578–13590. https://doi.org/10.1021/acs.inorgchem.9b00763.

Hagen, W. R. (2008). *Biomolecular EPR Spectroscopy* (1st ed.). Boca Raton, Florida, USA: CRC Press.

Hahn, E. L. (1950). Spin Echoes. *Physics Review*, *80*(4), 580–594. https://doi.org/10.1103/PhysRev.80.580.

Hai, Y., Dugery, R. J., Healy, D., & Christianson, D. W. (2013). Formiminoglutamase from *Trypanosoma Cruzi* is an arginase-like manganese metalloenzyme. *Biochemistry*, *52*(51), 9294–9309. https://doi.org/10.1021/bi401352h.

Hajnal, I., Łyskowski, A., Hanefeld, U., Gruber, K., Schwab, H., & Steiner, K. (2013). Biochemical and structural characterization of a novel bacterial manganese-dependent hydroxynitrile lyase. *The FEBS Journal*, *280*(22), 5815–5828. https://doi.org/10.1111/febs.12501.

Halkides, C. J., Farrar, C. T., Larsen, R. G., Redfield, A. G., & Singel, D. J. (1994). Characterization of the active site of p21 *ras* by electron spin-Echo envelope modulation spectroscopy with selective labeling: Comparisons between GDP and GTP forms. *Biochemistry*, *33*(13), 4019–4035. https://doi.org/10.1021/bi00179a031.

Harmer, J. R. (2017). Hyperfine spectroscopy—ENDOR. In D. Goldfarb, & S. Stoll (Eds.), *eMagRes* (pp. 1493–1514). John Wiley & Sons, Ltd.

Hayden, J. A., Brophy, M. B., Cunden, L. S., & Nolan, E. M. (2013). High-affinity manganese coordination by human calprotectin is calcium-dependent and requires the histidine-rich site formed at the dimer Interface. *Journal of the American Chemical Society*, *135*(2), 775–787. https://doi.org/10.1021/ja3096416.

Hearn, A. S., Stroupe, M. E., Cabelli, D. E., Ramilo, C. A., Luba, J. P., Tainer, J. A., et al. (2003). Catalytic and structural effects of amino acid substitution at histidine 30 in human manganese superoxide dismutase: Insertion of valine Cγ into the substrate Access Channel. *Biochemistry*, *42*(10), 2781–2789. https://doi.org/10.1021/bi0266481.

Hoffman, B. M. (2003). ENDOR of metalloenzymes. *Accounts of Chemical Research*, *36*(7), 522–529. https://doi.org/10.1021/ar0202565.

Hoffman, B. M., Dean, D. R., & Seefeldt, L. C. (2009). Climbing nitrogenase: Toward a mechanism of enzymatic nitrogen fixation. *Accounts of Chemical Research*, *42*(5), 609–619. https://doi.org/10.1021/ar8002128.

Högbom, M. (2010). The manganese/iron-carboxylate proteins: What is what, where are they, and what can the sequences tell us? *Journal of Biological Inorganic Chemistry*, *15*(3), 339–349. https://doi.org/10.1007/s00775-009-0606-5.

Hoogstraten, C. G., Grant, C. V., Horton, T. E., DeRose, V. J., & Britt, R. D. (2002). Structural analysis of metal ion ligation to nucleotides and nucleic acids using pulsed EPR spectroscopy. *Journal of the American Chemical Society*, *124*(5), 834–842. https://doi.org/10.1021/ja0112238.

Horitani, M., Offenbacher, A. R., Carr, C. A. M., Yu, T., Hoeke, V., Cutsail, G. E., III, et al. (2017). ^{13}C ENDOR spectroscopy of lipoxygenase–substrate complexes reveals the structural basis for C–H activation by tunneling. *Journal of the American Chemical Society*, *139*(5), 1984–1997. https://doi.org/10.1021/jacs.6b11856.

Horitani, M., Shisler, K., Broderick, W. E., Hutcheson, R. U., Duschene, K. S., Marts, A. R., et al. (2016). Radical SAM catalysis via an organometallic intermediate with an Fe-[5'-C]-deoxyadenosyl bond. *Science*, *352*(6287), 822–825. https://doi.org/10.1126/science.aaf5327.

Horitani, M., Yashiro, H., Hagiwara, M., & Hori, H. (2008). Multi-frequency and high-field EPR study of manganese(III) protoporphyrin IX reconstituted myoglobin with an $S=2$ integer electron spin. *Journal of Inorganic Biochemistry*, *102*(4), 781–788. https://doi.org/10.1016/j.jinorgbio.2007.11.015.

Horton, T. E., Clardy, D. R., & DeRose, V. J. (1998). Electron paramagnetic resonance spectroscopic measurement of Mn^{2+} binding affinities to the hammerhead ribozyme and correlation with cleavage activity. *Biochemistry*, *37*(51), 18094–18101. https://doi.org/10.1021/bi981425p.

Hu, Y., Lee, C. C., & Ribbe, M. W. (2012). Vanadium nitrogenase: A two-hit wonder? *Dalton Transactions*, *41*(4), 1118–1127. https://doi.org/10.1039/C1DT11535A.

Hunsicker-Wang, L., Vogt, M., & DeRose, V. J. (2009). Chapter 16 - EPR methods to study specific metal-ion binding sites in RNA. In *Vol. 468. Methods in enzymology* (pp. 335–367). Academic Press.

ICBIC. (2019). *19th international conference on biological inorganic chemistry (ICBIC-19)*. (Retrieved from https://www.chem.uzh.ch/en/icbic19.html).

Jackson, T. A., Gutman, C. T., Maliekal, J., Miller, A.-F., & Brunold, T. C. (2013). Geometric and electronic structures of manganese-substituted iron superoxide dismutase. *Inorganic Chemistry*, *52*(6), 3356–3367. https://doi.org/10.1021/ic302867y.

John Wiley & Sons, Ltd. (2007). *eMagRes*. https://doi.org/10.1002/9780470034590.

Just, V. J., Stevenson, C. E. M., Bowater, L., Tanner, A., Lawson, D. M., & Bornemann, S. (2004). A closed conformation of *bacillus subtilis* oxalate decarboxylase Oxdc provides evidence for the true identity of the active site. *The Journal of Biological Chemistry*, *279*, 19867. https://doi.org/10.1074/jbc.M313820200.

Juttukonda, L. J., & Skaar, E. P. (2015). Manganese homeostasis and utilization in pathogenic bacteria. *Molecular Microbiology*, *97*(2), 216–228. https://doi.org/10.1111/mmi.13034.

Kanyo, Z. F., Scolnick, L. R., Ash, D. E., & Christianson, D. W. (1996). Structure of a unique binuclear manganese cluster in arginase. *Nature*, *383*(6600), 554–557. https://doi.org/10.1038/383554a0.

Khangulov, S. V., Barynin, V. V., Voevodskaya, N. V., & Grebenko, A. I. (1990). ESR spectroscopy of the binuclear cluster of manganese ions in the active center of Mn-catalase from *Thermus thermophilus*. *Biochimica et Biophysica Acta (BBA) - Bioenergetics*, *1020*(3), 305–310. https://doi.org/10.1016/0005-2728(90)90161-V.

Kisgeropoulos, E. C., Gan, Y. J., Greer, S. M., Hazel, J. M., & Shafaat, H. S. (2022). Pulsed multifrequency EPR spectroscopy reveals key branch points for one- vs. two-electron reactivity in Mn/Fe proteins. (under review).

Kisgeropoulos, E. C., Griese, J. J., Smith, Z. R., Branca, R. M. M., Schneider, C. R., Högbom, M., et al. (2020). Key structural motifs balance metal binding and oxidative reactivity in a heterobimetallic Mn/Fe protein. *Journal of the American Chemical Society*, *142*(11), 5338–5354. https://doi.org/10.1021/jacs.0c00333.

Kositzki, R., Mebs, S., Marx, J., Griese, J. J., Schuth, N., Högbom, M., et al. (2016). Protonation state of MnFe and FeFe cofactors in a ligand-binding oxidase revealed by X-ray absorption, emission, and vibrational spectroscopy and QM/MM calculations. *Inorganic Chemistry*, *55*(19), 9869–9885. https://doi.org/10.1021/acs.inorgchem.6b01752.

Kostenko, A., Ray, K., Iavarone, A. T., & Offenbacher, A. R. (2019). Kinetic characterization of the C–H activation step for the lipoxygenase from the pathogenic fungus *Magnaporthe oryzae*: Impact of N-linked glycosylation. *Biochemistry, 58*(29), 3193–3203. https://doi.org/10.1021/acs.biochem.9b00467.

Koua, F. H. M., Umena, Y., Kawakami, K., & Shen, J.-R. (2013). Structure of Sr-substituted photosystem II at 2.1 Å resolution and its implications in the mechanism of water oxidation. *Proceedings of the National Academy of Sciences of the United States of America, 110*(10), 3889–3894. https://doi.org/10.1073/pnas.1219922110.

Kovacs, J. A. (2015). Tuning the relative stability and reactivity of manganese dioxygen and Peroxo intermediates via systematic ligand modification. *Accounts of Chemical Research, 48*(10), 2744–2753. https://doi.org/10.1021/acs.accounts.5b00260.

Krstić, I., Endeward, B., Margraf, D., Marko, A., & Prisner, T. F. (2012). Structure and dynamics of nucleic acids. In M. Drescher, & G. Jeschke (Eds.), *EPR spectroscopy: Applications in chemistry and biology* (pp. 159–198). Berlin, Heidelberg: Springer.

Krzystek, J., Ozarowski, A., & Telser, J. (2006). Multi-frequency, high-field EPR as a powerful tool to accurately determine zero-field splitting in high-spin transition metal coordination complexes. *Coordination Chemistry Reviews, 250*(17–18), 2308–2324. https://doi.org/10.1016/j.ccr.2006.03.016.

Krzystek, J., & Telser, J. (2016). Measuring giant anisotropy in paramagnetic transition metal complexes with relevance to single-ion magnetism. *Dalton Transactions, 45*(42), 16751–16763. https://doi.org/10.1039/C6DT01754A.

Lee, H. B., Marchiori, D. A., Chatterjee, R., Oyala, P. H., Yano, J., Britt, R. D., et al. (2020). $S = 3$ ground state for a tetranuclear $Mn^{IV4}O_4$ complex mimicking the S_3 state of the oxygen-evolving complex. *Journal of the American Chemical Society, 142*(8), 3753–3761. https://doi.org/10.1021/jacs.9b10371.

Lingappa, U. F., Yeager, C. M., Sharma, A., Lanza, N. L., Morales, D. P., Xie, G., et al. (2021). An ecophysiological explanation for manganese enrichment in rock varnish. *Proceedings of the National Academy of Sciences of the United States of America, 118*(25), e2025188118. https://doi.org/10.1073/pnas.2025188118.

Liu, J., Krzystek, J., Hill, S., Barrios, L., & Aromi, G. (2013). Elucidating magnetic exchange and anisotropy in weakly coupled MnIII dimers. *Inorganic Chemistry, 52*(2), 718–723. https://doi.org/10.1021/ic301849f.

Livada, J., Martinie, R. J., Dassama, L. M. K., Krebs, C., Bollinger, J. M., & Silakov, A. (2015). Direct measurement of the radical translocation distance in the class I ribonucleotide reductase from *chlamydia trachomatis*. *The Journal of Physical Chemistry. B, 119*(43), 13777–13784. https://doi.org/10.1021/acs.jpcb.5b04067.

Lockart, M. M., Edwards, K. C., Vincent, J. B., & Pierce, B. S. (2021). Electron paramagnetic spectrum of dimanganic human serum transferrin. *Polyhedron, 203*, 115224. https://doi.org/10.1016/j.poly.2021.115224.

Lohmiller, T., Krewald, V., Sedoud, A., Rutherford, A. W., Neese, F., Lubitz, W., et al. (2017). The first state in the catalytic cycle of the water-oxidizing enzyme: Identification of a water-derived μ-Hydroxo bridge. *Journal of the American Chemical Society, 139*(41), 14412–14424. https://doi.org/10.1021/jacs.7b05263.

Mailer, C., & Taylor, C. P. S. (1973). Rapid adiabatic passage EPR of ferricytochrome *c*. signal enhancement and determination of the spin-lattice relaxation time. *Biochimica et Biophysica Acta, 322*, 195–203. https://doi.org/10.1016/0005-2795(73)90293-6.

Marchiori, D. A., Debus, R. J., & Britt, R. D. (2020). Pulse EPR spectroscopic characterization of the S_3 state of the oxygen-evolving complex of photosystem II isolated from Synechocystis. *Biochemistry, 59*(51), 4864–4872. https://doi.org/10.1021/acs.biochem.0c00880.

Martin, J. E., & Giedroc, D. P. (2016). Functional determinants of metal ion transport and selectivity in paralogous cation diffusion facilitator transporters CzcD and MntE in *Streptococcus pneumoniae*. *Journal of Bacteriology*, *198*(7), 1066. https://doi.org/10.1128/JB.00975-15.

Martinie, R. J., Blaesi, E. J., Krebs, C., Bollinger, J. M., Silakov, A., & Pollock, C. J. (2017). Evidence for a Di-μ-oxo diamond Core in the Mn(IV)/Fe(IV) activation intermediate of ribonucleotide reductase from *chlamydia trachomatis*. *Journal of the American Chemical Society*, *139*(5), 1950–1957. https://doi.org/10.1021/jacs.6b11563.

Maugeri, P. T., Griese, J. J., Branca, R. M., Miller, E. K., Smith, Z. R., Eirich, J., et al. (2018). Driving protein conformational changes with light: Photoinduced structural rearrangement in a heterobimetallic oxidase. *Journal of the American Chemical Society*, *140*(4), 1471–1480. https://doi.org/10.1021/jacs.7b11966.

McConnell, I. L., Grigoryants, V. M., Scholes, C. P., Myers, W. K., Chen, P.-Y., Whittaker, J. W., et al. (2012). EPR–ENDOR characterization of (^{17}O, ^{1}H, ^{2}H) water in manganese catalase and its relevance to the oxygen-evolving complex of photosystem II. *Journal of the American Chemical Society*, *134*(3), 1504–1512. https://doi.org/10.1021/ja203465y.

McDevitt, C. A., Ogunniyi, A. D., Valkov, E., Lawrence, M. C., Kobe, B., McEwan, A. G., et al. (2011). A molecular mechanism for bacterial susceptibility to zinc. *PLoS Pathogens*, 7(11), e1002357. https://doi.org/10.1371/journal.ppat.1002357.

McGarvey, B. R. (1998). Survey of ligand field parameters of strong field d^5 complexes obtained from the g matrix. *Coordination Chemistry Reviews*, *170*(1), 75–92. https://doi.org/10.1016/S0010-8545(97)00073-8.

McNaughton, R. L., Reddi, A. R., Clement, M. H. S., Sharma, A., Barnese, K., Rosenfeld, L., et al. (2010). Probing in vivo Mn^{2+} speciation and oxidative stress resistance in yeast cells with electron-nuclear double resonance spectroscopy. *Proceedings of the National Academy of Sciences of the United States of America*, *107*(35), 15335–15339. https://doi.org/10.1073/pnas.1009648107.

Milikisiyants, S., Chatterjee, R., & Lakshmi, K. V. (2011). Two-dimensional ^{1}H HYSCORE spectroscopy of Dimanganese Di-μ-oxo mimics of the oxygen-evolving complex of photosystem II. *The Journal of Physical Chemistry. B*, *115*(42), 12220–12229. https://doi.org/10.1021/jp205629g.

Miller, E. K., Trivelas, N. E., Maugeri, P. T., Blaesi, E. J., & Shafaat, H. S. (2017). Time-resolved investigations of heterobimetallic cofactor assembly in R2lox reveal distinct Mn/Fe intermediates. *Biochemistry*, *56*(26), 3369–3379. https://doi.org/10.1021/acs.biochem.7b00403.

Moomaw, E. W., Angerhofer, A., Moussatche, P., Ozarowski, A., García-Rubio, I., & Richards, N. G. J. (2009). Metal dependence of oxalate decarboxylase activity. *Biochemistry*, *48*(26), 6116–6125. https://doi.org/10.1021/bi801856k.

Moomaw, E. W., Hoffer, E., Moussatche, P., Salerno, J. C., Grant, M., Immelman, B., et al. (2013). Kinetic and spectroscopic studies of Bicupin oxalate oxidase and putative active site mutants. *PLoS One*, *8*(3). https://doi.org/10.1371/journal.pone.0057933, e57933.

Morley, G. W. (2011). *GavinMorley, HahnEcho GWM, CC BY-SA 3.0*. (Retrieved from https://en.wikipedia.org/wiki/Spin_echo#/media/File:HahnEcho_GWM.gif).

Morrissey, S. R., Horton, T. E., & DeRose, V. J. (2000). Mn^{2+} sites in the hammerhead ribozyme investigated by EPR and continuous-wave Q-band ENDOR spectroscopies. *Journal of the American Chemical Society*, *122*(14), 3473–3481. https://doi.org/10.1021/ja992989z.

Morrissey, S. R., Horton, T. E., Grant, C. V., Hoogstraten, C. G., Britt, R. D., & DeRose, V. J. (1999). Mn^{2+}–nitrogen interactions in RNA probed by electron spin – Echo envelope modulation spectroscopy: Application to the hammerhead ribozyme. *Journal of the American Chemical Society*, *121*(39), 9215–9218. https://doi.org/10.1021/ja9921571.

Nguyen, A. I., Suess, D. L. M., Darago, L. E., Oyala, P. H., Levine, D. S., Ziegler, M. S., et al. (2017). Manganese–cobalt Oxido Cubanes relevant to manganese-doped water oxidation catalysts. *Journal of the American Chemical Society, 139*(15), 5579–5587. https://doi.org/10.1021/jacs.7b01792.

Offenbacher, A. R., Sharma, A., Doan, P. E., Klinman, J. P., & Hoffman, B. M. (2020). The soybean lipoxygenase–substrate complex: Correlation between the properties of tunneling-ready states and ENDOR-detected structures of ground states. *Biochemistry, 59*(7), 901–910. https://doi.org/10.1021/acs.biochem.9b00861.

Pake, G. E., & Estle, T. L. (1973). *The physical principles of electron paramagnetic resonance.* London, ON, Canada: W. A. Benjamin, Advanced Book Program.

Papp-Wallace, K. M., & Maguire, M. E. (2006). Manganese transport and the role of manganese in virulence. *Annual Review of Microbiology, 60*(1), 187–209. https://doi.org/10.1146/annurev.micro.60.080805.142149.

Peloquin, J. M., Campbell, K. A., Randall, D. W., Evanchik, M. A., Pecoraro, V. L., Armstrong, W. H., et al. (2000). ^{55}Mn ENDOR of the S_2-state multiline EPR signal of photosystem II: Implications on the structure of the tetranuclear Mn cluster. *Journal of the American Chemical Society, 122*(44), 10926–10942. https://doi.org/10.1021/ja002104f.

Petersen, J., Gessner, C., Fisher, K., Mitchell, C. J., Lowe, D. J., & Lubitz, W. (2005). Mn^{2+}-adenosine nucleotide complexes in the presence of the nitrogenase iron-protein: Detection of conformational rearrangements directly at the nucleotide binding site by EPR and 2D-ESEEM (two-dimensional electron spin-echo envelope modulation spectroscopy). *The Biochemical Journal, 391*(3), 527–539. https://doi.org/10.1042/BJ20050226.

Petrie, S., Stranger, R., & Pace, R. J. (2017). What Mn Kβ spectroscopy reveals concerning the oxidation states of the Mn cluster in photosystem II. *Physical Chemistry Chemical Physics, 19*(40), 27682–27693. https://doi.org/10.1039/C7CP04797E.

Potapov, A., & Goldfarb, D. (2006). Quantitative characterization of the Mn^{2+} complexes of ADP and ATPγS by W-band ENDOR. *Applied Magnetic Resonance, 30*(3), 461. https://doi.org/10.1007/BF03166212.

Rajakovich, L. J., Zhang, B., McBride, M. J., Boal, A. K., Krebs, C., & Martin Bollinger, J. (2020). 5.10 - emerging structural and functional diversity in proteins with dioxygen-reactive Dinuclear transition metal cofactors. In H.-W. Liu, & T. P. Begley (Eds.), *Comprehensive natural products III* (pp. 215–250). Oxford: Elsevier.

Rapatskiy, L., Ames, W. M., Pérez-Navarro, M., Savitsky, A., Griese, J. J., Weyhermüller, T., et al. (2015). Characterization of oxygen bridged manganese model complexes using multifrequency ^{17}O-hyperfine EPR spectroscopies and density functional theory. *The Journal of Physical Chemistry. B, 119*(43), 13904–13921. https://doi.org/10.1021/acs.jpcb.5b04614.

Retegan, M., Collomb, M.-N., Neese, F., & Duboc, C. (2013). A combined high-field EPR and quantum chemical study on a weakly ferromagnetically coupled dinuclear Mn(III) complex. A complete analysis of the EPR spectrum beyond the strong coupling limit. *Physical Chemistry Chemical Physics, 15*(1), 223–234. https://doi.org/10.1039/C2CP42955A.

Rice, D. B., Massie, A. A., & Jackson, T. A. (2017). Manganese–oxygen intermediates in O–O bond activation and hydrogen-atom transfer reactions. *Accounts of Chemical Research, 50*(11), 2706–2717. https://doi.org/10.1021/acs.accounts.7b00343.

Rife, C. L., Pharris, R. E., Newcomer, M. E., & Armstrong, R. N. (2002). Crystal structure of a Genomically encoded Fosfomycin resistance protein (FosA) at 1.19 Å resolution by MAD phasing off the L-III edge of Tl^+. *Journal of the American Chemical Society, 124*(37), 11001–11003. https://doi.org/10.1021/ja026879v.

Roth, A., & Plass, W. (2008). Carboxylate-bridged Dinuclear active sites in Oxygenases: Diiron, Dimanganese, or is Heterodinuclear better? *Angewandte Chemie, International Edition, 47*(40), 7588–7591. https://doi.org/10.1002/anie.200802366.

Ruskoski, T. B., & Boal, A. K. (2021). The periodic table of ribonucleotide reductases. *The Journal of Biological Chemistry, 297*(4), 101137. https://doi.org/10.1016/j.jbc.2021.101137.

Sahu, S., & Goldberg, D. P. (2016). Activation of dioxygen by iron and manganese complexes: A heme and nonheme perspective. *Journal of the American Chemical Society, 138*(36), 11410–11428. https://doi.org/10.1021/jacs.6b05251.

Sauvé, V., Roversi, P., Leath, K. J., Garman, E. F., Antrobus, R., Lea, S. M., et al. (2009). Mechanism for the hydrolysis of a sulfur-sulfur bond based on the crystal structure of the Thiosulfohydrolase SoxB. *The Journal of Biological Chemistry, 284*(32), 21707–21718. https://doi.org/10.1074/jbc.M109.002709.

Schweiger, A., & Jeschke, G. (2001). *Principles of pulse electron paramagnetic resonance.* Oxford, UK: Oxford University Press.

Seeler, J. F., Sharma, A., Zaluzec, N. J., Bleher, R., Lai, B., Schultz, E. G., et al. (2021). Metal ion fluxes controlling amphibian fertilization. *Nature Chemistry, 13*(7), 683–691. https://doi.org/10.1038/s41557-021-00705-2.

Shafaat, H. S., Griese, J. J., Pantazis, D. A., Roos, K., Andersson, C. S., Popović-Bijelić, A., et al. (2014). Electronic structural flexibility of heterobimetallic Mn/Fe cofactors: R2lox and R2c proteins. *Journal of the American Chemical Society, 136*(38), 13399–13409. https://doi.org/10.1021/ja507435t.

Sharma, A., Gaidamakova, E. K., Grichenko, O., Matrosova, V. Y., Hoeke, V., Klimenkova, P., et al. (2017). Across the tree of life, radiation resistance is governed by antioxidant Mn^{2+}, gauged by paramagnetic resonance. *Proceedings of the National Academy of Sciences of the United States of America, 114*(44), E9253–E9260. https://doi.org/10.1073/pnas.1713608114.

Sharma, A., Gaidamakova, E. K., Matrosova, V. Y., Bennett, B., Daly, M. J., & Hoffman, B. M. (2013). Responses of Mn^{2+} speciation in *Deinococcus radiodurans* and *Escherichia coli* to γ-radiation by advanced paramagnetic resonance methods. *Proceedings of the National Academy of Sciences of the United States of America, 110*(15), 5945–5950. https://doi.org/10.1073/pnas.1303376110.

Sheng, Y., Abreu, I. A., Cabelli, D. E., Maroney, M. J., Miller, A.-F., Teixeira, M., et al. (2014). Superoxide Dismutases and Superoxide Reductases. *Chemical Reviews, 114*(7), 3854–3918. https://doi.org/10.1021/cr4005296.

Signorella, S., Palopoli, C., & Ledesma, G. (2018). Rationally designed mimics of antioxidant manganoenzymes: Role of structural features in the quest for catalysts with catalase and superoxide dismutase activity. *Coordination Chemistry Reviews, 365*, 75–102. https://doi.org/10.1016/j.ccr.2018.03.005.

Sladek, V., & Tvaroška, I. (2017). First-principles interaction analysis assessment of the manganese cation in the catalytic activity of glycosyltransferases. *The Journal of Physical Chemistry. B, 121*(25), 6148–6162. https://doi.org/10.1021/acs.jpcb.7b03714.

Smith, S. J., Hadler, K. S., Schenk, G., Hanson, G. R., & Mitić, N. (2010). Manganese Metalloproteins. In G. Hanson, & L. Berliner (Eds.), *Metals in biology: Applications of high-resolution EPR to metalloenzymes* (pp. 273–341). New York, NY: Springer.

Smoukov, S. K., Telser, J., Bernat, B. A., Rife, C. L., Armstrong, R. N., & Hoffman, B. M. (2002). EPR study of substrate binding to the Mn(II) active site of the bacterial antibiotic resistance enzyme FosA: A better way to examine Mn(II). *Journal of the American Chemical Society, 124*(10), 2318–2326. https://doi.org/10.1021/ja012480f.

Song, R., Doan, P. E., Gurbiel, R. J., Sturgeon, B. E., & Hoffman, B. M. (1999). Non-Kramers ENDOR and ESEEM of the $S = 2$ ferrous ion of $[Fe(II)EDTA]^{2-}$. *Journal of Magnetic Resonance, 141*(2), 291–300. https://doi.org/10.1006/jmre.1999.1887.

Stich, T. A., Lahiri, S., Yeagle, G., Dicus, M., Brynda, M., Gunn, A., et al. (2007). Multifrequency pulsed EPR studies of biologically relevant manganese(II) complexes. *Applied Magnetic Resonance, 31*(1), 321. https://doi.org/10.1007/BF03166263.

Stich, T. A., Whittaker, J. W., & Britt, R. D. (2010). Multifrequency EPR studies of manganese catalases provide a complete description of proteinaceous nitrogen coordination. *The Journal of Physical Chemistry. B*, *114*(45), 14178–14188. https://doi.org/10.1021/jp908064y.

Sturgeon, B. E., Doan, P. E., Liu, K. E., Burdi, D., Tong, W. H., Nocek, J. M., et al. (1997). Non-Kramers ESEEM of integer-spin Diferrous carboxylate-bridged clusters in proteins. *Journal of the American Chemical Society*, *119*(2), 375–386. https://doi.org/10.1021/ja9628157.

Suga, M., Akita, F., Yamashita, K., Nakajima, Y., Ueno, G., Li, H., et al. (2019). An oxyl/oxo mechanism for oxygen-oxygen coupling in PSII revealed by an x-ray free-electron laser. *Science*, *366*(6463), 334–338. https://doi.org/10.1126/science.aax6998.

Tabares, L. C., Gätjens, J., Hureau, C., Burrell, M. R., Bowater, L., Pecoraro, V. L., et al. (2009). pH-dependent structures of the manganese binding sites in oxalate decarboxylase as revealed by high-field electron paramagnetic resonance. *The Journal of Physical Chemistry. B*, *113*(26), 9016–9025. https://doi.org/10.1021/jp9021807.

Tan, X., Poyner, R., Reed, G. H., & Scholes, C. P. (1993). Electron nuclear double resonance study of the Mn^{2+} environs in the oxalate-ATP complex of pyruvate kinase. *Biochemistry*, *32*(30), 7799–7810. https://doi.org/10.1021/bi00081a027.

Tao, L., Stich, T. A., Soldatova, A. V., Tebo, B. M., Spiro, T. G., Casey, W. H., et al. (2018). Mn(III) species formed by the multi-copper oxidase MnxG investigated by electron paramagnetic resonance spectroscopy. *Journal of Biological Inorganic Chemistry*, *23*(7), 1093–1104. https://doi.org/10.1007/s00775-018-1587-z.

Telser, J. (2017). EPR interactions—Zero-field Splittings. In *eMagRes (Vol. 6)* (pp. 207–233). John Wiley & Sons, Ltd.

Telser, J., Krzystek, J., & Ozarowski, A. (2014). High-frequency and high-field electron paramagnetic resonance (HFEPR): A new spectroscopic tool for bioinorganic chemistry. *Journal of Biological Inorganic Chemistry*, *19*(3), 297–318. https://doi.org/10.1007/s00775-013-1084-3.

Thompson, M. K., Keithly, M. E., Goodman, M. C., Hammer, N. D., Cook, P. D., Jagessar, K. L., et al. (2014). Structure and function of the Genomically encoded Fosfomycin resistance enzyme, FosB, from *Staphylococcus aureus*. *Biochemistry*, *53*(4), 755–765. https://doi.org/10.1021/bi4015852.

Tipton, P. A., McCracken, J., Cornelius, J. B., & Peisach, J. (1989). Electron spin echo envelope modulation studies of pyruvate kinase active site complexes. *Biochemistry*, *28*(14), 5720–5728. https://doi.org/10.1021/bi00440a004.

Tsednee, M., Castruita, M., Salomé, P. A., Sharma, A., Lewis, B. E., Schmollinger, S. R., et al. (2019). Manganese co-localizes with calcium and phosphorus in *Chlamydomonas acidocalcisomes* and is mobilized in manganese-deficient conditions. *The Journal of Biological Chemistry*, *294*(46), 17626–17641. https://doi.org/10.1074/jbc.RA119.009130.

Twahir, U. T., Ozarowski, A., & Angerhofer, A. (2016). Redox cycling, pH dependence, and ligand effects of Mn(III) in oxalate decarboxylase from Bacillus subtilis. *Biochemistry*, *55*(47), 6505–6516. https://doi.org/10.1021/acs.biochem.6b00891.

Umena, Y., Kawakami, K., Shen, J.-R., & Kamiya, N. (2011). Crystal structure of oxygen-evolving photosystem II at a resolution of 1.9Å. *Nature*, *473*(7345), 55–60. https://doi.org/10.1038/nature09913.

Un, S., & Bruch, E. M. (2015). How bonding in manganous phosphates affects their Mn(II)–^{31}P hyperfine interactions. *Inorganic Chemistry*, *54*(21), 10422–10428. https://doi.org/10.1021/acs.inorgchem.5b01864.

Un, S., Tabares, L. C., Cortez, N., Hiraoka, B. Y., & Yamakura, F. (2004). Manganese(II) zero-field interaction in Cambialistic and manganese superoxide Dismutases and its relationship to the structure of the metal binding site. *Journal of the American Chemical Society*, *126*(9), 2720–2726. https://doi.org/10.1021/ja036503x.

Van Doorslaer, S. (2017). Hyperfine spectroscopy: ESEEM. In D. Goldfarb, & S. Stoll (Eds.), *eMagRes* (pp. 51–70). John Wiley & Sons, Ltd.

Vertregt, F., Torrelo, G., Trunk, S., Wiltsche, H., Hagen, W. R., Hanefeld, U., et al. (2016). EPR study of substrate binding to Mn(II) in Hydroxynitrile lyase from *Granulicella tundricola*. *ACS Catalysis*, *6*(8), 5081–5085. https://doi.org/10.1021/acscatal.6b01204.

Vetting, M. W., Wackett, L. P., Que, L., Jr., Lipscomb, J. D., & Ohlendorf, D. H. (2004). Crystallographic comparison of manganese- and iron-dependent Homoprotocatechuate 2,3-dioxygenases. *Journal of Bacteriology*, *186*(7), 1945–1958. https://doi.org/10.1128/JB.186.7.1945-1958.2004.

Vincent, J. B. (2021). Chromium. In *Reference module in food science*. Elsevier. https://doi.org/10.1016/B978-0-12-821848-8.00001-9.

Vogt, M., Lahiri, S., Hoogstraten, C. G., Britt, R. D., & DeRose, V. J. (2006). Coordination environment of a site-bound metal ion in the hammerhead ribozyme determined by ^{15}N and ^{2}H ESEEM spectroscopy. *Journal of the American Chemical Society*, *128*(51), 16764–16770. https://doi.org/10.1021/ja057035p.

Walsby, C. J., Telser, J., Rigsby, R. E., Armstrong, R. N., & Hoffman, B. M. (2005). Enzyme control of small-molecule coordination in FosA as revealed by ^{31}P pulsed ENDOR and ESE-EPR. *Journal of the American Chemical Society*, *127*(23), 8310–8319. https://doi.org/10.1021/ja044094e.

Waters, L. S. (2020). Bacterial manganese sensing and homeostasis. *Current Opinion in Chemical Biology*, *55*, 96–102. https://doi.org/10.1016/j.cbpa.2020.01.003.

Weil, J. A., & Bolton, J. R. (2007). *Electron paramagnetic resonance: Elementary theory and practical applications* (2nd ed.). Hoboken, NJ: John Wiley & Sons, Inc.

Wennman, A., Magnuson, A., Hamberg, M., & Oliw, E. H. (2015). Manganese lipoxygenase of *F. oxysporum* and the structural basis for biosynthesis of distinct 11-hydroperoxy stereoisomers. *Journal of Lipid Research*, *56*(8), 1606–1615. https://doi.org/10.1194/jlr.M060178.

Wiedner, R., Kothbauer, B., Pavkov-Keller, T., Gruber-Khadjawi, M., Gruber, K., Schwab, H., et al. (2015). Improving the properties of bacterial R-selective Hydroxynitrile lyases for industrial applications. *ChemCatChem*, 7(2), 325–332. https://doi.org/10.1002/cctc.201402742.

Williams, G. J., & Thorson, J. S. (2008). Natural product glycosyltransferases: Properties and applications. In *Advances in enzymology and related areas of molecular biology* (pp. 55–119).

Woo, E.-J., Dunwell, J. M., Goodenough, P. W., Marvier, A. C., & Pickersgill, R. W. (2000). Germin is a manganese containing homohexamer with oxalate oxidase and superoxide dismutase activities. *Nature Structural & Molecular Biology*, 7(11), 1036–1040. https://doi.org/10.1038/80954.

Wu, A. J., Penner-Hahn, J. E., & Pecoraro, V. L. (2004). Structural, spectroscopic, and reactivity models for the manganese catalases. *Chemical Reviews*, *104*(2), 903–938. https://doi.org/10.1021/cr020627v.

Zein, S., Duboc, C., Lubitz, W., & Neese, F. (2008). A systematic density functional study of the zero-field splitting in Mn(II) coordination compounds. *Inorganic Chemistry*, *47*(1), 134–142. https://doi.org/10.1021/ic701293n.

Zhang, Y., & Stubbe, J. (2011). *Bacillus subtilis* class Ib ribonucleotide reductase is a Dimanganese(III)-tyrosyl radical enzyme. *Biochemistry*, *50*(25), 5615–5623. https://doi.org/10.1021/bi200348q.

Zhu, W., Easthon, L. M., Reinhardt, L. A., Tu, C., Cohen, S. E., Silverman, D. N., et al. (2016). Substrate binding mode and molecular basis of a specificity switch in oxalate decarboxylase. *Biochemistry*, *55*(14), 2163–2173. https://doi.org/10.1021/acs.biochem.6b00043.

Zhu, W., Wilcoxen, J., Britt, R. D., & Richards, N. G. J. (2016). Formation of hexacoordinate Mn(III) in *Bacillus subtilis* oxalate decarboxylase requires catalytic turnover. *Biochemistry*, *55*(3), 429–434. https://doi.org/10.1021/acs.biochem.5b01340.

Zygiel, E. M., & Nolan, E. M. (2018). Transition metal sequestration by the host-defense protein calprotectin. *Annual Review of Biochemistry*, *87*(1), 621–643. https://doi.org/10.1146/annurev-biochem-062917-012312.

CHAPTER ELEVEN

Application of EPR and related methods to molybdenum-containing enzymes

Russ Hille* and Dimitri Niks

Department of Biochemistry, University of California, Riverside, CA, United States
*Corresponding author: e-mail address: russ.hille@ucr.edu

Contents

Abstract

A description is provided of the contributions made to our understanding of molybdenum-containing enzymes through the application of electron paramagnetic resonance spectroscopy and related methods, by way of illustrating how these can be applied to better understand enzyme structure and function. An emphasis is placed on the use of EPR to identify both the coordination environment of the molybdenum coordination sphere as well as the structures of paramagnetic intermediates observed transiently in the course of reaction that have led to the elucidation of reaction mechanism.

Methods in Enzymology, Volume 666
ISSN 0076-6879
https://doi.org/10.1016/bs.mie.2022.02.006

1. Introduction

Enzymes other than nitrogenase that possess a molybdenum center in their active sites can be divided into three families (Hille, 1996), as characterized by the eponymous enzymes xanthine oxidase, sulfite oxidase and DMSO reductase, each being historically the best-characterized member of its respective group. The vast majority of these enzymes catalyze oxidation-reduction reactions, with the molybdenum center cycling between Mo(VI) and Mo(IV) states, with the intermediate Mo(V) state forming to varying degree depending on the conditions. As shown in Fig. 1, the active site of xanthine oxidase (often referred to as xanthine oxidoreductase for reasons described below) can be represented as LMoOS(OH) in the oxidized Mo(VI) state, with terminal Mo=O and Mo=S ligands, Mo–OH and L representing a bidentate enedithiolate ligand contributed by a pyranopterin cofactor that is common to all molybdenum-(and tungsten-) containing enzyme—again excepting nitrogenase (also shown in Fig. 1). The coordination geometry is square-pyramidal, with the Mo=O occupying the apical position. Sulfite oxidase family members have $LMoO_2$-SCys active sites, with a thiolate ligand contributed by a highly conserved cysteine residue. Again, the coordination geometry is square pyramidal with one apical Mo=O and one equatorial Mo=O. The active sites seen in members of the DMSO reductase family are the most diverse, but all have molybdenum centers possessing two equivalents of the pyranopterin and can be represented as L_2MoXY, where X is a terminal Mo=O, Mo=S or Mo=Se ligand and Y an O-Ser, S-Cys, Se-Cys, Asp

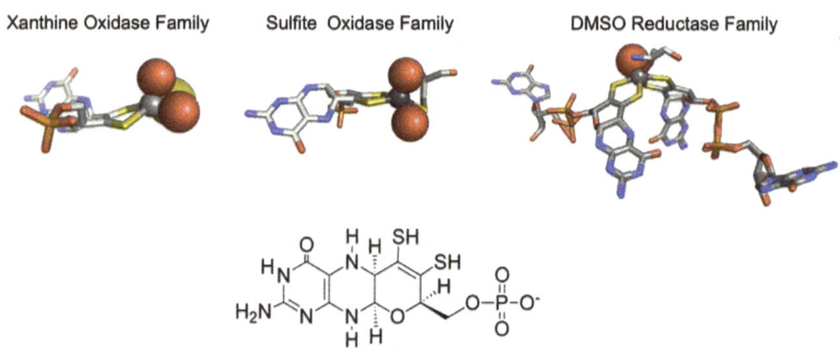

Fig. 1 *Structures of the active sites of the three families of molybdenum-containing enzymes.* From *Left* to *Right*: the xanthine oxidase family, the sulfite oxidase family and the DMSO reductase family. The structure of the pyranopterin coordinated to the molybdenum via its enedithiolate side chain is also shown.

or -OH ligand; the coordination geometry in the oxidized state is trigonal prismatic and the pyranopterin cofactor is elaborated as a dinucleotide (most commonly of guanosine, as shown). It is now well-recognized that, minimally, the xanthine oxidase and DMSO reductase families are evolutionarily ancient, tracing back to the Last Universal Common Ancestor to all extant life on earth (Schoepp-Cothenet et al., 2012). Xanthine oxidase family enzymes carry out the hydroxylation of carbon centers in the absence of O_2 (bearing in mind that life arose in a highly anoxic environment). Members of the DMSO reductase family are involved in a wide variety of energy-conserving reactions as well as the reduction of CO_2 to formate, which constitutes the first step each branch of the Wood-Ljungdahl pathway that the earliest life forms utilized to fix carbon. The following account focuses on work utilizing EPR, ENDOR and related methodologies leading to our present understanding of the structure and function of these enzymes, considering each family in turn.

2. Xanthine oxidase family enzymes

Xanthine oxidase and related enzymes, including xanthine dehydrogenases, aldehyde oxidases, CO dehydrogenase and 4-hydroxybenzoylCoA reductase, catalyze the oxidative hydroxylation of a wide variety of aromatic heterocycles as well as the oxidation of aldehydes to carboxylic acids (Hille, 1996; Hille, Hall, & Basu, 2014). They are fundamentally distinct from flavin-, pterin- or metal-containing monooxygenases in utilizing H_2O rather than O_2 as the source of the oxygen atom incorporated into substrate. Physiologically, these enzymes catalyze the final two steps in purine metabolism, the conversion of hypoxanthine to xanthine and xanthine to uric acid, and are extremely broadly distributed in biology. The enzyme from non-mammalian eukaryotes is a constitutive dehydrogenase (Amaya, Yamazaki, Sato, Noda, & Nishino, 1990; Massey, Schopfer, & Nishino, 1989; Nishino, 1994; Nishino, Schopfer, & Massey, 1989; Saito & Nishino, 1989; Schopfer, Massey, & Nishino, 1988), passing the pair of reducing equivalents from the hydroxylated substrate to NAD^+. In mammals, the enzyme is initially expressed as a dehydrogenase but can be converted to an oxidase form that passes the reducing equivalents to O_2, forming either peroxide or superoxide depending on the conditions (the dehydrogenase forms of these enzymes also react to varying degree with O_2 (Nishino et al., 1989)). For this reason the enzyme is frequently referred to as xanthine oxidoreductase. The enzyme from cow's milk, originally referred to as the Shardinger enzyme, is typically isolated as the oxidase form and in being first purified in 1924 (Dixon & Thurlow, 1924a, 1924b) is one of the longest studied metalloenzymes of any kind.

2.1 Xanthine oxidase

Certainly the signal achievement of EPR in the study of molybdenum-containing enzymes was the demonstration in 1966 by Bray and Meriwether (Bray & Meriwether, 1966) that cow's milk xanthine oxidase possessed molybdenum in its active site, and that the metal became reduced in the course of the reaction of enzyme with substrate. The key experiment involved the demonstration that enzyme isolated from the milk of cows injected with ^{95}Mo-molybdate ($I95_{Mo}=5/2$) afforded previously characterized Mo(V) EPR signals split into the expects sextets. It was a tour de force isotopic substitution experiment in a (very large) eukaryote. ^{95}Mo and ^{97}Mo (also $I=5/2$) in fact constitute 25% of natural abundance molybdenum, and the tell-tale sextets first identified by Bray and Meriwether are evident even in the EPR signals from natural-abundance enzyme samples; their observation presently constitutes definitive evidence that a given EPR signal is indeed derived from molybdenum.

In previous work with Beinert and Palmer (Bray, Palmer, & Beinert, 1964; Palmer, Bray, & Beinert, 1964), Bray had developed a freeze-quench protocol (variations of which remain widely used today) that was used to identify, in addition to signals attributable to two discrete [2Fe–2S] iron-sulfur clusters and FADH•, two discrete Mo(V) EPR signals designated αβ and δγ; on the basis of their relative rates of formation and decay in the course of reduction of enzyme by xanthine, these were subsequently termed "Rapid" and "Very Rapid" (Bray & Vanngard, 1969). The "Very Rapid" signal was nearly axial in lineshape and was devoid of proton hyperfine, while the "Rapid" signal was more uniformly rhombic and exhibited coupling to two inequivalent protons (one strongly coupled, the second more weakly). Bray and coworkers subsequently demonstrated that the more strongly coupled of the two protons in the species giving rise to the "Rapid" signal was derived from the C-8 position of xanthine, the position hydroxylated in the course of oxidation of xanthine to uric acid (Bray & Knowles, 1968). A second type of "Rapid" signal was subsequently characterized, exhibiting strong coupling to two equivalent protons and designated "Rapid Type II" (Bray & Vanngard, 1969), the earlier signal now being designated "Rapid Type I". A fourth "Slow" form was subsequently found to arise from nonfunctional enzyme that had lost the catalytically essential

Mo=S, being replaced by a second Mo=O (Bray & Vanngard, 1969). Simulations of each of these four spectra are shown in Fig. 2, including the ^{95,97}Mo coupling referred to above. The parameters used for these and all other simulations to be discussed are given in Table 1.

The relative amounts of "Very Rapid," "Rapid Type I" and "Rapid Type II" signals varied with the reaction conditions and substrate used (xanthine oxidase being able to oxidatively hydroxylate a wide variety of purines and other aromatic heterocycles, as well as aldehydes to carboxylic acids). 2-hydroxy-6-methylpurine was found to transiently form particularly high levels of the species giving rise to the "Very Rapid" signal (Bray & George, 1985), whose structure was recognized as being key to understanding the chemical mechanism by which substrates become oxidatively hydroxylated. It was subsequent work with this substrate that demonstrated that formation of the "Very Rapid" species was an oxidative event, resulting in the loss of one electron from a Mo^{IV} species that preceded it in the reaction mechanism (McWhirter & Hille, 1991); the Mo^{IV} species exhibited an absorbance difference spectrum relative to oxidized enzyme with a maximum at 470 nm, while the Mo^{V} "Very Rapid" species exhibited a maximum in the difference spectrum at 540 nm. Interestingly, no "Rapid" signal of either type was seen in the course of the reaction of enzyme with substoichiometric 2-hydroxy-6-methylpurine. ESEEM work tracking weak ^{1}H coupling (Lorigan, Britt, Kim, & Hille, 1994), ENDOR with 8-^{13}C-2-hydroxy-6-methylpurine (Manikandan, Choi, Hille, & Hoffman, 2001) and ultimately protein crystallography (Pauff, Zhang, Bell, & Hille, 2008) established that the species giving rise to the "Very Rapid" EPR signal has the 8-hydroxylated product coordinated equatorially to the molybdenum via the catalytically introduced hydroxyl group in a simple end-on fashion.

It has been demonstrated that the "Rapid" signals simply arises from the orientation-dependent binding of substrate to enzyme that had been partially reduced by reaction with a prior equivalent of substrate, with the molybdenum center in the Mo(V) state (Hille, Kim, & Hemann, 1993). The important implication is that the signal-giving species represented a paramagnetic analog of the Michaelis complex that could not proceed in the reaction mechanism because the molybdenum center was partially reduced. The counterintuitive conclusion is that, the relative rates of formation and decay of "Very Rapid" and "Rapid" signal notwithstanding, the species giving rise to the "Rapid" signal effectively lay upstream rather than downstream to that giving rise to the "Very Rapid" EPR signal. Indeed, partial reduction of the enzyme prior to mixing with the slow substrate

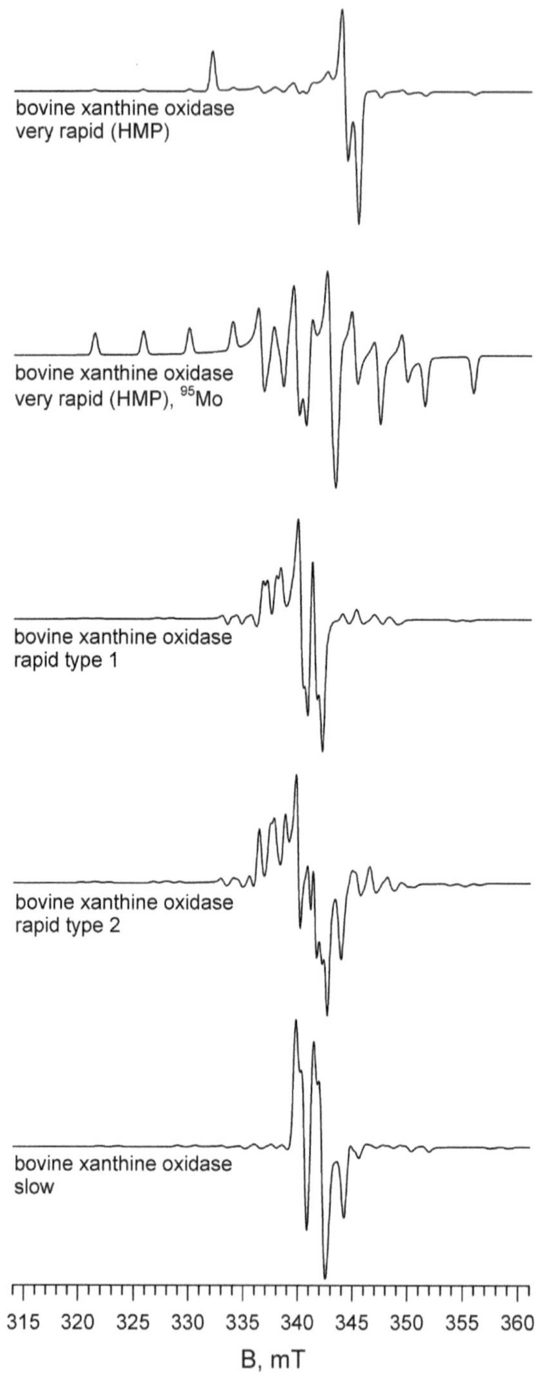

Fig. 2 See figure legend on opposite page.

2-hydroxy-6-methylpurine results in extremely rapid appearance of the "Rapid Type 1" signal that precedes the eventual accumulation of the "Very Rapid" signal.

The final EPR experiment critical to understanding the reaction mechanism of xanthine oxidase came from an examination of the reaction of enzyme with xanthine under single-turnover conditions in $H_2^{17}O$ in which it was demonstrated that ^{17}O was incorporated into a site in which the oxygen was strongly and anisotropically coupled to the Mo(V) after completion of a single turnover (Xia, Dempski, & Hille, 1999). Comparison with model compound studies (Greenwood, Wilson, Pilbrow, & Wedd, 1993) clearly indicated that the oxygen was incorporated into an equatorial Mo-OH group in the signal-giving species. Earlier work (Hille & Sprecher, 1987) had demonstrated that a catalytically labile oxygen site in the molybdenum center was the source of the oxygen incorporated into product, subsequently regenerated by oxygen from solvent water in the present experiment. The cumulative body of evidence has led to the now widely accepted mechanism for the reaction catalyzed by xanthine oxidase in which the equatorial Mo–OH undertakes base-assisted nucleophilic attack (using Glu 1261) on the C8 position of substrate. Concomitant hydride transfer of the C8–H to the catalytically essential Mo=S group of the molybdenum center yields an $LMo^{IV}O(SH)(OR)$ species, with product now coordinated to the molybdenum via the catalytically introduced oxygen of the erstwhile Mo–OH group as shown in Fig. 3.

In addition to the above signals relevant to the reaction mechanism of xanthine oxidase, a wide variety of Mo(V) signals arising from nonfunctional forms of the enzyme under various conditions have been observed. These include the "Slow" signal mentioned above, due to the so-called desulfo form of the enzyme in which the catalytically essential equatorial Mo=S has been lost and replaced with a second Mo=O, as well as inhibited forms

Fig. 2 *Simulations of the principal EPR signals exhibited by the molybdenum center of xanthine oxidase.* From *Top* to *Bottom*: the "very rapid" signal seen with natural abundance molybdenum and with 100% ^{95}Mo (using the simulation parameters in both cases given in (George & Bray, 1988)); the "Rapid Type 1" signal (using the simulation parameters in both cases given in (George & Bray, 1988)); the "Rapid Type 2" signal as originally reported (Malthouse, Gutteridge, & Bray, 1980), using the simulation parameters of (Wilson, Greenwood, Pilbrow, Spence, & Wedd, 1991); and the "Slow" signal originally as originally described by (Bray & Vanngard, 1969), using the simulation parameters of (Wilson et al., 1991). Simulations here and throughout were performed using the EasySpin software package (Stoll & Schweiger, 2006).

Table 1 Simulation parameters for various Mo^V EPR signals.[a]

					^{95,97}Mo (MHz)			Euler Angles (°)[b]			^{1}H (MHz)		
	g_1	g_2	g_3	g_{av}	A_1	A_2	A_3	α	β	γ	A_1	A_2	A_3
Fig. 2: bovine xanthine oxidase (XO)													
very rapid (HMP)	2.0229	1.9518	1.9446	1.9731	141.6	60.1	63.4	7	42	0			
rapid type 1	1.9901	1.9710	1.9666	1.9759	184	74	77	0	18	0	34 12.8	34.8 4.7	36.3 4.4
rapid type 2	1.9897	1.9681	1.9617	1.9732	181.2	74.1	74.4	0	20	0	39 27.9	42 29.4	45.3 37.5
slow	1.9706	1.9655	1.9542	1.9634	196	78.5	81.2	0	33	0	44.7 3.9	44.4 4.5	42.6 6.3
Fig. 4: *Oc* CO dehydrogenase (CODH) ^{63,65}Cu or ^{107,109}Ag (MHz)													
CO_{red}	2.001	1.9604	1.9549	1.9721	117	164	132	0	40	40			
$H_{2\ red}$	2.0127	1.9676	1.9594	1.9799	169.1	200.3	170.2	−31.5	−112	111	80 80	20 20	130 120
$dithionite_{red}$ (HCO_3^-)	2.001	1.9604	1.9549	1.9721	192.9	190.9	221.3	−11.3	−59	159			
Ag sub., CO_{red}	2.0043	1.9595	1.9540	1.9726	82	78.85	81.9						
Fig. 5: chicken/plant sulfite oxidase (SO)													
cSO, low pH	2.007	1.974	1.968	1.983	170.0	74.9	50.1	0	18	0	23.8	22.1	35.8
cSO, high pH	1.990	1.966	1.954	1.970	163.1	63.0	33.9	0	14	22			
cSO, PO_4-inhibited	1.9926	1.9701	1.9625	1.9751							0.8	3.3	0.8

At SO, blocked (2 species, ~7:3 ratio)	2.005 2.000	1.974 1.974	1.963 1.963	1.981 1.979									
Fig. 6: assimilatory nitrate reductase (NR)													
Cv low pH	1.994	1.969	1.967	1.977							27.9	30.3	35.6
Cv high pH	1.984	1.951	1.947	1.961									
Fig. 6: sulfite dehydrogenase (SDH)													
Tt low pH	1.9965	1.9700	1.9620	1.9762	167	80	66	0	27	0	22	21	29
Sn high pH	1.9914	1.9661	1.9541	1.9705									
Fig. 7: human mitochondrial amidoxime-reducing component (mARC)													
hmARC-1	1.9983	1.9691	1.9585	1.9753							23.9	0	0
hmARC-2	1.9994	1.9658	1.9616	1.9756							24.6	39.0	31.0
Fig. 7: methionine sulfoxide reductase (MsrP), formerly YedY													
Ec MsrP (2 species, ~1:1 ratio)	2.033 2.030	1.977 1.976	1.972 1.970	1.994 1.992	155.3	61.4	59	78.6	19.5	−68.8			
Fig. 8: *Rc*/*Rs* DMSO reductase (DMSOR)													
high-g split	1.9922	1.9818	1.9673	1.9804							40.4	29.1	13.7
high-g unsplit type 1	1.9906	1.9833	1.9611	1.9783									
low-g type 1	1.9702	1.9678	1.9560	1.9647							35.6	31.7	34.8
low-g type 2	1.9759	1.9540	1.9359	1.9553							29.6	33.4	36.3

Continued

Table 1 Simulation parameters for various Mo^{V} EPR signals.[a]—cont'd

					^{95,97}Mo (MHz)			Euler Angles (°)[b]			^{1}H (MHz)		
	g_1	g_2	g_3	g_{av}	A_1	A_2	A_3	α	β	γ	A_1	A_2	A_3
Rs high-g split[c]	1.9988	1.9885	1.9722	1.9865	155.9	40.0	112.0	14.3	85.9	5.7	38.1	25.6	30.1
Fig. 9: dissimilatory nitrate reductase													
Ec NAR, low pH	1.9989	1.9855	1.9628	1.9827							31.1	23.3	22.3
Ec NAR, high pH	1.9875	1.9807	1.9617	1.9766	96	124.4	95.7				10.6	8.9	9.1
Pd NAP, very high-g	2.0222	1.9993	1.9935	2.005							20.9	20.7	18.4
Pd NAP, high-g	1.9985	1.9902	1.9806	1.9898							17.9 7.8	14.5 0	15.3 0
Pd NAP, low-g	1.9957	1.9688	1.9607	1.9751							36.3	37.5	42
Fig. 10: formate dehydrogenase													
Ec FdhF[d]	2.094	2.001	1.990	2.028	118	71	71				7.5	18.9	21
Cn FdsDABG	2.009	2.001	1.992	2.000	138	82	45	10.8	0.2	152.5	18	21	18
Cn FdsDABG ($^{2}H_2O$)	2.009	2.001	1.992	2.000	139	84	49	10.8	0.2	152.5			
Ds Fdh	2.012	1.996	1.985	1.998							23.1 35.1	30 0	27.9 0
Ds Fdh (azide)	2.092	2.000	1.989	2.027							0	21	21

[a]All simulations were performed using $\nu = 9.4$ GHz with EasySpin (Stoll & Schweiger, 2006).
[b]Euler angles correspond to successive rotations as follows; α about z, β about the new y, and γ about the new z.
[c]Euler angles and A(^{1}H) values were provided courtesy of Professor Martin Kirk.
[d]Estimated A(^{95,97}Mo) values were not used in the simulation; in the absence of Euler angles they do not well describe the experimental spectrum.

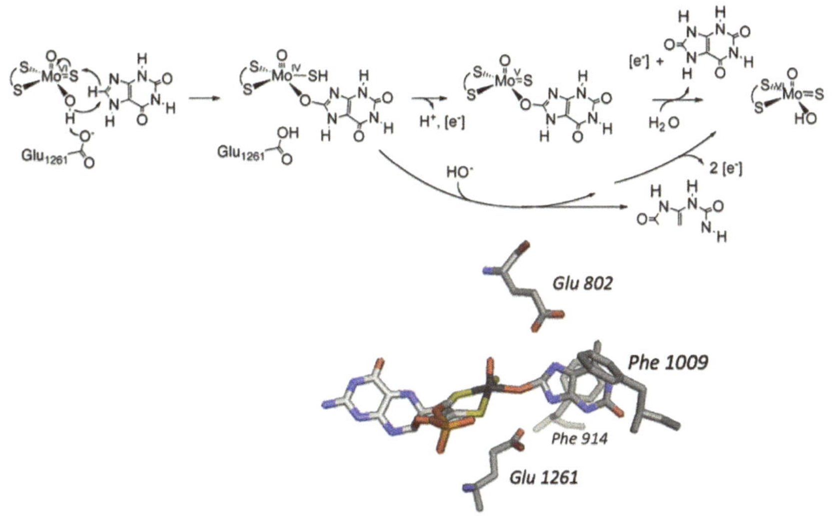

Fig. 3 *The reaction mechanism of xanthine oxidase.* The reaction proceeds by nucleophilic attack on the C8 carbon by the equatorial Mo–OH, with Glu 1261 acting as a general base. Concomitant hydride transfer to the Mo═S results in reduction of the molybdenum from Mo(VI) to Mo(IV), followed by sequential transfer of the electrons to other redox-active centers in the enzyme. When electron transfer occurs prior to product dissociation, the Mo(V) species generating the "Very Rapid" is formed, with product coordinated to the molybdenum center via the catalytically introduced hydroxyl group. Also shown is the X-ray crystal structure of xanthine oxidase in complex with the slow substrate 2-hydroxy-6-methylpurine, demonstrating the coordination of product to the molybdenum(Pauff et al., 2008). On the basis of the short Mo═S distance seen in the crystal structure (<2.2 Å), the molybdenum is most likely in the MO(V) valence state.

in complex with alloxanthine, formaldehyde, ethylene glycol, p-CMB, formamide and arsenite. These latter signals have been summarized elsewhere (Hille, 2010) and are not considered further here.

2.2 Aldehyde oxidases

Aldehyde oxidases from eukaryotes are closely related to the xanthine oxidoreductases and share extensive overlap in substrate specificity with them. They are constitutive oxidases that do not utilize NAD^+ but do have the same cofactor makeup as the xanthine oxidoreductases, containing two [2Fe–2S] clusters and FAD in addition to the molybdenum center that is the site of substrate hydroxylation. As expected on the basis of the broad overlap in substrate specificity, the active site molybdenum centers of these enzyme are very similar to that of the xanthine oxidoreductases. The rabbit liver enzyme exhibits both Rapid Type 1 and Rapid Type 2 Mo(V) EPR signals, for example (the latter seen in the presence of borate), (Turner, Doyle, Ventom, & Bray, 1995) as well as the "Slow" and other signals due to one or another form of the inactive desulfo form of the enzyme (Bray et al., 1982). No signal comparable to the "Very Rapid" signal seen with xanthine oxidase has been observed with an aldehyde oxidase, presumably because of unfavorable rates of formation and decay.

The four murine aldehyde oxidases have been studied extensively on the basis of their substrate specificity (specifically as regards drugs and xenobiotics) and tissue distribution (Hartmann et al., 2012; Kurosaki et al., 2004; Marelja et al., 2014; Schumann et al., 2009; M. Terao et al., 2001), and the topic has recently been reviewed (Mota et al., 2018; Mineko Terao, Garattini, Romao, & Leimkuehler, 2020). The X-ray crystal structure of murine aldehyde oxidase 3 has been reported (Coelho et al., 2012), confirming the near-identity of the two molybdenum centers, including the conservation of the two phenylalanine residues proximal to the molybdenum center between which substrate binds. Unfortunately, none of the mouse isozymes have to date been examined extensively by EPR.

Plant aldehyde oxidases have been known for many years, one of which is responsible for the oxidation of abscisic aldehyde to abscisic acid, an important plant hormone responsible for the plant's response to environmental stress and plant pathogens. Like the mammalian aldehyde oxidases, most of the work on these systems has been molecular and cell biological in nature (Mendel, 2011; Mendel & Kruse, 2012; Mendel & Schwarz, 1999; Sauer & Frebort, 2003; Schwarz & Mendel, 2006) and none has been examined extensively by EPR. It has long been assumed, however, that the basic chemical mechanism of the aldehyde oxidases is analogous to that seen with xanthine oxidase, with the equatorial Mo–OH of the molybdenum center undertaking nucleophilic attack on the carbonyl carbon of the aldehyde, with concomitant hydride transfer to the Mo=S group to simultaneously bring about the two-electron reduction of the molybdenum and creation of the C–O bond of the carboxylic acid product.

Although bacterial aldehyde oxidases differ from the eukaryotic enzymes in typically lacking the FAD (Rebelo et al., 2000; Romao et al., 1995), their molybdenum centers are considered to be very similar to those of the eukaryotic enzyme (Barata, Legall, & Moura, 1993; Bray, Turner, Legall, Barata, & Moura, 1991; Marangon et al., 2013; Romao et al., 1995). The enzyme from *Desulfovibrio gigas* exhibits both Slow and Rapid Type 2

Mo(V) signals that are both virtually identical to the corresponding signals seen with xanthine oxidase. This extends to the hyperfine coupling to two equivalent and solvent-exchangeable protons in the Rapid Type 2 signal (Bray et al., 1991).

2.3 CO dehydrogenase

The CO dehydrogenase from aerobes such as *Oligotropha carboxydovorans* (Haenzelmann, Hofmann, Meisen, & Meyer, 1999; Haenzelmann & Meyer, 1998) is clearly a member of the xanthine oxidase family on the basis of extensive sequence homology, but is unique in possessing a binuclear Mo–S–Cu active site, with the copper bridged to the molybdenum by what would otherwise be the equatorial Mo=S group (Dobbek, Gremer, Kiefersauer, Huber, & Meyer, 2002). The molybdenum coordination environment is otherwise similar to that seen in other members of the xanthine oxidase family, with an overall square-pyramidal coordination geometry; the copper coordination sphere is completed by a Cys ligand in an approximately linear coordination geometry. In the oxidized state, the molybdenum is Mo(VI) and the copper Cu(I) but one-electron reduction yields an EPR signal with extremely strong coupling to the $^{63/65}Cu$ nucleus (Zhang, Hemann, & Hille, 2010), indicating by analogy to binuclear Mo/Cu models considerable spin delocalization onto the copper (Fig. 4) (Gourlay et al., 2006). An ENDOR analysis utilizing ^{13}CO (Shanmugam et al., 2013) has provided evidence for a discrete Mo–S–Cu(CO) species in the one-electron reduced state that is presumed to represent a paramagnetic analog of the enzyme-substrate complex. On the basis of this interpretation, a mechanism has been proposed as shown in Fig. 4 (Shanmugam et al., 2013) in which CO initially coordinates to the Cu of the binuclear center and is then nucleophilically attacked by the equatorial Mo–OH to result ultimately in the incorporation of the oxygen from the Mo–OH into CO_2 and its regeneration with oxygen derived from solvent in a manner reminiscent of the mechanism of xanthine oxidase. Mechanisms that involve bicarbonate as an intermediate are unlikely, as bicarbonate binding to enzyme partially reduced with dithionite yields an EPRsignal that is distinct from that seen upon reduction of enzyme with CO (Wilcoxen & Hille, 2013). The *O. carboxydovorans* CO dehydrogenase is also reduced by H_2 (Wilcoxen & Hille, 2013), and yields an EPR signal that still possessing strong copper hyperfine but as shown in Fig. 4 is distinct from that seen on reduction of the enzyme by CO. The proposed mechanism involves polarization of the H_2 bond after side-on binding to the copper,

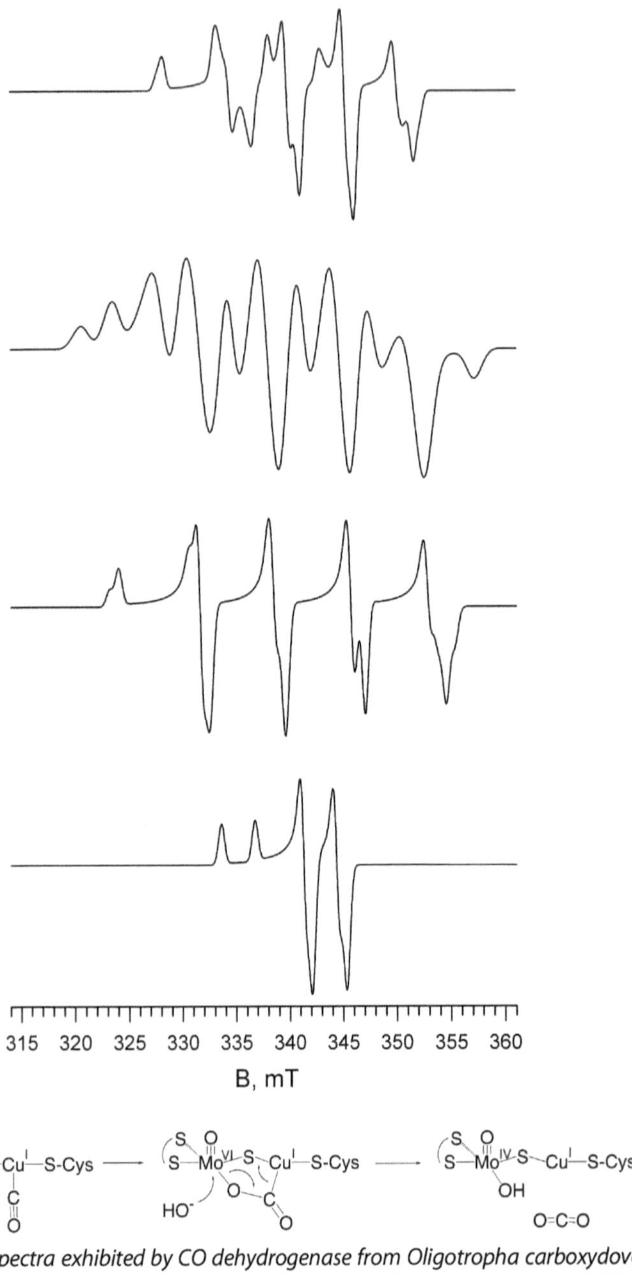

Fig. 4 *The EPR spectra exhibited by CO dehydrogenase from Oligotropha carboxydovorans.* From *Top* to *Bottom*: the spectrum seen on reduction of wild-type enzyme with CO (using the simulation parameters reported in (Zhang, Hemann, & Hille, 2010); the spectra seen on reduction of wild- Cu–CODH with H_2 and dithionite (using the simulation parameters reported in (Wilcoxen & Hille, 2013); and the signal on reduction of the silver-substituted enzyme with CO using the simulation parameters reported in (Wilcoxen, Snider, & Hille, 2011). At bottom is the proposed reaction mechanism (Shanmugam et al., 2013).

followed by deprotonation to give a transient Cu(I) hydride and a second deprotonation to give the fully reduced binuclear center.

As-isolated CODH is often partially depleted of Cu and inactive, but a protocol has been developed utilizing Cu(I) that enables reconstitution of the binuclear (Resch, Dobbek, & Meyer, 2005). Interestingly, activity is also recovered when Ag(I) is substituted for the copper in the reconstitution, although the silver-substituted enzyme reacts with CO at only 1/6 the rate of native, copper-containing enzyme (Wilcoxen et al., 2011). The lower activity is readily rationalized by the known lower backbonding capacity of Ag relative to Cu, which would render coordinated CO less susceptible to nucleophilic attack in the silver-substituted enzyme. The EPR signal seen with this silver-substituted enzyme exhibits strong hyperfine coupling to Ag ($I = ½$ for $^{107/109}Ag$) analogous to that seen with copper in the native enzyme, as shown in Fig. 4 (Wilcoxen et al., 2011).

2.4 4-HydroxybenzoylCoA reductase

Finally, there is the bacterial 4-hydroxybenzoylCoA reductase which in addition to the full complement of redox-active centers seen in the eukaryotic enzymes possesses an insertion domain that contains a [4Fe–4S] cluster unique to this enzyme (Unciuleac, Warkentin, Page, Boll, & Ermler, 2004). There has been relatively little EPR work done with 4-hydroxybenzoylCoA dehydrogenase, although "Rapid type 2" like and "Slow" like signals have been reported for as-isolated and cyanide-treated enzyme, respectively (Johannes et al., 2008). The former signal exhibits the characteristic strong coupling to two-solvent-exchangeable protons that typifies the signal-giving species.

3. Sulfite oxidase family enzymes

3.1 Sulfite oxidase

Bray undertook the initial EPR work with sulfite oxidase (Gutteridge, Lamy, & Bray, 1980; Lamy, Gutteridge, & Bray, 1980), demonstrating that three discrete Mo(V) signals could be seen, depending on conditions: "high-pH," "low-pH" and "phosphate-inhibited" (Fig. 5). This early work was facilitated by the fact that the chicken liver enzyme that was being used possesses a high-potential *b*-type cytochrome in addition to the molybdenum center, and high levels of Mo(V) were achieved after electron transfer to the heme after initial two-electron reduction of the molybdenum from Mo(VI) to Mo(IV). The "low-pH" form exhibits strong coupling to one

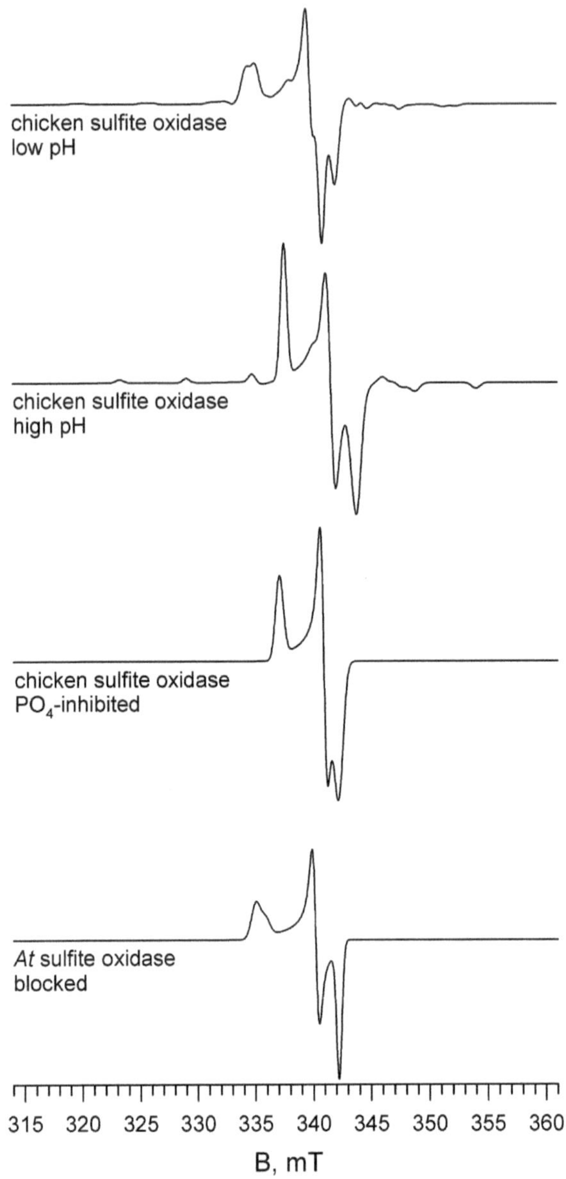

Fig. 5 *The EPR signals seen with chicken liver and A. thaliana sulfite oxidases.* From *Top* to *Bottom*: the "low-pH" and "high-pH" signals originally reported by (Lamy et al., 1980) using the simulation parameters reported by (Dhawan & Enemark, 1996); the "phosphate-inhibited" signal (George, Prince, Kipke, Sunde, & Enemark, 1988); and the "blocked" signal seen with the sulfite oxidase from *A. thaliana* as originally reported by (Hemann et al., 2005)using the simulation parameters reported by (Astashkin et al., 2005).

solvent-exchangeable proton, while neither the "high-pH" or "phosphate-inhibited" signals exhibit proton hyperfine in continuous-wave EPR spectra. Subsequent pulsed EPR work by Enemark and coworkers has clearly demonstrated, however, that the "high-pH" form does in fact exhibit coupling to a weakly coupled proton (Astashkin, Mader, Pacheco, Enemark, & Raitsimring, 2000; Raitsimring, Pacheco, & Enemark, 1998). It is now well-accepted that the proton coupling in both cases is due to an equatorial Mo–OH (the equatorial Mo=O of the oxidized molybdenum center becoming protonated upon reduction of the metal by one electron), with the magnitude of the proton coupling being extremely sensitive to the orientation of the Mo–O–H bond relative to the metal xy plane (the redox-active orbital being d_{xy}) (Enemark, 2017; John H. Enemark, Astashkin, & Raitsimring, 2010). In the species giving rise to the "low-pH" signal, the Mo–O–H bond lies in the xy plane, while in the species giving rise to the "high-pH" signal the proton is rotated out of the xy plane. The lack of proton coupling in the "phosphate-inhibited" signal is due to the displacement of the equatorial Mo–OH by phosphate. Although no ^{31}P coupling seen was seen in the continuous-wave EPR, subsequent 2D ESEEM with ^{31}P-labeled phosphate provided clear evidence for anisotropic ^{31}P coupling in the range of 4–11 MHz (Pacheco, Basu, Borbat, Raitsimring, & Enemark, 1996). Chloride was also found to perturb the "low-pH" EPR spectrum of chicken sulfite oxidase (Bray, Gutteridge, Lamy, & Wilkinson, 1983; Doonan et al., 2008), and it was subsequently shown by both XAS (Pushie, Doonan, Wison, Rajagopalan, & George, 2011) and X-ray crystallography (Karakas et al., 2005) that the anion binds in the (anionic) substrate binding pocket adjacent to the molybdenum center rather than coordinating directly to it.

Unlike the enzyme from vertebrate sources, that from plants lacks a heme group (the enzyme utilizes O_2 rather than cytochrome *c* as the oxidizing substrate (Byrne, Haensch, Mendel, & Hille, 2009; Hansch et al., 2006). As shown in Fig. 5, the wild-type enzyme from *Arabidopsis thaliana* exhibits a signal under turnover conditions that has a g-tensor resembling the canonical "low-pH" signal but, like the "phosphate-inhibited" signal seen with the chicken liver enzyme, is devoid of proton coupling (Astashkin et al., 2007; Astashkin, Hood, et al., 2005). Although originally thought to represent a complex of enzyme with the product sulfate, subsequent DFT simulations of the observed coupling when ^{33}S-labeled sulfite is used to generate the signal suggest that the large quadrupole coupling seen by pulsed EPR (36–40 MHz) is likelier due to sulfite rather than sulfate (Enemark, 2017).

Pulsed EPR has shown that when generated in ^{17}O-labeled H_2O, the "low-pH" signal exhibits coupling to two inequivalent oxygens, one more strongly coupled than the other (Astashkin, Feng, Raitsimring, & Enemark, 2005). Although the strongly and weakly coupled oxygens were originally assigned to the equatorial and axial oxygens of the molybdenum coordination (implying that both exchanged rapidly with solvent), a reanalysis of the data has led to the conclusion that the strongly coupled oxygen was the equatorial oxygen bridged between the molybdenum and sulfur of bound sulfite while the more weakly coupled oxygen was due to non-coordinating oxygen(s) on the sulfite, sulfite (but not sulfate) being known to rapidly exchange its oxygens with solvent water (Enemark, 2017). That this is the correct assignment for the oxygen coupling is reflected in the fact that addition of phosphate results in the loss of both strongly and weakly coupled oxygens upon displacing the bound sulfite (Enemark, 2017).

3.2 Bacterial sulfite dehydrogenases

A variety of bacteria possess a sulfite: cytochrome *c* oxidoreductase, commonly referred to as sulfite dehydrogenase, that is also members of the sulfite oxidase family. As typified by the enzyme from *Starkeya novella* (previously termed *Thiobaciillus novellus*) (Kappler et al., 2000);(Kappler & Bailey, 2004), these enzymes are heterodimers with separate molybdenum- and cytochrome *c*-containing subunits that are much more intimately associated with one another than the only loosely tethered heme domain of the vertebrate sulfite oxidases. Like the plant sulfite oxidase, the *S. novella* sulfite dehydrogenases exhibit only a high-pH spectrum upon reduction by sulfite (Kappler et al., 2000), and a Y236F variant exhibits a low-pH form below pH7 (Kappler et al., 2006), although the high-pH form of the variant has somewhat higher g_2 and g_3 than the corresponding signal seen with wild-type enzyme (Fig. 6. The tyrosine mutation does disrupt a hydrogen-bonding network in the vicinity of the molybdenum center, and results in both k_{cat} and K_m being compromised by a factor of $\sim$6 compared to wild-type enzyme (Kappler et al., 2006).

3.3 Assimilatory nitrate reductases

The assimilatory nitrate reductase from algae and higher plants are also members of the sulfite oxidase family of molybdenum enzymes (and are not to be confused with the *dissimilatory*—or respiratory—nitrate reductases that are members of the DMSO reductase family, considered further below).

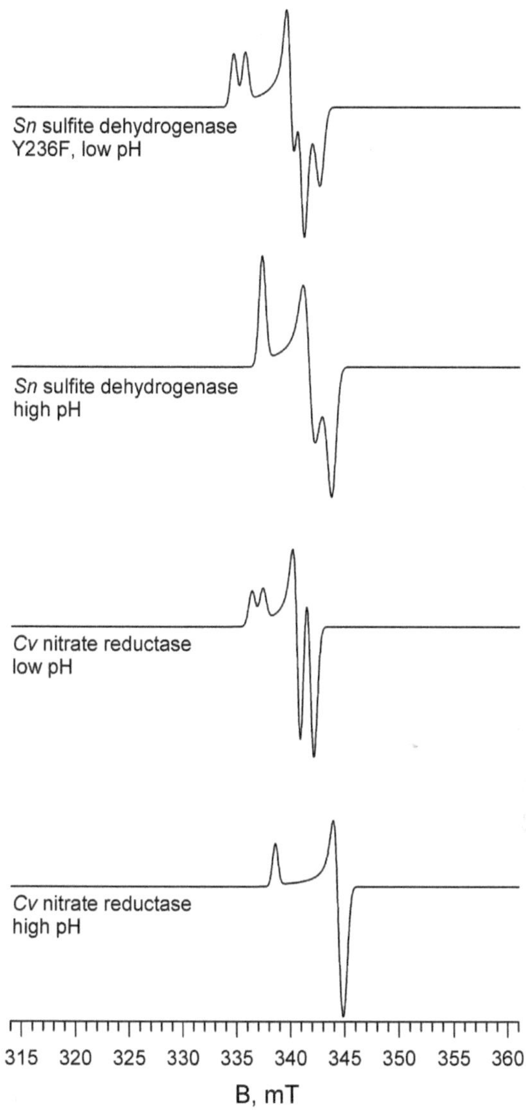

Fig. 6 *The EPR signals seen with bacterial sulfite dehydrogenases and vertebrate assimilatory nitrate reductases.* From *Top* to *Bottom*: The "low-pH signal seen with the Y238F mutant of the *S. novella* sulfite dehydrogenase, using the simulation parameters reported by (Kappler et al., 2006); the "high-pH" signal ween with the wild-type *S. novella* enzyme using the simulation parameters reported in (Kappler et al., 2000); the "low-pH" and "high-pH" signals seen with the *C. vulgaris* nitrate reductase, using the simulation parameters reported by (Solomonson, Barber, Howard, Johnson, & Rajagopalan, 1984).

As with the vertebrate sulfite oxidases, "low-pH" and "high-pH" forms of the EPR spectra exhibited by both the spinach (Gutteridge, Bray, Notton, Fido, & Hewitt, 1983) and *Chlorella* (Solomonson et al., 1984) enzymes are observed, with the "low-pH" signal showing strong coupling to a solvent-exchangeable proton (Fig. 6). The basis for sulfite oxidase being an oxidase and nitrate reductase being a reductase has little to do with the intrinsic properties of the molybdenum center, but rather the intrinsic thermodynamic stabilities of S=O versus N=O bonds (Holm & Donahue, 1993).

3.4 mARC and methionine sulfoxide reductase

The sulfite oxidase family was for many years thought to be represented only by the eponymous sulfite oxidases (and related bacterial sulfite dehydrogenases) and the assimilatory nitrate reductases. More recently it has been recognized that the mitochondrial amidoxime reductase component (mARC) and the C-terminal domain of the molybdenum cofactor sulfurase (MOSC) protein and its homologs are also members of the sulfite oxidase family of molybdenum enzymes, and in particular possess the highly conserved cysteine residue that coordinates the molybdenum center. The mARC system consists of three proteins, a cytochrome b_5, a b_5 reductase and a molybdenum-containing component that catalyzes the reduction of amidoximes and related compounds (Gruenewald et al., 2008; Havemeyer et al., 2006; Havemeyer et al., 2010; Krompholz et al., 2012). Both isoforms of human mARC, hmARC-1 and hmARC-2, exhibit EPR properties characteristic of the sulfite oxidase family, with proton coupling to a solvent-exchangeable proton (Wahl et al., 2010), although the g-tensor is somewhat more axial in symmetry than is seen with other members of the family; a simulation of the spectrum is shown in Fig. 7. Subsequent pulsed EPR work probing ^1H and ^{17}O coupling, in conjunction with a density functional theory treatment, has identified the proton as arising from an equatorial Mo–OH ligand (Rajapakshe et al., 2009).

Another member of the sulfite oxidase family is the methionine sulfoxide reductase (MsrP) of *E. coli* that is encoded by the *yedYZ* operon (the enzyme is also termed YedY in the literature). Most unusually for members of this family, MsrP was isolated primarily in the Mo(V) state, and exhibited a nearly axial EPR signal that, as shown in Fig. 7, exhibits no coupling to protons (Yang et al., 2009). Subsequent electrochemical studies showed that while the molybdenum could be reduced to the Mo(IV) state with an unusually high reduction potential of +174 mV, there was no evidence that

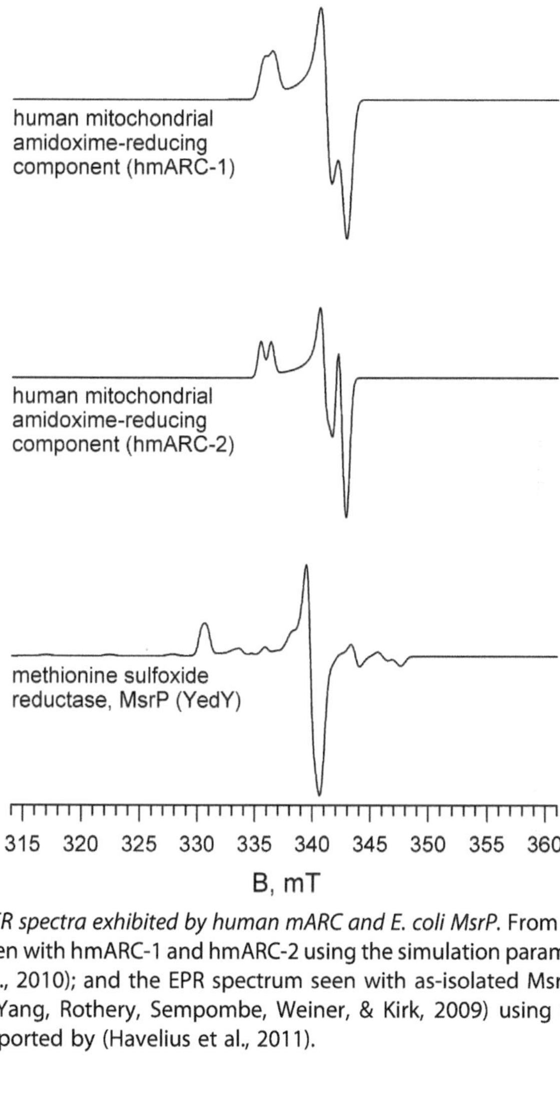

Fig. 7 *The EPER spectra exhibited by human mARC and E. coli MsrP.* From *Top* to *Bottom*: the spectra seen with hmARC-1 and hmARC-2 using the simulation parameters reported by (Wahl et al., 2010); and the EPR spectrum seen with as-isolated MsrP, as originally reported by (Yang, Rothery, Sempombe, Weiner, & Kirk, 2009) using the simulation parameters reported by (Havelius et al., 2011).

the Mo(V) state could be oxidized to Mo(VI) (Adamson et al., 2015). A second two-electron wave at $\sim -250\,\text{mV}$ in cyclic voltammograms was attributed to reduction of the pyranopterin ligand to the molybdenum from the di- to tetrahydropterin forms, and it was concluded that the enzyme cycled in the course of catalysis between these two states, with one of the two electrons required coming from the Mo(V). More recently, however, the EPR and XAS spectroscopic features of MsrP have been reassessed with the conclusion that the as-isolated enzyme does not possess the hallmark

MoO_2 structure of the sulfite oxidase family, but instead is found as an MoO form with a sulfur-containing inhibitory species displacing the catalytically labile equatorial Mo=O (Ingersol et al., 2020). The nature of this inhibitor is not at present known, although poorly resolved electron density at the equatorial position, extending into the substrate binding site, is observed in the original crystallographic work (modeled at the time as water and urea molecules) and is consistent with tightly bound homocysteine (Ingersol et al., 2020).

4. DMSO reductase family enzymes

The DMSO reductase family is the most diverse, both structurally and functionally, of the molybdenum-containing enzymes. The present focus is on those enzymes that have been best characterized by EPR and include the DMSO reductases, dissimilatory nitrate reductases and formate dehydrogenases, which are considered in turn below.

4.1 DMSO reductases

The simplest members of the DMSO reductase (and first to be characterized crystallographically (Li, Temple, Rajagopalan, & Schindelin, 2000; Schindelin, Kisker, Hilton, Rajagopalan, & Rees, 1996)) are the dissimilatory DMSO reductases from *Rhodobacter sphaeroides* or *R. capsulatus* for which the family is named. These are periplasmic proteins whose sole redox-active site is the molybdenum center, with terminal Mo=O and serinate ligands to the molybdenum in addition to the two equivalents of the bidentate pyranopterin cofactor. Working with the *R. capsulatus* protein, Bray and coworkers (Bennett, Benson, McEwan, & Bray, 1994) identified and carefully characterized five discrete and well-simulated Mo(V) EPR signals exhibited by the enzyme, as shown in Fig. 8. One of these exhibited a strong resemblance to the "slow" signal manifested by xanthine oxidase, that appears to arise from a non-functional form of the enzyme in which one of the two equivalents of pyranopterin has dissociated from the molybdenum, being replaced by a second Mo=O group; this form has since been characterized crystallographically (Li et al., 2000). Two of the remaining signals were termed "high-g" with g_{av} ~1.98, with one exhibiting strong coupling to a solvent-exchangeable proton and the other no; a second pair of "low-g" signals, with g_{av} below 1.96, exhibited proton coupling while the second did not. These signals are shown in Fig. 8. The high-g_{av} signal exhibiting proton splitting, frequently referred to as the "high-g split" signal, was considered to

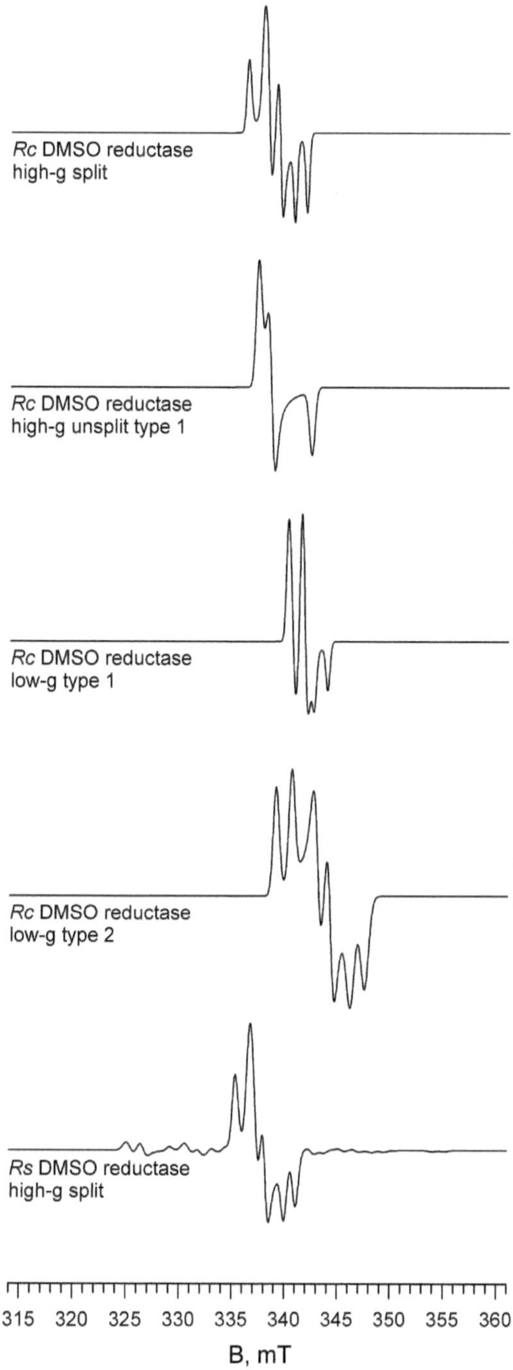

Fig. 8 *The EPER spectra exhibited by DMSO reductase from R. capsulatus and R. sphaeroides.* From *Top* to *Bottom*: the "high-g split," high-unsplit type 1," "low-g type 1" and "low-g type 2" signals seen with the *R; capsulatus* enzyme as reported by (Bennett, Benson, et al., 1994); and the "high-g split" signal with the *R. sphaeroides* enzyme as reported by (Mtei et al., 2011).

represent the most mechanistically relevant of the signals observed, a conclusion subsequently demonstrated in enzyme-monitored turnover experiments with the *R. sphaeroides* enzyme, using both DMSO and trimethylamine-N-oxide as substrates (Cobb, Conrads, & Hille, 2005). In this study, the UV/visible absorption spectrum of enzyme was followed as a function of time during turnover using sodium dithionite as an artificial electron donor. Using multicomponent analysis, the observed spectrum at any point in the course of turnover with DMSO as substrate could be fit to a weighted sum of the absorption spectra of five species: fully oxidized enzyme, fully reduced enzyme, an E_{red}•DMSO complex and the Mo(V) species giving rise to the "high-g split" EPR signal; with TMAO as substrate, no intermediate corresponding to the E_{red}•DMSO complex was observed, but the "high-g split" species was generated essentially quantitatively in the course of turnover, reflecting the fact that its reduction on to fully reduced enzyme was principally rate-limiting to turnover under the experimental conditions. With DMSO as substrate, the "high-g" split initially accounted for ~60% of the total enzyme population during turnover, but decreased at the expense of the E_{red}•DMSO complex as product accumulated in the course of the reaction and rebound to the oxidized enzyme to reverse the product dissociation step. This process accounted for the product inhibition to which the enzyme was known to be subject. Using a similar enzyme-monitored turnover approach, active site variants in which active site residues Trp 116 and Tyr 114 were mutated to Phe were examined (Cobb et al., 2007). The Y114F mutant showed significantly lower levels of accumulation of the E_{red}•DMSO complex in the course of turnover, consistent with its proposed role in hydrogen-bonding to bound DMSO in the reduced enzyme. The as-isolated W116F mutant was found to have one of the pyranopterin ligands dissociated, but this could be made to rebind by reduction of the enzyme with dithionite followed by reoxidation with DMSO (so-called "annealing"(Garton, Garrett, Rajagopalan, & Johnson, 1997)). Both W114F and Y116F mutants exhibited large amounts of the "high-g split" EPR signal in the course of turnover that were essentially indistinguishable from that seen with wild-type enzyme.

It is now generally understood that the "high-g split" signal arises from catalytically functional enzyme, whereas the "low-g" signals arise from non-functional forms in which one of the pyranopterins (that designated Q in the crystal structure of the *R. sphaeroides* enzyme, or "proximal" in the nomenclature proposed by Weiner and coworkers (Rothery, Workun, & Weiner, 2008), being located between the molybdenum and [4Fe–4S] cluster also present in the Mo-containing subunit of the *E. coli* DMSO reductase, see below) has dissociated from the molybdenum. This being the case, it is somewhat surprising that a "high-g split" signal has not been reported to date with the membrane-integral DMSO reductase from *E. coli*, which has a molybdenum-containing subunit closely homologous to the *Rhodobacter* proteins. Only a single signal clearly attributable to Mo(V) has been reported, which has a g_{av} even lower than is seen with the "low-g" signals described above (Cammack & Weiner, 1990); it does not manifest any coupling to protons and presumably arises from a nonfunctional form of the protein. It is important to recognize that the *E. coli* DMSO reductase, the product of the *dmsABC* operon, has two other subunits not present in the *Rhodobacter* enzymes in addition to the DmsA that contains the molybdenum center (and proximal [4Fe–4S] cluster): DmsB has four [4Fe–4S] clusters, and the cofactorless DmsC constitutes an integral membrane anchor; in vivo this minimal protomeric structure is probably further aggregated as a dimer or trimer as seen with formate dehydrogenase and nitrate reductase (discussed below) with which it shares considerable overall homology.

4.2 Dissimilatory nitrate reductases

To date, four different bacterial nitrate reductases have been characterized. The first two are the dissimilatory, membrane-integral NarGHI and NarYWV systems that function in conjunction with the FdhGHI and FdoGHI formate dehydrogenases (described below) as energy-conserving formate:nitrate oxidoreductases. These systems are dimeric, with each protomer containing a cytosolically-oriented subunit with the active site molybdenum and a single [4Fe–4S] cluster, a medial domain with three [4Fe–4S] clusters and a [3Fe–4S] cluster distal to the molybdenum center, and a membrane anchoring domain possessing two *b*-type cytochromes (Jormakka, Tornroth, Byrne, & Iwata, 2002). The third nitrate reductase is the periplasmically-localized, soluble and dissimilatory Nap protein, which is either a monomeric NapA with molybdenum and [4Fe–4S] centers (Dias et al., 1999) or a dimeric NapAB (with B subunit containing a pair of *b*-type cytochromes (Thomas, Potter, & Cole, 1999)) depending on the source organism. Finally, there are the *assimilatory* NasA (or NasAB) enzymes, coexpressed with nitrite reductases to from a nitrogen uptake system, which are cytosolic (Gangeswaran, Lowe, & Eady, 1993). The various nitrate reductases differ from the DMSO reductases (described above) in two important ways: the protein ligand to the molybdenum is either an Asp (e.g., in the *E. coli* NarGHI enzyme) or a Cys (e.g., in the periplasmic Nap enzymes from a variety of species) rather than the Ser found in the DMSO reductases, and at least in the case of the Nap proteins the terminal ligand is a Mo═S rather than Mo═O. Both substitutions are expected to give rise to EPR signals with higher rather than lower g_{av} values.

The *E. coli* NarGHI was the first to be investigated by EPR (Bray, Vincent, Lowe, Clegg, & Garland, 1976), work that was done prior to the recognition that members of the DMSO reductase family of enzymes were fundamentally distinct from those of the sulfite oxidase (or xanthine oxidase) families. The first signals to be characterized (Bray et al., 1976), shown in Fig. 9, closely resembled the "high pH" and "low pH" signals seen with sulfite oxidase). Comparable signals are seen with the dissimilatory nitrate reductase from *Pseudomonas aeruginosa* (Godfrey, Greenwood, Thomson, Bray, & George, 1984) and *Paracoccus denitrificans* (N. Turner, Ballard, Bray, & Ferguson, 1988) the; *P. denitrificans* enzyme also exhibits a third, "very high-g," signal (Turner et al., 1988). Nitrate and nitrite perturb the "high-g split" signal in the enzyme from *Pseudomonas aeruginosa*, constituting the first (if indirect) evidence that this signal was the more mechanistically relevant. Subsequently, again with the *E. coli* enzyme, a variety of signals arising from complexes with anions (including Cl^- and F^-, the latter exhibiting strong coupling to the ^{19}F nucleus) have been characterized (George et al., 1989). As with the DMSO reductases, it has been concluded that the "high-g split" signal seen with the periplasmic Nap arises from the most directly catalytically relevant form of the enzyme, several varieties of which are observed depending on the experimental conditions (specifically, the presence or absence of substrate). The relationship between the "high-g split" "very high-g" signal remains unresolved, however. The most convincing analysis to date, involving a combined EPR and DFT approach (Rendon et al., 2017), suggests that the "high-g" split signal arises from a catalytically relevant species in which a Mo═S ligand is partially bonded to the sulfur of the metal-coordinated Cys residue, forming what is in essence a cysteine persulfide that is coordinated side-on to the metal—indeed, in the best crystal structures of the *Desulfovibrio desulfuricans* NapA enzyme, the two sulfurs are only some 2.4 Å apart (i.e., closer than van der Waals contact). The "very high g" signal is thought to arise by reduction of the S—S bond to give frank terminal Mo═S and Cys thiolate ligands to the molybdenum. The "low g" signal, which exists in split and unsplit forms that resemble the EPR spectra exhibited by sulfite oxidase, is believed to arise from an inactive form of the enzyme in which, as in the case of the DMSO reductases, one of the pyranopterins has dissociated from the

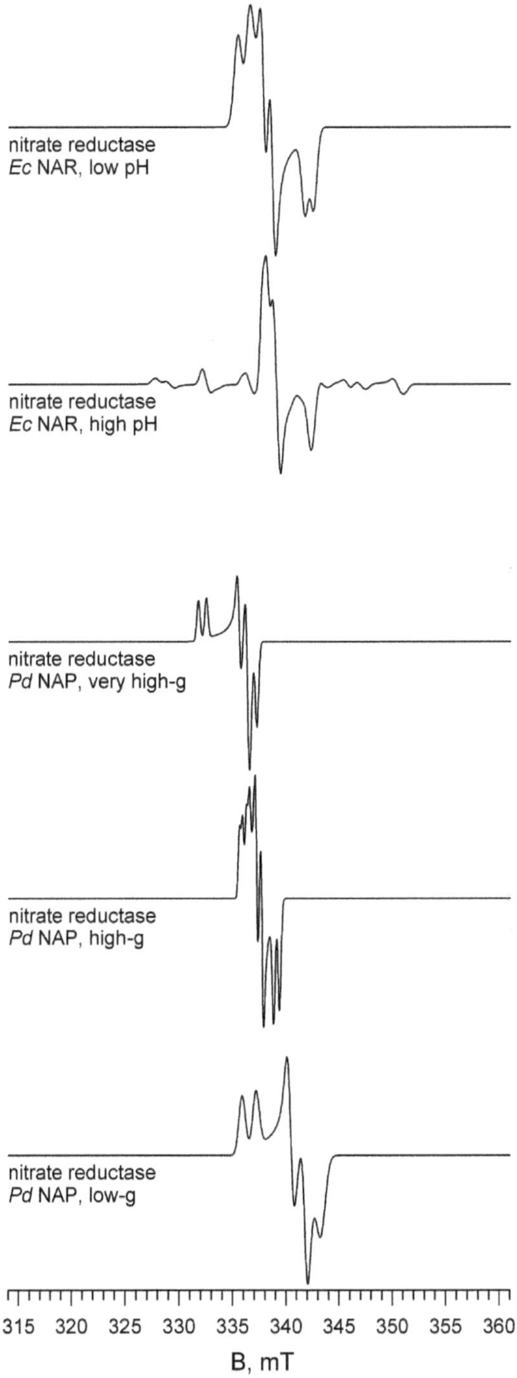

Fig. 9 *The EPR signals seen with various bacterial nitrate* reductases. From Top to Bottom: the "low-pH" signal seen with the *E. coli* NarGHI enzyme as reported by (Vincent & Bray, 1978); the "high-pH" signal seen with the *E. coli* NarGHI as reported by (George, Bray, Morpeth, & Boxer, 1985); the "very high-g" and "high-g" signals seen with the *P. denitrificans* NapA enzyme as reported by (Butler et al., 1999); and the "low-g" signal seen with the *P. denitrificans* NapA as reported by (Bennett, Berks, Ferguson, Thomson, & Richardson, 1994).

molybdenum, replaced by a solvent-derived Mo–OH; the terminal Mo=S may also have been lost as well, replaced by a solvent-derived Mo=O.

4.3 Formate dehydrogenases

A final group of enzyme systems in the DMSO reductase family for which EPR has proven extremely useful are the formate dehydrogenases, molybdenum- (and occasionally tungsten-) containing enzymes that catalyze the reversible interconversion of formate and CO_2. As in the case of the nitrate reductases, there are multiple varieties of formate dehydrogenase (Hille et al., 2014). The FdnGHI and FdoGHI systems are trimeric, membrane-integral complexes that work in concert with the NarGHI and NarYWV systems to generate a transmembrane potential. They are oriented oppositely to the Nar systems in the membrane, however, with the subunit containing the active site molybdenum center oriented toward the periplasm rather than the cytosol. In addition to the molybdenum- and [4Fe–4S]-containing FdnG subunit, the medial FdnH subunit has a four [4Fe–4S] clusters, and the membrane-integral FdnI anchor, like the Nar enzymes (but different from the *E. coli* DMSO reductase) contains a pair of *b*-type cytochromes (Jormakka et al., 2002). Altogether, the Nar and Fdh/Fdo systems share a great deal of structural homology, their opposite orientations in the membrane notwithstanding. By contrast, the FdhF formate dehydrogenase is a dissociable, periplasmically oriented subunit of the formate:H_2 lyase complex (Axley, Grahame, & Stadtman, 1990). It is monomeric with molybdenum and [4Fe–4S] centers, and exhibits a high degree of structural homology to the molybdenum-containing subunits of the DMSO reductase and nitrate reductase systems described above. Finally, there is the cytosolic FdsDABG enzyme from organisms such as *Cupriavidus necator* (formerly *Ralstonia eutropha*) (Friedebold & Bowien, 1993; Friedebold, Mayer, Bill, Trautwein, & Bowien, 1995; Oh & Bowien, 1998) or *Rhodobacter capsulatus* (Schrapers et al., 2015). In addition to the molybdenum-containing FdhA subunit, which has an FdhF-like C-terminal domain and an N-terminal domain with five additional iron-sulfur clusters, there is an FMN-containing FdsB subunit and an FdsG subunit possessing a [4Fe–4S] and [2Fe–2S] cluster; the small FdsD subunit, identified unambiguously by cryoelectron microscopy (Radon et al., 2020) is cofactorless and appears to be involved in "capping" the molybdenum center of the FdsA subunit. The overall topology closely resembles that of the matrix-exposed portion of the mitochondrial Complex I (Hille et al., 2014; Young et al., 2020). All the formate dehydrogenases possess a catalytically essential terminal Mo=S ligand to the metal that is incorporated into the maturing cofactor by a specific sulfurase prior to insertion into the apoenzyme (Boehmer, Hartmann, & Leimkuehler, 2014). Fdn/FdoGHI and FdhF have selenocysteine as the molybdenum ligand contributed by the polypeptide, while FdsDABG has a cysteine residue instead (it is worth noting parenthetically that the FdhF from *Pelobacter atrosepticum* also has cysteine rather than selenocysteine coordinated to the molybdenum).

The FdhF from *E. coli* exhibits a well-resolved Mo(V) EPR signal upon reduction of the enzyme by formate (the Mo(V) state being formed after the initial two-electron reduction of the molybdenum followed by transfer of one electron to the [4Fe–4S] cluster of the protein), as shown in Fig. 10. The signal exhibits poorly resolved ^{1}H hyperfine and, when ^{77}Se is used, strong coupling to selenium that clearly demonstrates direct SeCys coordination to the molybdenum (Gladyshev, Khangulov, Axley, & Stadtman, 1994); the proton coupling is lost on going to D_2O (Khangulov et al., 1998). A virtually identical signal is seen with the Cys-containing FdsDABG from *C. necator* (Niks et al., 2016). With both systems, when enzyme is reduced with ^{2}H-labeled formate in H_2O, no proton coupling is observed at short times after mixing, but grows in with time (Khangulov et al., 1998; Niks et al., 2016). The conclusion in both cases is that the hydrogen of substrate formate is transferred directly to the molybdenum center. The earlier work with FdhF was done prior to the recognition that the molybdenum center possessed a Mo=S group, and the interpretation with FdsDABG that the reaction mechanism involves direct hydride transfer to the Mo^{VI}=S to give a Mo^{IV}-SH (Niks et al., 2016) has now received general acceptance in the field. Such a hydride transfer mechanism is reminiscent of the reaction mechanism of the oxidative hydroxylation mechanism of xanthine oxidase, which also involves transfer of a hydride (or, more properly, a hydride equivalent) from substrate to the molybdenum center. The functionally related molybdenum-containing formylmethanofuran dehydrogenases from methanogenic archaea (Hochheimer, Hedderich, & Thauer, 1998; Hochheimer, Linder, Thauer, & Hedderich, 1996) likely operate via the same mechanism, as do their tungsten-containing orthologs (Hochheimer et al., 1996; Hochheimer, Schmitz, Thauer, & Hedderich, 1995).

The formate dehydrogenase *D. desulfuricans* is substantially less air-sensitive than is the enzyme from most other organisms. The resting enzyme exhibits a rhombic signal (Fig. 10) that is devoid of proton hyperfine; this and a second signal with similar anisotropy but somewhat lower g values is seen in the course of reductive titrations and a third signal that is reminiscent of the signals seen with the *E. coli* and *C. necator* enzymes upon reduction with

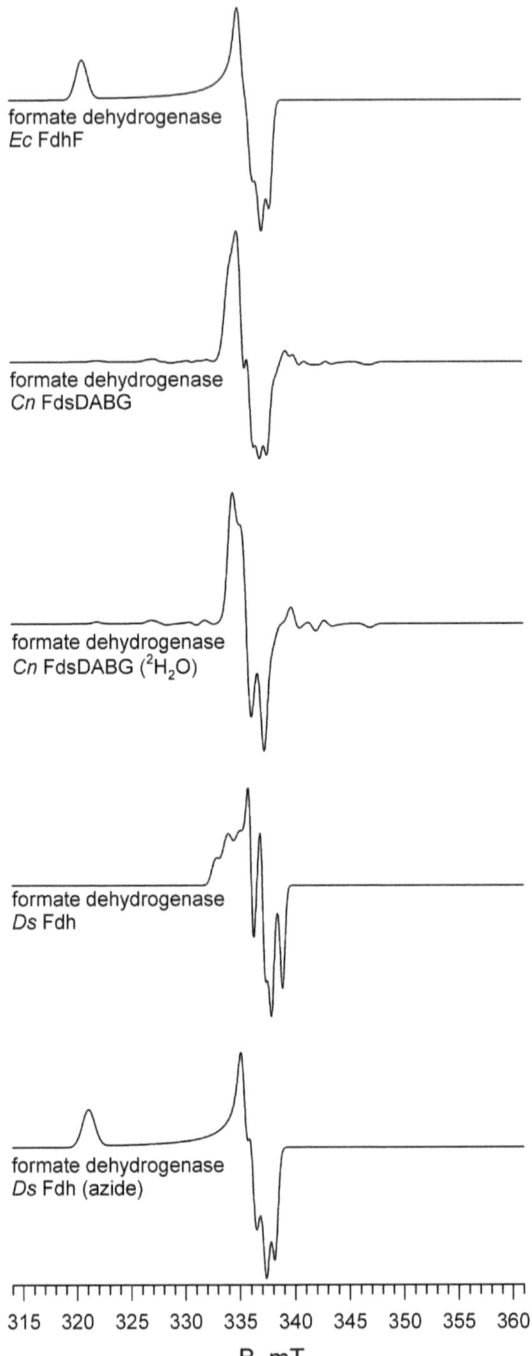

Fig. 10 *The EPR signals seen with formate dehydrogenases from various organisms.* From Top to Bottom: the signal seen with the *E. coli* FdhF enzyme as reported by (Khangulov, Gladyshev, Dismukes, & Stadtman, 1998); the signals seen with the *C. necator* FdsDABG enzyme in H2O and D2O using the simulation parameters reported by (Niks, Duvvuru, Escalona, & Hille, 2016); and the EPR signals seen with the *D. desulfuricans* formate dehydrogenase in the absence and presence of azide using the simulation parameters reported by (Rivas, Gonzalez, Brondino, Moura, & Moura, 2007).

formate, including coupling to at least one $I = 1/2$ nucleus, presumably a proton (Costa, Moura, Moura, Wang, & Huynh, 1996). More recent work on the signal seen on reduction with formate (Rivas et al., 2007) has shown that there is a single solvent-exchangeable proton coupled to the formate-reduced signal, with coupling to a second $I = 1/2$ nucleus remaining when the sample is prepared in D_2O. In the presence of azide, a much more anisotropic signal is observed with a particularly high g_1 of 2.094 with coupling to a proton that is initially absent when deuterated formate is used to reduced enzyme, but which grows in when the experiment is performed in H_2O. It was concluded that the signal seen in the absence of azide was due to an L_2Mo^V(SeCys)XH species, with it now recognized that the X is the catalytically essential Mo=S of the oxidized enzyme that has been protonated upon reduction with formate. The azide signal has been proposed to be due to a species in which the selenocysteine has dissociated from the metal (for which crystallographic evidence exists with the *E. coli* enzyme (Boyington, Gladyshev, Khangulov, Stadtman, & Sun, 1997)).

5. Summary

Electron paramagnetic resonance has played a historically critical role in advancing our understanding of molybdenum in biology, and for each of the three major categories of molybdenum enzyme has provided key information in the elucidation of reaction mechanism. One common element is the use of a Mo^{VI}=S group to accept (or donate) a hydride equivalent to yield a Mo^{IV}-SH moiety. EPR, ESEEM, ENDOR, HYSCORE and related magnetic methods have proven to be extremely powerful tools to gain insight into the detailed chemical environment of the molybdenum, and in particular its first and second coordination sphere.

Acknowledgments

The authors thank Dr. Jarett Wilcoxen, Department of Chemistry, University of Wisconsin, Milwaukee for helpful discussions. Work in the authors' laboratory has been funded by th e Department of Energy (DE-SC0010666 to RH), the National Institutes of Health (GM 135088 to RH) and the National Science Foundation (CHE 2101672 to RH).

References

Adamson, H., Simonov, A. N., Kierzek, M., Rothery, R. A., Weiner, J. H., Bond, A. M., et al. (2015). Electrochemical evidence that pyranopterin redox chemistry controls the catalysis of YedY, a mononuclear Mo enzyme. *Proceedings of the National Academy of Sciences of the United States of America, 112*(47), 14506–14511. https://doi.org/10.1073/pnas.1516869112.

Amaya, Y., Yamazaki, K., Sato, M., Noda, K., & Nishino, T. (1990). Proteolytic conversion of xanthine dehydrogenase from the NAD+-dependent type to the O2-dependent type—Amino acid sequence of rate liver xanthine dehydrogeanse and identification of cleavage sites of the enzyme protein during irreversible conversion by trypsin. *Journal of Biological Chemistry*, *265*(24), 14170–14175.

Astashkin, A. V., Feng, C. J., Raitsimring, A. M., & Enemark, J. H. (2005). O-17 ESEEM evidence for exchange of the axial oxo ligand in the molybdenum center of the high pH form of sulfite oxidase. *Journal of the American Chemical Society*, *127*(2), 502–503. https://doi.org/10.1021/ja0461466.

Astashkin, A. V., Hood, B. L., Feng, C. J., Hille, R., Mendel, R. R., Raitsimring, A. M., et al. (2005). Structures of the Mo(V) forms of sulfite oxidase from Arabidopsis thaliana by pulsed EPR spectroscopy. *Biochemistry*, *44*(40), 13274–13281. https://doi.org/10.1021/bi051220y.

Astashkin, A. V., Johnson-Winters, K., Klein, E. L., Byrne, R. S., Hille, R., Raitsimring, A. M., et al. (2007). Direct demonstration of the presence of coordinated sulfate in the reaction pathway of Arabidopsis thaliana sulfite oxidase using S-33 labeling and ESEEM spectroscopy. *Journal of the American Chemical Society*, *129*(47), 14800–14810.

Astashkin, A. V., Mader, M. L., Pacheco, A., Enemark, J. H., & Raitsimring, A. M. (2000). Direct detection of the proton-containing group coordinated to Mo(V) in the high pH form of chicken liver sulfite oxidase by refocused primary ESEEM spectroscopy: Structural and mechanistic implications. *Journal of the American Chemical Society*, *122*(22), 5294–5302.

Axley, M. J., Grahame, D. A., & Stadtman, T. C. (1990). *Escherichia coli* formate-hydrogenlyase—purification and properties of the selenium-dependent formate dehydrogenase component. *Journal of Biological Chemistry*, *265*(30), 18213–18218.

Barata, B. A. S., Legall, J., & Moura, J. J. G. (1993). Aldehyde oxidoreductase activity in *Desulfovibrio gigas*—*in vitro* reconstitution of an electron transfer chain from aldehydes to the production of molecular hydrogen. *Biochemistry*, *32*(43), 11559–11568.

Bennett, B., Benson, N., McEwan, A. G., & Bray, R. C. (1994). Multiple states of the molybdenum center of dimethylsulfoxide reductase from *Rhodobacter capsulatus* revealed by EPR spectroscopy. *European Journal of Biochemistry*, *225*(1), 321–331.

Bennett, B., Berks, B. C., Ferguson, S. J., Thomson, A. J., & Richardson, D. J. (1994). Mo(V) Electron-paramagnetic resonance signals from the periplasmic nitrater reductase of *Thiosphaera pantotropha*. *European Journal of Biochemistry*, *226*(3), 789–798. https://doi.org/10.1111/j.1432-1033.1994.00789.x.

Boehmer, N., Hartmann, T., & Leimkuehler, S. (2014). The chaperone FdsC for Rhodobacter capsulatus formate dehydrogenase binds the bis-molybdopterin guanine dinucleotide cofactor. *FEBS Letters*, *588*(4), 531–537. https://doi.org/10.1016/j.febslet.2013.12.033.

Boyington, J. C., Gladyshev, V. N., Khangulov, S. V., Stadtman, T. C., & Sun, P. D. (1997). Crystal structure of formate dehydrogenase H: Catalysis involving Mo, molybdopterin, selenocysteine, and an Fe4S4 cluster. *Science*, *275*(5304), 1305–1308. https://doi.org/10.1126/science.275.5304.1305.

Bray, R. C., & George, G. N. (1985). Electron paramagnetic resonance studies using pre-steady-state kinetics and substitution with stable isotopes on the mecahnism of action of molybdoenzymes. *Biochemical Society Transactions*, *13*(3), 560–567.

Bray, R. C., George, G. N., Gutteridge, S., Norlander, L., Stell, J. G. P., & Stubley, C. (1982). Studies by electron paramagnetic resonance spectroscopy of the molybdenum center of aldehyde oxidase. *Biochemical Journal*, *203*(1), 263–267.

Bray, R. C., Gutteridge, S., Lamy, M. T., & Wilkinson, T. (1983). Equilibria amongst different molybdenum(V)-containing species from sulfite oxidase—Evidence for a halide ligand of molybdenum in the low-pH species. *Biochemical Journal*, *211*(1), 227–236.

Bray, R. C., & Knowles, P. F. (1968). Electron paramagnetic resonance in enzyme chemistry—mechanism of action of xanthine oxidase. *Proceedings of the Royal Society of London Series a-Mathematical and Physical Sciences*, *302*(1470), 351–359. https://doi.org/10.1098/rspa.1968.0020.

Bray, R. C., & Meriwether, L. S. (1966). Electron spin resonance of xanthine oxidase substituted with molybdenum-95. *Nature*, *212*(5061), 467–469. https://doi.org/10.1038/212467a0.

Bray, R. C., Palmer, G., & Beinert, H. (1964). Direct studies on electron transfer sequence in xanthine oxidase by electron paramagnetic resonance spectroscopy. 2. Kinetic studies employing rapid freezing. *Journal of Biological Chemistry*, *239*(8), 2667–2676.

Bray, R. C., Turner, N. A., Legall, J., Barata, B. A. S., & Moura, J. J. G. (1991). Information from EPR spectroscopy on the iron-sulfur centers fo the iron-molybdenum protein (aldehyde oxidoreductase) of *Desulfovibrio gigas*. *Biochemical Journal*, *280*, 817–820.

Bray, R. C., & Vanngard, T. (1969). "Rapidly appearing" molybdenum elecrton paramagnetic resonance signals from reduced xanthine oxidase. *Biochemical Journal*, *114*(4), 725–734.

Bray, R. C., Vincent, S. P., Lowe, D. J., Clegg, R. A., & Garland, P. B. (1976). Electron-paramagnetic-resonance studies on molybdneum of nitrate reductase from *Escherichia coli* K12. *Biochemical Journal*, *155*(1), 201–203. https://doi.org/10.1042/bj1550201.

Butler, C. S., Charnock, J. M., Bennett, B., Sears, H. J., Reilly, A. J., Ferguson, S. J., et al. (1999). Models for molybdenum coordination during the catalytic cycle of periplasmic nitrate reductase from Paracoccus denitrificans derived from EPR and EXAFS spectroscopy. *Biochemistry*, *38*(28), 9000–9012. https://doi.org/10.1021/bi990402n.

Byrne, R. S., Haensch, R., Mendel, R. R., & Hille, R. (2009). Oxidative half-reaction of Arabidopsis thaliana sulfite oxidase. Generation of superoxide by a peroxisomal enzyme. *Journal of Biological Chemistry*, *284*(51), 35479–35484. https://doi.org/10.1074/jbc.M109.067355.

Cammack, R., & Weiner, J. H. (1990). Electron-paramagnetic-resonance sectroscopic characterization of dimethylsulfoxide reductase from *Escherichia coli*. *Biochemistry*, *29*(36), 8410–8416.

Cobb, N., Conrads, T., & Hille, R. (2005). Mechanistic studies of Rhodobacter sphaeroides Me2SO reductase. *Journal of Biological Chemistry*, *280*(12), 11007–11017. https://doi.org/10.1074/jbc.M412050200.

Cobb, N., Hemann, C., Polsinelli, G. A., Ridge, J. P., McEwan, A. G., & Hille, R. (2007). Spectroscopic and kinetic studies of Y114F and W116F mutants of Me2SO reductase from Rhodobacter capsulatus. *Journal of Biological Chemistry*, *282*(49), 35519–35529. https://doi.org/10.1074/jbc.M704458200.

Coelho, C., Mahro, M., Trincao, J., Carvalho, A. T. P., Ramos, M. J., Terao, M., et al. (2012). The first mammalian aldehyde oxidase crystal structure. Insights into substrate specificity. *Journal of Biological Chemistry*, *287*(48), 40690–40702. https://doi.org/10.1074/jbc.M112.390419.

Costa, C., Moura, J. J. G., Moura, I., Wang, Y. N., & Huynh, B. H. (1996). Redox properties of cytochrome c nitrite reductase from Desulfovibrio desulfuricans ATCC 27774. *Journal of Biological Chemistry*, *271*(38), 23191–23196.

Dhawan, I. K., & Enemark, J. H. (1996). EPR studies of oxo-molybdenum(V) complexes with sulfur donor ligands: Implications for the molybdenum center of sulfite oxidase. *Inorganic Chemistry*, *35*(17), 4873–4882.

Dias, J. M., Than, M. E., Humm, A., Huber, R., Bourenkov, G. P., Bartunik, H. D., et al. (1999). Crystal structure of the first dissimilatory nitrate reductase at 1.9 angstrom solved by MAD methods. *Structure with Folding & Design*, 7(1), 65–79. https://doi.org/10.1016/s0969-2126(99)80010-0.

Dixon, M., & Thurlow, S. (1924a). Studies on xanthine oxidase. I. Preparation and properties of the active material. *Biochemical Journal*, *18*(5), 971–975.

Dixon, M., & Thurlow, S. (1924b). Studies on xanthine oxidase. II. The dynamics of the oxidase system. *Biochemical Journal*, *18*(5), 976–988.

Dobbek, H., Gremer, L., Kiefersauer, R., Huber, R., & Meyer, O. (2002). Catalysis at a dinuclear CuSMo(==O)OH cluster in a CO dehydrogenase resolved at 1.1-angstrom resolution. *Proceedings of the National Academy of Sciences of the United States of America*, *99*(25), 15971–15976. https://doi.org/10.1073/pnas.212640899.

Doonan, C. J., Wilson, H. L., Bennett, B., Prince, R. C., Rajagopalan, K. V., & George, G. N. (2008). Mo-V electron paramagnetic resonance of sulfite oxidase revisited: The low-pH chloride signal. *Inorganic Chemistry*, *47*(6), 2033–2038. https://doi.org/10.1021/ic7017083.

Enemark, J. H. (2017). Consensus structures of the Mo(v) sites of sulfite-oxidizing enzymes derived from variable frequency pulsed EPR spectroscopy, isotopic labelling and DFT calculations. *Dalton Transactions*, *46*(39), 13202–13210. https://doi.org/10.1039/c7dt01731f.

Enemark, J. H., Astashkin, A. V., & Raitsimring, A. M. (2010). High-resolution EPER spectroscopy of molybdenum enzymes. Sulfite oxidases: Structural and functional implications. In G. Hanson, & L. Berliner (Eds.), *Vol. 29. Metals in biology: Applications of high-resolution EPR to metalloenzymes* (pp. 121–168).

Friedebold, J., & Bowien, B. (1993). Physiological and biochemical characterization of the soluble formate dehydrogenase, a molybdoenzyme from *Alcaligenes eutrophus*. *Journal of Bacteriology*, *175*(15), 4719–4728.

Friedebold, J., Mayer, F., Bill, E., Trautwein, A. X., & Bowien, B. (1995). Structural and immunological studies on the soluble formate dehydrogenase from *Alcaligenes eutrophus*. *Biological Chemistry Hoppe-Seyler*, *376*(9), 561–568. https://doi.org/10.1515/bchm3.1995.376.9.561.

Gangeswaran, R., Lowe, D. J., & Eady, R. R. (1993). Purification and characterization of the assimilatory nitrate reductase of *Azotobacter vinelandii*. *Biochemical Journal*, *289*, 335–342.

Garton, S. D., Garrett, R. M., Rajagopalan, K. V., & Johnson, M. K. (1997). Resonance Raman characterization of the molybdenum center in sulfite oxidase: Identification of Mo=O stretching modes. *Journal of the American Chemical Society*, *119*(10), 2590–2591.

George, G. N., & Bray, R. C. (1988). Studies by electron-paramagnetic-resoannce spectroscopy of xanthine oxidase enriched with Mo-95 and Mo97. *Biochemistry*, *27*(10), 3603–3609.

George, G. N., Bray, R. C., Morpeth, F. F., & Boxer, D. H. (1985). Complexes with halide and other anions of teh molybdenum center of nitrate reductase from *Escherichia coli*. *Biochemical Journal*, *227*(3), 925–931.

George, G. N., Prince, R. C., Kipke, C. A., Sunde, R. A., & Enemark, J. H. (1988). The nature of the phosphate complex of sulfite oxidase from electron-paramagnetic-resonance studies. *Biochemical Journal*, *256*(1), 307–309.

George, G. N., Turner, N. A., Bray, R. C., Morpeth, F. F., Boxer, D. H., & Cramer, S. P. (1989). X-ray absorption and electron paramagnetic resonance spectroscopic studies of the environment of moluybdenum in high-pH and low-pH forms of *Escherichia coli* nitratte reductase. *Biochemical Journal*, *259*(3), 693–700.

Gladyshev, V. N., Khangulov, S. V., Axley, M. J., & Stadtman, T. C. (1994). Coordination of selenium to molybdenum in formate dehydrogenase H from *Escherichia coli*. *Proceedings of the National Academy of Sciences of the United States of America*, *91*(16), 7708–7711. https://doi.org/10.1073/pnas.91.16.7708.

Godfrey, C., Greenwood, C., Thomson, A. J., Bray, R. C., & George, G. N. (1984). Electron paramagnetic resonance spectroscopy studies on the dissimilatory nitrate reductase from *Pseudomonas aeruginosa*. *Biochemical Journal*, *224*(2), 601–608.

Gourlay, C., Nielsen, D. J., White, J. M., Knottenbelt, S. Z., Kirk, M. L., & Young, C. G. (2006). Paramagnetic active site models for the molybdenum-copper carbon monoxide

dehydrogenase. *Journal of the American Chemical Society, 128*(7), 2164–2165. https://doi.org/10.1021/ja056500f.
Greenwood, R. J., Wilson, G. L., Pilbrow, J. R., & Wedd, A. G. (1993). Molybdenum(V) sites in xanthine oxidase and relevant analog complexes—Comparison of O-17 hyperfine coupling. *Journal of the American Chemical Society, 115*(13), 5385–5392.
Gruenewald, S., Wahl, B., Bittner, F., Hungeling, H., Kanzow, S., Kotthaus, J., et al. (2008). The fourth molybdenum containing enzyme mARC: Cloning and involvement in the activation of N-hydroxylated prodrugs. *Journal of Medicinal Chemistry, 51*(24), 8173–8177. https://doi.org/10.1021/jm8010417.
Gutteridge, S., Bray, R. C., Notton, B. A., Fido, R. J., & Hewitt, E. J. (1983). Studieso by electron paramagnetic resoance of teh molybdenum center of spinach (*Spinacia oleracea*) nitrate reductase. *Biochemical Journal, 213*(1), 137–142.
Gutteridge, S., Lamy, M. T., & Bray, R. C. (1980). The nature of the phosphate inhibitor complex of sulfite oxidase from electron paramagnetic resoannce studies using O-17. *Biochemical Journal, 191*(1), 285–288.
Haenzelmann, P., Hofmann, B., Meisen, S., & Meyer, O. (1999). The redox centers in the molybdo iron-sulfur flavoprotein CO dehydrogenase from the thermophilic carboxidotrophic bacterium Pseudomonas thermocarboxydovorans. *FEMS Microbiology Letters, 176*(1), 139–145. https://doi.org/10.1111/j.1574-6968.1999.tb13654.x.
Haenzelmann, P., & Meyer, O. (1998). Effect of molybdate and tungstate on the biosynthesis of CO dehydrogenase and the molybdopterin cytosine-dinucleotide-type of molybdenum cofactor in Hydrogenophaga pseudoflava. *European Journal of Biochemistry, 255*(3), 755–765. https://doi.org/10.1046/j.1432-1327.1998.2550755.x.
Hansch, R., Lang, C., Riebeseel, E., Lindigkeit, R., Gessler, A., Rennenberg, H., et al. (2006). Plant sulfite oxidase as novel producer of H2O2—Combination of enzyme catalysis with a subsequent non-enzymatic reaction step. *Journal of Biological Chemistry, 281*(10), 6884–6888. https://doi.org/10.1074/jbc.M513054200.
Hartmann, T., Terao, M., Garattini, E., Teutloff, C., Alfaro, J. F., Jones, J. P., et al. (2012). The impact of single nucleotide polymorphisms on human aldehyde oxidase. *Drug Metabolism and Disposition, 40*(5), 856–864. https://doi.org/10.1124/dmd.111.043828.
Havelius, K. G. V., Reschke, S., Horn, S., Doerlng, A., Niks, D., Hille, R., et al. (2011). Structure of the molybdenum site in YedY, a sulfite oxidase homologue from Escherichia coli. *Inorganic Chemistry, 50*(3), 741–748. https://doi.org/10.1021/ic101291j.
Havemeyer, A., Bittner, F., Wollers, S., Mendel, R., Kunze, T., & Clement, B. (2006). Identification of the missing component in the mitochondrial benzamidoxime prodrug-converting system as a novel molybdenum enzyme. *Journal of Biological Chemistry, 281*(46), 34796–34802. https://doi.org/10.1074/jbc.M607697200.
Havemeyer, A., Grunewald, S., Wahl, B., Bittner, F., Mendel, R., Erdelyi, P., et al. (2010). Reduction of N-hydroxy-sulfonamides, including N-hydroxy-valdecoxib, by the molybdenum-containing enzyme mARC. *Drug Metabolism and Disposition, 38*(11), 1917–1921. https://doi.org/10.1124/dmd.110.032813.
Hemann, C., Hood, B. L., Fulton, M., Hansch, R., Schwarz, G., Mendel, R. R., et al. (2005). Spectroscopic and kinetic studies of Arabidopsis thaliana sulfite oxidase: Nature of the redox-active orbital and electronic structure contributions to catalysis. *Journal of the American Chemical Society, 127*(47), 16567–16577. https://doi.org/10.1021/ja0530873.
Hille, R. (1996). The mononuclear molybdenum enzymes. *Chemical Reviews, 96*(7), 2757–2816.
Hille, R. (2010). EPR studies of xanthine oxidoreductase and other molybdenum-containing hydroxylases. In G. Hanson, & L. Berliner (Eds.), *Vol. 29. Metals in Biology: Applications of high-resolution EPR to Metalloenzymes* (pp. 91–120).

Hille, R., Hall, J., & Basu, P. (2014). The mononuclear molybdenum enzymes. *Chemical Reviews, 114*(7), 3963–4038.

Hille, R., Kim, J. H., & Hemann, C. (1993). The reductive half-reaction of xanthine oxidase—mechanistic role of the rapid type-1 Mo(V) electronparamagnetic resonance signal. *Biochemistry, 32*(15), 3973–3980.

Hille, R., & Sprecher, H. (1987). On the mechanism of xanthine oxidase—Evidence in support of an oxo transfer mechanism in the molybdenum-containing hydroxylases. *Journal of Biological Chemistry, 262*(23), 10914–10917.

Hochheimer, A., Hedderich, R., & Thauer, R. K. (1998). The formylmethanofuran dehydrogenase isoenzymes in Methanobacterium wolfei and Methanobacterium thermoautotrophicum: Induction of the molybdenum isoenzyme by molybdate and constitutive synthesis of the tungsten isoenzyme. *Archives of Microbiology, 170*(5), 389–393. https://doi.org/10.1007/s002030050658.

Hochheimer, A., Linder, D., Thauer, R. K., & Hedderich, R. (1996). The molybdenum formylmethanofuran dehydrogenase operon and the tungsten formylmethanofuran dehydrogenase operon from Methanobacterium thermoautotrophicum—Structures and transcriptional regulation. *European Journal of Biochemistry, 242*(1), 156–162. https://doi.org/10.1111/j.1432-1033.1996.0156r.x.

Hochheimer, A., Schmitz, R. A., Thauer, R. K., & Hedderich, R. (1995). The tungsten formylmethanofuran dehydrogenase from Methanobacterium thermoautotrophicum contains sequence motifs characteristic for enzymes containing molybdopterin dinucleotide. *European Journal of Biochemistry, 234*(3), 910–920. https://doi.org/10.1111/j.1432-1033.1995.910_a.x.

Holm, R. H., & Donahue, J. P. (1993). A thermodynamic scale for oxygen atom transfer reactions. *Polyhedron, 12*(6), 571–589.

Ingersol, L. J., Yang, J., Khadanand, K. C., Pokhrel, A., Astashkin, A. V., Weiner, J. H., et al. (2020). Addressing ligand-based redox in molybdenum-dependent methionine sulfoxide reductase. *Journal of the American Chemical Society, 142*(6), 2721–2725. https://doi.org/10.1021/jacs.9b11762.

Johannes, J., Unciuleac, M. C., Friedrich, T., Warkentin, E., Ermler, U., & Boll, M. (2008). Inhibitors of the molybdenum cofactor containing 4-Hydroxybenzoyl-CoA reductase. *Biochemistry, 47*(17), 4964–4972. https://doi.org/10.1021/bi800137v.

Jormakka, M., Tornroth, S., Byrne, B., & Iwata, S. (2002). Molecular basis of proton motive force generation: Structure of formate dehydrogenase-N. *Science, 295*(5561), 1863–1868. https://doi.org/10.1126/science.1068186.

Kappler, U., & Bailey, S. (2004). Crystallization and preliminary X-ray analysis of sulfite dehydrogenase from Starkeya novella. *Acta Crystallographica Section D-Biological Crystallography, 60*, 2070–2072. https://doi.org/10.1107/s0907444904022395.

Kappler, U., Bailey, S., Feng, C. J., Honeychurch, M. J., Hanson, G. R., Bernhardt, P. V., et al. (2006). Kinetic and structural evidence for the importance of Tyr236 for the integrity of the Mo active site in a bacterial sulfite dehydrogenase. *Biochemistry, 45*(32), 9696–9705. https://doi.org/10.1021/bi060058b.

Kappler, U., Bennett, B., Rethmeier, J., Schwarz, G., Deutzmann, R., McEwan, A. G., et al. (2000). Sulfite : Cytochrome c oxidoreductase from Thiobacillus novellus—Purification, characterization, and molecular biology of a heterodimeric member of the sulfite oxidase family. *Journal of Biological Chemistry, 275*(18), 13202–13212. https://doi.org/10.1074/jbc.275.18.13202.

Karakas, E., Wilson, H. L., Graf, T. N., Xiang, S., Jaramillo-Buswuets, S., Rajagopalan, K. V., et al. (2005). Structural insights into sulfite oxidase deficiency. *Journal of Biological Chemistry, 280*(39), 33506–33515. https://doi.org/10.1074/jbc.M505035200.

Khangulov, S. V., Gladyshev, V. N., Dismukes, G. C., & Stadtman, T. C. (1998). Selenium-containing formate dehydrogenase H from Escherichia coli: A molybdopterin enzyme that catalyzes formate oxidation without oxygen transfer. *Biochemistry*, *37*(10), 3518–3528. https://doi.org/10.1021/bi972177k.

Krompholz, N., Krischkowski, C., Reichmann, D., Garbe-Schoenberg, D., Mendel, R.-R., Bittner, F., et al. (2012). The mitochondrial amidoxime reducing component (mARC) is involved in detoxification of N-hydroxylated base analogues. *Chemical Research in Toxicology*, *25*(11), 2443–2450. https://doi.org/10.1021/tx300298m.

Kurosaki, M., Terao, M., Barzago, M. M., Bastone, A., Bernardinello, D., Salmona, M., et al. (2004). The aldehyde oxidase gene cluster in mice and rats—Aldehyde oxidase homologue 3, a novel member of the molybdo-flavoenzyme family with selective expression in the olfactory mucosa. *Journal of Biological Chemistry*, *279*(48), 50482–50498. https://doi.org/10.1074/jbc.M408734200.

Lamy, M. T., Gutteridge, S., & Bray, R. C. (1980). Electron paramagnetic resonance paramaters om molybdenum(V) in sulfite oxidase from chicken liver. *Biochemical Journal*, *185*(2), 397–403.

Li, H. K., Temple, C., Rajagopalan, K. V., & Schindelin, H. (2000). The 1.3 angstrom crystal structure of Rhodobacter sphaeroides dimethyl sulfoxide reductase reveals two distinct molybdenum coordination environments. *Journal of the American Chemical Society*, *122*(32), 7673–7680. https://doi.org/10.1021/ja000643e.

Lorigan, G. A., Britt, R. D., Kim, J. H., & Hille, R. (1994). Electron-spin echo envelope modulation spectroscopy of the molybdenum center of xanthine oxidase. *Biochimica et Biophysica Acta-Bioenergetics*, *1185*(3), 284–294.

Malthouse, J. P. G., Gutteridge, S., & Bray, R. C. (1980). Rapid type-2 molybdenum(V) electron paramagnetic resonance signals from xanthine oxidase and the structure of the active center of the enzyme. *Biochemical Journal*, *185*(3), 767–770.

Manikandan, P., Choi, E. Y., Hille, R., & Hoffman, B. M. (2001). 35 GHz ENDOR characterization of the "very rapid" signal of xanthine oxidase reacted with 2-hydroxy-6-methylpurine ((13)C8): Evidence against direct Mo-C8 interaction. *Journal of the American Chemical Society*, *123*(11), 2658–2663. https://doi.org/10.1021/ja003894w.

Marangon, J., Correia, H. D., Brondino, C. D., Moura, J. J. G., Romao, M. J., Gonzalez, P. J., et al. (2013). Kinetic and structural studies of aldehyde oxidoreductase from Desulfovibrio gigas reveal a Dithiolene-based chemistry for enzyme activation and inhibition by H2O2. *PLoS One*, *8*(12). https://doi.org/10.1371/journal.pone.0083234.

Marelja, Z., Dambowsky, M., Bolis, M., Georgiou, M. L., Garattini, E., Missirlis, F., et al. (2014). The four aldehyde oxidases of Drosophila melanogaster have different gene expression patterns and enzyme substrate specificities. *Journal of Experimental Biology*, *217*(12), 2201–2211. https://doi.org/10.1242/jeb.102129.

Massey, V., Schopfer, L. M., & Nishino, T. (1989). Differences in protein structure of xanthine dehydrogenase and xanthine oxidase revealed by reconstitution with flavin active site probes. *Journal of Biological Chemistry*, *264*(18), 10567–10573.

McWhirter, R. B., & Hille, R. (1991). The reductive half-reaction of xanthine oxidase—Identification of spectral intermediates in the hydroxylation of 2-hydroxy-6-methylpurine. *Journal of Biological Chemistry*, *266*(35), 23724–23731. Retrieved from <Go to ISI>://WOS:A1991GV31900033.

Mendel, R. R. (2011). Cell biology of molybdenum in plants. *Plant Cell Reports*, *30*(10), 1787–1797. https://doi.org/10.1007/s00299-011-1100-4.

Mendel, R. R., & Kruse, T. (2012). Cell biology of molybdenum in plants and humans. *Biochimica et Biophysica Acta-Molecular Cell Research*, *1823*(9), 1568–1579. https://doi.org/10.1016/j.bbamcr.2012.02.007.

Mendel, R. R., & Schwarz, G. (1999). Molybdoenzymes and molybdenum cofactor in plants. *Critical Reviews in Plant Sciences*, *18*(1), 33–69. https://doi.org/10.1016/s0735-2689(99)00386-x.

Mota, C., Coelho, C., Leimkuehler, S., Garattini, E., Terao, M., Santos-Silva, T., et al. (2018). Critical overview on the structure and metabolism of human aldehyde oxidase and its role in pharmacokinetics. *Coordination Chemistry Reviews*, *368*, 35–59. https://doi.org/10.1016/j.ccr.2018.04.006.

Mtei, R. P., Lyashenko, G., Stein, B., Rubie, N., Hille, R., & Kirk, M. L. (2011). Spectroscopic and electronic structure studies of a dimethyl sulfoxide reductase catalytic intermediate: Implications for electron- and atom-transfer reactivity. *Journal of the American Chemical Society*, *133*(25), 9762–9774. https://doi.org/10.1021/ja109178q.

Niks, D., Duvvuru, J., Escalona, M., & Hille, R. (2016). Spectroscopic and kinetic properties of the molybdenum-containing, NAD(+)—Dependent formate dehydrogenase from Ralstonia eutropha. *Journal of Biological Chemistry*, *291*(3), 1162–1174. https://doi.org/10.1074/jbc.M115.688457.

Nishino, T. (1994). The conversion of xanthine dehydrogenase to xanthine oxidase and the role of the enzyme in reperfusion injury. *Journal of Biochemistry*, *116*(1), 1–6.

Nishino, T., Schopfer, L. M., & Massey, V. (1989). The reactivity of chicken liver xanthine dehydrogeanes with molecular oxygen. *Journal of Biological Chemistry*, *264*(5), 2518–2527.

Oh, J. I., & Bowien, B. (1998). Structural analysis of the fds operon encoding the NAD(+)-linked formate dehydrogenase of Ralstonia eutropha. *Journal of Biological Chemistry*, *273*(41), 26349–26360. https://doi.org/10.1074/jbc.273.41.26349.

Pacheco, A., Basu, P., Borbat, P., Raitsimring, A. M., & Enemark, J. H. (1996). Multifrequency ESEEM spectroscopy of sulfite oxidase in phosphate buffer: Direct evidence for coordinated phosphate. *Inorganic Chemistry*, *35*(24), 7001–7008.

Palmer, G., Bray, R. C., & Beinert, H. (1964). Direct studies on electron trnasfer in xanthine oxidase by electron paramagnetic resonance. I. Techniques and description of spectra. *Journal of Biological Chemistry*, *239*(8), 2657–2665.

Pauff, J. M., Zhang, J., Bell, C. E., & Hille, R. (2008). Substrate orientation in xanthine oxidase—Crystal structure of enzyme in reaction with 2-hydroxy-6-methylpurine. *Journal of Biological Chemistry*, *283*(8), 4818–4824. https://doi.org/10.1074/jbc.M707918200.

Pushie, M. J., Doonan, C. J., Wison, H. L., Rajagopalan, K. V., & George, G. N. (2011). Nature of halide binding to the molybdenum site of sulfite oxidase. *Inorganic Chemistry*, *50*(19), 9406–9413. https://doi.org/10.1021/ic201030u.

Radon, C., Mittelstadt, G., Duffus, B. R., Burger, J., Hartmann, T., Mielke, T., et al. (2020). Cryo-EM structures reveal intricate Fe-S cluster arrangement and charging in Rhodobacter capsulatus formate dehydrogenase. *Nature Communications*, *11*(1), 1912. https://doi.org/10.1038/s41467-020-15614-0.

Raitsimring, A. M., Pacheco, A., & Enemark, J. H. (1998). ESEEM investigations of the high pH and low pH forms of chicken liver sulfite oxidase. *Journal of the American Chemical Society*, *120*(44), 11263–11278.

Rajapakshe, A., Snyder, R. A., Astashkin, A. V., Bernardson, P., Evans, D. J., Young, C. G., et al. (2009). Insights into the nature of Mo(V) species in solution: Modeling catalytic cycles for molybdenum enzymes. *Inorganica Chimica Acta*, *362*(12), 4603–4608. https://doi.org/10.1016/j.ica.2009.05.040.

Rebelo, J., Macieira, S., Dias, J. M., Huber, R., Ascenso, C. S., Rusnak, F., et al. (2000). Gene sequence and crystal structure of the aldehyde oxidoreductase from Desulfovibrio desulfuricans ATCC 27774. *Journal of Molecular Biology*, *297*(1), 135–146.

Rendon, J., Biaso, F., Ceccaldi, P., Toci, R., Seduk, F., Magalon, A., et al. (2017). Elucidating the structures of the low- and high-pH Mo(V) species in respiratory nitrate reductase: A combined EPR, N-14,N-15 HYSCORE, and DFT Study. *Inorganic Chemistry*, *56*(8), 4422–4434. https://doi.org/10.1021/acs.inorgchem.6b03129.

Resch, M., Dobbek, H., & Meyer, O. (2005). Structural and functional reconstruction in situ of the CuSMoO2 active site of carbon monoxide dehydrogenase from the carbon monoxide oxidizing eubacterium Oligotropha carboxidovorans. *Journal of Biological Inorganic Chemistry, 10*(5), 518–528. https://doi.org/10.1007/s00775-005-0006-4.

Rivas, M. G., Gonzalez, P. J., Brondino, C. D., Moura, J. J. G., & Moura, I. (2007). EPR characterization of the molybdenum(V) forms of formate dehydrogenase from Desulfovibrio desulfuricans ATCC 27774 upon formate reduction. *Journal of Inorganic Biochemistry, 101*(11 – 12), 1617–1622. https://doi.org/10.1016/j.jinorgbio.2007.04.011.

Romao, M. J., Archer, M., Moura, I., Moura, J. J. G., Legall, J., Engh, R., et al. (1995). Crystal structure of the xanthine oxidase-related aldehyde oxidoreductase from *Desulfovibrio gigas*. *Science, 270*(5239), 1170–1176.

Rothery, R. A., Workun, G. J., & Weiner, J. H. (2008). The prokaryotic complex iron-sulfur molybdoenzyme family. *Biochimica et Biophysica Acta-Biomembranes, 1778*(9), 1897–1929. https://doi.org/10.1016/j.bbamem.2007.09.002.

Saito, T., & Nishino, T. (1989). Differences in environment of FAD between NAD +-dependent and O2-dependent types of rat liver xanthine dehydrogenase. *Journal of Biological Chemistry, 264*(17), 10015–10022.

Sauer, P., & Frebort, I. (2003). Molybdenum cofactor-containing oxidoreductase family in plants. *Biologia Plantarum, 46*(4), 481–490. https://doi.org/10.1023/a:1024814007027.

Schindelin, H., Kisker, C., Hilton, J., Rajagopalan, K. V., & Rees, D. C. (1996). Crystal structure of DMSO reductase: Redox-linked changes in molybdopterin coordination. *Science, 272*(5268), 1615–1621.

Schoepp-Cothenet, B., van Lis, R., Philippot, P., Magalon, A., Russell, M. J., & Nitschke, W. (2012). The ineluctable requirement for the trans-iron elements molybdenum and/or tungsten in the origin of life. *Scientific Reports, 2*. https://doi.org/10.1038/srep00263.

Schopfer, L. M., Massey, V., & Nishino, T. (1988). Rapid reaction studies on the reduction and oxidation of chicken liver xanthine dehydrogenase by the xanthine/urate and NAD +/NADH couples. *Journal of Biological Chemistry, 263*(27), 13528–13538.

Schrapers, P., Hartmann, T., Kositzki, R., Dau, H., Reschke, S., Schulzke, C., et al. (2015). Sulfido and cysteine ligation changes at the molybdenum cofactor during substrate conversion by formate dehydrogenase (FDH) from Rhodobacter capsulatus. *Inorganic Chemistry, 54*(7), 3260–3271. https://doi.org/10.1021/ic502880y.

Schumann, S., Terao, M., Garattini, E., Saggu, M., Lendzian, F., Hildebrandt, P., et al. (2009). Site directed mutagenesis of amino acid residues at the active site of mouse aldehyde oxidase AOX1. *PLoS One, 4*(4). e534810.1371/journal.pone.0005348.

Schwarz, G., & Mendel, R. R. (2006). Molybdenum cofactor biosynthesis and molybdenum enzymes. *Annual Review of Plant Biology, 57*, 623–647. https://doi.org/10.1146/annurev.arplant.57.032905.105437.

Shanmugam, M., Wilcoxen, J., Habel-Rodriguez, D., Cutsail, G. E., III, Kirk, M. L., Hoffman, B. M., et al. (2013). C-13 and Cu-63,Cu-65 ENDOR studies of CO Dehydrogenase from *Oligotropha carboxidovorans*. Experimental Evidence in Support of a Copper-Carbonyl Intermediate. *Journal of the American Chemical Society, 135*(47), 17775–17782. https://doi.org/10.1021/ja406136f.

Solomonson, L. P., Barber, M. J., Howard, W. D., Johnson, J. L., & Rajagopalan, K. V. (1984). Electron paramagnetic resonance studies of the molybdenum center of assimilatory nitrate reductase from *Chlorella vulgaris*. *Journal of Biological Chemistry, 259*(2), 849–853.

Stoll, S., & Schweiger, A. (2006). EasySpin, a comprehensive software package for spectral simulation and analysis in EPR. *Journal of Magnetic Resonance, 178*(1), 42–55. https://doi.org/10.1016/j.jmr.2005.08.013.

Terao, M., Garattini, E., Romao, M. J., & Leimkuehler, S. (2020). Evolution, expression, and substrate specificities of aldehyde oxidase enzymes in eukaryotes. *Journal of Biological Chemistry, 295*(16), 5377–5389. https://doi.org/10.1074/jbc.REV119.007741.

Terao, M., Kurosaki, M., Marini, M., Vanoni, M. A., Saltini, G., Bonetto, V., et al. (2001). Purification of the aldehyde oxidase homolog 1 (AOH1) protein and cloning of the AOH1 and aldehyde oxidase homolog 2 (AOH2) genes—Identification of a novel molybdo-flavoprotein gene cluster on mouse chromosome 1. *Journal of Biological Chemistry*, *276*(49), 46347–46363. https://doi.org/10.1074/jbc.M105744200.

Thomas, G., Potter, L., & Cole, J. A. (1999). The periplasmic nitrate reductase from Escherichia coli: A heterodimeric molybdoprotein with a double-arginine signal sequence and an unusual leader peptide cleavage site. *FEMS Microbiology Letters*, *174*(1), 167–171. https://doi.org/10.1111/j.1574-6968.1999.tb13564.x.

Turner, N., Ballard, A. L., Bray, R. C., & Ferguson, S. (1988). Investigation by electron paramagnetic resonance spectroscopy of the molybdenum center of respiratory nitrtate reductase from *Paracoccus denitrificans*. *Biochemical Journal*, *252*(3), 925–926.

Turner, N. A., Doyle, W. A., Ventom, A. M., & Bray, R. C. (1995). Properties of rabbit liver aldehyde oxidase and the relationship of the enzyme to xanthine oxidase and dehydrogenase. *European Journal of Biochemistry*, *232*(2), 646–657.

Unciuleac, M., Warkentin, E., Page, C. C., Boll, M., & Ermler, U. (2004). Structure of a xanthine oxidase-related 4-hydroxybenzoyl-CoA reductase with an additional 4Fe-4S cluster and an inverted electron flow. *Structure*, *12*(12), 2249–2256. https://doi.org/10.1016/j.str.2004.10.008.

Vincent, S. P., & Bray, R. C. (1978). Electron paramagnetic resonance studies on nitrate reductase from *Escherichia coli* K12. *Biochemical Journal*, *171*(3), 639–647.

Wahl, B., Reichmann, D., Niks, D., Krompholz, N., Havemeyer, A., Clement, B., et al. (2010). Biochemical and spectroscopic characterization of the human mitochondrial Amidoxime reducing components hmARC-1 and hmARC-2 suggests the existence of a new molybdenum enzyme family in eukaryotes. *Journal of Biological Chemistry*, *285*(48), 37847–37859. https://doi.org/10.1074/jbc.M110.169532.

Wilcoxen, J., & Hille, R. (2013). The hydrogenase activity of the molybdenum/copper-containing carbon monoxide dehydrogenase of Oligotropha carboxidovorans. *Journal of Biological Chemistry*, *288*(50), 36052–36060. https://doi.org/10.1074/jbc.M113.522441.

Wilcoxen, J., Snider, S., & Hille, R. (2011). Substitution of silver for copper in the binuclear Mo/Cu Center of carbon monoxide dehydrogenase from oligotropha carboxidovorans. *Journal of the American Chemical Society*, *133*(33), 12934–12936. https://doi.org/10.1021/ja205073j.

Wilson, G. L., Greenwood, R. J., Pilbrow, J. R., Spence, J. T., & Wedd, A. G. (1991). Molybdenum(V) sites in xanthine oxidase and relevant analog complexes. Comparison of Mo-95 and S-33 hyperfine coupling. *Journal of the American Chemical Society*, *113*(18), 6803–6812. https://doi.org/10.1021/ja00018a014.

Xia, M., Dempski, R., & Hille, R. (1999). The reductive half-reaction of xanthine oxidase—Reaction with aldehyde substrates and identification of the catalytically labile oxygen. *Journal of Biological Chemistry*, *274*(6), 3323–3330.

Yang, J., Rothery, R., Sempombe, J., Weiner, J. H., & Kirk, M. L. (2009). Spectroscopic characterization of YedY: The role of sulfur coordination in a Mo(V) sulfite oxidase family enzyme form. *Journal of the American Chemical Society*, *131*(43), 15612-+. https://doi.org/10.1021/ja903087k.

Young, T., Niks, D., Hakopian, S., Tam, T. K., Yu, X., Hille, R., et al. (2020). Crystallographic and kinetic analyses of the FdsBG subcomplex of the cytosolic formate dehydrogenase FdsABG from Cupriavidus necator. *Journal of Biological Chemistry*, *295*(19), 6570–6585. https://doi.org/10.1074/jbc.RA120.013264.

Zhang, B., Hemann, C. F., & Hille, R. (2010). Kinetic and spectroscopic studies of the molybdenum-copper CO dehydrogenase from Oligotropha carboxidovorans. *Journal of Biological Chemistry*, *285*(17), 12571–12578. https://doi.org/10.1074/jbc.M109.076851.

CHAPTER TWELVE

EPR of Type I photosynthetic reaction centers

John H. Golbeck[a,*] and Art van der Est[b]

[a]Department of Biochemistry and Molecular Biology, and Department of Chemistry, The Pennsylvania State University, University Park, PA, United States
[b]Department of Chemistry, Brock University, St. Catharines, ON, Canada
[*]Corresponding author: e-mail address: jhg5@psu.edu

Contents

Abstract

Light-induced reactions in photosynthetic reaction centers are initiated by the absorption of a photon, which results in the transfer of a single electron and the generation of radical ions in the donor and acceptor molecules involved in the charge-separated state. Electron paramagnetic resonance (EPR) spectroscopy is the ideal method for the study of such reactions. In addition to measuring spectra of the electron transfer cofactors in continuous light, reactions can be initiated by brief flashes of light, thereby allowing the kinetics of forward electron transfer as well as recombination reactions to be obtained. Because the donor and acceptor pairs are so closely spaced and because light induced charge separation is so rapid, the donor and early acceptors are in a quantum mechanically spin entangled state, which confers properties such as increased sensitivity, the ability to measure reactions on the nanosecond timescale, and the determination of bond angles between cofactors. Additionally, distances between pairs of cofactors

Methods in Enzymology, Volume 666
ISSN 0076-6879
https://doi.org/10.1016/bs.mie.2022.02.007

can be measured by detecting the modulation of a phase shifted "out-of-phase" electron spin echo signal. In this methods article, we will describe how continuous wave EPR, time resolved EPR, and pulsed EPR can be used to measure these properties in Type I photosynthetic reaction centers. Methods of analysis are described for the bound electron transfer cofactors in the heterodimeric Photosystem I reaction center of plants and cyanobacteria and in the homodimeric reaction centers found in phototrophic members of the phyla Bacillota, Chlorobiota, and Acidobacteriota.

1. Introduction

1.1 Heterodimeric Type I reaction centers

Photosystem I (PS I) is a heterodimeric Type I reaction center (RC) present in oxygenic organisms such as plants and cyanobacteria (comprehensively reviewed in Golbeck, 2006). The X-ray crystal structure of PS I from the cyanobacterium *Thermosynechococcus elongatus* (Jordan et al., 2001) revealed a central core of six chlorophyll (Chl) *a* molecules (named Chl_{1A}, Chl_{1B}, Chl_{2A}, Chl_{2B}, Chl_{3A}, Chl_{3B}) arranged in two quasi-symmetrical branches (Fig. 1, top). Chl_{1A}/Chl_{1B} form a dimeric "special pair" termed P_{700} that has been treated historically as the primary electron donor (Kok, 1956; Norris, Uphaus, Crespi, & Katz, 1971). Chl_{2A}/Chl_{3A} and Chl_{2B}/Chl_{3B} each form additional dimeric "special pairs" (Gorka, Charles, et al., 2021) named A_{0A} and A_{0B}, which are the spectroscopic names for the primary electron acceptor(s) (Mathis, Ikegami, & Setif, 1988; Shuvalov, Nuijs, Van Gorkom, Smit, & Duysens, 1986). The substituted menaquinones A_{1A} and A_{1B} function as secondary electron acceptors (Biggins & Mathis, 1988; Sauer, Acker, Mathis, & Van Best, 1977), after which the two branches converge to transfer electrons serially through three iron-sulfur clusters known as F_X, F_B and F_A. F_X is an interpolypeptide [4Fe-4S] cluster ligated between the PsaA and PsaB subunits, whereas F_A and F_B are [4Fe-4S] clusters (Malkin & Bearden, 1971) ligated by the tightly-bound PsaC subunit (Bordier, Svendsen, Scheller, & Møller, 1987). An electron is capable of being transferred through both A- and B-branches (Guergova-Kuras, Boudreaux, Joliot, Joliot, & Redding, 2001), although in cyanobacteria the A-branch is dominant (Dashdorj, Xu, Cohen, Golbeck, & Savikhin, 2005). The electron is handed off to either ferredoxin, which contains a [2Fe-2S] cluster or to flavodoxin, which contains a flavin mononucleotide, for entry into the cellular metabolic pool.

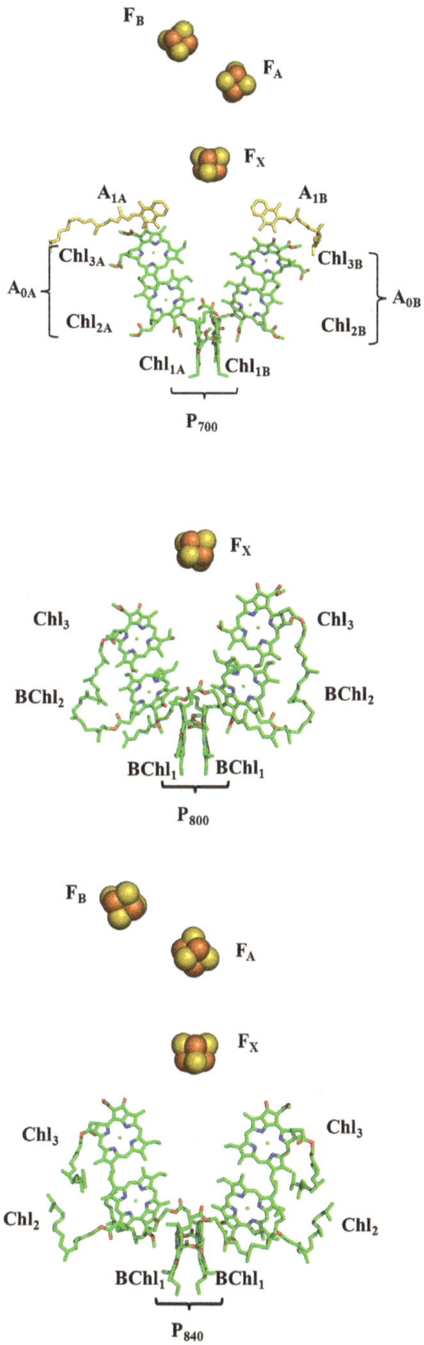

Fig. 1 Electron transfer cofactors from the X-ray crystal structures of the heterodimeric PS I RC from *Thermosynechococcus elongatus* (1JB0, top) and the homodimeric RC from *Heliomicrobium modesticaldum* (5V8K, middle), and the cryoEM structure of the homodimeric RC from *Chlorobaculum tepidum* (6M32, bottom). Chl molecules are depicted in green, quinones in yellow, iron atoms as red spheres, and sulfide ions as yellow spheres. The dimeric Chl_{1A}/Chl_{1B}, Chl_1/Chl_1 and $BChl_1/BChl_1$ special pairs are the primary donors in

(Continued)

1.2 Homodimeric Type I reaction centers

Homodimeric Type-1 RCs are found in the phototrophic members of the phyla Bacillota (e.g., *Heliomicrobium modesticaldum*), Chlorobiota (e.g., *Chlorobaculum tepidum*), and Acidobacteriota (*Chloracidobacterium thermophilum*)[a]. The X-ray crystal structure of the *H. modesticaldum* RC (HbRC) shows similarities to that of PS I (Gisriel et al., 2017), although there are a number of significant differences (Fig. 1, middle). The common feature is the presence of six chlorin molecules that are arranged in two perfectly symmetrical branches, forming the central core of the HbRC. Unlike PS I, $BChl_1$ and $BChl_2$ are bacteriochlorophyll *g* molecules and Chl_3 is a chlorophyll *a* derivative, 8^1-hydroxy-chlorophyll *a*. Chl_1/Chl_1 form a perfectly dimeric "special pair" termed P_{800}, which is considered the primary electron donor. The most significant difference with PS I is that the HbRC does not contain bound quinones, nor does it contain a tightly bound extrinsic subunit containing the F_A and F_B clusters. It does, however, harbor F_X, an interpolypeptide [4Fe-4S] cluster that serves as the tertiary electron acceptor (Romberger & Golbeck, 2012), handing off the electron to soluble, low molecular mass, bacterial-like dicluster ferredoxins. In the green sulfur bacterial RC (GsbRC) from *C. tepidum*, $BChl_1$ is a bacteriochlorophyll *a* molecule. The quinones are similarly missing (Fig. 1, bottom), although F_A and F_B are present due to the retention of the PscB subunit during sample preparation (Chen et al., 2020).

The above-described cofactors transfer a single electron and thus are EPR-visible in one of their oxidation states. In the following sections we will outline how continuous wave EPR, time resolved EPR, and pulsed EPR and can be used to detect, identify, quantify, and localize cofactor-bound radicals generated in heterodimeric and homodimeric Type I RCs.

Fig. 4—Cont'd PS I the HbRC, and the GsbRC respectively, and are specified by their spectroscopic names P_{700}, P_{800}, and P_{840}. The Chl_{2A}/Chl_{3A} and Chl_{2B}/Chl_{3B} special pairs in PS I, the two special pairs $BChl_2/Chl_3$ in the HbRC and the two special pairs Chl_2/Chl_3 are specified by their spectroscopic name A_0. Note that the A_{1A}/A_{1B} quinones and the F_A and F_B iron-sulfur clusters are absent in the ancestral *H. modesticaldum* RC. The A_{1A}/A_{1B} quinones are also absent in the *C. tepidum* RC but the F_A and F_B iron-sulfur clusters are present due to the retention of the loosely bound PscB subunit.

[a] At the time of writing, the nomenclature is in a state of flux. The following name changes have either been approved or proposed: 'Bacillota' for 'Firmicutes'; 'Chlorobiota' for 'Chlorobi'; 'Acidobacteriota' for 'Acidobacteria'; '*Heliomicrobium*' for '*Heliobacterium*'; and '*Chlorobaculum*' for '*Chlorobium*'.

2. Continuous wave (CW) EPR spectroscopy of Type I reaction centers

2.1 Methodology

2.1.1 Fundamentals of CW EPR spectroscopy

A brief review of the theory behind biological EPR spectroscopy is provided in Golbeck and van der Est (2013); readers interested in a more comprehensive treatment of the method should refer to Goldfarb and Stoll (2018). In this article, we provide only the bare essentials needed to understand why experimental condition are chosen for study of a given cofactor. Because most researchers in the field are familiar with optical absorption spectroscopy, we refer to it at times to underscore the fact that EPR differs in aspects that have practical consequences for the measurement of the RC cofactors.

- Whereas optical spectroscopy uses the electric vector of electromagnetic radiation to promote transitions between the electronic energy levels in atoms and molecules without the necessity of an external magnetic field, EPR spectroscopy uses the magnetic vector of electromagnetic radiation to promote transitions between spin sublevels in the presence of an external magnetic field. In the absence of the field, the m_s sublevels for S = ½ systems are degenerate, and thus electromagnetic radiation cannot be absorbed during transitions between them. The sample must be placed in an external field (Fig. 2, top left), which is provided either by an iron core electromagnet or, if fields above 1.5 mT are required, by a superconducting magnet.
- The energy difference between the spin sublevels that can be attained by the available magnetic fields are a fraction of the energy differences that are probed in optical spectroscopy. For example, at the magnetic field of 330 mT used in X-band (9.4 GHz) EPR, the difference in energy between the two levels is only 6.13×10^{-24} J, whereas an optical transition at 700 nm is 2.84×10^{-19} J. This has practical consequences: because the kinetic energy of molecules at room temperature is 4×10^{-21} J, and because the occupancy of the two energy levels is governed by the Boltzmann equation

$$N_\alpha/N_\beta = exp\left(-\Delta E/kT\right) \quad (1)$$

the ratio N_α/N_β is 0.9985 in EPR spectroscopy whereas it is $<10^{-29}$ in optical spectroscopy. Thus, the two energy levels are nearly equally populated in the former, whereas only the lower energy level is effectively

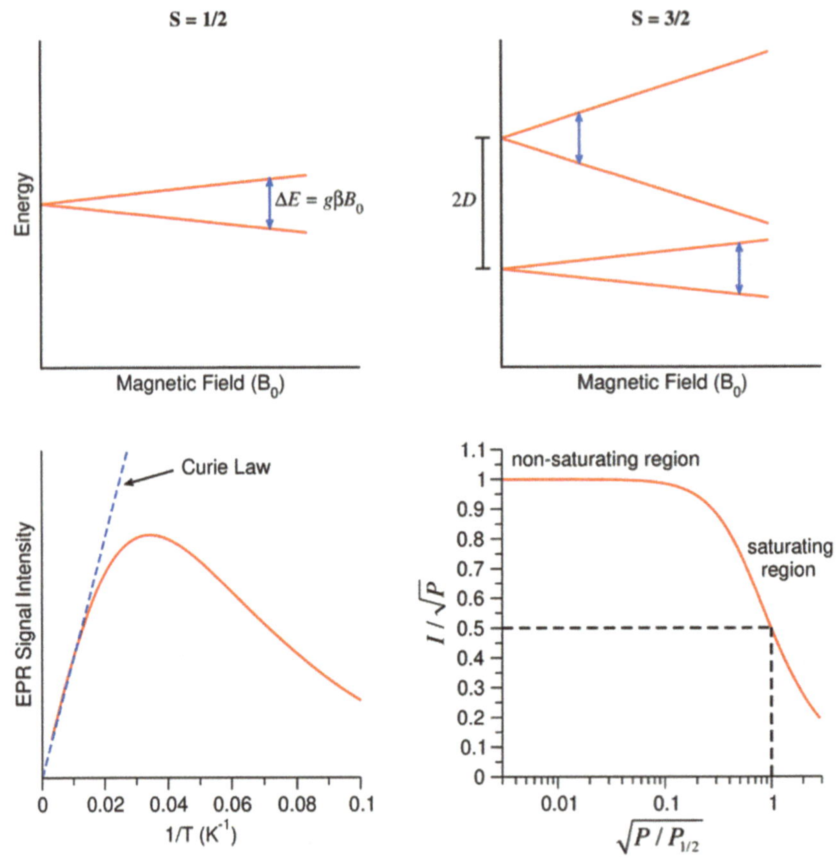

Fig. 2 Energy level diagrams and effect of temperature and power on signal intensity. Top: Energy level diagram for a S = ½ system (left) and for a S = 3/2 system (right) with zero-field splitting parameter $D = +0.5\,\mathrm{cm}^{-1}$ and $E = 0$. The arrows indicate the field positions at which X-band transitions $\Delta E = g\beta B_0$ occur (g is 2.0023193 for a free electron, β is 9.274096×10^{-24} J/T for the Bohr magneton, and B_0 is the magnetic field strength in Tesla). The population of the energy levels is governed by the Boltzmann distribution $N^+/N^- = e^{(-\Delta E/kT)}$, where ΔE is the energy difference indicated above, k is the Boltzmann constant $1.3806 \times 10^{-23}\,\mathrm{J\,K^{-1}}$ and T is the temperature in K. Bottom, left: The intensity of an EPR signal as a function of temperature. As the temperature is lowered (to the right on the X-axis), the intensity follows Curie law behavior (blue dotted line), but below a certain temperature, the intensity decreases due to saturation. Bottom, right: The intensity of an EPR signal as a function of microwave power. As the power is raised, the signal of a homogeneously broadened line transitions from the non-saturating region to the saturating region.

populated in the latter. Because photon-induced transitions between the ground state and the excited state (absorption), and transitions between the excited state and ground state (stimulated emission), have the same probability, only the difference between the upper and lower populations can be detected. Hence, EPR spectroscopy is much less sensitive than optical spectroscopy, and this translates to the need for high sample concentrations and the use of a resonator to focus the microwave energy. One additional consideration is that lower temperatures can be used to increase sensitivity (see Eq. 1); for example, at 77 K, $N_\alpha/N_\beta=0.994$, at 10 K, $N_\alpha/N_\beta=0.957$, and at 4.2 K $N_\alpha/N_\beta=0.900$. Such low temperatures require the use of liquid gases such as nitrogen or helium to cool the sample.

- Because both CW EPR and optical spectroscopy are steady-state methods, i.e., the microwave field and the optical source are continuously on during measurement, the population measured represents a balance between the rates of excitation and relaxation. Accordingly, the excited state must relax to the ground state before the sample can absorb another microwave or optical photon. In optical spectroscopy, relaxation is typically on the order of nanoseconds or shorter (the exception is a sample that phosphoresces), and this condition is easily met except at extremely high photon fluxes that are rarely encountered in the laboratory. However, in EPR spectroscopy, the relaxation can be either fast or slow, depending on the cofactor under study and on its immediate environment. The practical consequence is that for fast relaxing systems, low temperatures and high microwave powers are required to achieve an adequate S/N. Examples of fast relaxing systems in Type I RCs include reduced iron-sulfur clusters F_X^-, F_B^- and F_A^-. For slow relaxing systems, low temperatures cause slower relaxation rates, resulting in an increasingly equalized N_α/N_β ratio, making the detected signal smaller. The practical consequence is that for slow relaxing systems, higher temperatures and lower microwave powers are used to achieve an adequate S/N. Examples of slow relaxing systems in Type I RCs include the organic radicals A_{1A}^-, A_{1B}^-, A_{0A}^-, A_{0B}^- and P_{700}^+.
- The relationship between temperature, power, and signal strength is formalized in a concept termed saturation. As shown in Fig. 2, bottom left, as the temperature is lowered at a given microwave power, the signal becomes increasingly larger because of the greater population difference between the ground and excited states (see Eq. 1). This is known as the Curie law region. However, as the temperature is lowered beyond a certain point, the rate of relaxation slows relative to the constant rate of absorption of the microwaves, leading to a smaller difference in population, and hence, an increasingly smaller signal. Similarly, as the microwave power is raised at a given temperature (Fig. 2, bottom right), the rate of absorption of the microwaves outcompetes the rate of

relaxation, leading to an increased difference in population; as the two processes become competitive, the difference in population decreases. Temperature and microwave power are therefore interrelated, and both must be varied by trial-and-error to achieve the highest signal-to-noise ratio for a given sample.

- Because magnetic dipole transitions are considerably weaker than electric dipole transitions, a resonator is required to separate the magnetic component of the microwave radiation from the electric component and to concentrate the magnetic component at the position of the sample tube. It does this by setting up standing waves in the resonator such that the maxima of the magnetic and electric fields are out of phase and thus spatially separated. A change in the impedance of the resonator caused by the absorption of microwave energy in the sample is the measurable in an EPR experiment. If low temperatures are required, a cryostat incorporating a quartz insert Dewar maintains the temperature of the sample using either liquid nitrogen or liquid helium as the coolant. A heater in the gas stream allows precision temperature control from 298 to 4.2 K, the latter being the temperature of boiling helium.

One additional feature of CW EPR is the use of phase sensitive lock-in detection in which a small AC-modulated magnetic field is superimposed on the larger, slowly moving "static" magnetic field. The alternating field is applied by means of modulation coils that are attached to the resonator, causing the magnetic field to oscillate over a small range, typically at a frequency 100 kHz. The oscillating signal is then mixed with a reference signal and only the DC component of the mixed signal is retained which eliminates a large fraction of the noise. The modulation amplitude can be varied; the larger the value, the more sensitive the detection; however, the signal will become distorted if the magnitude of the sweep exceeds the linewidth of the signal. The analog in optical spectroscopy is the slit width of the monochromator; if it is wider than the width of the peak, the peak will be distorted. As a result of phase sensitive detection, the output of an EPR experiment is a derivative rather than an absorption spectrum. The derivative is used directly because it makes small features visible in that would be otherwise difficult to observe in broad EPR spectra.

In summary, to achieve an acceptable signal-to-noise ratio, high sample concentrations are needed; fast relaxing systems require low temperatures and high microwave powers; and slow relaxing systems require higher temperatures and lower microwave powers. These and other parameters are explicitly specified in the examples shown below.

2.1.2 Quantifying spin concentrations

In EPR, a large number of parameters must be taken into account when determining spin concentration. Because CW EPR outputs a derivative spectrum, a double integration is performed to produce a signal height. The baseline must be corrected, especially in low temperature measurements, due to the presence of contaminating paramagnetic species, notably triplet-state dioxygen. The sample and standard must be measured under non-saturating conditions (see Fig. 2), which requires careful attention to power settings and temperature, especially for slow relaxing systems. Three common standards used for spin concentration are 1 mM $CuSO_4$ in 10 mM EDTA (ethylenediaminetetraacetic acid), 1 mM MetMbCN (where Mb is myoglobin) in pH 7.0 buffer, and 0.1 mM TEMPO (2,2,6,6-tetramethyl-1-piperidinyloxyl) in benzene. Whereas $CuSO_4$ and MetMbCN can be obtained in purified forms, the concentration of TEMPO solutions should be calculated by optical spectroscopy using an extinction coefficient of 10.5 M^{-1} cm^{-1} at 476 nm (Chateauneuf, Lusztyk, & Ingold, 1990). Because the linewidths and saturation properties must be similar (see Eq. 2), the rule of thumb is to use an organic standard for an organic sample and a metal standard for a metal containing sample. EPR tubes prepared with these standards can be stored at low temperature and reused indefinitely. Also, if a light-induced radical is under study, the light must be intense enough to completely penetrate the sample to quantitative oxidize or reduce the cofactor.

The following equation illustrates the parameters that must be taken into account to allow the spin concentration of a S = ½ spin system to be determined:

$$C_u = C_s \cdot \frac{\iint I_u \, dB_0^2}{\iint I_s \, dB_0^2} \cdot \frac{g_s^{av}}{g_u^{av}} \cdot \frac{T_u}{T_s} \cdot \frac{MA_s}{MA_u} \sqrt{\frac{P_s}{P_u}} \cdot \frac{f_s^2}{f_u^2} \cdot \frac{G_s}{G_u} \tag{2}$$

where C is the concentration of spins, u and s refer to the unknown and the standard, $\iint IdB_0^2$ is the double integral of the signal I over the field B_0, g^{av} is the average g-value and is given by $g^{av} = \frac{2}{3}\sqrt{\left(g_x^2 + g_y^2 + g_x^2\right)/2 +} \left(g_x + g_y + g_z\right)/3$, T is the temperature in Kelvin, MA is the modulation amplitude, P is the microwave power in mW, f is the inner diameter of the sample tube, and G is the receiver gain. If the resonances occur within a reasonably narrow magnetic field range (typically 20 mT) the g^{av} term can be neglected, and if the sample and standard are measured at the same temperature, modulation amplitude, microwave power, sample tube and receiver gain, these factors can similarly be neglected, simplifying the equation to:

$$C_u = C_s \cdot \frac{\iint I_u \, dB_0^2}{\iint I_s \, dB_0^2} \tag{3}$$

All measurements are made in the Curie law region under non-saturating conditions, i.e., the doubly integrated signal intensity is proportional to the square root of the microwave power. Eq. (2) is most useful if a spectrum has been taken previously of a sample under a different set of conditions. It should be noted that the accuracy of the spin concentration is rarely better than 10–15%, hence, the value obtained should be used with that caveat in mind.

2.2 Room temperature measurements

2.2.1 Dynamic measurement of $P_{700}{}^{+}$ and $A_1{}^{-}$ in cyanobacterial cells

$P_{700}{}^{+}$ and $A_1{}^{-}$ are organic radicals, hence, they relax slowly and can be detected at room temperature. The ability to measure these radicals in whole cells allows the dynamics of the onset and decay of $P_{700}{}^{+}$ and $A_1{}^{-}$ to be visualized on the tens-of-ms to *sec* timescale in wild-type and variants (Klughammer & Pace, 1997; Sun & Golbeck, 2015). The problem with using cells is that water has a dielectric constant of ~80 at 20 °C, and thus quenches the standing waves in the resonator. The workaround is to use a flat cell and a specially designed cylindrical TM_{110} cavity. A flat cell can be visualized as an EPR tube that has been pinched midway so that it is thin in one dimension and widened in a second dimension. This geometry allows the sample to be positioned in the resonator at the anti-node of the magnetic field component in a nodal plane of the electric field component. The top and bottom of the flat cell are connected with tubing to a sample reservoir, and a programmable peristaltic pump is used to introduce a fresh sample into the cell prior to illumination. The pump is stopped during the measurement to allow the sample to be exposed to light and the measurement to be made. A computer is programmed to cycle through the following steps: (1) set the magnetic field, (2) stop the pump, (3) open the shutter to illuminate the sample, (4) record the EPR signal as a function of time, (5) close the shutter, and (6) start the pump to introduce fresh sample into the cell.

By repeating the experiment stepwise over a range of fixed field values, a three-dimensional plot is generated of signal intensity vs magnetic field vs time.

As shown in the (Fig. 3, top), the derivative signal of P_{700}^{+} appears at the onset of illumination due to rapid electron donation to the bound electron acceptors. This is followed by a decline due to electron donation from the plastoquinone pool to P_{700}^{+} followed by its gradual reappearance as a result of electron exhaustion in the plastoquinone pool. Superimposed is the signal from A_1^{-}, which shows a characteristic low field shoulder due to its greater anisotropy. The A_1^{-} radical appears only after the bound iron-sulfur clusters are filled and electrons are backed up into the quinone pool, and it disappears when rate limitations on the acceptor side are relieved, thereby allowing the electrons stored in A_1, F_X, F_B and F_A to flow into the cellular metabolic pool (Sun & Golbeck, 2015). Slicing the 3D dataset along signal intensity vs magnetic field at a given time provides a typical field swept EPR spectrum (Fig. 3, middle) and slicing the 3D dataset along signal intensity vs time at a given magnetic field provides a time course of the radical (Fig. 3, bottom). The spectrum of A_1^{-} is characterized by an Hpp (peak-to-peak) of $\sim$ 1.08 mT and a crossover at $g = 2.0049$ and because P_{700}^{+} has zero intensity at 347.1 and 348.4 mT ($g = 2.0100$ and 2.0025), the time-course of A_1^{-} can be followed independently of P_{700}^{+}. Conversely, the spectrum of P_{700}^{+} is characterized by an H_{pp} of $\sim$ 0.78 mT and a crossover at $g = 2.0025$ and because the A_1^{-} radical has zero intensity at 348.2 mT ($g = 2.0049$), P_{700}^{+} can be followed independently of A_1^{-}. Because of these differences, changes in donor side and acceptor side dynamics can be readily monitored under different physiological conditions. For example, when the optical cross section of PS I is enlarged by the presence of the IsiA antenna protein (Bibby, Nield, & Barber, 2001), a greater number of photons are processed per unit time, resulting in a faster onset of A_1^{-} followed by a more rapid rise in the reappearance of P_{700}^{+} due to faster depletion of the electrons in the plastoquinone pool (Sun & Golbeck, 2015).

Equipment

(1) X-band CW EPR spectrometer: Bruker ESP-300E or Bruker ELEXSYS-II E500

(2) TM_{110} resonator: Bruker ER-4103TM

(3) Flat cell: Wilmad WG-813-Q or Bruker Aqua-X capillary cell

(4) Cycling flow system and computer controlled peristaltic pump: ISCO WIZ

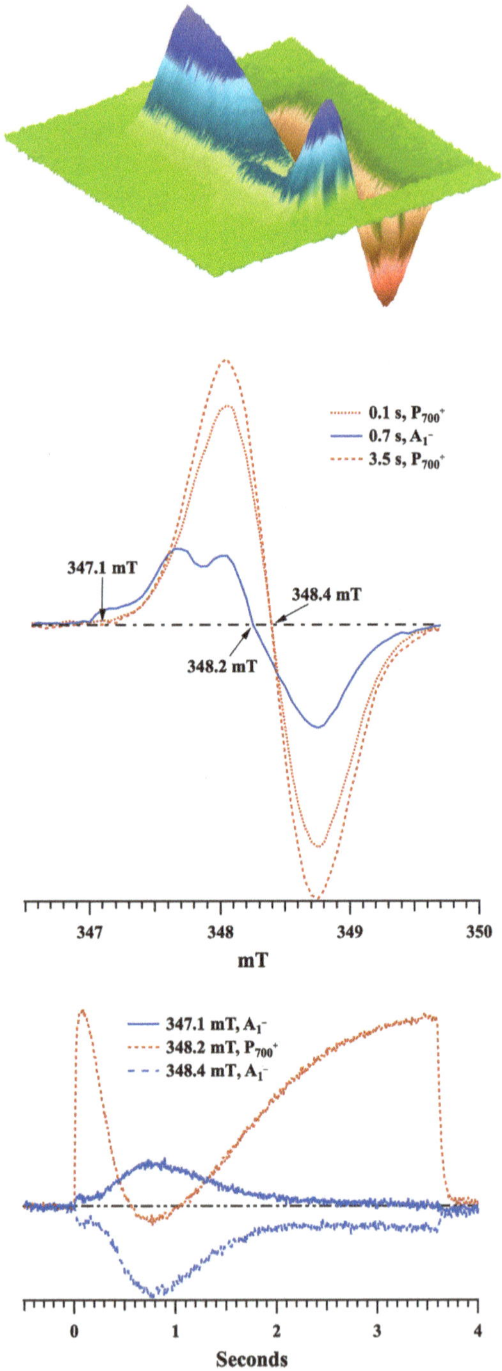

Fig. 3 Dynamic detection of $P_{700}{}^{+}A_1{}^{-}$ in cyanobacterial PS I. Top: three-dimensional plot of magnetic field (X-axis from 346.5 to 350 mT), signal intensity (Y-axis in arbitrary units) and time (*Z*-axis from −0.5 to 4 s). Middle: Two-dimensional slice of amplitude vs

(5) Xenon lamp (200 W) with water and IR-blocking filters or frequency-doubled CW Nd-YAG laser (3-W) equipped with a threefold beam expander
(6) Mechanical shutter: Thor Labs SH1 and SC10 shutter controller
(7) Computer to control the magnetic field, shutter, and peristaltic pump

Protocol and Experimental Conditions: Live cyanobacterial cells are placed in a 10 mL reservoir, which is attached to an aqueous flat cell using a short, closed loop of tubing and a microprocessor-controlled peristaltic pump. The reservoir, tubing, and the TM_{110} optical window are shielded from light to keep cells dark-adapted. The computer is programmed to begin data acquisition at $t=0$, open the shutter at $t=0.5$ s, close the shutter at $t=4.1$ s, and pump a fresh sample into the flat cell at $t=4.5$ s. This protocol is repeated at 7 s intervals at a given magnetic field to obtain signals with an acceptable signal-to-noise ratio. The magnetic field is stepped 0.05 mT and the process is repeated until data has been obtained at 60 different magnetic field settings. The data is downloaded to a computer and plotted as a three-dimensional data set that can be later sliced as a function of time or magnetic field.

2.2.2 Static room temperature detection of P^+ in Type I reaction centers

Protocol and Experimental Conditions: The detection and quantitation of the primary electron donor, P^+, is relatively straightforward. P_{700} in PS I has midpoint potential of +430 mV and can be oxidized to a radical with the use of chemical agents such as potassium ferricyanide. Alternately, samples can be illuminated in the EPR resonator under conditions similar to that described above for the dynamic measurement of $P_{700}{}^+$ and $A_1{}^-$ except that the flow system is discarded and only a standard scan is made of signal intensity vs magnetic field. Samples can consist of whole cells, membranes, or purified RCs. Because the light is turned on before the field is swept, any $A_1{}^-$ will have decayed by the time the measurement has started (see Fig. 3 top) and only $P_{700}{}^+$ will be present. One advantage of using an illumination protocol is that a light-minus-dark difference spectrum eliminates background signals, thereby yielding a pure P^+ spectrum.

magnetic field at 0.1 s (dotted red line), 0.7 s (solid blue line) and 3.5 s (dashed red line). Note that at 0.1 s and 3.5 s, the trace is entirely from $P_{700}{}^+$ and that at 0.7 s, the trace is entirely from $A_1{}^-$. Bottom: Two-dimensional slice of signal amplitude vs time. At 347.1 mT (solid blue line) and 348.4 (dotted blue line), the kinetics are due to $A_1{}^-$ and at 348.2 mT (dashed red line), the kinetics are due to $P_{700}{}^+$. EPR conditions: microwave power, 50 mW; modulation frequency, 100 kHz; modulation amplitude, 0.1 mT.

2.3 Low temperature measurements

2.3.1 Quantification of P^+ in Type I reaction centers

The quantification of P^+ in Type I RCs is best carried out at 77 K rather than at room temperature. The reason is that a standard 3 mm quartz EPR sample tube accommodates a higher sample amount than a flat cell, leading to a higher S/N (see Section 2.1.2 above). Due to the low dielectric constant of ice at 77 K, a standard EPR tube and a standard TE_{102} cavity can be used for the measurement. Determination of the spin concentration of P^+ can provide an accurate measurement of the RC concentration in whole cells and membranes as well as in purified RCs.

Equipment

(1) X-band CW EPR spectrometer: Bruker ESP-300E or Bruker ELEXSYS-II E500

(2) TM_{102} resonator: Bruker ER-4102ST or Bruker SuperX high Q resonator

(3) Liquid nitrogen flow cryostat: Oxford Instruments ER4118CV

(4) Temperature control unit: Oxford Instruments ER4121

(5) Quartz 3 mm EPR sample tube: Wilmad 707-SQ-250M

(6) Xenon lamp (200 W) with water and IR-blocking filters or frequency-doubled CW Nd-YAG laser (3-W) equipped with a threefold beam expander

Experimental Protocol: PS I RCs are placed in a standard EPR tube at a Chl concentration of 1 mg mL^{-1} and $P_{700}{}^+$ is generated either by oxidation with potassium ferricyanide or by illumination. It is important to use just enough potassium ferricyanide to oxidize $P_{700}{}^+$ because an excess may lead to the oxidation of Chl molecules in the antenna. Although the linewidth of $P_{700}{}^+$ is narrower than that of a monomeric Chl^+ cation, both radicals have similar g-values and the two signals cannot be easily separated. The amount of ferricyanide used is determined by trial-and-error; repeated measurements are made until the signal intensity asymptotically approaches a limit. If light is used to generate P^+, the intensity must be sufficiently intense to penetrate the sample and completely oxidize P_{700}, but not so intense as to cause heating of the sample. Similar conditions can be used to measure $P_{800}{}^+$ in *H. modesticaldum* and $P_{840}{}^+$ in *C. tepidum* (Vassiliev, Ronan, Hauska, & Golbeck, 2000) and $P_{840}{}^+$ in *A. thermophilum* (Charles et al., 2020).

2.3.2 Measurement of $F_A{}^-$, $F_B{}^-$, and $F_X{}^-$

The terminal iron-sulfur clusters F_A and F_B were the first of the bound electron acceptors to be discovered in PS I (Malkin & Bearden, 1971). They are coordinated by Cys residues to the tightly bound PsaC subunit and are considered an evolutionary adaptation to a heliobacterial-like RC at the onset of oxygenic photosynthesis (Jagannathan, Shen, & Golbeck, 2012). F_A and F_B are [4Fe-4S] clusters, hence they are fast-relaxing spin systems, and require very low temperatures and high microwave powers for detection. Because the local protein environment is unique to F_A and F_B, the two clusters have distinguishable characteristics (reviewed in Vassiliev, Antonkine, & Golbeck, 2001). F_A has a midpoint potential of −520 mV and exhibits a rhombic lineshape with g-values of 2.05, 1.95 and 1.85 whereas F_B has a midpoint potential of −580 mV and exhibits a rhombic lineshape with g-values of 2.07, 1.93 and 1.88. Because the two reduced clusters are close to each other, they undergo spin-spin interaction; the resulting dipolar coupling introduces a new term in the so-called spin Hamiltonian, resulting in a new set of resonances with g-values of 2.05, 1.94, 1.92 and 1.88. Depending on the experimental protocol, it is possible to reduce (i) F_A or F_B within a given PsaC to produce the individual resonances or (ii) F_A and F_B within a given PsaC to produce an interaction spectrum. F_X is an interpolypeptide [4Fe-4S] cluster coordinated between the PsaA and PsaB subunits. It is even faster relaxing, requiring lower temperatures and higher microwave powers for detection. Because the midpoint potential of F_X is −705 mV (Chamorovsky & Cammack, 1982), only a small percentage can be reduced chemically; hence, F_X^- must be generated by a process termed photoaccumulation. To accomplish this, F_A and F_B are first reduced with sodium hydrosulfite after which the sample is illuminated during freezing to generate the state $P_{700}^+ F_X^- F_A^- F_B^-$.

Equipment

(1) X-band CW EPR spectrometer: Bruker ESP300E or ELEXSYS-II E500
(2) TM_{102} resonator: Bruker ER-4102ST or Bruker SuperX high Q
(3) Liquid helium cryostat: Oxford Instruments ESR900
(4) Temperature control unit: Oxford Instruments ITC4
(5) Quartz 3 mm EPR sample tube: Wilmad 707-SQ-250M
(6) Xenon lamp (200 W) with water and IR filters or frequency-doubled CW Nd-YAG laser (3-W) equipped with a threefold beam expander

2.3.2.1 One electron reduction: F_A^- or F_B^- in photosystem I

PS I RCs at a Chl concentration of 300 μg mL^{-1} are placed in a 3 mm quartz EPR tube and dark adapted in the presence of 1 mM ascorbic acid and 50 μM dichlorophenolindophenol (DCPIP) for 5 min. The tube is transferred to the resonator, cooled to 15 K, and the magnetic field is swept from 325 to 360 mT to record a background scan. The sample is illuminated in the cavity with either a xenon lamp or a CW laser and the magnetic field is swept to record a sample scan. The light should remain on during the scan, otherwise the P_{700}^{+} $[F_A/F_B]^-$ signal will decrease in amplitude due to the recombination of ~50% of the population after cessation of illumination. Care must be taken that the sample temperature does not rise during illumination, otherwise the signal may decrease in amplitude due to the onset of saturation (see Fig. 2). The sample scan is subtracted from the background scan to obtain a light-minus-dark difference spectrum of F_A^- and a light-minus-dark difference spectrum F_B^- in a ratio of ~3:1 (Fig. 4, top). The ability to detect F_A^- and F_B^- is not limited to purified PS I; whole cyanobacterial cells as well as membranes can be packed in an EPR tube and the difference spectrum of F_A^- and F_B^- can be measured albeit with a lower S/N.

2.3.2.2 Two electron reduction: F_A^- and F_B^- in photosystem I

Two methods can be used to generate an interaction spectrum of $[F_A^-/F_B^-]$. The first is to use illumination to promote the transfer of multiple electrons to the bound iron-sulfur clusters. A background scan is recorded similar to that described in Section 2.3.2.1. The EPR tube is removed from the resonator and lowered slowly into liquid nitrogen in a transparent Dewar while focusing the light of either a Xenon lamp or a CW Nd-YAG laser at the solid-liquid interface. It should take about 2 min to carry out the process for a sample height of 2.5 cm. *It is necessary to wear eye protection while carrying out the photoreduction protocol, as the heat of the lamp can lead to stresses in the glass that may cause the Dewar to implode.* This protocol locks-in the state P_{700} F_A^- F_B^- although there may be a small population of P_{700}^{+} F_A^- F_B^- present due to incomplete reduction of P_{700}^{+}. The sample is repositioned in the resonator and the spectrum is again recorded. The sample scan is subtracted from the background scan to obtain a light-minus-dark difference spectrum of F_A^-/F_B^- with g-values of 2.05, 1.94, 1.92 and 1.88 (Fig. 4, middle). The second method is to raise the pH of the sample to 10.0 using 1 M glycine buffer and to add sodium hydrosulfite to a concentration of 1 mM to reduce F_A and F_B. Because protons are involved in the reaction, and because the redox potential of the solution is related to the pH through the Nernst equation, a pH of 10.0 is required to achieve a solution potential of −600 mV. A similar protocol is used to detect the resonances of F_A^- and F_B^- in membranes and RCs from *Heliobacterium chlorum* (Nitschke, Setif, Liebl, Feiler, & Rutherford, 1990), *H. modesticaldum*, *C. tepidum* (Nitschke, Feiler, & Rutherford, 1990) and *C. thermophilum* (He et al., 2019).

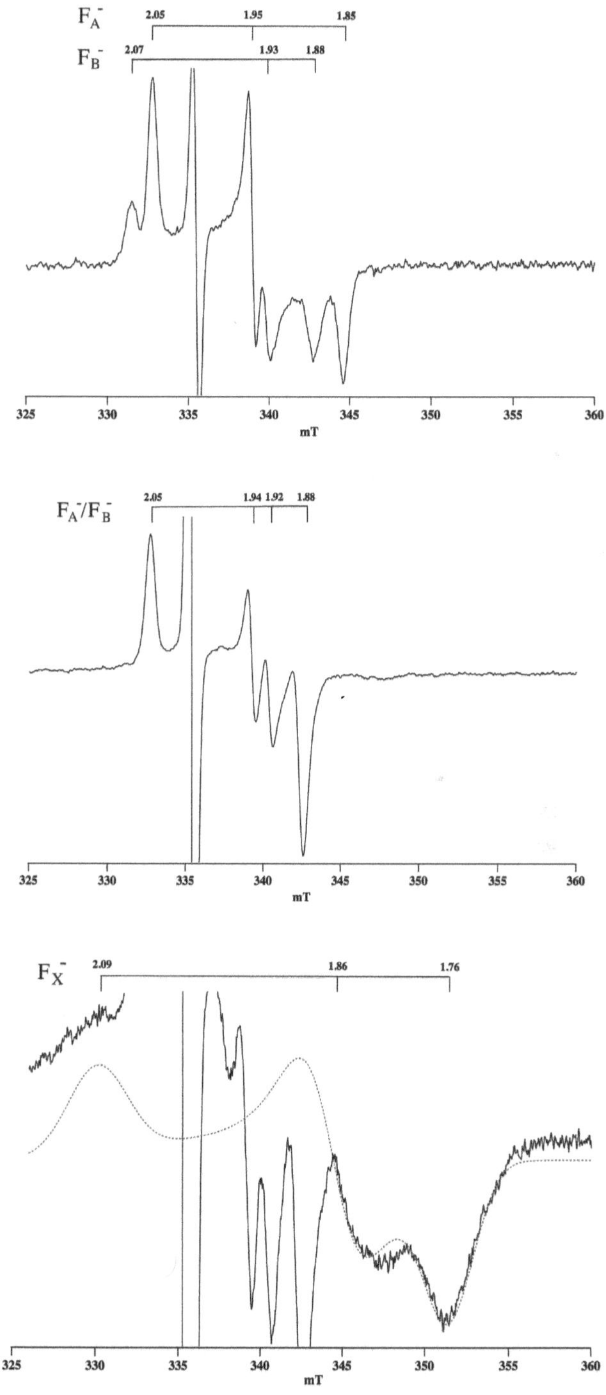

Fig. 4 CW X-band EPR spectra of the F_A, F_B and F_X iron-sulfur clusters in PS I. Top: PS I sample frozen in the dark to 18 K and illuminated during measurement. The spectrum

(Continued)

2.3.2.3 Three electron reduction: F_A^- F_B^- F_X^- in photosystem I

The ability to detect F_X^- requires the chemical reduction of F_A and F_B prior to carrying out the photoaccumulation protocol. The resulting recombination rate between P_{700}^+ and F_X^- is so fast that it typically outcompetes the rate of reduction of P_{700}^+ by forward electron donation from sulfite ion. However, every so often the latter outcompetes the former, and over time, the P_{700} F_X^- state becomes locked in. PS I RCs at a Chl concentration of 500 μg mL^{-1} are placed in a 3 mm quartz EPR tube in the presence of 1 M glycine buffer, pH 10.0, and 25 mM sodium hydrosulfite. The sample is treated with sodium hydrosulfite to reduce F_A and F_B as described above in Section 2.3.2.2. The tube is then transferred to the resonator, illuminated with a CW Nd-YAG laser, and scanned while under illumination at 6 K. Two factors must be taken into account when making the measurement. First, because F_X^- is a very fast-relaxing system, the temperature should be between 6 and 8 K to maximize the signal intensity. Second, because F_X^- has a greater anisotropy than F_A^- and F_B^-, the S/N will be lower at a given spin concentration; however, the loss in sensitivity can be partly compensated by increasing the modulation amplitude. The highfield resonance is the only feature of F_X^- fully resolved; the midfield and low field resonances are partly obscured by those from F_A^- and F_B^-, which are in saturation at this temperature and microwave power (Fig. 4, bottom, black line). The experimental spectrum can be simulated with g-values of 2.09, 1.86 and 1.76 (Fig. 4, bottom, dotted red line). The complete spectrum of F_X^- can be observed independent of F_A^- and F_B^- by removing PsaC either with the use of chaotropic agents (Golbeck, Parrett, Mehari, Jones, & Brand, 1988) or by genetic interruption of the *psaC* gene (Yu, Smart, Jung, Golbeck, & McIntosh, 1995).

Fig. 4—Cont'd shows F_A^- with g-values of 2.05, 1.95 and 1.85 and F_B^- with g-values of 2.07, 1.93 and 1.88 in a ca. 3:1 ratio. The sharp radical at 385.5 mT is P_{700}^+, which is in saturation at this temperature. EPR conditions: temperature 15 K; microwave power 100 mW; modulation frequency 100 kHz; modulation amplitude 1 mT. Middle: PS I sample illuminated while freezing and measured in the dark at 18 K. The spectrum shows an interaction spectrum of F_A^-/F_B^- with g-values of 2.05, 1.94, 1.92 and 1.88. The sharp radical at 385.5 mT is P_{700}^+, which is in saturation. EPR conditions: same as top spectrum. Bottom: PS I sample reduced with sodium hydrosulfite at pH 10.0, frozen in the dark, and illuminated at 6 K. The spectrum shows F_X^- as well as resonances from the interaction spectrum of F_A^-/F_B^-; the latter are in saturation at this temperature and are therefore reduced in amplitude relative to F_X^-. The sharp radical at 385.5 mT is P_{700}^+, which is also in saturation at this temperature. The simulated spectrum of F_X^- (dotted red line) shows a highly anisotropic, rhombic set of resonances with g-values of 2.09, 1.86 and 1.76. EPR conditions: temperature 6 K; microwave power 200 mW; modulation frequency 100 kHz; modulation amplitude 3.2 mT.

2.3.2.4 Measurement of F_X^- in homodimeric Type I reaction centers

F_X is an interpolypeptide [4Fe-4S] cluster coordinated between the identical PshA subunits in the HbRC. Its midpoint potential is −504 mV, making it possible for reductants such as sodium hydrosulfite to reduce the majority of the clusters chemically. In PS I, two of the four iron atoms of F_X are in the ferric (S=5/2) state and two are in the ferrous (S=2) state. The four iron atoms couple magnetically to generate a ground spin state of S=0. Hence, an oxidized [4Fe-4S] cluster in PS I is EPR silent. When one of the two ferric iron atoms is reduced, the magnetic coupling between the resulting three ferrous atoms and one ferric iron generates a ground spin state of S=½. This is fortunate for EPR spectroscopists, as a S=½ spin system is straightforward to detect and quantify. However, in certain instances, the three ferrous and one ferric iron atoms couple magnetically to generate a ground spin state of S=3/2. This frequently occurs when a [4Fe-4S] cluster bridges two identical proteins, as is the case with the F_X cluster in the homodimeric RC in *H. modesticaldum* (Ferlez et al., 2016; Heinnickel, Agalarov, Svensen, Krebs, & Golbeck, 2006) and *C. tepidum* (Jagannathan & Golbeck, 2008). The F_X cluster in the GsbRC appears to have at least some S=½ ground spin state character (Vassiliev et al., 2000); it therefore appears to exist in a mixed population of S=3/2 and S=½ ground spin states (Romberger & Golbeck, 2010). (F_X has not yet been detected by EPR in *A. ther-mophilum* (He et al., 2019).) High spin S=3/2 clusters show two sets of Zeeman splittings termed Kramers doublets separated by 2D, the axial zero-field splitting parameter (see Fig. 2, top right). As the magnetic field increases, each of the doublets splits and can be treated as if it were an isolated S=½ system. Both the lower and upper pseudo S=½ doublets are populated according to a Boltzmann distribution; hence, the larger the value of D, the smaller will be the total spin population in the upper doublet. An additional parameter E is the rhombic splitting parameter, which governs degree of rhombicity; when E/D=0, the resonances will have axial symmetry, and when E/D=1/3, the resonances will have rhombic symmetry. The two pseudo S=½ systems can be distinguished by their temperature dependence; as the temperature is lowered, the upper Kramers doublet will depopulate relative to the lower Kramers doublet provided the value of D is <0; the opposite is true if $D>0$.

Equipment

Same as for detection of F_X^- in PS I.

Protocol and Experimental Conditions: HbRC cores at a BChl *g* concentration of 500 μg mL^{-1} are placed in a 3 mm quartz EPR tube in the presence of 1 M glycine buffer, pH 10.0, and 25 mM sodium hydrosulfite and incubated for

10 min to reduce F_A and F_B. The tube is positioned in the resonator, illuminated with a Nd-YAG laser, and the magnetic field is swept from 50 to 450 mT at temperatures of 12, 10, 8, 6, and 4.2 K. High spin $[4Fe\text{-}4S]^+$ clusters with a $S=3/2$ ground state exhibit temperature-dependent resonances in the range $g\approx4$ to 6 emanating from the two Kramers doublets. At 4.2 K, the low-field resonance shows two distinctive features: a peak at g = 5.4 and a shoulder at g = 4.4 (Fig. 5). As the temperature is increased to 12 K, the g = 5.4 peak decreases in intensity, while the g = 4.4 shoulder increases in intensity. The temperature dependence indicates that the g = 5.4 signal originates from the ground Kramers doublet, whereas the g = 4.4 feature originates from the excited Kramers doublet. The value for the zero-field splitting parameter D is estimated by measuring the population

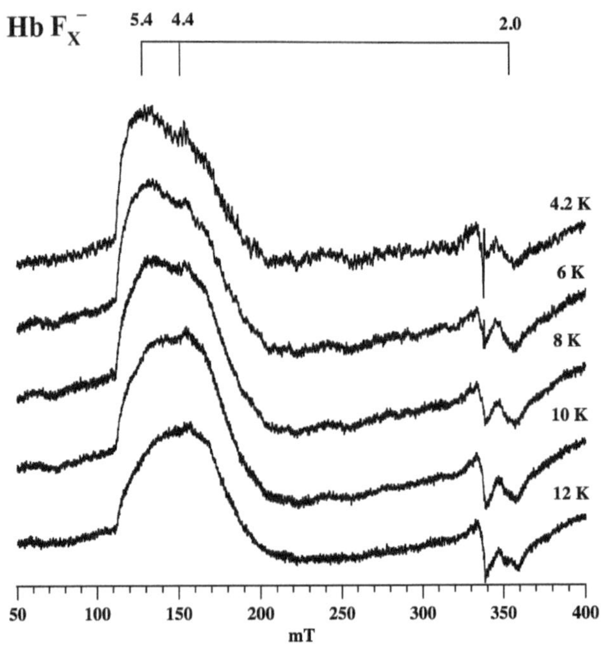

Fig. 5 CW X-band EPR spectra of the F_X^- iron-sulfur cluster in the homodimeric HbRC. The spectra were measured at temperatures of 4.2, 6, 8, 10, and 12 K; each trace represents four averages of 8000 total data points. The lowfield resonances are assigned to a $[4Fe\text{-}4S]^+$ cluster with an S = 3/2 ground spin state (see Fig. 2, top right). The temperature dependence indicates that the g = 5.4 feature is derived from the ground Kramers doublet, whereas the g = 4.4 feature is derived from with the excited Kramers doublet of the S = 3/2 spin manifold. EPR conditions: microwave power, 126 mW; modulation frequency 100 kHz; modulation amplitude, 1 mT.

of the two Kramers doublets at g-values of 4.4 and 5.4. A plot of the natural logarithm of the ratio of the peak intensities vs the inverse of the absolute temperature is fit to a straight line, the slope of which yields the energy separation of the two Kramers doublets (Duderstadt, Brereton, Adams, & Johnson, 1999). In this instance, D was found to be $-1.5\,cm^{-1}$; given that $D<0$, an $S=3/2$ spin system exhibiting a rhombicity, E/D, of approximately 0.2 is expected to exhibit g-values of 5.8, 1.3, and 1.0 for the ground state doublet and 5.0, 2.7, and 1.8 for the excited state doublet. The low field resonances are reasonably close to those measured experimentally; the other resonances are too broad to be detected.

2.3.3 Measurement of A_1^- and A_0^- in photosystem I

A_{1A} and A_{1B} are bound menaquinones located between A_0 and F_X and are capable of being singly reduced to the semiquinone radical state and doubly reduced to the neutral hydroquinone state. The quinone binding sites in PS I typically allows only a single reduction to occur, although a double reduction can be forced under certain conditions. Only the semiquinone state is EPR visible. The midpoint potentials of A_{1A} and A_{1B} are sufficiently electronegative that they are incapable of being chemically reduced, hence, the semiquinone radicals must be generated by photoaccumulation. Because the CW EPR spectra of A_{1A}^- and A_{1B}^- are indistinguishable, they will be referred to collectively as A_1^-. A_{0A} and A_{0B} are Chl molecules and can only be singly reduced. They have midpoint potentials even more reducing than that of A_{1A} and A_{1B}, and hence A_{0A}^- and A_{0B}^- must also be generated by photoaccumulation. Because the spectra of A_{0A}^- and A_{0B}^- are also indistinguishable by CW EPR, they will be referred to collectively as A_0^-. A_0^- and A_1^- are organic radicals, hence, they relax slowly; however, measurements are typically made at 77 K because of the need for low temperature in the photoaccumulation process. A PS I sample is illuminated at cryogenic temperatures, and over time the P_{700} A_0 A_1^- state accumulates plus a small amount of P_{700}^+ A_0 A_1^-. Once this state is achieved in any given RC, continued illumination begins to generate A_0^-, leading to the P_{700} A_0^- A_1^- state plus a small amount of P_{700}^+ A_0^- A_1^-. Hence a statistical mixture of states results in the ensemble of RCs, the ratio of which depends on the temperature and illumination time. In one example, the A_1^- to A_0^- ratio was reported to be 1.0:0.05 after 90 min of illumination at 205 K, but this rose to 1.0:0.43 after 30 min of illumination at 212 K (Yang et al., 1998).

The A_1^- and A_0^- radicals can be distinguished from each other by CW EPR. The presence of electron density at the highly electronegative oxygen atom of A_1 enables greater spin-orbit coupling and hence a larger anisotropy than that of a Chl anion or cation radical. Even so, at X-band, the g-tensor of A_1^- is obscured by overlapping hyperfine couplings from the ring protons as well as unresolved splittings from the three equivalent hydrogens on the 2-methyl group. The result is a spectrum showing a distorted derivative lineshape with a g-value of ca. 2.0050 at the crossover point (Fig. 6 top). For comparison, the spectrum of P_{700}^+ is also shown. The pure spectrum of A_0^- can only be obtained by photoaccumulating a PS I sample that lacks the quinone, typically by chemical extraction (Mathis et al., 1988) or by genetic manipulation (Gorka, Gruszecki, et al., 2021). The resulting spectrum of A_0^- is a featureless derivative at X-band with a g-value of 2.0034 at the crossover point and a linewidth of 1.4 mT (Fig. 6 top). At 34 GHz the resolution is enhanced by a factor of ~3.8, leading to a partially resolved set of four hyperfine lines from the 2-methyl group of A_1^- (Fig. 6 middle). At this frequency, the A_0^- spectrum continues to be dominated by unresolved hyperfine couplings from ring protons and remains a featureless derivative (Fig. 6 middle) but it can be distinguished from A_1^-. It should be noted that the simulated spectrum of A_1^- (Fig. 6 middle, dotted blue line) differs somewhat from that of the experimental spectrum; the difference between the two is a small amount of A_0^- in the sample. One additional detail is that even at Q-band, the g-tensor of A_1^- is obscured by unresolved hyperfine couplings from ring protons and the 2-methyl group (Fig. 6 middle, hatches above spectrum). Deuterium has a spin 1 nucleus and a relatively small magnetic moment, hence, these couplings can be removed by growing the cyanobacterium in D_2O. The resulting Q-band spectrum of A_1^- (Fig. 6 bottom) can be simulated with near-rhombic set of resonances at $g = 2.0062$, 2.0050, and 2.0021.

Equipment

(1) Q-band CW EPR spectrometer: Bruker ELEXSYS-II E500
(2) Q-band resonator: Bruker ER 5106 QT-W
(3) Liquid nitrogen cryostat: Bruker ER4118CV
(4) Temperature controller: Oxford Instruments ITC-4
(5) Quartz 1 mm EPR sample tube: Wilmad WG-221T-RB
(6) Xenon lamp (200 W) with water and IR filters or frequency-doubled CW Nd-YAG laser (3 W) equipped with a threefold beam expander
(7) Glass walled transparent Dewar

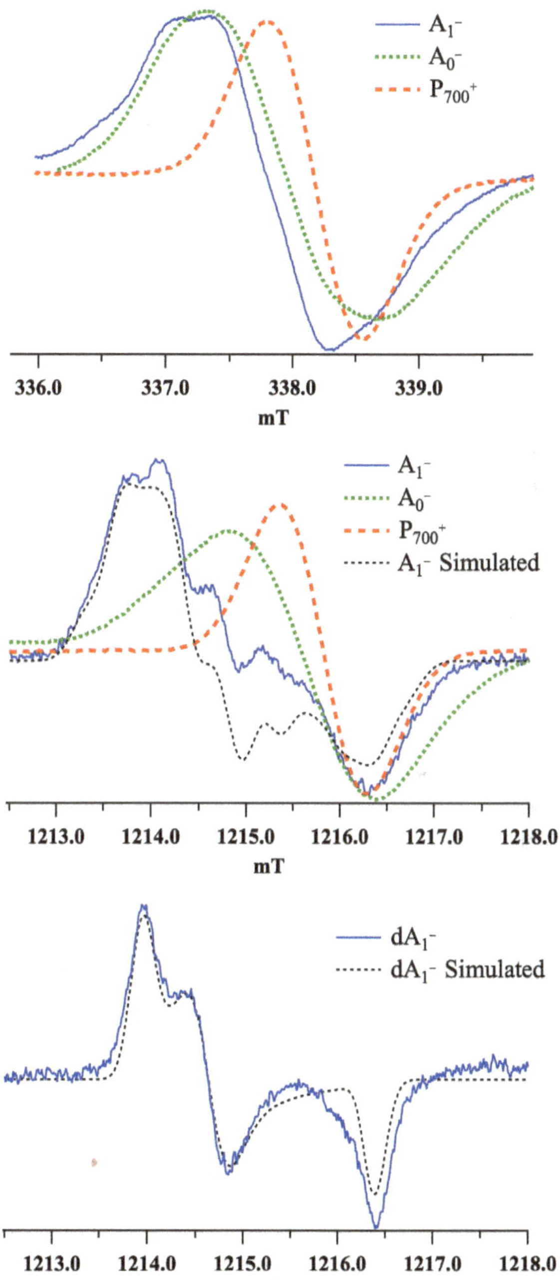

Fig. 6 CW X-band EPR spectra of A_1^-, A_0^- and P_{700}^+ in cyanobacterial PS I. A_1^- was generated by photoaccumulation of PS I RCs; A_0^- was generated by photoaccumulation

(Continued)

Protocol and Experimental Conditions: PS I RCs at a Chl concentration of $1.5\,mg\,mL^{-1}$ are placed in a 1 mm quartz EPR tube in the presence of 1 M glycine buffer, pH 10.0, and 25 mM sodium hydrosulfite and incubated for 10 min to reduce F_A and F_B. The tube is lowered into a glass walled transparent Dewar filled with ethanol and cooled with dry ice to a temperature of 205 K. The beam from either the xenon lamp or the CW Nd-YAG laser is focused on the sample over the course of 1.5 h, a procedure that allows F_X to become reduced and, over time allows A_1 to become trapped in the singly reduced state, leading to the P_{700} $F_A^-F_B^-F_X^-A_1^-A_0$ state along with a small amount of P_{700} $F_A^-F_B^-F_X^-A_1^-A_0^-$. The spectrum of A_1^- is acquired by sweeping the field from 1212 to 1218 mT at a temperature of 77 K. The large anisotropy of F_A^-, F_B^-, and F_X^- means that their resonances do not overlap with the considerably narrower resonances of A_1^-. At 34 GHz, the field-dependent g-anisotropy dominates the spectrum of A_1^- (Fig. 6 middle), revealing two of the three principal values as $g_{xx}=2.0062$ and $g_{zz}=2.0021$ (the g_{yy} component is partially obscured by the hyperfine lines from the 2-methyl group). Because it is difficult to generate a spectrum of A_1^- without generating some A_0^-, the spectrum of A_1^- should simulated before attempting double integration and spin quantitation (Fig. 6 middle). A matrix of microwave powers and temperature is run on both the sample and the standard to ensure that the signal heights obey the Curie law under non-saturating conditions (see Fig. 2).

Fig. 6—Cont'd of PS I RCs in which the A_{1A} and A_{1B} quinones were chemically extracted; P_{700}^+ was generated by oxidation of PS I complexes with potassium ferricyanide. Top: X-band EPR spectra of A_1^- (blue solid line) A_0^- (green dotted line) and P_{700}^+ (red dashed line). EPR conditions for A_1^-: microwave power, 20 μW; modulation amplitude, 0.2 mT; modulation frequency, 100 kHz; temperature, 100 K. EPR conditions for A_0^- and P_{700}^+: same except microwave power, 63 μW; temperature, 35 K. Middle: Q-band EPR spectra of A_1^- (blue solid line) A_0^- (green dotted line) and P_{700}^+ (red dashed line). EPR conditions for A_1^-, A_0^- and P_{700}^+: microwave power, 10 μW; modulation amplitude, 0.1 mT; modulation frequency, 100 kHz; temperature, 100 K. The A_1^- spectrum was simulated (dotted black line) using the g-tensor extracted from the deuterated A_1^- spectrum (see below), an A-tensor of 9.0, 12.9, and 9.0 MHz, and line widths of 0.335 mT for g_{xx}, g_{yy}, and g_{zz}. The lines above the spectrum indicate the field positions of the four hyperfine couplings from the 2-methyl group of phylloquinone. Bottom: Q-band EPR spectrum of deuterated A_1^- (dA_1^-, blue solid line) in PS I complexes isolated from *Synechocystis* sp. PCC 7002 grown in 99% D_2O. The dA_1^- spectrum was simulated (dotted black line) using $g_{xx}=2.0062$, $g_{yy}=2.0050$, and $g_{zz}=2.0021$ and line widths of 0.24, 0.37, and 0.24 mT, respectively. EPR conditions: microwave power, 10 μW, modulation amplitude, 0.1 mT, modulation frequency, 100 kHz, temperature, 100 K.

3. Time-resolved EPR of Type I reaction centers

In addition to measurements of the steady-state populations of radicals and reduced iron-sulfur clusters described in the preceding sections, EPR spectroscopy can also be used in a time-resolved mode to study the paramagnetic species generated in Type I RCs after a single turnover flash (van der Est, 2001). Using time-resolved EPR (TREPR) methods, the light-induced radical pairs generated by the charge separation event can be studied and both the forward and reverse electron transfer rates can be obtained. If chlorophyll triplet states are formed, they can also be detected. The spin polarization of the EPR signals, observed during the first several microseconds after the laser flash greatly enhances their information content compared to steady-state measurements. A unique feature of these data is that that the sign and intensity of the spin polarized signals depends on the pathway by which a given radical pair or triplet state has been formed. Thus, the same species can have different TREPR spectra, depending on how it was generated. Moreover, the magnetic properties of the radical pairs and triplet states depend on the local interactions of the cofactors with their surroundings, the distance between magnetic species and their relative orientation. Hence, information about the electron transfer kinetics, structure and protein-cofactor interactions is obtained in a single experiment.

3.1 Methodology

TREPR experiments can be broken into two main classes, transient EPR and pulsed EPR, that differ in the way the signal is detected. In transient EPR experiments, continuous microwave irradiation is used, while in pulsed EPR, the signal is detected as a spin echo. In both cases, the signal is collected at a fixed magnetic field as a function of the time after a laser flash. Spectral information is obtained by repeating the experiment stepwise over a range of fixed field values or by taking the Fourier transform of the time dependent data.

3.1.1 Transient EPR on different time scales

The basic procedure for a transient EPR experiment is similar to that described in Section 2.2.1 for dynamic measurement of P^+ and A_1^-, except that a pulsed laser is used and a photodiode triggers a digital oscilloscope to collect the signal after the laser flash. The time traces are collected over a range of magnetic field values to create a time/field dataset from which the spectrum at a given delay time can be extracted and the kinetics of the signal evolution and decay can be determined. A general overview of the transient EPR setup is given in the recent book chapter by Weber (2017). Often modified CW EPR instruments with an external digital oscilloscope and home-written software are used for such experiments but they can also be carried out on Bruker Elexsys spectrometers operating in transient mode.

A key aspect of the experiment is that the sample must flashed thousands of times during the data collection and the system being measured must return to its initial state before the next laser flash. For liquid samples, a circulation pump can be used if necessary to allow more time for the system to relax. For isolated Type I RCs at room temperature, the charge recombination occurs in less than ~100 ms, and within this time window a fast electron donor such as phenazine methosulfate is also able to re-reduce P^+ in any RCs in which electrons have been lost to oxygen. At low temperature, reversible electron transfer occurs more rapidly than at room temperature. Thus, for both temperature regimes a flash repetition rate of 10 Hz can be used, and circulation of the sample is not necessary.

A crucial factor in any time-resolved spectroscopy experiment is the time resolution that is required and the response time of the spectrometer. In CW experiments there is a similar trade-off between sensitivity and response time. For transient EPR experiments, the response time is limited by the bandwidth of the resonator, microwave detector and amplifier and detection scheme used.

3.1.1.1 Field modulation detection

In standard CW EPR experiments, the signal is detected using field modulation and a lock-in amplifier (see Section 1.1.1). The modulation frequency is usually 100 kHz and since about 10 oscillations of the modulation are required to detect the signal, the response time of the detection is on the order of 100 μs. This timescale is longer than the lifetime of electron transfer to the iron-sulfur clusters and is similar to the low temperature back reaction lifetime from phylloquinone in PS I. Because the spin-lattice relaxation of the iron-sulfur clusters is extremely fast, they cannot be detected at temperatures above about 20 K. Thus, with field modulation detection usually only P^+ is observed. Indeed, the spectrum of $P_{700}{}^+$ from PS I, originally known as signal 1, was one of the first pieces of evidence for light-induced electron transfer in photosynthesis (Commoner, Heise, & Townsend, 1956). The decay of the EPR signals from P^+ can be used to determine the back reaction lifetimes for electrons returning from the iron-sulfur clusters.

3.1.1.2 Direct detection

Because the field modulation limits the time resolution of an EPR experiment the response time can be decreased considerably by using so-called direct detection in which no modulation is employed, rather, the signal taken directly from the detector is amplified and digitized. With a sufficiently fast detector and amplifier the response time is limited by the bandwidth of the resonator and can be pushed as low as ~10 ns. This also results in a dramatic decrease in the sensitivity of the spectrometer. However, it is compensated for by the strong spin polarization that is inherent in photogenerated species. In Type I RCs this allows detection of $P^+A_1^-$, $P^+(FeS)^-$ and the triplet states of the primary donor and antenna chlorophyll. The low sensitivity of direct detection means that the signals are no longer detectable once the populations of the spin states return to thermal equilibrium. This restriction places an upper detection limit of a few tens of microseconds on the method.

3.1.2 Pulsed EPR

Pulsed EPR encompasses a very wide range of methods and allows weak interactions to be measured that are not easily extracted directly from EPR spectra. Here, we will restrict the discussion to the measurement of the radical pairs generated by pulsed laser excitation of PS I. We will not include any discussion of methods such as ENDOR and ESEEM, which are used to measure hyperfine couplings. In pulsed EPR experiments, the signal is generated by applying one or more microwave pulses to the sample. Because the EPR spectra from PS I are relatively broad, the FID generated after a single pulse typically decays within the ring down time of the resonator and is difficult to detect. To circumvent this problem, two or more pulses are applied, and the signal is detected as a spin echo. The height of the echo or the area under it is then measured as one of the experimental parameters is varied. A field swept echo spectrum is obtained if the magnetic field is varied. Echo modulation curves are acquired by varying a pulse length or delay time.

3.2 Room temperature measurements

As discussed in Section 2.2.1, EPR measurements of aqueous solutions is complicated by the strong damping of the resonator by the solvent. A flat cell or capillary that places the aqueous sample in a region of maximum amplitude of the magnetic field component and minimum amplitude of the electric component of the microwaves in the resonator can be used to overcome the damping. However, this becomes increasingly difficult as the dimensions of the resonator become smaller. As a result, essentially all room temperature experiments on Type I RCs have been carried out at X-band using a flat cell and a TM110 or TE102 resonator. Pulsed EPR measurements, which require a low Q resonator with a high filling factor are difficult to perform on liquid aqueous samples. Thus, virtually all room temperature time resolved EPR experiments on Type I RCs in the literature have been performed as transient EPR measurements.

Fig. 7 shows typical transient EPR data obtained from isolated PS I complexes at room temperature. The signals are spin polarized and are presented with positive signals as absorption (A) and negative signals as emission (E). Slices through the dataset allow transient spectra at different times or time traces at different magnetic field values to be extracted as shown in the bottom of Fig. 7. For PS I samples an emission/absorption/emission (E/A/E) pattern due to $P_{700}^{+}A_1^{-}$ is observed at early time, which then evolves into a primarily emissive spectrum from $P_{700}^{+}(FeS)^{-}$ at later time. Because the iron-sulfur cluster contribution to the spectrum is not observed, the actual location of the transferred electron on the iron-sulfur clusters is ambiguous and FeS is used to mean any of the three iron-sulfur clusters. The lifetime of A_{1A}^{-} due forward electron transfer ($\tau \approx 250\,ns$) can be determined from such data by fitting the transients with an appropriate function (van der Est et al., 2004). The low field end of the data set is mostly due to the emissive A_{1A}^{-} contribution to the $P_{700}^{+}A_1^{-}$ charge separated state and the decay of the signal (transient a) is dominated by the decay of A_{1A}^{-} due to electron transfer to F_X. In the center of the dataset the $P_{700}^{+}A_1^{-}$ signal is absorptive and that of FeS^{-} is emissive as seen in transient b.

The same experiment can be carried out using field modulation detection similar to that described in Section 2.1.1 except that a pulsed laser replaces the CW light source and the magnetic field is fixed at a single value. In this case, the response time of the spectrometer is generally slower than the decay of the spin polarization, but the spectrometer sensitivity is much higher. Thus, the P_{700}^{+} signal at thermal equilibrium is seen and the spin polarized signal is only observed as a spike in the first few data points after the laser flash. From the decay of thermal equilibrium signal, the lifetime of the backreaction from the iron-sulfur clusters can be obtained.

Equipment

(1) X-band Transient EPR spectrometer: modified CW EPR spectrometer with 500 MHz digital oscilloscope and higher bandwidth preamplifier or Bruker ELEXSYS-II E580

(2) TM_{110} or TE_{102} resonator with optical access: Bruker ER-4103TM or Bruker ER-4102ST

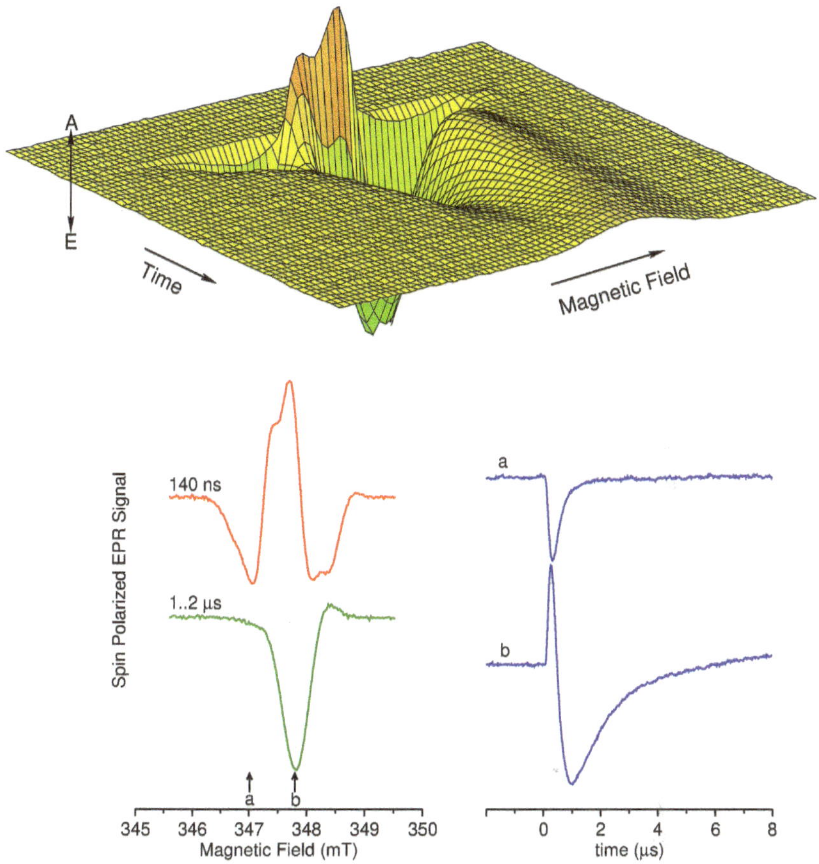

Fig. 7 X-band spin polarized transient EPR data of PS I at 293 K. Positive signals are in absorption (A) and negative signals are in emission (E). Top: full time/field dataset. The time axis spans a range of 10 μs and the magnetic field sweep width is 4 mT. Bottom: Spectra and time traces extracted from the full data set at the indicated times and magnetic field positions. The spectra and transients have been offset vertically from one another to simplify the figure. EPR conditions: microwave frequency, 9.75 GHz; microwave power 10 mW.

(3) Quartz flat cell: Wilmad WG-813-Q or WG-812-Q

(4) Pulsed, frequency doubled Nd-YAG laser (pulse width < 10 ns, repetition rate 10 Hz or less)

(5) Photodiode trigger

Protocol and Experimental Conditions: The flat cell is filled with a solution of RCs at a Chl concentration of ~1 mg/mL containing 5 μM phenazine methosulfate and 20 mM sodium ascorbate. The redox mediators should be added from freshly prepared stock solutions immediately prior to the experiments. The laser power is set to ~5 mJ/pulse. Transients are collected at a fixed magnetic field value using the digital oscilloscope, which is triggered by a photodiode that detects stray light from the laser flashes. Typically, 50 to 100 transients are averaged and stored. The magnetic field is then stepped, and the process is repeated. For measurements of $P_{700}^{+} A_1^{-}$ a field range of 4–5 mT and 100–200 field values are used. The stored time traces form a time/field data set that can be analyzed as described above.

3.3 Low temperature measurements

Frozen solutions do not cause the severe damping of the microwave resonance that occurs with liquid aqueous samples. Because of this, a much wider range of TREPR experiments become feasible, in particular, high field/frequency and pulsed EPR measurements. However, a drawback of working at low temperature is that the electron transfer kinetics may be altered compared to native conditions. In PS I, both reversible electron transfer between P_{700} and A_1 and irreversible transfer between P_{700} and F_A/F_B transfer occur, and this complicates the data analysis. In homodimeric Type I RCs only reversible electron transfer between P and F_X takes place.

3.3.1 *Radical pair states*

For PS I samples transient EPR experiments give the spin polarized spectrum of $P_{700}^{+} A_{1A}^{-}$. This spectrum can also be measured by the field swept echo method. However, because irreversible charge separation occurs and spin echo measurements detect static paramagnetic species, a light-minus-dark spectrum must be taken to remove the contribution from the static P_{700}^{+} signal. In homodimeric Type I RCs only the P^{+} contribution to the spectrum of $P^{+} F_X^{-}$ is observed due to the absence of bound quinones.

3.3.1.1 Spin polarization patterns of $P_{700}^{+} A_1^{-}$ and $P^{+} F_X^{-}$

Fig. 8 shows spin polarized spectra of $P_{700}^{+} A_1^{-}$ in PS I and $P_{800}^{+} F_X^{-}$ in HbRCs measured at two different microwave frequencies. For the spectra of $P_{700}^{+} A_1^{-}$ in PS I, the g-anisotropy of the phylloquinone anion leads to a wider and better resolved spectrum at Q-band. Because the $P_{700}^{+} A_1^{-}$ radical pair is generated from a singlet state on a time scale that is fast compared to the motions of the spins it has pure singlet character, which leads to the absorptive and emissive parts of the spectrum having equal overall intensity. For HbRCs only the P_{800}^{+} contribution to the $P_{800}^{+} F_X^{-}$ spectrum is observed. Because the g-anisotropy of P_{800}^{+} is small there is little difference in the widths of the X-band and Q-band spectra. In contrast to the situation

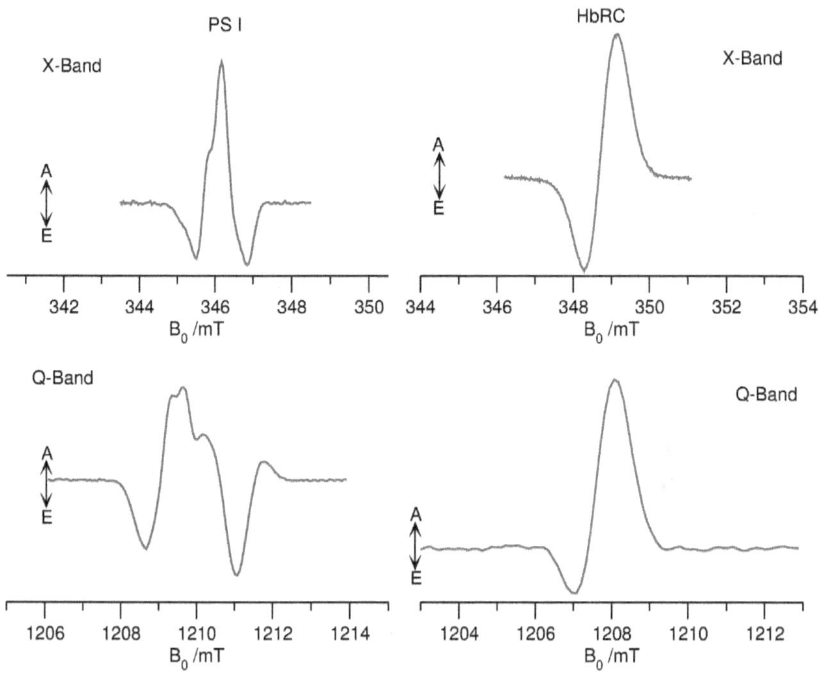

Fig. 8 X-band and Q-band spin polarized transient EPR data of PS I and HbRCs at 80 K taken at 1 μs after the laser flash. Positive signals are in absorption (A) and negative signals are in emission (E). EPR conditions: PS I: X-band microwave frequency 9.71 GHz; microwave power, 0.1 mW Q-band: microwave frequency 33.93 GHz microwave power 2.5 mW; HbRC's: X-band microwave frequency 9.769 GHz, microwave power, 0.1 mW; Q-band: microwave frequency, 33.84 GHz; microwave power 2.5 mW.

in PS I, in HbRCs the lifetime of $P_{800}{}^{+}A_0{}^{-}$ is sufficiently long (~600 ps) for singlet-triplet mixing to occur. This leads to net absorptive polarization of $P_{800}{}^{+}F_X{}^{-}$. At higher microwave frequencies the rate of the singlet-triplet mixing increases and as a result the Q-band spectrum has stronger net polarization (a larger difference in intensities of the absorptive and emissive components) than the X-band spectrum.

Equipment

(1) X-band and Q-band transient or pulsed EPR spectrometers

(2) Dielectric ring, split ring or cylindrical resonators with optical access: Bruker ER 4118X-MD5, Bruker ER 4118X-MS5, or ER5106QTW

(3) Quartz 3 or 4 mm EPR tubes

(4) He gas flow cryostat and controller: Oxford Instruments CF935

(5) Pulsed, frequency doubled Nd-YAG laser (pulse width < 10 ns, repetition rate 10 Hz or less)

(6) Photodiode or laser trigger

Protocol and Experimental Conditions: The sample concentrations and redox mediators are as described in Section 3.2. The cryostat is first cooled to the desired temperature, typically 80 or 50 K. The sample is placed in the EPR tube and then dark adapted for ~20 min before being frozen in the dark and transferred to the resonator. To avoid having the sample tube break due to the expansion of the water during freezing, the EPR tube can be cooled slowly by placing a centrifuge tube filled with isopropanol in liquid nitrogen and then freezing the EPR sample in the cold isopropanol. For transient EPR measurements the time/field dataset is collected as described in Section 3.2. For field-swept echo EPR data collection, the pulse generator is triggered by the laser and generates a 90°-τ-180° pulse sequence. The excitation bandwidth of the pulses must be set to be smaller than twice the spin-spin coupling of ~5 MHz. This typically means a 90° pulse length of 50 to 100 ns. Two field swept echo spectra are then collected with the trigger delay set so that the microwave pulses are applied either before or after the laser flash. The difference of the two spectra gives the light-induced spin polarized spectrum.

3.3.1.2 Out-of-phase echo modulation

If the excitation bandwidth of the pulses used in a spin echo measurement of $P_{700}{}^{+} A_1{}^{-}$ in PS I is larger than the spin-spin coupling, the phase of the echo is shifted by 90° so that it has the same phase as the first pulse in the echo sequence. The phase shift of the echo arises because the radical pair is generated from a singlet state and the two unpaired electrons are spin correlated. The amplitude of this so-called out-of-phase echo is a maximum when it is generated using a 45°-τ-180° pulse sequence (Bittl & Zech, 2001) and it exhibits very deep amplitude modulation as the spacing between the two pulses is varied. Fig. 9 shows an echo modulation curve obtained from PS I particles from *C. reinhardtii*. The intensity of the echo is given by Salikhov, Kandrashkin, and Salikhov (1992) and Tan, Thurnauer, and Norris (1994)

$$I(\tau) = \int_0^{2\pi}\int_0^{\pi} \Delta\omega^2 (2J + d)^2 4\Omega^4 \cdot sin(\Gamma\tau) sin\theta d\theta d\phi \tag{4}$$

where τ is the spacing between pulses, $\Delta\omega$ is the difference between the resonance frequencies of the two radicals, J and d and the exchange and dipolar coupling between the unpaired electrons, respectively. $\Omega^2 = \Delta\omega^2 + (J + d/2)^2$ and $\Gamma = 2(J - d)$. The integration in Eq. (4) is over all orientations of the radical pair and θ is the angle between the vector connecting

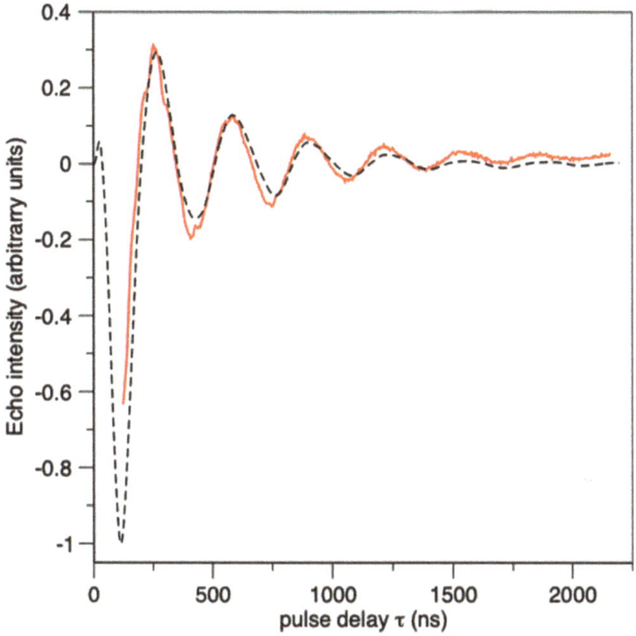

Fig. 9 X-band out-of-phase echo modulation curve of PS I particles from *Chlamydomonas reinhardtii* at 80 K. Red: experimental modulation curve. Black: theoretical echo modulation curve calculated as described in Mula et al. (2012) with D = −170 μT and J = 1 μT. EPR conditions: microwave frequency 9.777 GHz, magnetic field 348.5 mT, delay after flash 200 ns, π/2 pulse length 8 ns.

the two spins and the magnetic field and it is assumed that the microwave pulses are not orientation selective. This assumption is valid at X-band with the power level set to give a 90° pulse length of 8 ns. At higher field/frequency, the pulses are orientation selective, and this selectivity must be taken into account when calculating the echo intensity. The dashed curve in Fig. 9 is a fit of Eq. (4) to the experimental curve. The dipolar part of the spin-spin coupling $d = D(3cos^2\theta - 1)/3$ depends on the distance between the two spins and using the point dipole approximation $D = -2786\, r^{-3}$ mT $\mathring{A}^3$. Thus, the distance between P^+ and A_1^- can be obtained from the fit of the calculated modulation curve to the experimental data.

Equipment

(1) X-band pulsed EPR spectrometer: Bruker Elexsys E580

(2) Dielectric ring or split ring resonator with optical access: Bruker ER 4118X-MD5, Bruker ER 4118X-MS5 or ER5106QTW

(3) Quartz 3 or 4 mm EPR tubes

(4) He gas flow cryostat and controller: Oxford Instruments CF935
(5) Pulsed, frequency doubled Nd-YAG laser (pulse width < 10 ns, repetition rate 10 Hz or less)
(6) Photodiode or laser trigger

Protocol and Experimental Conditions: Sample preparation is the same as in Section 3.3.1.

3.3.2 Triplet states

Light-induced triplet states in Type I RCs can also be studied using TREPR. If energy transfer to the primary donor is inefficient, spin-orbit coupling mediated intersystem crossing can occur generating Chl triplet states in the antenna. Similarly, if electron transfer beyond the primary acceptor becomes slow, charge recombination to ^3P can occur. The TREPR spectra of chlorophyll triplet states are approximately 60 mT wide and are thus much broader than the spectra of radical pair states, which are ~2 mT wide. An advantage of TREPR methods over other spectroscopic techniques for studying triplet states is that the spin polarization patterns depend on the process by which the triplet state was generated. This means that the charge recombination triplet states can be easily distinguished from the antenna triplet states formed by intersystem crossing. Fig. 10 shows characteristic spin polarized transient EPR spectra of triplet states in PS I complexes. The top spectrum is from a PS I sample from *C. reinhardtii* containing a significant amount of Chl that is unable to transfer its excitation energy to P_{700}. The broad wings of the spectrum are from the triplet states of these Chl molecules that are formed by intersystem crossing, while the sharp peaks at the center of the spectrum are from $P_{700}^+ A_1^-$. The bottom part of Fig. 10 shows the spectrum of PS I complexes from spinach in which phylloquinone has been removed by lyophilization and ether extraction. This treatment eliminates the signal contribution from $P_{700}^+ A_1^-$ and results in a triplet state spectrum with the polarization pattern characteristic of selective population of the T_0 sublevel due to S-T_0 mixing in $P_{700}^+ A_0^-$. The absence of these spectra can be used as an indicator for the purity and integrity of a RC preparation. In heliobacteria the energy transfer is less efficient and antenna triplet states are generated at low temperature even in samples of good quality. Moreover, the zero-field-splitting parameters of the triplet state of the BChl *g* found in heliobacteria are such that when the triplet state is formed by charge recombination, the absorptive and emissive contributions cancel each other, and no spectrum is observed (Ferlez, Agostini, Carbonera, Golbeck, & Van Der Est, 2017).

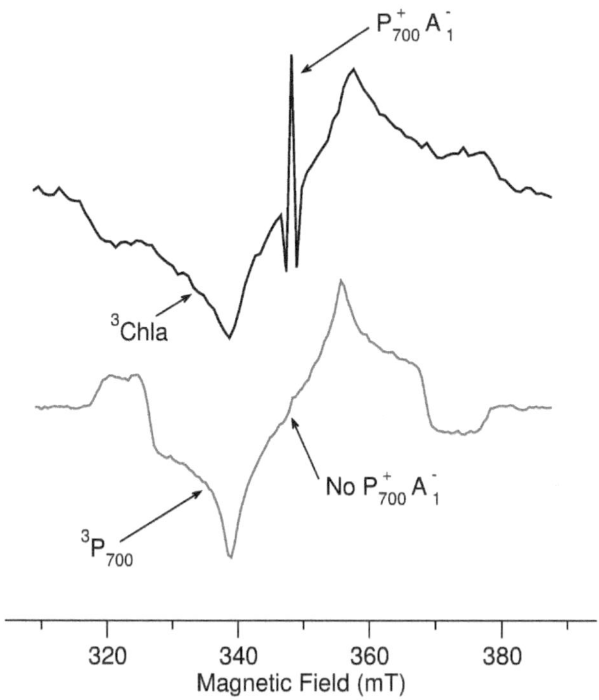

Fig. 10 X-band transient EPR spectra of triplet states in PS I at 80 K. Top: PS I particles from *Chlamydomonas reinhardtii* containing a significant amount of antenna complex fragments. Spectrum at 1.0 μs after the laser flash. EPR conditions: microwave power 0.2 mW, microwave frequency 9.77 GHz. Bottom: PS I particles from spinach following lyophilization and ether extraction to remove phylloquinone. Spectrum at 1.1 μs after the laser flash. EPR conditions: microwave power 1 mW, microwave frequency: 9.75 GHz.

Equipment Protocol and Experimental Conditions

The equipment and sample preparation are the same as in Section 3.3.1.

4. Summary

In this article, we have outlined how continuous wave EPR, time resolved EPR, and pulsed EPR can be used to measure field-swept spectra as well as spatial and time-dependent properties of cofactor-bound radicals generated by light in heterodimeric and homodimeric Type I RCs. A comprehensive research program will supplement these techniques with optical absorbance, fluorescence, Mössbauer, magnetic circular dichroism, infrared, resonance Raman EXAFS, and NMR studies to obtain a complete understanding of the photobiological reaction under study.

Acknowledgments

The work was supported by a grant from the US National Science Foundation (MCB-1613022) to J.H.G., by a grant from the Photosynthetic Systems Program, Chemical Sciences, Geosciences, & Biosciences (CSGB) Division, US Department of Energy (DESC0010575) to J.H.G. and by the NSERC in the form of a Discovery Grant to A.v.d.E.

References

Bibby, T. S., Nield, J., & Barber, J. (2001). Iron deficiency induces the formation of an antenna ring around trimeric photosystem I in cyanobacteria. *Nature*, *412*(6848), 743–745.

Biggins, J., & Mathis, P. (1988). Functional role of vitamin K_1 in photosystem I of the cyanobacterium *Synechocystis* 6803. *Biochemistry*, *27*(5), 1494–1500.

Bittl, R., & Zech, S. (2001). Pulsed EPR spectroscopy on short-lived intermediates in photosystem I. *Biochimica et Biophysica Acta, Bioenergetics*, *1507*(1–3), 194–211.

Bordier, P., Svendsen, I., Scheller, H., & Møller, B. (1987). Identification of a chloroplast-encoded 9-kDa polypeptide as a 2[4Fe-4S] protein carrying centers a and B of photosystem I. *The Journal of Biological Chemistry*, *262*(26), 12676–12684.

Chamorovsky, S. K., & Cammack, R. (1982). Direct determination of the midpoint potential of the acceptor X in chloroplast photosystem I by electrochemical reduction and ESR spectroscopy. *Photobiochemistry and Photobiophysics*, *4*(4), 195–200.

Charles, P., Kalendra, V., He, Z., Khatami, M. H., Golbeck, J. H., van der Est, A., et al. (2020). Two-dimensional ^{67}Zn HYSCORE spectroscopy reveals that a Zn-bacteriochlorophyll aP' dimer is the primary donor (P840) in the type-1 reaction centers of *Chloracidobacterium thermophilum*. *Physical Chemistry Chemical Physics*, *22*(11), 6457–6467.

Chateauneuf, J., Lusztyk, J., & Ingold, K. U. (1990). Photoinduced electron transfer from dialkyl nitroxides to halogenated solvents. *The Journal of Organic Chemistry*, *55*, 1061–1065.

Chen, J. H., Wu, H., Xu, C., Liu, X. C., Huang, Z., Chang, S., et al. (2020). Architecture of the photosynthetic complex from a green sulfur bacterium. *Science*, *370*(6519).

Commoner, B., Heise, J. J., & Townsend, J. (1956). Light-induced paramagnetism in chloroplasts. *Proceedings of the National Academy of Sciences of the United States of America*, *42*(10), 710.

Dashdorj, N., Xu, W., Cohen, R. O., Golbeck, J. H., & Savikhin, S. (2005). Asymmetric electron transfer in cyanobacterial photosystem I: Charge separation and secondary electron transfer dynamics of mutations near the primary electron acceptor A_0. *Biophysical Journal*, *88*(2), 1238–1249.

Duderstadt, R. E., Brereton, P. S., Adams, M. W., & Johnson, M. K. (1999). A pure $S = 3/2$ $[Fe_4S_4]^+$ cluster in the A33Y variant of *Pyrococcus furiosus* ferredoxin. *FEBS Letters*, *454*(1–2), 21–26.

Ferlez, B., Agostini, A., Carbonera, D., Golbeck, J. H., & Van Der Est, A. (2017). Triplet charge recombination in heliobacterial reaction centers does not produce a spin-polarized EPR spectrum. *Zeitschrift für Physikalische Chemie*, *231*(3), 593–607.

Ferlez, B., Cowgill, J., Dong, W., Gisriel, C., Lin, S., Flores, M., et al. (2016). Thermodynamics of the electron acceptors in *Heliobacterium modesticaldum*: An exemplar of an early homodimeric type I photosynthetic reaction center. *Biochemistry*, *55*(16), 2358–2370.

Gisriel, C., Sarrou, I., Ferlez, B., Golbeck, J. H., Redding, K. E., & Fromme, R. (2017). Structure of a symmetric photosynthetic reaction center-photosystem. *Science*, *357*(6355), 1021–1025.

Golbeck, J. (2006). *Photosystem I: The light-driven plastocyanin:ferredoxin oxidoreductase. Vol. 24.* Dordrecht: Springer.

Golbeck, J. H., Parrett, K. G., Mehari, T., Jones, K. L., & Brand, J. J. (1988). Isolation of the intact photosystem I reaction center core containing P700 and iron-sulfur center F_X. *FEBS Letters, 228*(2), 268–272.

Golbeck, J., & van der Est, A. (2013). Electron paramagnetic resonance spectroscopy. In N. Allewell, L. O. Narhi, & I. Rayment (Eds.), *Molecular biophysics for the life sciences* (pp. 175–214). New York: Springer.

Goldfarb, D., & Stoll, S. (2018). *EPR spectroscopy: Fundamentals and methods.* John Wiley and Sons.

Gorka, M., Charles, P., Kalendra, V., Baldansuren, A., Lakshmi, K. V., & Golbeck, J. H. (2021). A dimeric chlorophyll electron acceptor differentiates type I from type II photosynthetic reaction centers. *iScience, 24*(7), 102719. https://doi.org/10.1016/j.isci.2021.102719.

Gorka, M., Gruszecki, E., Charles, P., Kalendra, V., Lakshmi, K. V., & Golbeck, J. H. (2021). Two-dimensional HYSCORE spectroscopy reveals a histidine imidazole as the axial ligand to Chl_{3A} in the $M688H_{PsaA}$ genetic variant of photosystem I. *Biochimica et Biophysica Acta, Bioenergetics, 1862*(7), 148424.

Guergova-Kuras, M., Boudreaux, B., Joliot, A., Joliot, P., & Redding, K. (2001). Evidence for two active branches for electron transfer in photosystem I. *Proceedings of the National Academy of Sciences of the United States of America, 98*(8), 4437–4442.

He, Z., Ferlez, B., Kurashov, V., Tank, M., Golbeck, J. H., & Bryant, D. A. (2019). Reaction centers of the thermophilic microaerophile, *Chloracidobacterium thermophilum* (Acidobacteria) I: Biochemical and biophysical characterization. *Photosynthesis Research, 142*(1), 87–103.

Heinnickel, M., Agalarov, R., Svensen, N., Krebs, C., & Golbeck, J. H. (2006). Identification of F_X in the heliobacterial reaction center as a [4Fe-4S] cluster with an $S = 3/2$ ground spin state. *Biochemistry, 45*(21), 6756–6764.

Jagannathan, B., & Golbeck, J. H. (2008). Unifying principles in homodimeric type I photosynthetic reaction centers: Properties of PscB and the F_A, F_B and F_X iron-sulfur clusters in green sulfur bacteria. *Biochimica et Biophysica Acta, Bioenergetics, 1777*(12), 1535–1544.

Jagannathan, B., Shen, G. Z., & Golbeck, J. H. (2012). The evolution of type I reaction centers: The response to oxygenic photosynthesis. In *Vol. 33. Functional genomics and evolution of photosynthetic systems* (pp. 285–316). Springer.

Jordan, P., Fromme, P., Witt, H. T., Klukas, O., Saenger, W., & Krauß, N. (2001). Three dimensional structure of photosystem I at 2.5 Å resolution. *Nature, 411*, 909–917.

Klughammer, C., & Pace, R. J. (1997). Photoreduction of the secondary photosystem I acceptor vitamin K_1 in intact spinach chloroplasts and cyanobacteria in vivo. *Biochimica et Biophysica Acta, Bioenergetics, 1318*, 133–144.

Kok, B. (1956). On the reversible absorption change at 705 mu in photosynthetic organisms. *Biochimica et Biophysica Acta, Bioenergetics, 22*(2), 399–401.

Malkin, R., & Bearden, A. J. (1971). Primary reactions of photosynthesis: Photoreduction of a bound chloroplast ferredoxin at low temperature as detected by EPR spectroscopy. *Proceedings of the National Academy of Sciences of the United States of America, 68*(1), 16–19 (Retrieved from).

Mathis, P., Ikegami, I., & Setif, P. (1988). Nanosecond flash studies of the absorption spectrum of the photosystem I primary acceptor A_0. *Photosynthesis Research, 16*(3), 203–210.

Mula, S., McConnell, M. D., Ching, A., Zhao, N., Gordon, H. L., Hastings, G., et al. (2012). Introduction of a hydrogen bond between phylloquinone PhQA and a threonine side-chain OH group in photosystem I. *The Journal of Physical Chemistry. B, 116*(48), 14008–14016.

Nitschke, W., Feiler, U., & Rutherford, A. W. (1990). Photosynthetic reaction center of green sulfur bacteria studied by EPR. *Biochemistry, 29*(16), 3834–3842.

Nitschke, W., Setif, P., Liebl, U., Feiler, U., & Rutherford, A. W. (1990). Reaction center photochemistry of *Heliobacterium chlorum. Biochemistry, 29*(50), 11079–11088.

Norris, R. J., Uphaus, R. A., Crespi, H. L., & Katz, J. J. (1971). Electron spin resonance of chlorophyll and the origin of signal I in photosynthesis. *Proceedings of the National Academy of Sciences of the United States of America, 68*, 625–628.

Romberger, S. P., & Golbeck, J. H. (2010). The bound iron-sulfur clusters of type-I homodimeric reaction centers. *Photosynthesis Research, 104*(2–3), 333–346. https://doi.org/10.1007/s11120-010-9543-y.

Romberger, S. P., & Golbeck, J. H. (2012). The F_X iron-sulfur cluster serves as the terminal bound electron acceptor in heliobacterial reaction centers. *Photosynthesis Research, 111*(3), 285–290.

Salikhov, K., Kandrashkin, Y., & Salikhov, A. (1992). Peculiarities of free induction and primary spin echo signals for spin-correlated radical pairs. *Applied Magnetic Resonance, 3*(1), 199–216.

Sauer, K., Acker, S., Mathis, P., & Van Best, J. A. (1977). Optical studies of photosystem I particles: Evidence for the presence of multiple electron acceptors. *Developments in Bioenergetics and Biomembranes, 1*, 351–359.

Shuvalov, V. A., Nuijs, A. M., Van Gorkom, H. J., Smit, H. W. J., & Duysens, L. N. M. (1986). Picosecond absorbance changes upon selective excitation of the primary electron donor P-700 in photosystem I. *Biochimica et Biophysica Acta, Bioenergetics, 850*(2), 319–323.

Sun, J., & Golbeck, J. H. (2015). The presence of the IsiA-PSI supercomplex leads to enhanced photosystem I electron throughput in iron-starved cells of *Synechococcus sp.* PCC 7002. *The Journal of Physical Chemistry. B, 119*(43), 13549–13559.

Tan, J., Thurnauer, M. C., & Norris, J. R. (1994). Electron spin echo envelope modulation due to exchange and dipolar interactions in a spin-correlated radical pair. *Chemical Physics Letters, 219*(3–4), 283–290.

van der Est, A. (2001). Light-induced spin polarization in type I photosynthetic reaction centres. *Biochimica et Biophysica Acta, Bioenergetics, 1507*(1–3), 212–225.

van der Est, A., Valieva, A. I., Kandrashkin, Y. E., Shen, G., Bryant, D. A., & Golbeck, J. H. (2004). Removal of PsaF alters forward electron transfer in photosystem I: Evidence for fast reoxidation of Q_K-A in subunit deletion mutants of *Synechococcus sp.* PCC 7002. *Biochemistry, 43*(5), 1264–1275.

Vassiliev, I. R., Antonkine, M. L., & Golbeck, J. H. (2001). Iron-sulfur clusters in type I reaction centers. *Biochimica et Biophysica Acta, Bioenergetics, 1507*(1–3), 139–160.

Vassiliev, I. R., Ronan, M. T., Hauska, G., & Golbeck, J. H. (2000). The bound electron acceptors in green sulfur bacteria: Resolution of the *g*-tensor for the F_X iron-sulfur cluster in *Chlorobium tepidum. Biophysical Journal, 78*(6), 3160–3169.

Weber, S. (2017). Transient EPR. In *Vol. 6. eMagRes books* (pp. 255–270). John Wiley & Sons.

Yang, F., Shen, G., Schluchter, W. M., Zybailov, B., Ganago, A. O., Vassiliev, I. R., et al. (1998). Deletion of the PsaF polypeptide modifies the environment of the redox-active phylloquinone. Evidence for unidirectionality of electron transfer in Photosytem I. *The Journal of Physical Chemistry. B, 102*, 8288–8299.

Yu, J. P., Smart, L. B., Jung, Y. S., Golbeck, J., & McIntosh, L. (1995). Absence of PsaC subunit allows assembly of photosystem I core but prevents the binding of PsaD and PsaE in *Synechocystis* sp PCC 6803. *Plant Molecular Biology, 29*(2), 331–342.

CHAPTER THIRTEEN

Experimental guidelines for trapping paramagnetic reaction intermediates in radical *S*-adenosylmethionine enzymes

Aidin R. Balo and R. David Britt*

Department of Chemistry, University of California, Davis, CA, United States
*Corresponding author: e-mail address: rdbritt@ucdavis.edu

Contents

Abstract

Due to their biological importance and functional diversity, radical *S*-adenosylmethionine (rSAM) enzymes have become popular targets for electron paramagnetic resonance (EPR) spectroscopic studies. EPR spectroscopy is a powerful tool that allows for the observation of the iron–sulfur clusters as well as paramagnetic reaction intermediates, thus providing insight into their catalytic mechanisms. While the iron–sulfur clusters may be readily observable by EPR spectroscopy in the enzymes' resting states, radical intermediates are often elusive and must be trapped. Here, we describe a protocol for trapping and analyzing the Lys-Trp$^•$ intermediate of the Lys-Trp-crosslinking rSAM enzyme SuiB, including modified expression and purification steps. This protocol is also intended to serve as a primer for trapping paramagnetic intermediates in other rSAM enzymes for studying by EPR spectroscopy.

Methods in Enzymology, Volume 666
ISSN 0076-6879
https://doi.org/10.1016/bs.mie.2022.02.015

1. Introduction

With over 570,000 known members spanning all domains of life, the radical *S*-adenosyl-L-methionine (rSAM) enzyme superfamily is the largest known, with most of its members still uncharacterized (Caruso, Martinie, Bushin, & Seyedsayamdost, 2019; Landgraf, McCarthy, & Booker, 2016; Seyedsayamdost, Caruso, & Davis, 2020; Sofia, Chen, Hetzler, Reyes-Spindola, & Miller, 2001). The common feature of rSAM enzymes is a conserved [4Fe-4S] cluster, called the rSAM (RS) cluster, usually bound by a CX_3CX_2C motif. The RS cluster catalyzes the reductive cleavage of SAM to form L-methionine and a strongly oxidizing 5′-deoxyadenosyl radical (5′-dA$^•$), which can rebind to the RS cluster to form an organometallic intermediate **Ω** (Broderick, Duffus, Duschene, & Shepard, 2014; Byer et al., 2018; Horitani et al., 2016; Sayler et al., 2019; Yang et al., 2019). 5′-dA$^•$ usually abstracts a hydrogen atom (H-atom) from substrate to initiate an extremely diverse set of chemical reactions, including the biosynthesis of DNA, cofactors, vitamins, and antibiotics (Berkovitch, Nicolet, Wan, Jarrett, & Drennan, 2004; Flühe et al., 2012; Grove et al., 2011; Klinman & Bonnot, 2014).

Of the 570,000 rSAM enzyme superfamily members known to date, ~165,000 genes (from the Enzyme Function Initiative—Enzyme Similarity Tool) possess C-terminal extensions, called SPASM and twitch domains, which bind auxiliary [4Fe-4S] clusters (Barr et al., 2018; Davis et al., 2017; Grell et al., 2018; Grell, Goldman, & Drennan, 2015; Grove et al., 2017; Haft & Basu, 2011; Seyedsayamdost et al., 2020). The SPASM domain typically binds two auxiliary [4Fe-4S] clusters (AuxI and AuxII) and is named after the rSAM enzymes involved in the synthesis of Subtilosin, Pyrroloquinoline quinone, Anaerobic Sulfatase, and Mycofactocin. The twitch domain is a truncated SPASM domain and only binds one auxiliary cluster (Aux) (Grell et al., 2015). Direct evidence for the catalytic function(s) of any SPASM/twitch domain auxiliary clusters has long remained elusive. While these clusters have been proposed to serve as electron acceptors (Grove, Ahlum, Sharma, Krebs, & Booker, 2010; Lanz & Booker, 2015), the AuxI cluster in the SPASM domain-containing rSAM enzyme of focus in this chapter, SuiB, was observed by electron paramagnetic resonance (EPR) spectroscopy to accept an electron directly from the radical intermediate (Balo et al., 2021).

EPR spectroscopy is a powerful tool for studying the catalytic mechanisms of rSAM enzymes, as it allows for direct observation of reaction intermediates that can sometimes be trapped. Reviews have been written about the studies of rSAM enzymes by continuous wave and pulse EPR spectroscopy, discussing the spectroscopic features of trapped paramagnetic intermediates and iron–sulfur clusters in various rSAM enzymes (Balo, Tao, & Britt, 2021; Broderick et al., 2014; Broderick, Hoffman, & Broderick, 2018; Stich, Myers, & Britt, 2014). While iron–sulfur clusters can be spectroscopically observed in most cases in rSAM enzymes, it is not always straightforward to trap radical intermediates and may require automated approaches such as rapid freeze–quench (RFQ) techniques for subsecond freezing (Burdi, Sturgeon, Tong, Stubbe, & Hoffman, 1996; Kim, Perera, Hager, Dawson, & Hoffman, 2006).

Streptide is a natural product synthesized in *Streptococcus suis* and its streptococcal relatives by a gene cluster encoding three proteins: a short precursor peptide SuiA, a rSAM enzyme SuiB and a protease/transporter SuiC (Fig. 1A) (Schramma, Bushin, & Seyedsayamdost, 2015; Schramma & Seyedsayamdost, 2017). SuiB is a SPASM domain-containing enzyme (Fig. 1B) that catalyzes the C—C bond crosslink formation between a Lys and a Trp residue on its precursor peptide SuiA (Fig. 1C). The crosslinked product is cleaved from its leader sequence and secreted by SuiC. Recently, we described a detailed EPR spectroscopic study of SuiB, with three of its paramagnetic intermediates characterized (Fig. 2): (i) the SAM-based organometallic intermediate **Ω**, (ii) the organic radical intermediate Lys-Trp$^{\bullet}$, and (iii) The AuxI cluster after accepting an electron from Lys-Trp$^{\bullet}$.

Here, we use SuiB as an example for trapping and characterizing a radical intermediate by EPR spectroscopy. The purpose of this chapter is to serve as a model for trapping paramagnetic intermediates in rSAM enzymes. In enzymes with multiple iron–sulfur clusters, such as twitch or SPASM domain-containing rSAM enzymes, extra precautions must be taken to ensure high-integrity incorporation of iron–sulfur clusters. In this chapter, we describe a step-by-step guide to express and purify SuiB, as well as trap its radical intermediate Lys-Trp$^{\bullet}$ and analyze Lys-Trp$^{\bullet}$ by EPR spectroscopy.

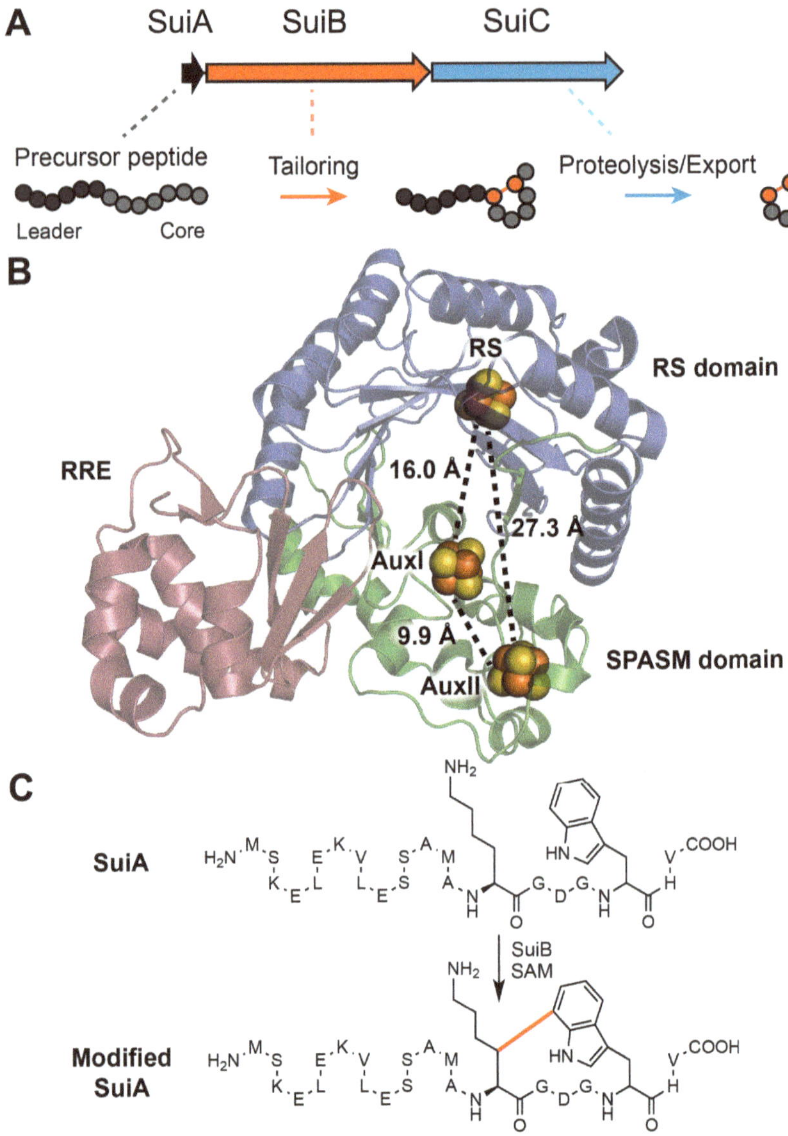

Fig. 1 Synthesis of streptide in *Streptococcus suis* (Schramma & Seyedsayamdost, 2017). (A) Streptide-forming gene cluster composed of a short precursor peptide (SuiA, black), an rSAM enzyme that installs the C—C bond between Lys2 and Trp6 (SuiB, red) and an ABC transporter that cleaves the N-terminal leader sequence and secretes the maturated streptide (SuiC, blue). (B) X-ray crystal structure of SuiB (PDB ID: 5V1T) (Davis et al., 2017). The RS domain, SPASM domain, and RiPP recognition element (RRE) are rendered blue, green, and pink, respectively. (C) Lys-Trp crosslink formation catalyzed by SuiB. The C—C bond installed by SuiB is shown in red.

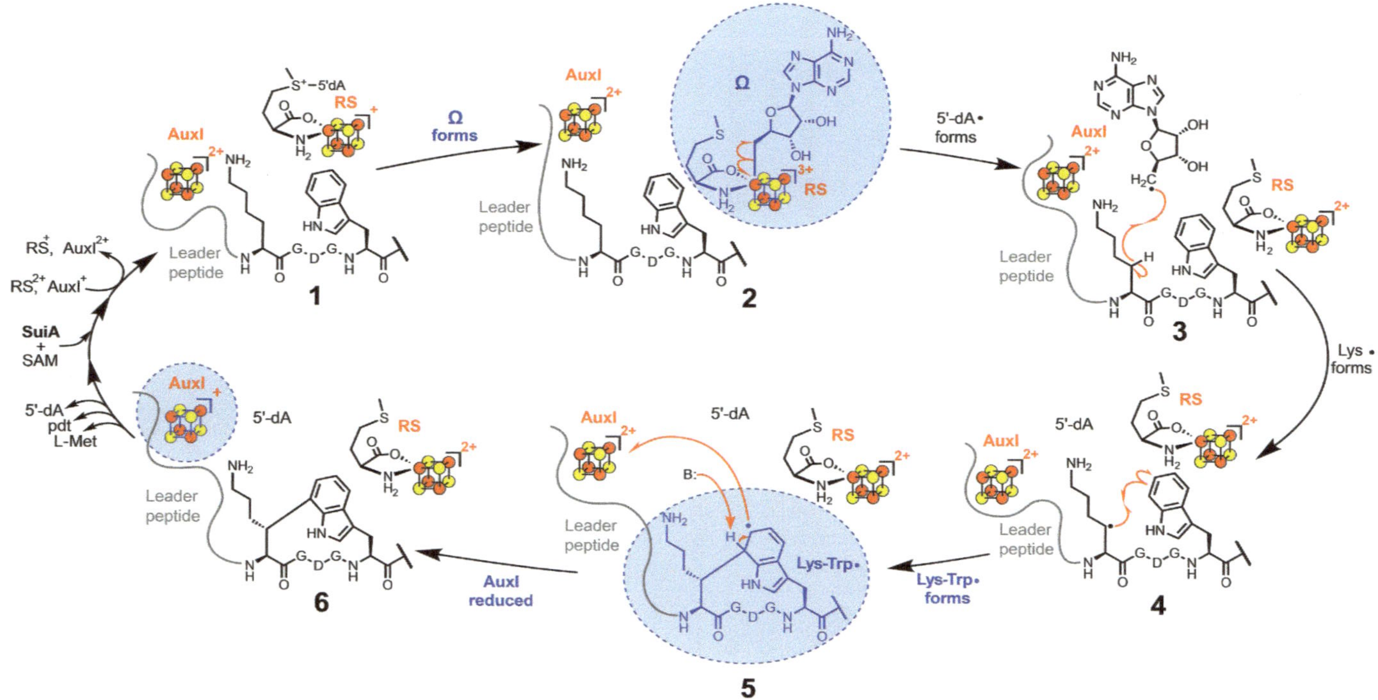

Fig. 2 Current rEAS mechanism for the SuiB-catalyzed carbon–carbon bond formation between the Lys2 and Trp6 sidechains (Balo et al., 2021). The spectroscopically observed paramagnetic intermediates, Ω, Lys-Trp$^•$, and AuxI, are highlighted in blue. In this previously published model, the catalytic cycle begins with binding of SuiA to SuiB, which results in a decrease of reduced AuxI signal and an increase of reduced RS signal, thus readying AuxI as an oxidant during the reaction and RS as a reductant to initiate SAM cleavage (state 1). The organometallic Ω is an observable intermediate (state 2) followed by the Lys-Trp$^•$ via two transient radical intermediates: 5′-dA$^•$ (state 3) and Lys$^•$ (state 4). The Lys$^•$ then reacts with Trp6 to generate the Lys-Trp$^•$, which under our conditions is stable in the protonated form (state 5). H7 is likely the gatekeeper of the ensuing reduction reaction, as the deprotonated radical anion Lys-Trp$^•$—is highly reducing and can reduce the low-potential AuxI cluster to generate rearomatized product and reduced AuxI (state 6). Release of product and binding of substrate would reset the redox states of the RS and AuxI clusters (state 1), thereby permitting multiple turnovers.

2. Overview of changes to original SuiB expression and purification protocol

A detailed protocol for expressing and purifying SuiB capable of catalysis and suitable for structural determination by X-ray crystallography has been published by the Seyedsayamdost lab (Bushin & Seyedsayamdost, 2018). While we were able to produce functional enzyme through this protocol, we found it challenging to trap paramagnetic intermediates in significant quantities. Trapping quantities of elusive radical intermediates large enough to be detectable by EPR spectroscopy is facilitated by not only a high concentration of protein, but also high-quality iron–sulfur cluster incorporation that is sometimes difficult to achieve through conventional biochemical methods. Thus, we worked to optimize this protocol for EPR spectroscopic studies (Balo et al., 2021). The principle was to maximize the [4Fe—4S] incorporation by *E. coli*, while minimizing the damage done to them by using gentler purification methods, such as the more metalloprotein-compatible Strep tag over the more standard His_6 tag, which has been successfully applied on other iron–sulfur proteins (Kuchenreuther, George, Grady-Smith, Cramer, & Swartz, 2011; Tao, Zhu, Klinman, & Britt, 2019). We highlight the following differences between the original SuiB preparation protocol and the modified one presented here (Fig. 3). The protocol described in this chapter:

- Includes higher concentrations of $(NH_4)_2Fe(SO_4)_2$ and L-cysteine in the *E. coli* expression medium, which are sources of bioavailable Fe and S for iron–sulfur cluster biosynthesis, respectively
- Does not include components in the buffer that have not been shown to be necessary for stabilizing this system, such as glycerol or reducing agents during the purification
- Utilizes the chemical protein extraction reagent BugBuster instead of sonication, the former of which is presumed to be gentler on the [4Fe—4S] clusters
- Avoids imidazole altogether, and instead utilizes the much gentler Strep tag purification method over the more standard His_6 tag method
- Skips the chemical iron–sulfur cluster reconstitution step, as this was found to be unnecessary when using the Strep tag purification approach
- Avoids buffer exchanges (e.g., to remove imidazole or chemical iron–sulfur cluster reconstitution reagents)

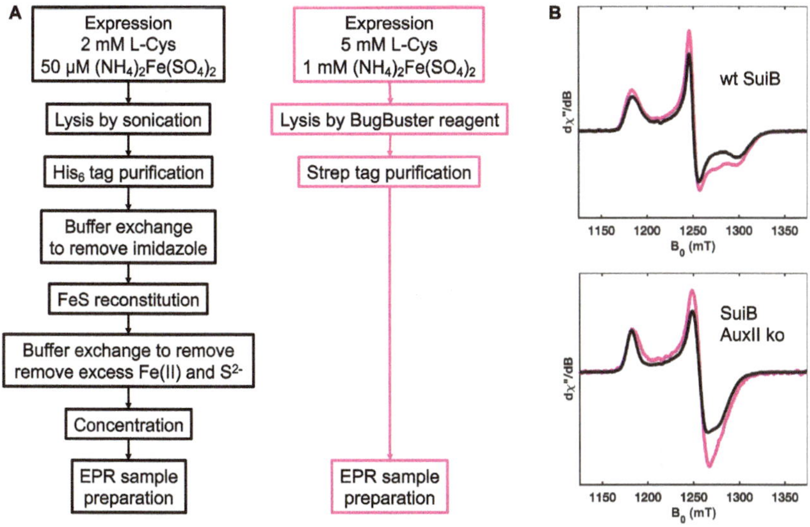

Fig. 3 (A) Outline of previous published SuiB preparation protocol (black) (Bushin & Seyedsayamdost, 2018) and the protocol described in this chapter (magenta). (B) Comparison of EPR spectra collected at Q-band of wt SuiB and the SuiB AuxII knock-out mutant reveals an increase in $[4Fe–4S]^+$ EPR signals when using the protocol reported here (Balo et al., 2021).

- Avoids concentration steps by centrifugation, as the protein eluted from the Strep-Tactin resin is concentrated enough for high-quality EPR sample preparation and
- Executes the entire purification protocol over a few hours to minimize cluster degradation, such that EPR samples are always made with freshly prepared protein

3. Protocol

3.1 Substitution of a His_6 tag with a Strep tag at the genetic level in the SuiB-encoding pET-28a plasmid

Cloning a synthesized gene encoding a Strep-tagged rSAM enzyme into an expression vector is a way to acquire a plasmid for expressing a Strep-tagged rSAM enzyme. However, when a plasmid encoding a His_6-tagged version of the rSAM enzyme is available, it is usually more cost effective and faster to replace the nucleotides encoding the His_6 tag (HHHHHH) for the Strep tag residues (WSHPQFEK) by the polymerase chain reaction (PCR).

```
                                  M   G   S   S   W   S   H
5' ...CT TTA AGA AGG AGA TAT ACC ATG GGC AGC AGC TGG TCT CAT
3' ...GA AAT TCT TCC TCT ATA TGG TAC CCG TCG TCG ACC AGA GTA

P   Q   F   E   K   S   S   G   L   V   P   R
CCG CAG TTC GAA AAA AGC AGC GGC CTG GTG CCG CGC GG... 3'
GGC GTC AAG GTT TTT TCG TCG CCG GAC CAC GGC GCG CC... 5'
```

Fig. 4 Partially overlapping primers used to substitute the nucleotides encoding the His_6 tag for the Strep tag by PCR. The long primer is shown in red and the short primer in blue. The *Nco*I restriction site used to clone the original synthesized SuiB gene encoding an N-terminal His_6 into a pET-28a vector is underlined.

We use the Agilent QuikChange Lightning Site-Directed Mutagenesis Kit along with its associated protocol (https://www.agilent.com/cs/library/usermanuals/Public/210518.pdf). However, we use a particular primer design to encompass this large modification (Fig. 4). In the case of N-terminally His_6-tagged SuiB cloned into a pET-28a vector (Bushin & Seyedsayamdost, 2018; Davis et al., 2017; Schramma & Seyedsayamdost, 2017), we use a partially overlapping pair of primers consisting of a long primer (5′-GG AGA TAT ACC ATG GGC AGC AGC TGG TCT CAT CCG CAG TTC GAA AAA AGC AGC GGC CTG GTG CCG CGC GG-3′) and a short primer (5′-GGT ATA TCT CCT TCT TAA AG-3′). It is better that the primers overlap slightly on their 5′ ends rather than the 3′ ends so that the PCR extension cannot occur between them. Due to the necessary length of the long primer, it is better that the overlapping region with the short primer not be too long. We use 11 nucleotides of overlap in this case. If PCR reagents from a different source are used, follow the optimal site-directed mutagenesis protocol for those reagents, but with this primer design.

3.2 Expression of Strep-SuiB

Once the plasmid encoding the Strep-tagged rSAM enzyme is acquired, we follow the following steps to express N-terminally Strep-tagged SuiB.

1. Co-transform the pET-28a plasmid bearing the Strep-SuiB gene and the pDB1282 plasmid encoding the six genes (iscS, iscU, iscA, hscA, hscB, and fdx) from the *Azotobacter vinelandii* operon (Lanz et al., 2012) into 50 μL of chemically component *E. coli* BL21(DE3) through standard heat shock
2. Add 200 μL of SOC Outgrowth Medium (New England BioLabs) and shake at 220 rpm for 1 h at 37 °C

3. Spread the mixture on an LB Miller agar plate containing 100 μg/mL of ampicillin and 50 μg/mL of kanamycin and incubate overnight at 37 °C
4. On the following day, use a single colony from the agar plate to inoculate 100 mL of LB Miller containing 100 μg/mL of ampicillin and 50 μg/mL of kanamycin in a small (250 mL) Erlenmeyer flask and shake at 180 rpm overnight at 37 °C. If no colonies appear on the agar plate, transform by electroporation
5. On the following day, use 15 mL of the pre-culture to inoculate 1.5 L of LB Miller containing 100 μg/mL of ampicillin, 50 μg/mL of kanamycin, and 1 mM $(NH4)_2Fe(SO4)_2$ in a large (2.8 L) Erlenmeyer flask. For large-scale protein preparations, grow six 1.5 L cultures for a total of 9 L
6. Shake the culture(s) at 200 rpm at 37 °C. Once $OD_{600} \sim 0.3$, add 5 mM L-cysteine and 0.2% L-arabinose
7. Continue to shake the culture(s) at 200 rpm at 37 °C. Once $OD_{600} \sim 0.6$, lower the shaking speed to 90 rpm and the temperature to 18 °C for 1 h, after which add 250 μM of IPTG and leave incubated at 90 rpm and 18 °C overnight
8. On the following day, harvest cells by centrifugation at 6000 rpm for 15 min. Transfer cell pellets to a sterile conical vial. Keep vials on ice, and minimize the time the cell pellets are exposed to open atmosphere
9. Flash freeze cell pellets in liquid nitrogen. If cell pellets are to be used within 3 weeks, store in −80 °C freezer. For long-term storage, store in liquid nitrogen

3.3 Purification of Strep-SuiB

To ensure high and consistent integrity of iron–sulfur clusters in the enzyme, the following protocol should be executed on a single day in an anaerobic chamber with pre-deoxygenated solutions. For best results, it is important to allocate enough time to perform both the purification and the EPR sample preparation consecutively on the same day with minimal gaps in the preparation.

1. Bring a frozen ~4 g cell pellet (from 1.5 L of cell culture) into an anaerobic chamber and resuspend into 40 mL of buffer A (300 mL KCl, 50 mM HEPES pH 7.5) supplemented with 15 μL of Benzonase Nuclease (Millipore), 1 μL of rLysozyme (Millipore) and 1 complete Protein Inhibitor Cocktail tablet (Millipore).

2. Add 5 mL of BugBuster 10X Protein Extraction Reagent (Millipore) to the mixture to induce lysis, and let stir for at room temperature for 1 h
3. Transfer lysate into ultracentrifuge tubes and sealed prior to bringing them out of the anaerobic chamber. Ultracentrifuge the samples at 35,000 *g* for 1 h at 4 °C
4. Transfer sample back into the anaerobic chamber without agitating and load the supernatant onto 4 mL of Strep-Tactin XT resin (IBA Lifesciences) pre-equilibrated with at least 3 column volumes of buffer A. If another streptavidin-based purification resin is used, it is important to note whether to elute with D-desthiobiotin or D-biotin for that particular resin
5. Wash the resin with 10 column volumes of buffer A. The resin should retain the dark-brown color after washing
6. Elute the protein by slow and gentle addition of 3 column volumes of buffer A + 5 mM D-biotin. Be careful not to resuspend significant amounts of resin. The protein should travel down the resin as a single, dark band with the D-biotin-containing buffer. Collect the darkest fractions of the eluant separately, typically consisting of ~1 mL of ~250–300 μM purified SuiB. In our experience, the highest integrity of iron–sulfur clusters was obtained when subsequent concentration steps via centrifugation were avoided

3.4 Synthesis of substrate peptide SuiA

The precursor peptide SuiA along with its isotopologue variants were synthesized as described by Seyedsayamdost and coworkers (Bushin & Seyedsayamdost, 2018; Schramma & Seyedsayamdost, 2017). The protocol used for the synthesis of protected isotopologue amino acids is also described.

3.5 Freeze trapping the Lys-Trp$^{\bullet}$ radical intermediate

EPR spectroscopy is the ideal technique for observing paramagnetic intermediates produced during enzymatic catalysis. Enzymatic reactions can be initiated and trapped in EPR sample tubes at various time points by plunging into liquid nitrogen by hand for slower enzymatic reactions, such as the one performed by SuiB ($k_{cat} = 0.18 \pm 0.05\ \mathrm{min}^{-1}$) (Schramma, Forneris, Caruso, & Seyedsayamdost, 2018). Alternatively, automated rapid freeze–quench (RFQ) strategies can achieve freeze trapping at sub-second time points. At some time points, new EPR signals in addition to the $[4Fe—4S]^{+}$ signals can be observed in trapped SuiB reaction samples.

1. Our EPR sample holder is equipped with an adapter which can support relatively short EPR tubes, and we use standard, disposable, 23 cm glass pipettes to rapidly deliver liquids directly to the bottom of the tubes. To ensure that these pipettes can reach the bottom of the EPR sample tube, we cut 25 cm quartz X-band EPR tubes (Wilmad-LabGlass; 4.0 mm outer diameter, 3.0 mm inner diameter) into thirds (~8.3 cm), and seal one end of the open tubes with an oxygen–natural gas torch. This step is recommended to maximize the amount of sample that is delivered to the bottom of the EPR tube and minimize the time to deliver sample, but ultimately not essential
2. Prepare fresh stock solutions of $Na_2S_2O_4$ (~250 μL of 1 M dissolved in water), SAM (~500 μL of 105 mM, dissolved and diluted in a 0.5 M KOH solution until pH ~7) and SuiA, the substrate peptide SuiA with sequence MSKELEKVLESSAMAKGDGWHV (~50 μL of 20 mM dissolved in buffer A) in an anaerobic chamber
3. Add 100 μL of ~250–300 μM freshly purified SuiB to an Eppendorf tube on ice. At the same time, pre-cool an EPR tube. We recommend placing the EPR tube inside of an open Eppendorf tube resting on ice to minimize water buildup on the outside of the EPR tube
4. Add 1 μL of the freshly prepared 1 M $Na_2S_2O_4$ stock solution to the Eppendorf tube containing SuiB (~10 mM final $Na_2S_2O_4$ concentration). Mix thoroughly and rapidly by stirring with the pipette tip
5. Add 10 μL of the 20 mM SuiA stock solution to the SuiB/$Na_2S_2O_4$ mixture (~2 mM final SuiA concentration). Mix thoroughly and rapidly by stirring with the pipette tip. Incubate the mixture on ice for at least 1 additional minute. During this time, transfer a small sample dewar filled with liquid nitrogen into the anaerobic chamber
6. To initiate the reaction, add 4 μL of the 105 mM SAM stock solution to the SuiB/$Na_2S_2O_4$/SuiA mixture (4 mM final SAM concentration). Mix thoroughly and rapidly by stirring with the pipette tip
7. Quickly transfer the sample to the bottom of a pre-cooled EPR tube with a standard, disposable glass 23 cm pipette. Plunge the sample into liquid nitrogen 20 s after the time that SAM is added. This was found to be the optimal timepoint for observing Lys-Trp$^{\bullet}$ (Fig. 5) (Balo et al., 2021).

3.6 EPR spectroscopy of SuiB reaction mixtures

As previously described (Balo et al., 2021), continuous-wave EPR experiments were performed at X-band (9.37 GHz) using a Bruker (Billerica, MA)

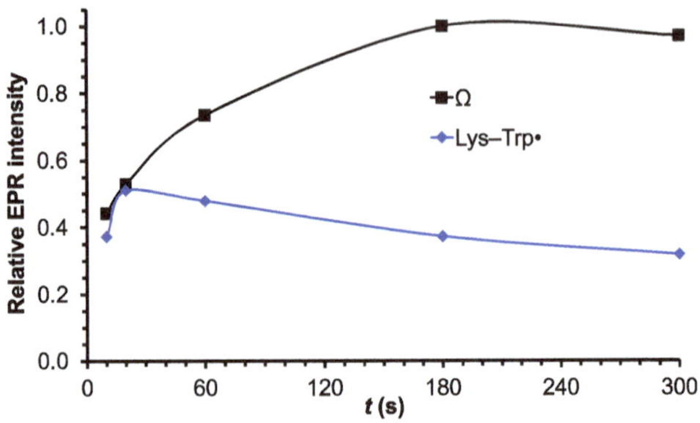

Fig. 5 Relative EPR signal intensities of paramagnetic intermediates Ω and Lys-Trp$^{\bullet}$ measured on samples frozen at various time points after initiation of reaction by SAM (*t*). While the Ω signal appears to increase as *t* approaches steady state, Lys-Trp$^{\bullet}$ reaches its maximum at *t*=20 s.

EleXsys E500 spectrometer equipped with a superhigh Q resonator (ER4122SHQE) with a conversion time of 60 ms and modulation frequency of 100 kHz. Cryogenic temperatures were achieved and controlled using an ESR900 liquid helium cryostat in conjunction with a temperature controller (Oxford Instruments ITC503) and a gas flow controller.

The EPR spectra of $[4Fe–4S]^{+}$ signals can be observed at cold temperatures (~10 K) using relatively low microwave powers (~2×10^{-2} mW). Generally, organic radicals are more clearly observed at higher temperatures (>40 K) and when applying higher microwave powers (~2 mW).

1. While the Strep-tagged protocol largely maintains the integrity of the [4Fe—4S] clusters, we recommend first collecting EPR spectra at 10 K with a modulation amplitude of 5 G to ensure that all [4Fe—4S] clusters are present. Degradation of [4Fe–4S] clusters in SuiB is common and may not be readily apparent by looking at the sample, but can explain the lack of trapped reaction intermediate signals. It may also be helpful to estimate the Fe and S content using the method developed by Barr et al. (Barr et al., 2016).
2. Power saturation can distort the shape of the EPR spectrum. While an applied microwave power of 2×10^{-2} mW will not saturate the $[4Fe—4S]^{+}$ spectral components of SuiB at 10 K, it is good practice to check when working on [4Fe–4S] cluster-containing proteins. Check collected EPR spectrum for saturation by collecting a second spectrum using a quarter of the applied microwave power (5×10^{-3} mW). The second spectrum should have half of the original signal intensity at each

field value, if the original spectrum is not saturated. If the signal intensity is greater than half of the original spectrum, repeat from the previous step using a lower applied microwave power

3. Collect EPR spectrum at 40 K using 2 mW of microwave power and a modulation amplitude of 1 G. Collecting EPR spectra at higher temperatures broadens $[4Fe–4S]^+$ signals such that the organic Lys-Trp$^\bullet$ radical signal is more apparent (Fig. 6). Moreover, because organic radicals yield much narrower spectra than $[4Fe–4S]^+$ clusters, we reduce the modulation amplitude to 1 G not to broaden the narrow Lys-Trp$^\bullet$ signal
4. Generally, CW or pulse EPR characterization of at least one effective substrate isotopologue is necessary to support a model of a hypothesized species. Using either SuiA-d_5-Trp6 or SuiA-d_8-Trp6 will remove the observed hyperfine splitting pattern (Fig. 7), thus showing that the hyperfine coupling observed in the Lys-Trp$^\bullet$ intermediate originates from the indole H atoms. A substrate analog not capable of turning over, SuiA-W6F, will remove the Lys-Trp$^\bullet$ signal, leaving only the precursor Ω signal (Fig. 7). Compared to substrate isotopologues, using substrate analogs is considered indirect evidence, as the disappearance of a signal

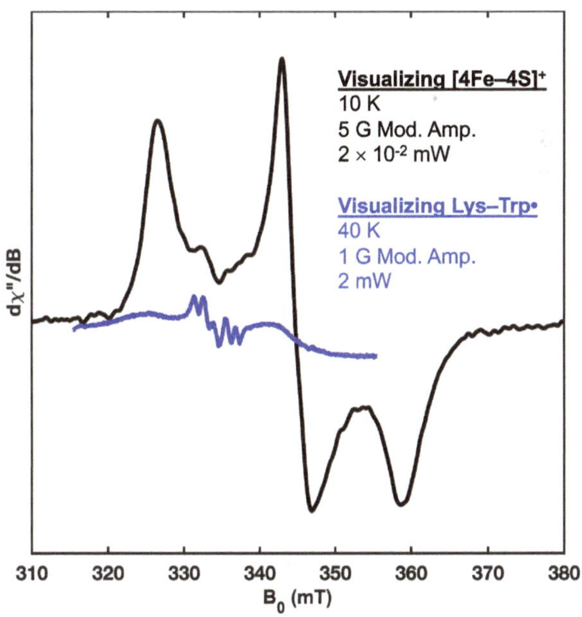

Fig. 6 Comparison of EPR spectra of a trapped SuiB reaction sample collected at conditions that highlight $[4Fe—4S]^+$ signals (black) and at conditions that highlight the Lys-Trp$^\bullet$ signal (blue).

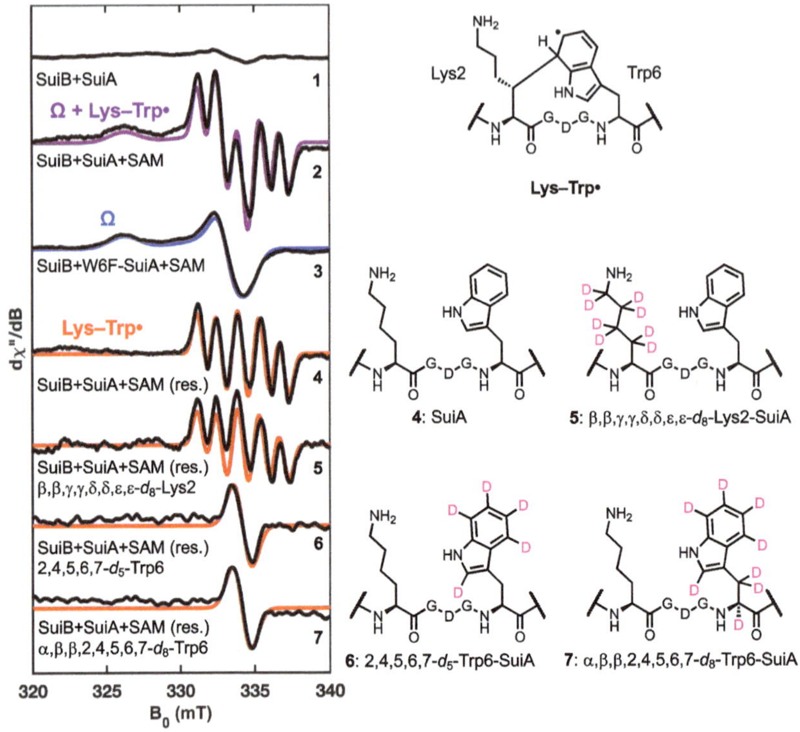

Fig. 7 Characterization of SuiB radical intermediates using X-band EPR at 50 K, 1 G modulation amplitude, and 2.0 mW of nonsaturating microwave power (Balo et al., 2021). (1) A sample containing 250 μM SuiB, 10 mM $Na_2S_2O_4$, and 2 mM SuiA yields no major signals. (2) Upon addition of 4 mM SAM, the sample was quenched in liquid nitrogen after 20 s on ice, which gave rise to two major signals (sum of components simulated in purple). These were deconvoluted using several substrate variants. (3) A 20 s reaction using the W6F-SuiA suppressed the formation of one of the two paramagnetic species observed with wt SuiA. We classify the remaining signal as the SAM-dependent organometallic species Ω. The blue trace is the simulated spectrum with $g = [2.054, 2.011, 2.002]$. (4) The organic radical component proposed to be Lys-Trp$^{\bullet}$ is the difference between the previous two spectra. The spectrum can be modeled (red trace) with strong hyperfine coupling to three protons ($g = [2.0042, 2.0042, 2.0042]$, $A_H = [80, 80, 80; 20, 50, 30; 50, 20, 30]$ MHz). (5) Use of the isotopologue d_8-Lys2-SuiA had no significant effect on the splitting pattern. (6, 7) However, with d_5-Trp6-SuiA or d_8-Trp6-SuiA, the splitting collapses entirely into a single line of equal line shape ($g = [2.0042, 2.0042, 2.0042]$).

by using an analog does not give insight into the hyperfine interactions or, by extension, the electronic structure of the observed EPR signal

3.7 Modeling the Lys-Trp$^{\bullet}$ intermediate using DFT

DFT calculations are a convenient way of simulating organic radicals such as Lys-Trp$^{\bullet}$ to estimate the hyperfine coupling constant to nearby atoms. It is

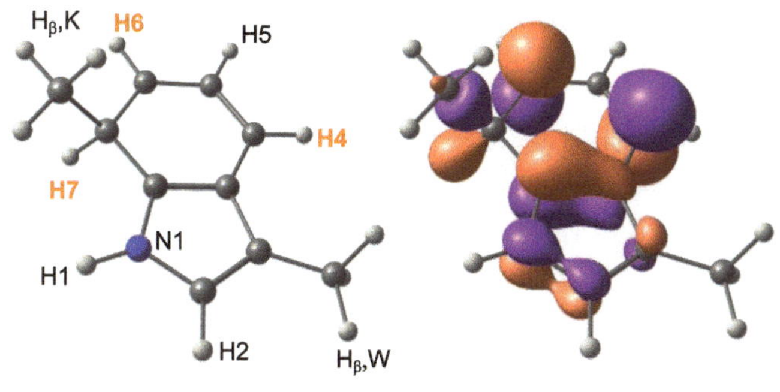

Atom	$\|A_{EPR}\|$ (MHz)	$\|A_{DFT}\|$ (MHz)
H4	[50, 20, 30]	[58, 16, 38]
H6	[20, 50, 30]	[13, 50, 33]
H7	[80, 80, 80]	[118, 122, 116]
N1, H1, H2, H_{β},W, H5, H_{β},K	Unresolved	$\|A_{iso}\|$ ~ 1–9

Fig. 8 Structure and numbering scheme (left) and the singly occupied molecular orbital (SOMO) generated through a DFT calculation of a doubly methyl-capped analog of the Lys-Trp• (right) (Balo et al., 2021). The three strongest EPR-based and DFT-predicted hyperfine coupling constants are highlighted and compared in the table. The hyperfine coupling constants of other protons and N1 are too small to resolve by EPR splitting patterns, as is consistent with DFT.

wise to perform these calculations and attempt to simulate the EPR spectra using the predicted hyperfine coupling constants before embarking on the synthesis of sometimes challenging and expensive substrate isotopologues, as observed radicals in reactions may not be part of the main catalytic cycle, such as tyrosyl radicals in the protein.

Using a doubly methyl-capped Lys-Trp• model, the geometry optimization and EPR parameter calculation were carried out using the ORCA 4.0.1 quantum chemistry program (Neese, Wennmohs, Becker, & Riplinger, 2020) using the ZORA-def2-TZVP basis set (Van Lenthe, Ehlers, & Baerends, 1999) and the BP86 density functional (Becke, 1988; Perdew, 1986) as previously described (Balo et al., 2021; Tao, Lai, Power, & Britt, 2019) (Fig. 8). From this, we can observe the three strong hyperfine coupling interactions of our EPR spectrum.

4. Conclusion

The sample preparation is often the bottleneck of EPR-based characterization studies, such that the advancement of our understanding of elusive reaction intermediates in enzymatic mechanisms depends on innovative protein biochemistry approaches. Here, we described how we trapped and characterized the organic radical intermediate Lys-Trp$^{\bullet}$ in the rSAM enzyme SuiB by EPR spectroscopy. While the discovery of this intermediate has shed light on the catalytic mechanism of SuiB, we believe that this protocol is a step towards a generalized approach for trapping radical intermediates in other rSAM enzymes and metalloenzymes in general, when kinetically favorable.

Acknowledgments

We thank Lizhi Tao and Alessio Caruso for impactful discussions leading to the development of this protocol. This work was funded through an NIH R35 Grant (1R35GM126961-01, to R.D.B.) and an NSERC Postdoctoral Fellowship (to A.R.B.).

References

Balo, A. R., Tao, L., & Britt, R. D. (2021). Characterizing SPASM/twitch domain-containing radical SAM enzymes by EPR spectroscopy. *Applied Magnetic Resonance*.

Balo, A. R., et al. (2021). Trapping a cross-linked lysine-tryptophan radical in the catalytic cycle of the radical SAM enzyme SuiB. *Proceedings of the National Academy of Sciences of the United States of America*, *118*.

Barr, I., et al. (2016). Demonstration that the radical S-adenosylmethionine (SAM) enzyme PqqE catalyzes de novo carbon-carbon cross-linking within a peptide substrate PqqA in the presence of the peptide chaperone PqqD. *The Journal of Biological Chemistry*, *291*, 8877–8884.

Barr, I., et al. (2018). X-ray and EPR characterization of the auxiliary Fe-S clusters in the radical SAM enzyme PqqE. *Biochemistry*, *57*, 1306–1315.

Becke, A. D. (1988). Density-functional exchange-energy approximation with correct asymptotic behavior. *Physical Review A: General Physics*, *38*, 3098–3100.

Berkovitch, F., Nicolet, Y., Wan, J. T., Jarrett, J. T., & Drennan, C. L. (2004). Crystal structure of biotin synthase, an S-adenosylmethionine-dependent radical enzyme. *Science*, *303*, 76–79.

Broderick, J. B., Duffus, B. R., Duschene, K. S., & Shepard, E. M. (2014). Radical S-adenosylmethionine enzymes. *Chemical Reviews*, *114*, 4229–4317.

Broderick, W. E., Hoffman, B. M., & Broderick, J. B. (2018). Mechanism of radical initiation in the radical S-adenosyl-l-methionine superfamily. *Accounts of Chemical Research*, *51*, 2611–2619.

Burdi, D., Sturgeon, B. E., Tong, W. H., Stubbe, J., & Hoffman, B. M. (1996). Rapid freeze – quench ENDOR of the radical X intermediate of Escherichia coli ribonucleotide reductase using 17O2, H217O, and 2H2O. *Journal of the American Chemical Society*, *118*, 281–282.

Bushin, L. B., & Seyedsayamdost, M. R. (2018). Guidelines for determining the structures of radical SAM enzyme-catalyzed modifications in the biosynthesis of RiPP natural products. *Methods in Enzymology*, *606*, 439–460.

Byer, A. S., et al. (2018). Paradigm shift for radical S-adenosyl-l-methionine reactions: The organometallic intermediate Ω is central to catalysis. *Journal of the American Chemical Society*, *140*, 8634–8638.

Caruso, A., Martinie, R. J., Bushin, L. B., & Seyedsayamdost, M. R. (2019). Macrocyclization via an arginine-tyrosine crosslink broadens the reaction scope of radical. *Journal of the American Chemical Society*, *141*, 16610–16614.

Davis, K. M., et al. (2017). Structures of the peptide-modifying radical SAM enzyme SuiB elucidate the basis of substrate recognition. *Proceedings of the National Academy of Sciences of the United States of America*, *114*, 10420–10425.

Flühe, L., et al. (2012). The radical SAM enzyme AlbA catalyzes thioether bond formation in subtilosin A. *Nature Chemical Biology*, *8*, 350–357.

Grell, T. A., Goldman, P. J., & Drennan, C. L. (2015). SPASM and twitch domains in S-adenosylmethionine (SAM) radical enzymes. *The Journal of Biological Chemistry*, *290*, 3964–3971.

Grell, T. A. J., et al. (2018). Structural and spectroscopic analyses of the sporulation killing factor biosynthetic enzyme SkfB, a bacterial AdoMet radical sactisynthase. *The Journal of Biological Chemistry*, *293*, 17349–17361.

Grove, T. L., Ahlum, J. H., Sharma, P., Krebs, C., & Booker, S. J. (2010). A consensus mechanism for radical SAM-dependent dehydrogenation? BtrN contains two [4Fe-4S] clusters. *Biochemistry*, *49*, 3783–3785.

Grove, T. L., et al. (2011). A radically different mechanism for S-adenosylmethionine-dependent methyltransferases. *Science*, *332*, 604–607.

Grove, T. L., et al. (2017). Structural insights into thioether bond formation in the biosynthesis of Sactipeptides. *Journal of the American Chemical Society*, *139*, 11734–11744.

Haft, D. H., & Basu, M. K. (2011). Biological systems discovery in silico: Radical S-adenosylmethionine protein families and their target peptides for posttranslational modification. *Journal of Bacteriology*, *193*, 2745–2755.

Horitani, M., et al. (2016). Radical SAM catalysis via an organometallic intermediate with an Fe-[5'-C]-deoxyadenosyl bond. *Science*, *352*, 822–825.

Kim, S. H., Perera, R., Hager, L. P., Dawson, J. H., & Hoffman, B. M. (2006). Rapid freeze-quench ENDOR study of chloroperoxidase compound I: The site of the radical. *Journal of the American Chemical Society*, *128*, 5598–5599.

Klinman, J. P., & Bonnot, F. (2014). Intrigues and intricacies of the biosynthetic pathways for the enzymatic quinocofactors: PQQ, TTQ, CTQ, TPQ, and LTQ. *Chemical Reviews*, *114*, 4343–4365.

Kuchenreuther, J. M., George, S. J., Grady-Smith, C. S., Cramer, S. P., & Swartz, J. R. (2011). Cell-free H-cluster synthesis and [FeFe] hydrogenase activation: All five CO and CN^- ligands derive from tyrosine. *PLoS One*, *6*, e20346.

Landgraf, B. J., McCarthy, E. L., & Booker, S. J. (2016). Radical S-adenosylmethionine enzymes in human health and disease. *Annual Review of Biochemistry*, *85*, 485–514.

Lanz, N. D., & Booker, S. J. (2015). Auxiliary iron-sulfur cofactors in radical SAM enzymes. *Biochimica et Biophysica Acta*, *1853*, 1316–1334.

Lanz, N. D., et al. (2012). RlmN and AtsB as models for the overproduction and characterization of radical SAM proteins. *Methods in Enzymology*, *516*, 125–152.

Neese, F., Wennmohs, F., Becker, U., & Riplinger, C. (2020). The ORCA quantum chemistry program package. *The Journal of Chemical Physics*, *152*, 224108.

Perdew, J. P. (1986). Density-functional approximation for the correlation energy of the inhomogeneous electron gas. *Physical Review B: Condensed Matter*, *33*, 8822–8824.

Sayler, R. I., et al. (2019). Trapping and electron paramagnetic resonance characterization of the 5'dAdo. *ACS Central Science*, *5*, 1777–1785.

Schramma, K. R., Bushin, L. B., & Seyedsayamdost, M. R. (2015). Structure and biosynthesis of a macrocyclic peptide containing an unprecedented lysine-to-tryptophan crosslink. *Nature Chemistry*, *7*, 431–437.

Schramma, K. R., Forneris, C. C., Caruso, A., & Seyedsayamdost, M. R. (2018). Mechanistic investigations of lysine-tryptophan cross-link formation catalyzed by streptococcal radical S-adenosylmethionine enzymes. *Biochemistry*, *57*, 461–468.

Schramma, K. R., & Seyedsayamdost, M. R. (2017). Lysine-tryptophan-crosslinked peptides produced by radical SAM enzymes in pathogenic streptococci. *ACS Chemical Biology*, *12*, 922–927.

Seyedsayamdost, M. R., Caruso, A., & Davis, K. M. (2020). The chemistry and structural enzymology of RiPP-modifying radical SAM metalloenzymes. In T. Begley, & B. Liu (Eds.), *Invited book chapter submission: "Comprehensive natural products III"* (pp. 49–64). Elsevier.

Sofia, H. J., Chen, G., Hetzler, B. G., Reyes-Spindola, J. F., & Miller, N. E. (2001). Radical SAM, a novel protein superfamily linking unresolved steps in familiar biosynthetic pathways with radical mechanisms: Functional characterization using new analysis and information visualization methods. *Nucleic Acids Research*, *29*, 1097–1106.

Stich, T. A., Myers, W. K., & Britt, R. D. (2014). Paramagnetic intermediates generated by radical S-adenosylmethionine (SAM) enzymes. *Accounts of Chemical Research*, *47*, 2235–2243.

Tao, L., Lai, T. Y., Power, P. P., & Britt, R. D. (2019). Germanium hydride radical trapped during the photolysis/thermolysis of Diarylgermylene. *Inorganic Chemistry*, *58*, 15034–15038.

Tao, L., Zhu, W., Klinman, J. P., & Britt, R. D. (2019). Electron paramagnetic resonance spectroscopic identification of the Fe-S clusters in the SPASM domain-containing radical SAM enzyme PqqE. *Biochemistry*, *58*, 5173–5187.

Van Lenthe, E., Ehlers, A., & Baerends, E.-J. (1999). Geometry optimizations in the zero order regular approximation for relativistic effects. *The Journal of Chemical Physics*, *110*, 8943–8953.

Yang, H., et al. (2019). The elusive 5'-Deoxyadenosyl radical: Captured and characterized by electron paramagnetic resonance and electron nuclear double resonance spectroscopies. *Journal of the American Chemical Society*, *141*, 12139–12146.

CHAPTER FOURTEEN

Using peptide substrate analogs to characterize a radical intermediate in NosN catalysis

Bo Wang[a,c], Alexey Silakov[a], and Squire J. Booker[a,b,c,*]
[a]Department of Chemistry, The Pennsylvania State University, University Park, PA, United States
[b]Department of Biochemistry and Molecular Biology, The Pennsylvania State University, University Park, PA, United States
[c]Howard Hughes Medical Institute, The Pennsylvania State University, University Park, PA, United States
*Corresponding author: e-mail address: squire@psu.edu

Contents

Abstract

Nosiheptide is a ribosomally produced and post-translationally modified thiopeptide antibiotic that displays potent antibacterial activity *in vitro*, especially against Gram-positive pathogens. It comprises a core peptide macrocycle that contains multiple thiazole rings, dehydrated serine and threonine residues, a tri-substituted 3-hydroxypyridine ring and several other modifications. Among these additional modifications includes a 3,4-dimethyl-2-indolic acid (DMIA) moiety that bridges Glu6 and Cys8 of the core peptide to form a second smaller ring system. This side-ring system is formed by the action of NosN, a radical *S*-adenosylmethionine (SAM) enzyme that falls within the class C radical SAM methylase (RSMT) family. However, the true function of NosN is to transfer a methylene group from the methyl moiety of SAM to C4 of 3-methylindolic acid (MIA) attached in a thioester linkage to Cys8 of the core peptide to set up a highly electrophilic species. This species is then trapped by the side chain of Glu6, resulting in formation of a lactone and the side-ring system. The NosN reaction requires two simultaneously bound molecules of SAM. The first, SAM[I], is cleaved to generate a 5′-deoxyadenosyl 5′-radical, which

Methods in Enzymology, Volume 666
ISSN 0076-6879
https://doi.org/10.1016/bs.mie.2022.02.008

abstracts a hydrogen atom from the methyl group of the second molecule of SAM, SAM^{II}. The resulting SAM^{II} radical is believed to add to C4 of MIA, affording a radical intermediate on the MIA substrate. Herein we describe synthetic approaches that allow detection of this radical by electron paramagnetic resonance (EPR) spectroscopy.

1. Introduction

Class C radical *S*-adenosylmethionine (SAM) methylases (RSMTs) operate on non-nucleophilic sp^2-hybridized carbons found most often during the biosynthesis of complex secondary metabolites. The chemical outcomes of the enzymes are one-carbon derived, and include methylation, lactonization, or cyclopropanation (Bauerle, Schwalm, & Booker, 2015; Jin, Wu, Xu, Yuan, & Tang, 2020; Zhang, van der Donk, & Liu, 2012). Recent mechanistic investigations on several class C RSMTs, including NosN (LaMattina et al., 2017; Wang, LaMattina, Marshall, & Booker, 2019), TbtI (Zhang, Mahanta, Hudson, Mitchell, & van der Donk, 2017), and C10P (Jin, Wu, Jian, Yuan, & Tang, 2018), detail the requirement for two simultaneously bound SAM molecules during catalysis, which appears to be common to radical SAM (RS) enzymes designated as HemN-like, a group into which class C RSMTs fall (Bauerle et al., 2015; Layer, Moser, Heinz, Jahn, & Schubert, 2003). During catalysis, the first SAM molecule (SAM^I), which binds to the unique iron of the $[Fe_4S_4]$ cluster, is reductively cleaved to methionine and a 5′-deoxyadenosyl 5′-radical (5′-dA•), which abstracts a hydrogen atom (H•) from the methyl group of the second SAM molecule (SAM^{II}). The resulting SAM^{II} methylene radical attacks the target sp^2-hybridized carbon of the substrate to yield a radical intermediate connected to SAM^{II} through its former methyl moiety. The formation of this radical intermediate is presumed to be common to all class C RSMTs, but it can be used to effect diverse outcomes, including methylation, lactonization or cyclopropanation, as dictated by the specific protein in which the adduct is formed.

NosN is an annotated class C RSMT that establishes the lactonic bond in the side-ring of nosiheptide (NOS, as shown in Fig. 1), a thiopeptide natural product with potent *in vitro* activity against a variety of clinically relevant Gram-positive bacteria (Haste et al., 2012). The biosynthetic gene cluster of NOS contains 16 genes, which include *nosA-nosP*. As shown in Fig. 1, the biosynthesis of NOS starts from NosM, a ribosomal precursor peptide of 50 amino acids. Qiu et al. showed that NosM is first processed by NosF, G, and H to give intermediate **1** (Qiu et al., 2017). Next, NosK,

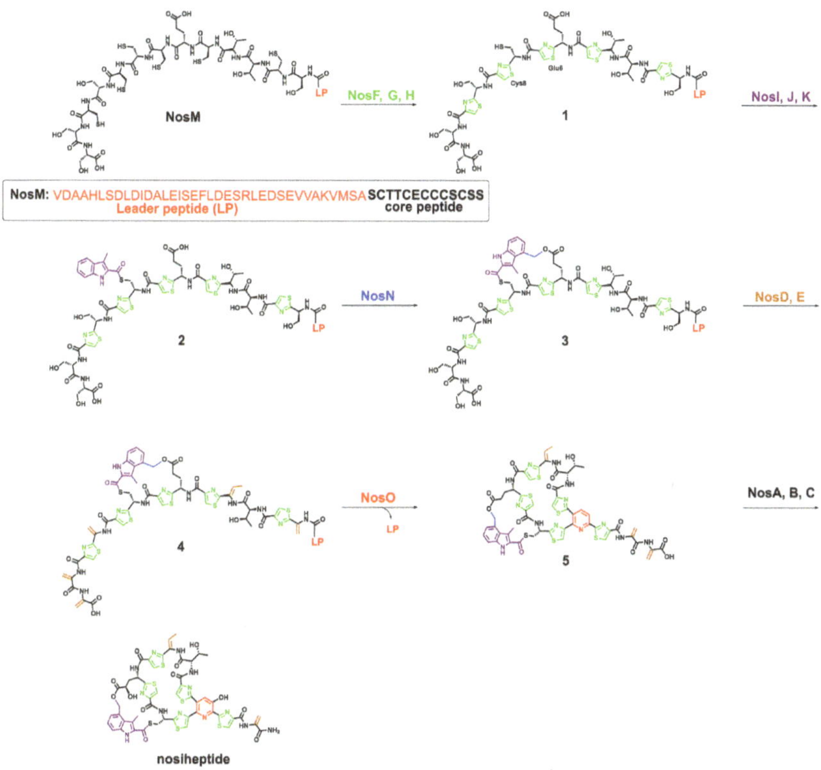

Fig. 1 Biosynthetic pathway of nosiheptide.

an α/β-hydrolase (Badding et al., 2017), recognizes this thiazole-bearing intermediate **1** and appends 3-methylindolic acid (MIA)—generated from tryptophan *via* the action of NosL (Bhandari, Fedoseyenko, & Begley, 2018)—to the thiol group of Cys8 to give intermediate **2** (Qiu et al., 2017). Our previous studies suggest that NosN then recognizes **2** as its substrate and generates the side ring to give intermediate **3** (LaMattina et al., 2017; Wang et al., 2019). Although not experimentally verified, intermediate **3** is expected to be processed by NosD and NosE to generate the dehydrated serine and threonine residues, and then by NosO to form the major macrocycle through a [4 + 2] cycloaddition reaction (Zheng, Fang, & Liu, 2017). Additional post-translational modifications give rise to the final nosiheptide structure.

As shown in Fig. 1, the substrate for NosN is complex. Therefore, to study the NosN reaction mechanism, we designed a synthetic tripeptide **6** that integrates the minimal fragments necessary for forming the side ring,

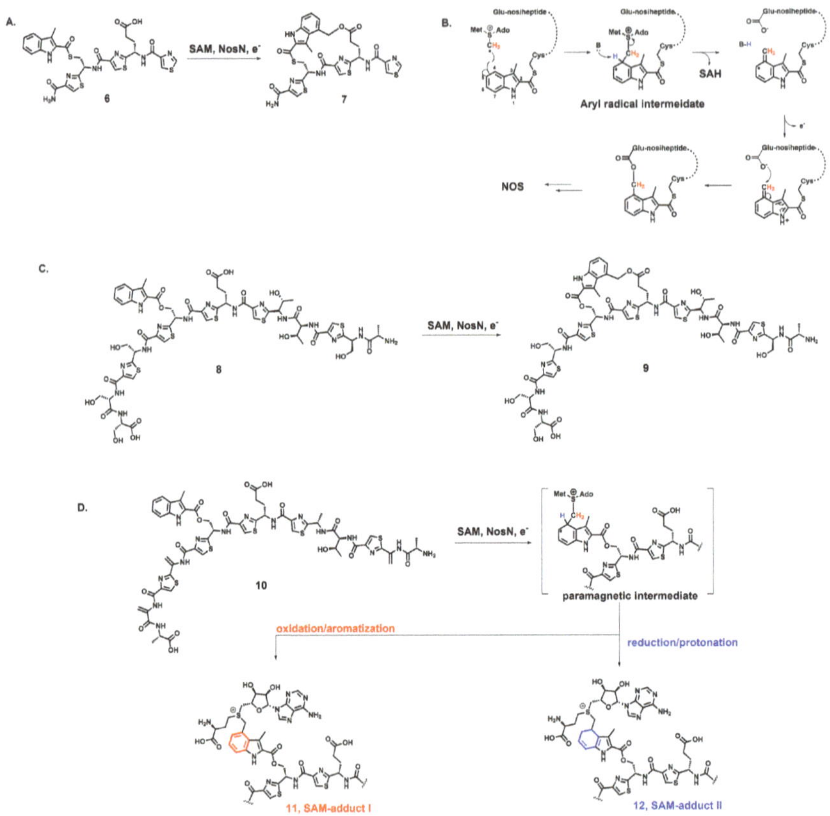

Fig. 2 (A) NosN catalyzes formation of the side-ring system on a tripeptide substrate surrogate; (B) Proposed NosN catalytic mechanism; (C) NosN catalyzed formation of the side-ring system on compound 8; (D) compound 10 gives two SAM adducts from a presumed common radical intermediate.

which includes the carboxylic side chain of Glu6 and MIA linked in a thioester bond to Cys8, as shown in Fig. 2A (LaMattina et al., 2017). Using this substrate surrogate, we established that the natural function of NosN is not to transfer a methyl group, but to transfer a methylene unit in such a way as to generate a highly electrophilic species that is trapped by the carboxylic group of Glu6. This step results in formation of the lactonic bond of nosiheptide concomitant with formation of the side ring. Fig. 2B shows the current working hypothesis for catalysis by NosN. After addition of the SAM^{II} methylene radical to C4 of MIA, an unidentified base from NosN removes the C4 proton of the MIA radical intermediate, which promotes elimination of *S*-adenosylhomocysteine (SAH) as a co-product.

The resulting radical intermediate then loses an electron, likely back to the iron-sulfur (Fe–S) cluster, affording an electrophilic methylene group that is attacked by the carboxylic group of Glu6 to close the ring.

To determine NosN's true substrate and when it acts during the biosynthesis of nosiheptide, we designed two substrate surrogates, compounds **8** and **10,** to be used as probes (Fig. 2C and D) (Wang et al., 2019). Compound **8** is a surrogate for intermediate **2** in Fig. 1, the species before dehydration by NosD and -E (with truncation of the leader peptide), while compound **10** is the dehydrated version of **8**. These two compounds were used to assess whether NosN acts before or after dehydration by NosD and -E. Use of **8** in NosN reactions afforded the side-ring closed species as the sole product, while the use of **10** afforded two MIA-SAM adducts as major products. We posited that the more rigid structure of **10,** due to the dehydrated serines, impedes the base from removing the C4 proton of the radical intermediate, a step that allows the normal reaction to advance forward to product. After quenching the reaction, the radical species would either be oxidized or reduced by one electron, which would be followed by re-aromatization to give **11** (SAM adduct I) or protonation to give **12** (SAM adduct II) (Fig. 2D). The observation of the two SAM adducts suggested the possibility that the presumed radical intermediate would accumulate sufficiently to be detected by electron paramagnetic resonance (EPR) spectroscopy (Fig. 2B). Although EPR spectroscopy has been a powerful tool to characterize radical intermediates in a variety of RS enzymes, no such characterization has been reported yet for class C RSMTs, mostly due to the complexity of their substrates. Herein, we provide synthetic strategies for generating substrate analogs, some with strategically placed isotopic labels, for studying the reaction mechanism of NosN.

2. Synthesis of NosN substrate surrogates and their deuterated analogs

Our EPR study of the NosN reaction requires compound **8**, compound **13** and its perdeuterated analogs **14** to assign the radical species observed. The chemical structures of compounds **13** and **14** are shown in Fig. 3. Compound **13** is an analog of compound **8** but with no methyl group on C3 of MIA. Although not the natural substrate, compound **13** was designed because it is more convenient to incorporate deuterium atoms into indolic acid than MIA. As shown in Fig. 3A, the synthesis of compound **8**, **13**, and **14** was accomplished by solid phase peptide synthesis (SPPS),

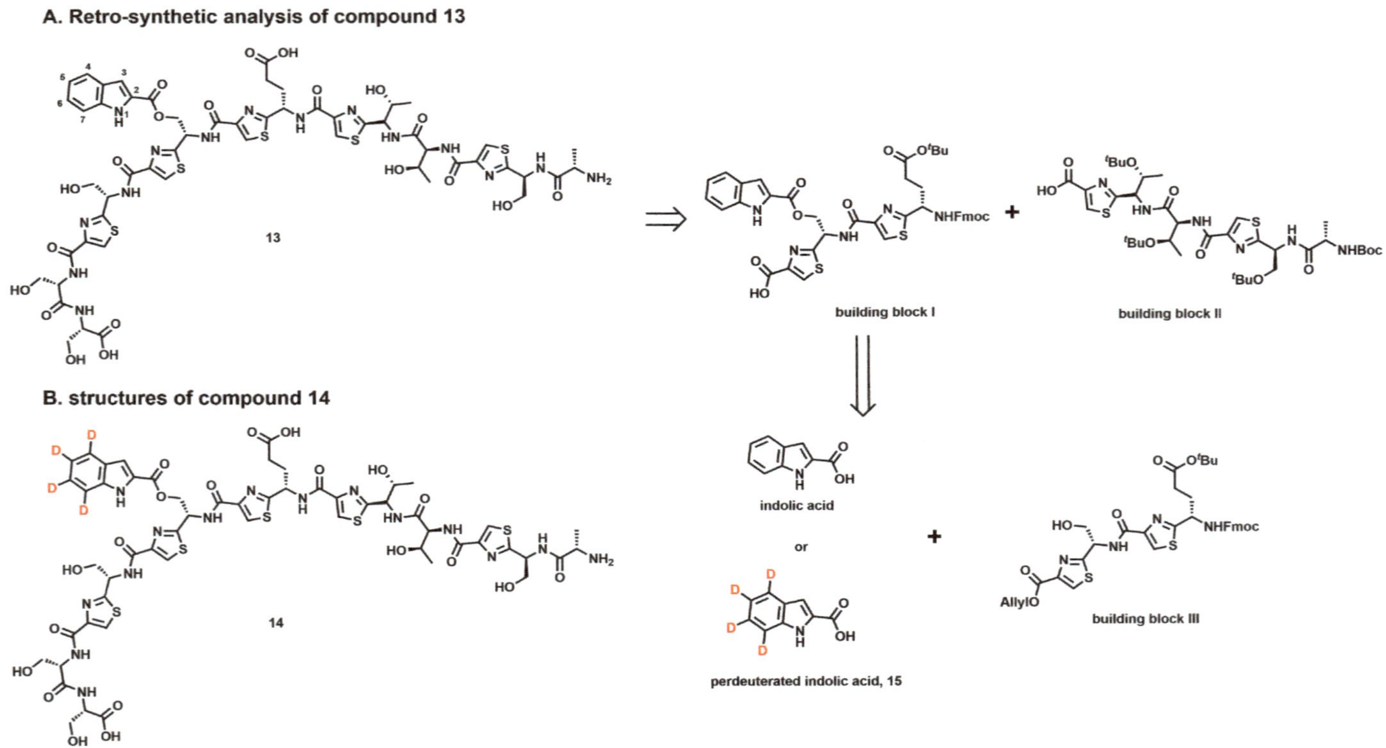

Fig. 3 (A) Retro-synthesis of peptide analogs; (B) Chemical structures of compound **14** with deuteration on indolic acid to study the structure of the radical species.

which requires two important building blocks, I and II. Herein, we detail our synthetic strategy in three sections: synthesis of deuterated indolic acid analogs **15**, synthesis of peptide building blocks, and solid phase peptide synthesis (SPPS).

2.1 Synthesis of 4,5,6,7-d_4 indolic acid 15

Indolic acid and MIA, which are used to synthesize compound **8** and **13** respectively, are commercially available. By contrast, 4,5,6,7-d_4 indolic acid **15** must be accessed synthetically. The construction of the indole structures is achieved by an aerobic cross-dehydrogenative coupling between aniline and ethyl pyruvate, as reported by Xiao (Ren, Nan, Wang, & Xiao, 2018). As shown in Fig. 4A, we start from the corresponding perdeuterated aniline, which is commercially available.

A.

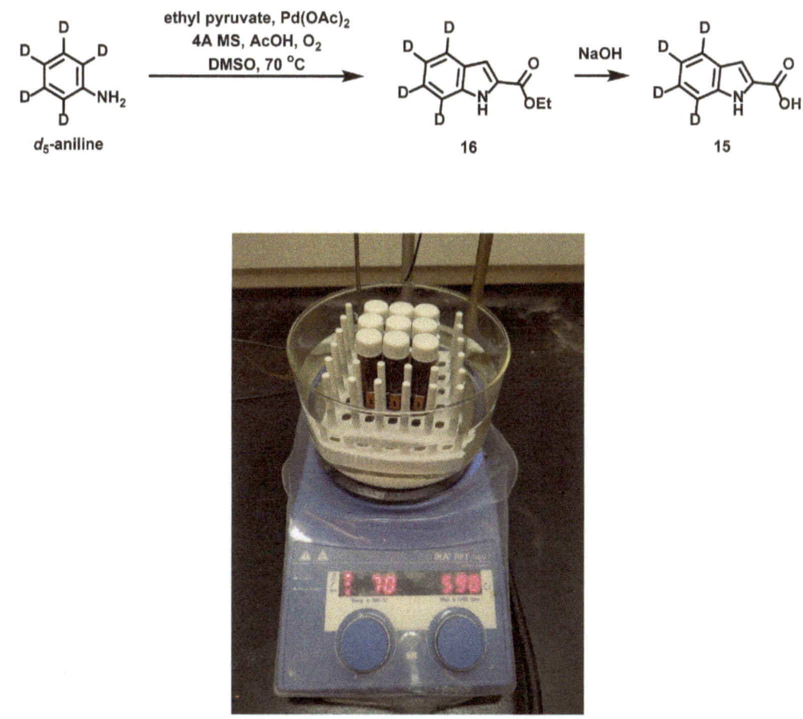

B.

Fig. 4 (A) Synthesis of compound **15**; (B) aerobic cross-dehydrogenative coupling set ups.

1. *We found that scaling up the coupling reaction results a lower yield.* Thus, to obtain sufficient product, we usually prepare 10–20 parallel reactions at the scale described below (Fig. 4B).
2. d_5-aniline (0.4 mmol, 1.0 equiv.), $Pd(OAc)_2$ (9.0 mg, 0.04 mmol, 0.1 equiv.), ethyl pyruvate (92.8 mg, 0.8 mmol, 2.0 equiv.), acetic acid (96.0 mg, 1.6 mmol, 4.0 equiv.), and 4A molecular sieve (80 mg) are added to DMSO (2.0 mL) in a 5 mL vial equipped with a stir bar. The reaction is bubbled with oxygen for 1 min then sealed. The reaction mixture is stirred at 70 °C for 12 h. After completion, the reaction mixtures are combined and diluted with ethyl acetate (250 mL) and washed thoroughly and sequentially with 0.1 M HCl (100 mL, twice), saturated aqueous sodium bicarbonate (100 mL, twice), water (100 mL), and brine (100 mL). The organic layer is dried over anhydrous sodium sulfate and concentrated *in vacuo*, and the resulting residue is purified by silica gel flash chromatography (hexanes: ethylacetate = 20:1–5:1), giving compound **16** as a yellow solid.
3. Sodium hydroxide (200 mg, 5.0 mmol, 5 equiv.) is added to the stirred solution of **16** (1.0 mmol, 1.0 equiv.) in THF (7 mL)-MeOH (7 mL)-water (7 mL). The reaction mixture is stirred at room temperature for 1 h. After completion, the reaction mixture is diluted with ethyl acetate (70 mL) and washed thoroughly and sequentially with 0.1 M HCl (100 mL) and brine (100 mL). The organic layer is dried over anhydrous sodium sulfate and concentrated *in vacuo* to give compound **15** as a yellow solid.

2.2 Synthesis of building block I for SPPS

O- to N- acyl transfer during deprotection of Fmoc, which occurs in SPPS upon treatment with 20% piperidine in DMF, is common when there is an ester bond on the side chain of serine or threonine. To avoid this side reaction, it is necessary to minimize the number of deprotection steps during SPPS cycles. We therefore designed and used building blocks I and II for SPPS. Indolic acid and its deuterated analogs, as well as MIA, are introduced into building block I by solution phase synthesis prior to SPPS, as shown in Fig. 5. Solution phase synthesis is also used to prepare building block II for SPPS, which is a tetra-peptide (Fig. 3). By using this strategy, we reduce the number of piperidine deprotections to one, which occurs between incorporation of building blocks I and II. We use Boc as the amino protecting group in building block II, because it can be removed concomitantly when

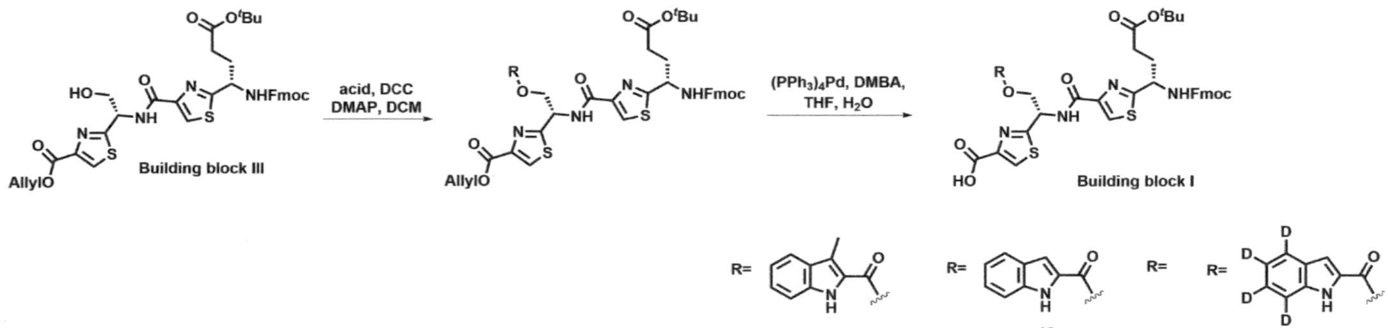

17 18 19

Fig. 5 Solution phase synthesis of building block I for SPPS. 1. Indolic acid (its deuterated analogs, or MIA (1 mmol, 2.0 equiv.)) and building block III (359 mg, 0.5 mmol, 1.0 equiv.) are dissolved in dichloromethane (10 mL) in a 50 mL flask with a magnetic stir bar.

cleaving the peptide from the resin. There are three unnatural amino acids used in solution phase synthesis and SPPS, including thiazolyl serine, thiazolyl glutamic acid, and thiazolyl threonine. The detailed preparation of these amino acids can be found in a previous publication, (Wang et al., 2018) while the preparation of building blocks II and III can be found in another of our previous publications. (Wang et al., 2019).

1. *N,N′*-Dicyclohexylcarbodiimide (DCC, 124 mg, 0.6 mmol, 1.2 equiv.) and 4-dimethylaminopyridine (DMAP, 6.1 mg, 0.05 mmol, 0.1 equiv.) are added to the reaction mixture.
2. The resulting solution is stirred at room temperature for 1 h. The reaction will become cloudy because of the formation of dicyclohexylurea.
3. After completion, the reaction mixture is diluted with dichloromethane (50 mL) and washed thoroughly and sequentially with 0.1 M HCl (100 mL), saturated aqueous sodium bicarbonate (100 mL), and brine (100 mL). The organic layer is dried over anhydrous sodium sulfate and concentrated in vacuo, and the resulting residue is used for the next step without further purification.
4. The residue is dissolved in THF (20 mL), and tetrakis(triphenylphosphine) palladium(0) ($Pd(PPh_3)_4$, 29 mg, 0.025 mmol, 0.05 equiv.) and 1,3-dimethylbarbituric acid (78 mg, 0.5 mmol, 1.0 equiv.) are added to the reaction. This reaction will remove the allyl ester to expose the carboxyl group for SPPS. The resulting reaction mixture is stirred at room temperature for 1 h. Next, the reaction mixture is diluted with ethyl acetate (100 mL) and washed thoroughly and sequentially with 0.1 M HCl (100 mL), saturated aqueous sodium bicarbonate (100 mL), water (50 mL), and brine (50 mL). The organic layer is dried over anhydrous sodium sulfate and concentrated in vacuo to give yellow foam which can be used for SPPS directly without further purification.

2.3 Solid phase peptide synthesis

We chose the Fmoc/*t*Bu strategy to perform our SPPS. In detail, the amino group of each building block is protected by Fmoc, which is removed by treatment with 20% piperidine/DMF after each cycle of amide coupling. The side chains of the building blocks are protected by acid labile groups, such as *t*Bu for hydroxyl and carboxyl groups (Ser, Thr, Asp, Glu), Pbf for guanidine groups (Arg), Boc for amino groups (Lys), and trityl for thiol groups (Cys), and these are concomitantly removed when cleaving the peptide from the resin using a TFA cocktail. We chose 2-chlorotrityl chloride resin because we want our final product to be a free acid on the C-terminus. Rink Amide Resin can be selected if a C-terminal amide is preferred. Here we will detail how our SPPS was performed.

2.3.1 Loading 2-chlorotrityl chloride resin

Weigh 2-chlorotrityl chloride resin (0.75 meq/g, 2 g, 1.5 mmol) in a 50 mL solid phase peptide synthesis vessel and swell the resin in dichloromethane (20 mL) for 1 h. Drain the dichloromethane and introduce a solution of Fmoc-L-Ser(*t*Bu)-OH (2.9 g, 7.5 mmol, 5 equiv.) and 2,4,6-collidine (2 mL) in dichloromethane (20 mL) into the vessel. Shake the vessel overnight, drain the reaction mixture and wash with dichloromethane three times.

2.3.2 Capping 2-chlorotrityl chloride resin

Introduce the capping solution (dichloromethane/methanol/*N*,*N*-diisopropylethylamine = 17 mL/2 mL/1 mL) into the vessel and shake for 1 h. Drain the capping solution and wash the resin with dichloromethane three times. Dry the resin under nitrogen flow. The obtained resin is 0.7 mmol/g.

2.3.3 Iterative/amide coupling/Fmoc removal

We started at a 0.1 mmol scale (equals to 145 mg resin from previous step). Removal of Fmoc is achieved by bubbling the resin in 20% piperidine in DMF (4 mL) for 10 min. Next, the resin is washed with DMF and dichloromethane. For amide coupling, a solution of the corresponding amino acid building block (0.4 mmol, 4.0 equiv.) and HATU (152 mg, 0.4 mmol, 4.0 equiv.) in DMF containing 20% *N*-methylmorpholine (5 mL) is added to the vessel and bubbled for 20 min. After completion, the reaction solution is removed and washed with DMF followed by dichloromethane. Repeat the deprotecting and amide coupling steps until all amino acids are incorporated. Generally, we do not recommend using a stir bar for the reaction because it may rupture the beads. Stirring can be achieved by bubbling, as shown in Fig. 6C.

2.3.4 Cleavage and deprotection

After all amino acid building blocks are incorporated, cleavage cocktail (trifluoroacetic acid/H_2O/triisopropylsilane = 9.5 mL/0.25 mL/0.25 mL) is added to the resin, and the resulting reaction mixture is shaken for 3 h. Next, the solution is collected in a 50 mL conical tube. The volume is

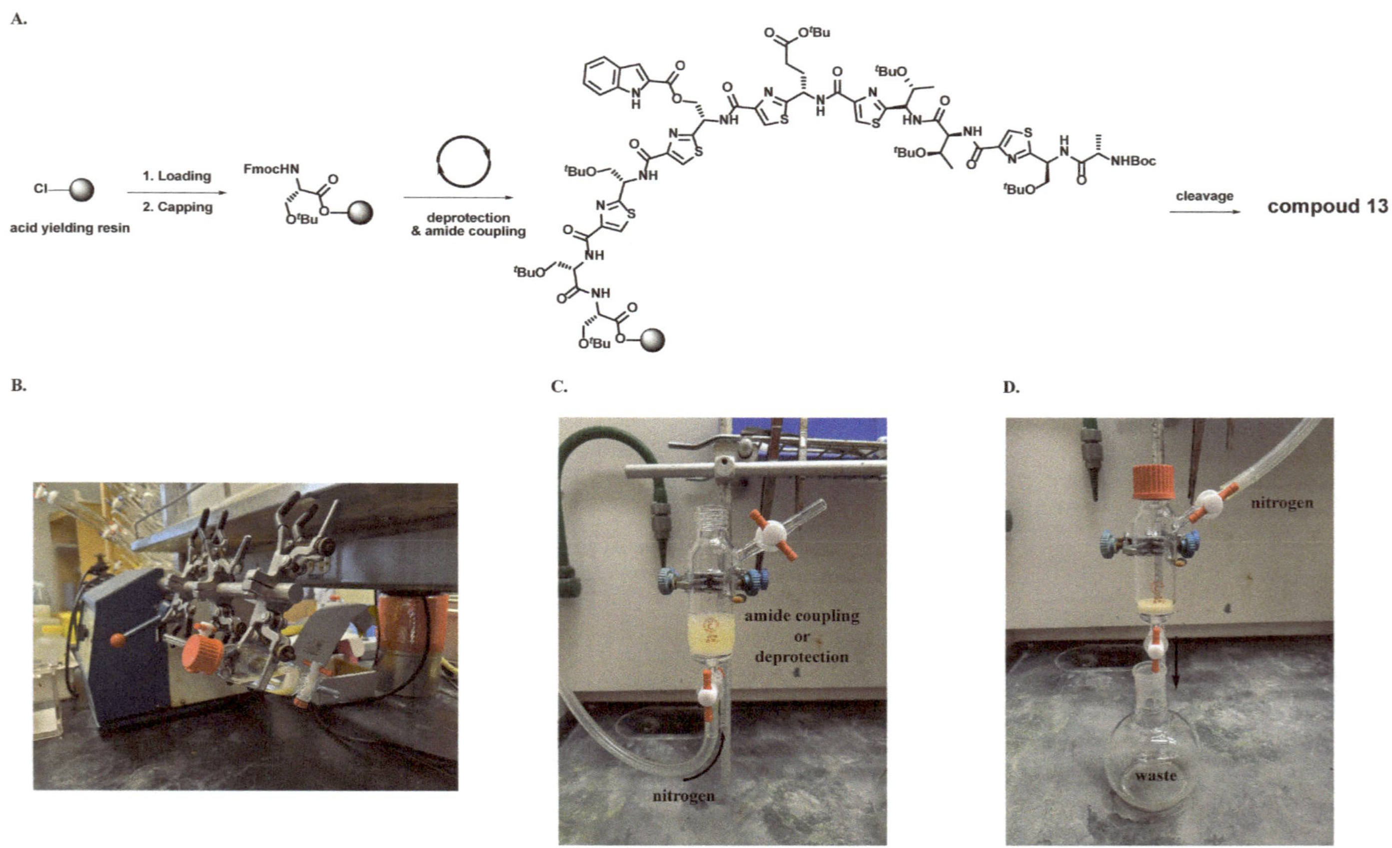

Fig. 6 (A) SPPS flow chart of compound 13; (B) resin swelling, 1st amino acid loading, capping, and cleavage are performed by shaking; (C) amide coupling and Fmoc removal reaction are stirred by bubbling; (D) liquid phase is pushed out of the SPPS vessel with air or nitrogen to separate it from beads.

reduced to ~4 mL under nitrogen flow and *t*-butyl methyl ether (40 mL, prechilled with ice) is added to precipitate the peptide. The white precipitate is pelleted by centrifugation and the residue is dried under a mild flow of nitrogen. The resulting residue is solubilized in water (10 mL), and the resulting solution is filtered before HPLC purification.

2.4 HPLC purification

The purification of crude peptides is performed on an Agilent 1260 Infinity II Preparative HPLC system using an Agilent 5 Prep-C18 (50 × 21.2 mm) column. The solvents for purification include 0.1% TFA in water (solvent A) and 0.1% TFA in acetonitrile (solvent B), and detection is performed by UV monitoring at 254 nm. The detailed methods are listed in Tables 1 and 2.

Table 1 HPLC conditions for purification of compound **8.**

Time (min)	A%	B%	Flow rate (mL/min)
0.00	75	25	30.00
2.00	75	25	30.00
5.00	60	40	30.00
5.20	20	80	30.00
6.00	20	80	30.00
6.20	75	25	30.00
7.00	75	25	30.00

Table 2 HPLC conditions for purification of compound **13** and **14.**

Time (min)	A%	B%	Flow rate (mL/min)
0.00	75	23	30.00
2.00	75	23	30.00
5.00	60	34.5	30.00
5.20	20	80	30.00
6.00	20	80	30.00
6.20	75	23	30.00
7.00	75	23	30.00

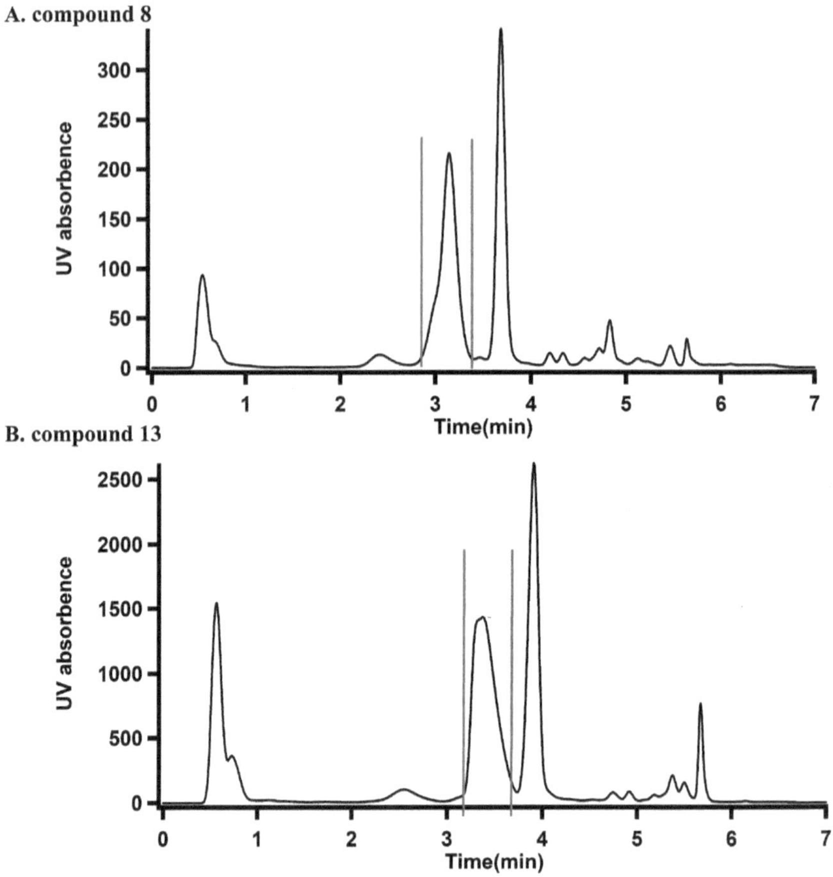

Fig. 7 HPLC trace of compound **8** (A) and compound **13** (B). The red lines indicate the desired peak that were collected.

As shown in Fig. 7, the peaks between the red lines are the desired compounds **8** or **13**. Iterative injections are performed to purify the entire crude peptide solution. Fractions are pooled and lyophilized to give a white powder as final products. The powder is re-dissolved in water to make a stock solution with a concentration of 5 mM. HPLC traces of compounds **14** are almost identical to that of compound **13** (data not shown).

3. EPR sample preparation and data collection

Our strategy to overproduce and purify NosN has been previously published (LaMattina et al., 2017).

1. Bring stocks of NosN, SAM, substrate peptides, and sodium dithionite (DT) powder, together with a bottle of liquid nitrogen-cooled isopentane for sample freezing, into an anaerobic glovebox.
2. Made a fresh stock solution of DT (100 mM in 50 mM HEPES, pH 7.5) each time EPR samples are prepared.
3. Recipe for making EPR samples is shown in Table 3:
4. Mix all reagents shown in the table (excluding DT) in an Eppendorf tube. Initiate the reaction by addition of DT to the reaction mixture and quickly transfer the reaction into an EPR tube. Incubate the reaction in the EPR tube for given periods of time and freeze the sample in liquid nitrogen-cooled isopentane.
5. For compound **8**, an incubation time between 5 and 10 min gives the strongest EPR signals.
6. It is recommended to explore the time course of each sample to identify the best time to freeze the sample for the strongest signal. For compound **8**, we prepared samples at 15 s, 30 s, 2 min, 5 min, 10 min, and 60 min. As shown in Fig. 8, samples prepared at 5 min and 10 min give the strongest signals. Thus, we chose 5 min as the standard reaction time for compound **8**.
7. The power dependence of the EPR signal was also explored, as shown in Fig. 9. Attenuation of 40 dB (20 μW) gave the best signal, which represents the upper limit for suitable microwave power that allows an undistorted signal of this radical to be obtained at 60 K.
8. For compound **13** and its ^{2}H-labeled analog **14**, incubation times of 2 min are used. The comparison of EPR spectra of the labeled and unlabeled samples together with the simulated spectra are shown in Fig. 10. Note that the modulation amplitude used to obtain these spectra is lowered to 0.2 mT to obtain the best spectroscopic resolution.

Table 3 EPR sample preparation recipe.

Reagent	Stock	Reaction	Volume to add
HEPES, pH 7.5	1000 mM	100 mM	25 μL
Substrate peptides	5 mM	1.5 mM	75 μL
SAM	28.8 mM	3 mM	26 μL
NosN	1200 μM	500 μM	104 μL
DT	100 mM	4 mM	10 μL
Water			10 μL
	Total volume:		250 μL

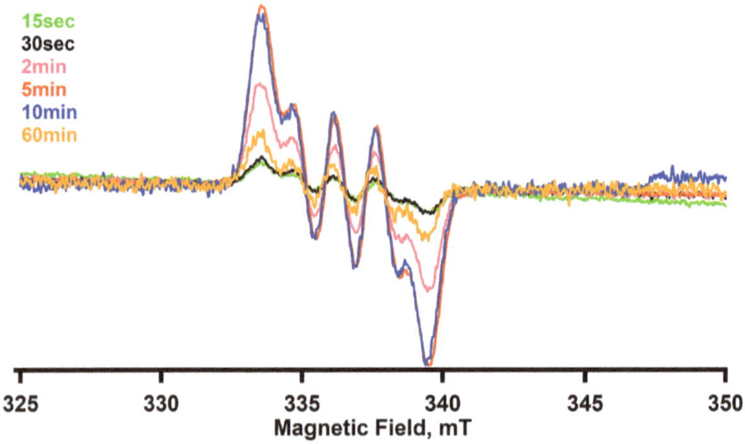

Fig. 8 Time course EPR signals of compound **8**. Experimental conditions: Temperature, 80 K; microwave frequency, 9.436 GHz; microwave power, 1 mW (20 dB attenuation); modulation amplitude, 1 mT; scan time of 60 s.

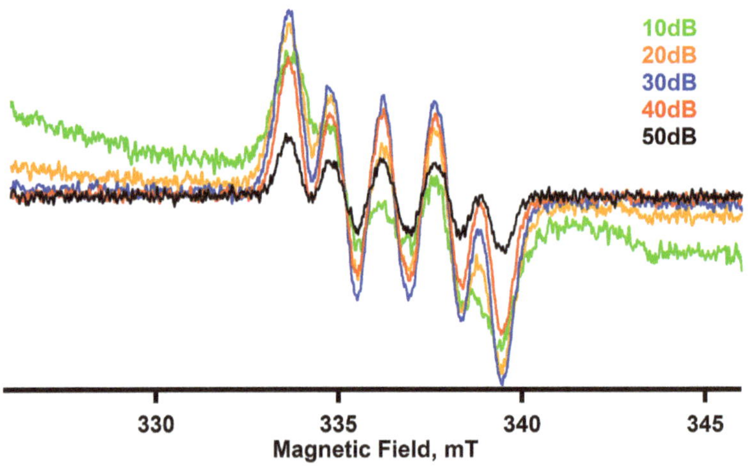

Fig. 9 EPR signals of compound **8** at 5 min using various attenuation numbers. Experimental conditions: Temperature, 60 K; microwave frequency, 9.436 GHz; modulation amplitude, 1 mT; scan time of 60 s.

4. EPR data simulation

Complete loss of hyperfine structure in the EPR spectra of the intermediate upon ^{2}H-labeling of C4,C5,C6 and C7 agrees with the assignment of the observed signal to the aryl radical intermediate (see Fig. 2).

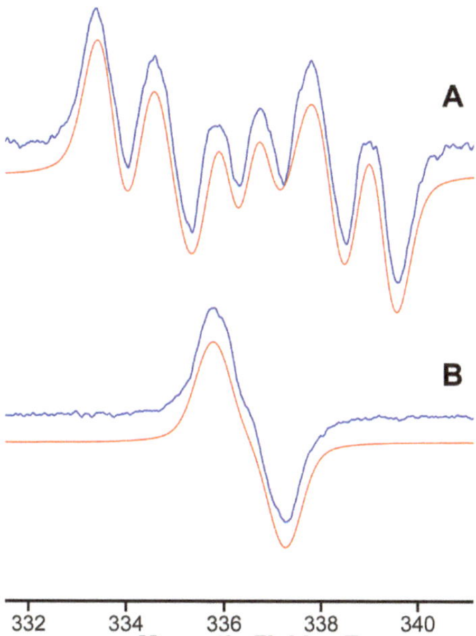

Fig. 10 Simulation of the EPR spectra of radical intermediate obtained using natural-abundance substrate **13** (A) and ^{2}H-isotope labeled compound **14** (B). Experimental conditions: Microwave frequency, 9.436 GHz, Microwave power, 0.1 mW (30 dB attenuation), modulation amplitude, 0.2 mT and scan time, 60 s.

To simulate the EPR spectra obtained we adopted a simple model in which the $^1H/^2H$ nuclei on sp^3-hybridized carbons (e.g., C4) exhibit a predominantly isotropic hyperfine (HF) interaction (i.e., with principal components of the HF coupling $a_1 \approx a_2 \approx a_3$) and $^1H/^2H$ nuclei on sp^2-hybridized carbons (C5, C6 and C7) exhibit strongly rhombic hyperfine anisotropy ($a_1 > a_2 > a_3$). Taking this model as a starting point we performed a fit of the simulation to the natural-abundance EPR spectrum using the "*esfit*" routine of the EasySpin package in MATLAB. During fitting, we allowed to vary HF coupling constants, three g-values and the unitary linewidth. "*pepper*" from EasySpin with the first-order perturbation theory ("perturb1") option was used as the core simulation routine. This procedure afforded a satisfactory fit to the experimental data with a pseudo-Voigt unitary lineshape (0.8 and 0.15 mT Gaussian and Lorentzian line-width parameters, respectively) and an almost isotropic g-tensor ($g_1 = 2.0029$, $g_2 = 2.0038$, $g_3 = 2.0041$). Inclusion of three strong ^1H HF couplings was sufficient to interpret the spectra: $A_1 = [84, 86, 93]$ MHz, $A_2 = [20, 40, 47]$ MHz, $A_3 = [13, 26, 38]$ MHz. Using these parameters, it was possible to interpret the ^{2}H spectra by scaling down the HF coupling constants by a factor of $g_n(H)/g_n(D)=6.51$ and slightly adjusting the linewidth parameters. It is worth noting, however, that a reasonable simulation of the natural-abundance EPR spectra can be also achieved with a set of almost completely isotropic HF couplings. While such an interpretation is not possible given that we have established the indole-based nature of the radical, it does highlight certain ambiguities associated with the interpretation of EPR spectra. Therefore, to fully characterize the electronic structure of the radical we need a comprehensive EPR analysis of various isotopologs obtained using synthetic procedures outlined above. This work is underway and will be published separately.

5. Conclusions

Class C RSMTs catalyze the formation of an adduct between the former methyl group of SAM and an sp^2-hybridized carbon of a substrate as a common intermediate. From this intermediate, reaction outcomes include methylation, lactonization, and cyclopropanation. Given the role that class C RSMTs play in the biosynthesis of bioactive natural products, determining the catalytic mechanism has the potential to shed light on how these molecules can be engineered to produce more effective therapeutics. The mechanistic study of NosN has been hindered by the complex nature of the substrate on which it acts. The methods described herein provide a facile synthetic approach to generate NosN's substrate as a means to probe its reaction as well as others in the nosiheptide biosynthetic pathway.

Acknowledgments

This research was supported by NIH (GM-122595 to S. J. B.) and NSF (CHE-1943748 to A.S.). S.J.B. is an Investigator of the Howard Hughes Medical Institute.

References

Badding, E. D., et al. (2017). Rerouting the pathway for the biosynthesis of the side ring system of Nosiheptide: The roles of NosI, NosJ, and NosK. *Journal of the American Chemical Society*, *139*, 5896–5905.

Bauerle, M. R., Schwalm, E. L., & Booker, S. J. (2015). Mechanistic diversity of radical S-adenosylmethionine (SAM)-dependent methylation. *The Journal of Biological Chemistry*, *290*, 3995–4002.

Bhandari, D. M., Fedoseyenko, D., & Begley, T. P. (2018). Mechanistic studies on tryptophan Lyase (NosL): Identification of cyanide as a reaction product. *Journal of the American Chemical Society*, *140*, 542–545.

Haste, N. M., et al. (2012). Activity of the thiopeptide antibiotic nosiheptide against contemporary strains of methicillin-resistant Staphylococcus aureus. *Journal of Antibiotics (Tokyo)*, *65*, 593–598.

Jin, W. B., Wu, S., Jian, X. H., Yuan, H., & Tang, G. L. (2018). A radical S-adenosyl-L-methionine enzyme and a methyltransferase catalyze cyclopropane formation in natural product biosynthesis. *Nature Communications*, *9*, 2771.

Jin, W. B., Wu, S., Xu, Y. F., Yuan, H., & Tang, G. L. (2020). Recent advances in HemN-like radical S-adenosyl-l-methionine enzyme-catalyzed reactions. *Natural Product Reports*, *37*, 17–28.

LaMattina, J. W., et al. (2017). NosN, a radical S-Adenosylmethionine Methylase, catalyzes both C1 transfer and formation of the Ester linkage of the side-ring system during the biosynthesis of Nosiheptide. *Journal of the American Chemical Society*, *139*, 17438–17445.

Layer, G., Moser, J., Heinz, D. W., Jahn, D., & Schubert, W. D. (2003). Crystal structure of coproporphyrinogen III oxidase reveals cofactor geometry of radical SAM enzymes. *The EMBO Journal*, *22*, 6214–6224.

Qiu, Y., et al. (2017). Thiolation protein-based transfer of Indolyl to a Ribosomally synthesized Polythiazolyl peptide intermediate during the biosynthesis of the side-ring system of Nosiheptide. *Journal of the American Chemical Society*, *139*, 18186–18189.

Ren, L., Nan, G., Wang, Y., & Xiao, Z. (2018). Carboxylic acid-promoted single-step indole construction from simple anilines and ketones via aerobic cross-dehydrogenative coupling. *The Journal of Organic Chemistry*, *83*, 14472–14488.

Wang, B., LaMattina, J. W., Marshall, S. L., & Booker, S. J. (2019). Capturing intermediates in the reaction catalyzed by NosN, a class C radical S-adenosylmethionine methylase involved in the biosynthesis of the Nosiheptide side-ring system. *Journal of the American Chemical Society*, *141*, 5788–5797.

Wang, B., et al. (2018). Using peptide mimics to study the biosynthesis of the side-ring system of Nosiheptide. *Methods in Enzymology*, *606*, 241–268.

Zhang, Z., Mahanta, N., Hudson, G. A., Mitchell, D. A., & van der Donk, W. A. (2017). Mechanism of a class C radical S-adenosyl-l-methionine thiazole methyl transferase. *Journal of the American Chemical Society*, *139*, 18623–18631.

Zhang, Q., van der Donk, W. A., & Liu, W. (2012). Radical-mediated enzymatic methylation: A tale of two SAMS. *Accounts of Chemical Research*, *45*, 555–564.

Zheng, Q., Fang, H., & Liu, W. (2017). Post-translational modifications involved in the biosynthesis of thiopeptide antibiotics. *Organic & Biomolecular Chemistry*, *15*, 3376–3390.

FSC
www.fsc.org
MIX
Papier | Fördert
gute Waldnutzung
FSC® C083411

Kolibri 360 GmbH
Ferdinand-Jühlke-Straße 7
99095 Erfurt, Deutschland
produktsicherheit@kolibri360.de